Edited by Evgeny Katz

Implantable Bioelectronics

Related Titles

Katz, E. (ed.)

Biomolecular Information Processing

From Logic Systems to Smart Sensors and Actuators

2012
ISBN: 978-3-527-33228-1

Katz, E. (ed.)

Molecular and Supramolecular Information Processing

From Molecular Switches to Logic Systems

2012
ISBN: 978-3-527-33195-6

Katz, E. (ed.)

Information Processing Set

2 Volumes
(consisting of "Biomolecular Information Processing" and "Molecular and Supramolecular Information Processing")

2012
ISBN: 978-3-527-33245-8

Wallace, G.G., Moulton, S., Kapsa, R.M.I., Higgins, M.

Organic Bionics

2012
ISBN: 978-3-527-32882-6

Cosnier, S., Karyakin, A. (Eds.)

Electropolymerization

Concepts, Materials and Applications

2010
ISBN: 978-3-527-32414-9

Alkire, R.C., Kolb, D.M., Lipkowski, J. (Eds.)

Bioelectrochemistry

Fundamentals, Applications and Recent Developments

2012
ISBN: 978-3-527-32885-7

Kumar, C.S. (ed.)

Nanotechnologies for the Life Sciences

10 Volume Set

2011
Print ISBN: 978-3-527-33114-7

Waser, R. (ed.)

Nanoelectronics and Information Technology

Advanced Electronic Materials and Novel Devices
Third, Completely Revised and Enlarged Edition

2012
Print ISBN: 978-3-527-40927-3

Edited by Evgeny Katz

Implantable Bioelectronics

Verlag GmbH & Co. KGaA

The Editor

Prof. Evgeny Katz
Clarkson University
Department of Chemistry
Clarkson Avenue 8
USA

All books published by **Wiley-VCH** are carefully produced. Nevertheless, authors, editors, and publisher do not warrant the information contained in these books, including this book, to be free of errors. Readers are advised to keep in mind that statements, data, illustrations, procedural details or other items may inadvertently be inaccurate.

Library of Congress Card No.: applied for

British Library Cataloguing-in-Publication Data
A catalogue record for this book is available from the British Library.

Bibliographic information published by the Deutsche Nationalbibliothek
The Deutsche Nationalbibliothek lists this publication in the Deutsche Nationalbibliografie; detailed bibliographic data are available on the Internet at <http://dnb.d-nb.de>.

© 2014 Wiley-VCH Verlag GmbH & Co. KGaA, Boschstr. 12, 69469 Weinheim, Germany

All rights reserved (including those of translation into other languages). No part of this book may be reproduced in any form – by photoprinting, microfilm, or any other means – nor transmitted or translated into a machine language without written permission from the publishers. Registered names, trademarks, etc. used in this book, even when not specifically marked as such, are not to be considered unprotected by law.

Print ISBN: 978-3-527-33525-1
ePDF ISBN: 978-3-527-67317-9
ePub ISBN: 978-3-527-67316-2
Mobi ISBN: 978-3-527-67315-5
oBook ISBN: 978-3-527-67314-8

Cover Design Bluesea Design, McLeese Lake, Canada
Typesetting Laserwords Private Ltd., Chennai, India
Printing and Binding Markono Print Media Pte Ltd, Singapore

Printed on acid-free paper

Contents

Preface *XV*
List of Contributors *XVII*

1	**Implantable Bioelectronics – Editorial Introduction** *1*	
	Evgeny Katz	
	References *5*	
2	**Magnetically Functionalized Cells: Fabrication, Characterization, and Biomedical Applications** *7*	
	Ekaterina A. Naumenko, Maria R. Dzamukova, and Rawil F. Fakhrullin	
2.1	Introduction *7*	
2.2	Magnetic Microbial Cells *8*	
2.2.1	Direct Deposition of MNPs onto Microbial Cells *8*	
2.2.2	Polymer-Mediated Deposition of MNPs onto Microbial Cells *9*	
2.2.2.1	Layer-by-Layer Magnetic Functionalization of Microbial Cells *9*	
2.2.2.2	Single-step Polymer-mediated Magnetic Functionalization of Microbial Cells *11*	
2.2.3	Applications of Magnetically Modified Microbial Cells *15*	
2.2.3.1	Biosorbents and Biocatalysts *15*	
2.2.3.2	Whole-Cell Biosensors and Microfluidic Devices *15*	
2.2.3.3	Remotely Controlled Organisms *16*	
2.3	Magnetic Labeling of Mammal (Human) Cells *18*	
2.3.1	Intracellular Labeling of Cells *18*	
2.3.1.1	Labeling with Anionic Magnetic Nanoparticles *18*	
2.3.1.2	Labeling with Cationic Magnetic Nanoparticles *19*	
2.3.2	Extracellular Labeling of Cells *20*	
2.3.3	Applications of Magnetically Labeled Cells in Biomedicine *20*	
2.3.3.1	MRI Imaging of MNPs-Labeled Cells *21*	
2.3.3.2	MNPs-Mediated Cell Delivery and Tissue Engineering *21*	

2.4	Conclusion 23	
	Acknowledgment 23	
	References 23	
3	**Untethered Insect Interfaces** 27	
	Amol Jadhav, Michel M. Maharbiz, and Hirotaka Sato	
3.1	Introduction 27	
3.2	Systems for Tetherless Insect Flight Control 30	
3.2.1	Various Approaches to Tetherless Flight Control 30	
3.2.2	Neurostimulation for Initiation of Wing Oscillations 30	
3.2.3	Extracellular Stimulation of the Muscles to Elicit Turns 32	
3.3	Implantable Bioelectronics in Insects 33	
3.3.1	Example: Insertion of Flexible Substrates into the Developing Eye 33	
3.4	Conclusions 39	
	References 39	
4	**Miniaturized Biomedical Implantable Devices** 45	
	Ada S.Y. Poon	
4.1	Introduction 45	
4.2	Energy Harvesting as a Pathway to Miniaturization 47	
4.3	Implementation of Implantable Devices 48	
4.3.1	RF Power Harvesting 49	
4.3.1.1	Matching Network 49	
4.3.1.2	Rectifier 49	
4.3.1.3	Regulator and Bandgap Reference 50	
4.3.1.4	Low-Power Controller and Auxiliary Circuits in the Implant Functional Block 50	
4.3.2	Wireless Communication Link 51	
4.3.2.1	Forward Data Link 51	
4.3.2.2	Reverse Data Link 54	
4.3.3	Payload and Applications: Locomotive Implant and Implantable Cardiac Probe 56	
4.3.3.1	Actuation for Therapeutics: Millimeter-Sized Wirelessly Powered and Remotely Controlled Locomotive Implant 56	
4.3.3.2	Low-Power Sensing for Diagnostics: Implantable Intracardiac Probe 59	
4.4	Conclusion 62	
	References 62	
5	**Cross-Hierarchy Design Exploration for Implantable Electronics** 65	
	Mrigank Sharad and Kaushik Roy	
5.1	Introduction 65	
5.2	System Overview of a Generic Bioelectronic Implant 65	
5.3	Circuit Design for Low-Power Signal Processing 67	
5.3.1	Design Challenges for Low-Power Bioelectronic Sensor Interface 67	

5.3.2	Analog Signal Processing Using Subthreshold Circuits	68
5.3.3	Analog-to-Digital Conversion	69
5.3.4	Low-Power Digital Signal Processing	71
5.3.4.1	V_{DD} Scaling and Parallel Processing	71
5.3.4.2	Dynamic Voltage and Frequency Scaling	72
5.3.4.3	Standby Mode Power Reduction	73
5.3.4.4	Minimum Energy Subthreshold Operation	73
5.3.5	FinFETs for Ultralow Voltage Subthreshold Circuits	74
5.4	Architecture-Level Optimizations for Low-Power Data Processing	76
5.4.1	Optimal Apportioning of Computation Task to Analog and Digital Blocks	76
5.4.2	Approximate Computing for Low Power	78
5.5	Design of Energy-Efficient Memory	79
5.5.1	Design Challenges with Subthreshold SRAM	79
5.5.1.1	On-Current to Off-Current Ratio	79
5.5.1.2	Sizing Constraints	79
5.5.1.3	Variability	80
5.5.2	Spin Transfer Torque MRAM (STT-MRAM) for Energy-Efficient Memory Design	80
5.6	Wireless Communication Power Delivery	81
5.6.1	Near-Field Electromagnetic Wireless Communication	82
5.6.2	Far-Field Electromagnetic Wireless Communication	82
5.6.3	Wireless Energy Transfer	83
5.7	Conclusion	83
	References	84
6	**Neural Interfaces: from Human Nerves to Electronics**	**87**
	Jessica D. Falcone, Joav Birjiniuk, Robert Kretschmar, and Ravi V. Bellamkonda	
6.1	Introduction	87
6.2	Fusing Robotics with the Human Body: Interfacing with the Peripheral Nervous System	87
6.2.1	The Anatomy of Peripheral Nerves	88
6.2.1.1	Glial Cells of the Peripheral Nervous System	88
6.2.1.2	Functional Afferent and Efferent Pathways	88
6.2.2	Interfacing with the Periphery for Recording and Stimulation	89
6.2.2.1	Noninvasive Electrodes	89
6.2.2.2	Extraneural Electrodes	90
6.2.2.3	Intrafascicular Electrodes	91
6.2.2.4	Regeneration-Based Electrodes	92
6.2.2.5	Research Designs and Challenges	92
6.3	Listening to the Brain: Interfacing with the Central Nervous System	93
6.3.1	Glial Cells of the Central Nervous System	93
6.3.1.1	Microglia – Sentinels of the Brain	93

6.3.1.2	Astrocytes – Cellular Support for Neurons	94
6.3.2	Interfacing with the Brain for Recordings	94
6.3.2.1	Noninvasive Electrodes	94
6.3.2.2	Extracortical Electrodes	95
6.3.2.3	Invasive Intracortical Electrodes	95
6.3.2.4	Research Designs and Challenges	97
6.4	Electrical Modulation of the Human Nervous System: Stimulation and Clinical Applications	99
6.4.1	Deep Brain Stimulation	100
6.4.1.1	Biological Mechanisms	100
6.4.1.2	Electrode Design and Stimulation	100
6.4.1.3	Research Designs and Challenges	101
6.4.2	Electrical Modulation of Nerve Regeneration	101
6.4.2.1	Biological Mechanisms	102
6.4.2.2	Electrode Stimulation	102
6.4.3	Pain Modulation	102
6.4.3.1	Biological Mechanisms	102
6.4.3.2	Clinical Outcomes	103
6.4.4	Electrical Modulation of Inflammation	103
6.4.4.1	The Vagus Nerve and Stimulation	103
6.4.4.2	Cholinergic Anti-Inflammatory Pathway	104
6.5	Future Directions for Neural Interfacing	105
	References	106
7	**Cyborgs – the Neuro-Tech Version**	**115**
	Kevin Warwick	
7.1	Introduction	115
7.2	Biological Brains in a Robot Body	116
7.3	Deep Brain Stimulation	120
7.4	General Purpose Brain Implants	123
7.5	Noninvasive Brain-Computer Interfaces	126
7.6	Subdermal Magnetic Implants	127
7.7	RFID Implants	128
7.8	Conclusions	130
	References	131
8	**Interaction with Implanted Devices through Implanted User Interfaces**	**133**
	Christian Holz, Tovi Grossman, George Fitzmaurice, and Anne Agur	
8.1	Implanted User Interfaces	135
8.1.1	Design Considerations	136
8.1.1.1	Input through Implanted Interfaces	136
8.1.1.2	Output through Implanted Interfaces	136
8.1.1.3	Communication and Synchronization	137
8.1.1.4	Power Supply through Implanted Interfaces	137

8.1.2	Summary 137
8.2	Evaluating Basic Implanted User Interfaces 137
8.2.1	Devices 138
8.2.2	Experimenters 139
8.2.3	Procedure 139
8.2.4	Medical Procedure 139
8.2.5	Study Procedure and Results 140
8.2.5.1	Touch Input Device (Pressure Sensor, Tap Sensor, Button) 140
8.2.5.2	Hover Input Device (Capacitive and Brightness Sensor) 141
8.2.5.3	Output Device (Red LED, Vibration Motor) 142
8.2.5.4	Audio Device (Speaker and Microphone) 144
8.2.5.5	Powering Device (Powermat Wireless Charger) 145
8.2.5.6	Wireless Communication Device (Bluetooth Chip) 146
8.2.6	Discussion 147
8.2.7	Exploring Exposed Components 147
8.3	Qualitative Evaluation 148
8.3.1	Simulating Implants: Artificial Skin 148
8.3.2	Task and Procedure 149
8.3.3	Participants 150
8.3.4	Results 150
8.4	Medical Considerations 150
8.4.1	Location 150
8.4.2	Device Parameters 151
8.4.3	Risks 151
8.4.4	Implications and Future Studies 152
8.5	Discussion and Limitations 152
8.5.1	Study Limitations 152
8.6	Conclusions 153
	References 153
9	**Ultralow Power and Robust On-Chip Digital Signal Processing for Closed-Loop Neuro-Prosthesis** 155
	Swarup Bhunia, Abhishek Basak, Seetharam Narasimhan, and Maryam Sadat Hashemian
9.1	Introduction 155
9.1.1	Neural Interfaces 158
9.1.2	Closing the Neural Loop: Significance of On-Chip DSP 160
9.2	Algorithm: a Vocabulary-Based Neural Signal 162
9.2.1	Analysis 162
9.2.2	Spike-Level Vocabulary 163
9.2.3	Spike Detection 164
9.2.4	Spike Characterization and Sorting 166
9.2.5	Burst-Level Vocabulary 167
9.2.6	Multichannel Vocabulary for Behavior-Specific Patterns 167
9.2.7	Output Packet Generation 169

9.3	Hardware Implementation	171
9.3.1	Wavelet Module	173
9.3.1.1	Vocabulary Module	177
9.3.2	Area, Power Reduction Methodologies	179
9.3.2.1	Subthreshold versus Super-Threshold Operation	181
9.3.3	Impact of Process Variations on Yield	183
9.3.3.1	Preferential Design	185
9.3.4	Overall Design Flow	188
9.4	Summary	191
	References	191
10	**Implantable CMOS Imaging Devices**	**195**
	Jun Ohta	
10.1	Introduction	195
10.2	Fundamentals of CMOS Imaging Devices	198
10.2.1	Photosensors	198
10.2.2	Active Pixel Sensor	199
10.2.3	Log Sensor	201
10.2.4	Pulse Width Modulation Sensor	202
10.2.5	SPAD Sensor	203
10.3	Artificial Retina	203
10.3.1	Principle of Artificial Retina	203
10.3.2	Artificial Retina Based on CMOS Imaging Device	204
10.4	Brain-Implantable CMOS Imaging Device	210
10.4.1	Measurement Methods for Brain Activities	210
10.4.2	Fiber Endoscope and Head-Mountable Device	211
10.4.3	Brain-Implantable CMOS Imaging Device	212
10.5	Summary and Future Directions	215
	Acknowledgments	217
	References	217
11	**Implanted Wireless Biotelemetry**	**221**
	Mehmet Rasit Yuce and Jean-Michel Redoute	
11.1	Introduction	221
11.2	Biotelemetry	223
11.2.1	Inductive Link for Forward Data	225
11.2.2	Wireless Power Link	226
11.2.3	Implantable Telemetry Links	228
11.2.3.1	Wideband Telemetry Link	228
11.2.3.2	Multichannel Neural Recording Systems	228
11.2.3.3	Wireless Endoscope	230
11.3	Microelectrode Arrays and Interface Electronics	232
11.3.1	Stimulation Front Ends	233
11.3.2	Recording Front-Ends	238

11.3.2.1	Instrumentation Amplifier *239*	
11.4	Conclusion *242*	
	References *242*	

12 Nano-Enabled Implantable Device for *In Vivo* Glucose Monitoring *247*
Esteve Juanola-Feliu, Jordi Colomer-Farrarons, Pere Miribel-Catalá, Manel González-Piñero, and Josep Samitier

12.1	Introduction *247*	
12.1.1	Nanotechnology *247*	
12.1.2	Nanomedicine *248*	
12.2	Biomedical Devices for *In Vivo* Analysis *249*	
12.2.1	State of the Art *249*	
12.2.2	The Innovative Biomedical Device *250*	
12.2.3	Architecture of the Implantable Device *251*	
12.2.4	Implantable Front-End Architecture for *In Vivo* Detection Biosensor Applications *254*	
12.2.4.1	Architecture of the Envisaged Subcutaneous Device *254*	
12.2.4.2	Implementation and Results *258*	
12.2.5	The Diabetes Care Devices Market *260*	
12.3	Conclusions and Final Recommendations *261*	
	References *262*	

13 Improving the Biocompatibility of Implantable Bioelectronics Devices *265*
Gymama Slaughter

13.1	Introduction *265*	
13.2	Implantable Bioelectronic Device Materials *267*	
13.3	Surface Composition *269*	
13.4	Response to Implantation *273*	
13.5	Conclusion *278*	
	References *279*	

14 Abiotic (Nonenzymatic) Implantable Biofuel Cells *285*
Sven Kerzenmacher

14.1	Introduction *285*	
14.1.1	The History of Implantable Abiotic Fuel Cells *285*	
14.2	Basic Principles *286*	
14.2.1	Electrode Reactions and Theoretical Potentials *287*	
14.2.2	Practical Fuel Cell Voltage, Power Density, and Efficiency *289*	
14.2.3	Reliable Characterization of Implantable Glucose Fuel Cells *291*	
14.3	Abiotic Catalyst Materials and Separator Membranes *292*	
14.3.1	Electrocatalysts for Glucose Oxidation *292*	
14.3.2	Electrocatalysts for Oxygen Reduction *293*	
14.3.3	Separator Membranes *294*	

14.4	Design Considerations 295	
14.4.1	Site of Implantation 295	
14.4.2	Strategies to Cope with the Presence of Mixed Reactants 297	
14.5	State-of-the-Art and Practical Examples 299	
14.5.1	Comparison of Fuel Cell Designs and Their Power Densities 299	
14.5.2	Factors Affecting Long-Term Operation 304	
14.6	Conclusion and Outlook 307	
14.6.1	State-of the-Art 307	
14.6.2	Applications 308	
14.6.3	Challenges and Future Trends 309	
	References 309	

15 Direct-Electron-Transfer-Based Enzymatic Fuel Cells *In Vitro*, *Ex Vivo*, and *In Vivo* 315

Magnus Falk, Dmitry Pankratov, Zoltan Blum, and Sergey Shleev

15.1	Introduction 315
15.2	Oxidoreductases for Direct-Electron-Transfer-Based Biodevices 316
15.2.1	Anodic Bioelements 317
15.2.2	Cathodic Bioelements 321
15.3	Design of Enzyme-Based Biodevices 323
15.3.1	Electrode Material 325
15.3.2	Electrode Function 327
15.4	Examples of Direct Electron Transfer Enzymatic Fuel Cells 329
15.4.1	Enzymatic Fuel Cells Operating *In Vitro* 329
15.4.2	Biodevices Operating *In Vivo* 334
15.4.3	Enzymatic Fuel Cells Operating *Ex Vivo* 336
15.4.4	Summary 339
15.5	Outlook 340
	References 341

16 Enzymatic Fuel Cells: From Design to Implantation in Mammals 347

Serge Cosnier, Alan Le Goff, and Michael Holzinger

16.1	Introduction 347
16.2	Design of Implantable Bioelectrodes of Glucose Biofuel Cells 352
16.3	Packaging of Implanted Biofuel Cells 356
16.4	Surgery 358
16.5	Implanted Biofuel Cell Performances 359
	References 361

17 Implanted Biofuel Cells Operating *In Vivo* 363

Evgeny Katz

17.1	Implanted Biofuel Cells 363
	Acknowledgment 377
	References 377

18	**Biomedical Implantable Systems – History, Design, and Trends** *381*	
	Wen H. Ko and Philip X.-L. Feng	
18.1	Introduction *381*	
18.2	History: Review of Implant Systems *383*	
18.2.1	Historical Review of Early Implant Systems, 1950–1970 *384*	
18.2.1.1	Historical Review of Early Implant *Telemetry* Systems, 1950–1970 *384*	
18.2.1.2	Historical Review of Early Implant *Stimulation* Systems, 1950–1970 *391*	
18.2.1.3	Historical Review of Early Implant *Control* Systems, 1950–1970 *391*	
18.2.2	Historical Review of Implant Systems – 1970–1990 *394*	
18.2.3	Historical Review of Implant Systems – 1990–2012 *395*	
18.3	Design of Implant Systems *396*	
18.3.1	Basic Considerations and Characteristics of RF MEMS Implantable Systems *397*	
18.3.1.1	Legal Considerations of the Radio Frequency (RF), Field Strength, and Power Levels *397*	
18.3.1.2	Biocompatibility and Protection of the Biomedical Implant Systems *398*	
18.3.1.3	Characteristics of Biological and Medical Signals *399*	
18.3.2	Design Considerations of Implantable Systems *400*	
18.3.3	Micropower Electronic Design Approaches and Samples *401*	
18.3.4	Power Supply Design *403*	
18.3.5	System Integration and Micro-Packaging *404*	
18.4	Present Challenges *405*	
18.5	Future Trends *406*	
	Acknowledgments *407*	
	References *407*	
19	**Brain–Computer Interfaces: Ethical and Policy Considerations** *411*	
	Ellen M. McGee	
19.1	Introduction *411*	
19.2	Neuroethics *412*	
19.3	Brain–Computer Interfaces *412*	
19.4	Noninvasive Interfaces *413*	
19.5	Partially Invasive Interfaces *413*	
19.6	Invasive Interfaces *413*	
19.7	Development of Brain–Computer Interfaces *414*	
19.8	Therapy/Enhancement *418*	
19.9	Ethical Issues *419*	
19.10	Brain Chips and Cloning *420*	
19.11	Regulatory Procedures *424*	
19.12	Principles and Standards for Adoption *426*	
	References *430*	

20 Conclusions and Perspectives *435*
Evgeny Katz
References *436*

Index *437*

Preface

Scientific research and engineering in the area of implantable bioelectronic devices have been progressing rapidly in the last two decades, greatly contributing to medical and technological advances, thus resulting in numerous applications. In addition, this research absorbs novel achievements and discoveries in microelectronics, computing, biotechnology, materials science, micromachinery, and many other science and technology areas. The present book overviews the multidisciplinary field of implantable bioelectronics, highlighting its key aspects and future perspectives. The chapters written by the leading experts cover different subareas of the science and technology related to implantable bioelectronics – together covering the multifaceted area and its applications. The different topics addressed in this book will be of high interest to the interdisciplinary community active in the area of implantable bioelectronics. It is hoped that the collection of the different chapters will be important and beneficial for researchers and students working in various areas related to bioelectronics, including microelectronics, biotechnology, materials science, computer science, medicine, and so on. Furthermore, the book is aimed at attracting young scientists and introducing them to the field, while providing newcomers with an enormous collection of literature references. I, indeed, hope that the book will spark the imagination of scientists to further develop the topic.

It should be noted that the field of implantable bioelectronics relates to some extent to the fascinating area of unconventional computing, the consideration of which is outside the scope of the present book. This complementary area of molecular/biomolecular computing was covered in two other recent books from Wiley-VCH: *Molecular and Supramolecular Information Processing: From Molecular Switches to Logic Systems*, E. Katz (Ed.), Wiley-VCH, Weinheim, Germany, **2012** and *Biomolecular Information Processing – From Logic Systems to Smart Sensors and Actuators*, E. Katz (Ed.), Wiley-VCH, Weinheim, Germany, **2012**.

Finally, the editor (E. Katz) and publisher (Wiley-VCH) express their gratitude to all authors of the chapters, whose dedication and hard work made this book possible, hoping that the book will be interesting and beneficial to researchers and students working in various areas related to bioelectronics. I would like to conclude this preface by thanking my wife Nina for her support in every respect in the past 40 years. Without her help, it would not have been possible to complete this work.

August 2013

Evgeny Katz
Potsdam, NY, USA

List of Contributors

Anne Agur
University of Toronto
Department of Surgery
Faculty of Medicine
Toronto, Medical Sciences
Building
Canada

Abhishek Basak
Case Western Reserve University
Department of Electrical
Engineering and Computer
Science
Nanoscape Research Laboratory
10900 Euclid Avenue
Cleveland, OH 44106
USA

Ravi V. Bellamkonda
WH Coulter Professor and
School Chair
GRA Distinguished Scholar
Georgia Institute of Technology
and Emory University School of
Medicine
Wallace H. Coulter Department
of Biomedical Engineering
313 Ferst Dr
Atlanta, GA 30332-0535
USA

Swarup Bhunia
Case Western Reserve University
Department of Electrical
Engineering & Computer Science
711 Glennan Building
Cleveland, OH 44106
USA

Joav Birjiniuk
Georgia Institute of Technology
and Emory University School of
Medicine
Wallace H. Coulter Department
of Biomedical Engineering
313 Ferst Dr
Atlanta, GA 30332
USA

Zoltan Blum
Malmö University
Biomedical Sciences
Health and Society
Jan Waldenströms gata 25
SE-20506 Malmö
Sweden

Jordi Colomer-Farrarons
University of Barcelona
Department of Electronics
Bioelectronics and
Nanobioengineering Research
Group (SIC-BIO)
Martí i Franquès 1
Planta 2
08028 Barcelona
Spain

Serge Cosnier
CNRS-Université Joseph Fourier
Grenoble 1
Département de Chimie
Moléculaire
UMR-5250
570 rue de la chimie
BP 53
38041 Grenoble cedex 9
France

Maria R. Dzamukova
Kazan (Idel buye\Volga region)
Federal University
Biomaterials and Nanomaterials
Group
Department of Microbiology
Institute of Fundamental
Medicine and Biology
Kreml urami 18
Kazan 420008
Republic of Tatarstan
Russian Federation

Rawil F. Fakhrullin
Kazan (Idel buye\Volga region)
Federal University
Biomaterials and Nanomaterials
Group
Department of Microbiology
Institute of Fundamental
Medicine and Biology
Kreml urami 18
Kazan 420008
Republic of Tatarstan
Russian Federation

Jessica D. Falcone
Georgia Institute of Technology
School of Electrical and
Computer Engineering
791 Atlantic Dr
Atlanta, GA 30332
USA

Magnus Falk
Malmö University
Biomedical Sciences
Health and Society
Jan Waldenströms gata 25
SE-20506 Malmö
Sweden

Philip X.-L. Feng
Case Western Reserve University
Department of Electrical
Engineering and Computer
Science
Case School of Engineering
10900 Euclid Avenue
Cleveland, OH 44106
USA

George Fitzmaurice
Autodesk Research
210 King Street East
Suite 600, Toronto
Ontario, M5A 1J7
Canada

Manel González-Piñero
University of Barcelona
Department of Public Economy
Political Economy and Spanish
Economy
Av. Diagonal 690-696
08034 Barcelona
Spain

and

Technical University of Catalonia
Biomedical Engineering Research
Centre
Pau Gargallo 5
08028 Barcelona
Spain

Tovi Grossman
Autodesk Research
210 King Street East
Suite 600, Toronto
Ontario, M5A 1J7
Canada

Maryam Sadat Hashemian
Case Western Reserve University
Department of Electrical
Engineering and Computer
Science
Nanoscape Research Laboratory
10900 Euclid Avenue
Cleveland, OH 44106
USA

Christian Holz
University of Potsdam
Hasso Plattner Institute
Prof.-Dr.-Helmert-Str. 2 - 3
Potsdam, 14482
Germany

Michael Holzinger
CNRS-Université Joseph Fourier
Grenoble 1
Département de Chimie
Moléculaire
UMR-5250
570 rue de la chimie, BP 53
38041 Grenoble cedex 9
France

Amol Jadhav
University of California at
Berkeley
Department of Electrical
Engineering and Computer
Science
Berkeley, CA
USA

Esteve Juanola-Feliu
University of Barcelona
Department of Electronics
Bioelectronics and
Nanobioengineering Research
Group (SIC-BIO)
Martí i Franquès 1
Planta 2
08028 Barcelona
Spain

Evgeny Katz
Clarkson University
Department of Chemistry and
Biomolecular Science
8 Clarkson Avenue
Potsdam, NY 13699
USA

Sven Kerzenmacher
University of Freiburg
Department of Microsystems
Engineering – IMTEK
Georges-Koehler-Allee 103
D-79110 Freiburg
Germany

Wen H. Ko
Case Western Reserve University
Department of Electrical
Engineering and Computer
Science
Case School of Engineering
10900 Euclid Avenue
Cleveland, Ohio 44106
USA

Robert Kretschmar
Georgia Institute of Technology
Wallace H. Coulter Department
of Biomedical Engineering
313 Ferst Dr
Atlanta, GA 30332
USA

Alan Le Goff
CNRS-Université Joseph Fourier
Grenoble 1
Département de Chimie
Moléculaire
UMR-5250
570 rue de la chimie
BP 53
38041 Grenoble cedex 9
France

Michel M. Maharbiz
University of California at
Berkeley
Department of Electrical
Engineering and Computer
Science
Berkeley, CA
USA

Ellen M. McGee
Retired Adjunct Philosophy
Professor
Retired Director, The Long Island
Center for Ethics
Long Island University
CW Post
720 Northern Blvd
Brookville, NY 11548
USA

Pere Miribel-Català
University of Barcelona
Department of Electronics
Bioelectronics and
Nanobioengineering Research
Group (SIC-BIO)
Martí i Franquès 1
Planta 2
08028 Barcelona
Spain

Seetharam Narasimhan
Intel Corporation
Portland, OR
USA

Ekaterina A. Naumenko
Kazan (Idel buye\Volga region)
Federal University
Biomaterials and Nanomaterials
Group
Department of Microbiology
Institute of Fundamental
Medicine and Biology
Kreml urami 18
Kazan 420008
Republic of Tatarstan
Russian Federation

Jun Ohta
Nara Institute of Science and Technology
Graduate School of Materials Science
8916-5 Takayama
Ikoma
Nara 630-0192
Japan

Dmitry Pankratov
Malmö University
Biomedical Sciences
Health and Society
Jan Waldenströms gata 25
SE-20506 Malmö
Sweden

Ada S. Y. Poon
Stanford University
Department of Electrical Engineering
David Packard Building
350 Serra Mall
Stanford, CA 94305-9505
USA

Jean-Michel Redoute
Monash University
Biomedical Integrated Circuits and Sensors Laboratory
Department of Electrical Engineering and Computer Systems Engineering
Clayton
Melbourne Vic 3800
Australia

Kaushik Roy
Purdue University
School of Electrical and Computer Engineering
Electrical Engineering Building
465 Northwestern Ave.
West Lafayette
Indiana 47907-2035
USA

Josep Samitier
University of Barcelona
Department of Electronics
Bioelectronics and Nanobioengineering Research Group (SIC-BIO)
Martí i Franquès 1
Planta 2
08028 Barcelona
Spain

and

IBEC-Institute for Bioengineering of Catalonia
µ nanosystems Engineering for Biomedical Applications Research Group
Baldiri Reixac 10-12
08028 Barcelona
Spain

and

CIBER-BBN-Biomedical Research Networking Center in Bioengineering
Biomaterials and Nanomedicine
María de Luna 11
Edificio CEEI
50018 Zaragoza
Spain

Hirotaka Sato
School of Mechanical and
Aerospace Engineering
3School of Electrical and
Electronic Engineering
Nanyang Technological
University
Singapore

Mrigank Sharad
Purdue University
School of Electrical and
Computer Engineering
Electrical Engineering Building
465 Northwestern Ave.
West Lafayette
Indiana 47907-2035
USA

Sergey Shleev
Malmö University
Biomedical Sciences
Health and Society
Jan Waldenströms gata 25
SE-20506 Malmö
Sweden

Gymama Slaughter
University of Maryland Baltimore
County
Department of Computer Science
and Electrical Engineering
Bioelectronics Laboratory
Baltimore, MD 21250
USA

Kevin Warwick
University of Reading
School of Systems Engineering
Whiteknights
Reading RG6 6AY
UK

Mehmet Rasit Yuce
Monash University
Biomedical Integrated Circuits
and Sensors Laboratory
Department of Electrical
Engineering and Computer
Systems Engineering
Clayton
Melbourne Vic 3800
Australia

1
Implantable Bioelectronics – Editorial Introduction
Evgeny Katz

The integration of electronic elements with biological systems, resulting in novel devices with unusual functionalities, attracts significant research efforts owing to fundamental scientific interest and the possible practical applications of such devices. The commonly used buzzword "bioelectronics" highlights the functional integration of two different areas of science and engineering – biology and electronics, to yield a novel subarea of biotechnology [1, 2]. Bioelectronics is a rapidly developing, multidisciplinary research direction, combining novel achievements from electronics miniaturization allowing devices to operate with ultralow power consumption [3], the development of flexible devices for interfacing with biological tissue via advances within materials science [4], bio-inspired unconventional computing for mimicking biological information processing [5], and many other highly innovative science and technology areas. One of the most advanced applications benefiting from the development of bioelectronics is the rapidly progressing area of biosensors technology [6]. The use of novel nanostructured materials integrated with biomolecular systems [7–9] tremendously contributes to the rapid progress of bioelectronics, especially in regard to biosensor applications [10]. The novel electronic systems based on flexible supports [11] for direct interfacing with biological tissues are very promising for use in implantable bioelectronic devices [12] (Figure 1.1).

The most challenging developments in bioelectronics are related to biomedical applications, particularly advancing the direct coupling of electronic devices/machines with living organisms, where electronics operates in a biological environment implanted within a living body. This technology is already highly advanced, at least in some medical applications such as implantable cardiostimulators [13, 14] and various other implantable prosthetic devices [15, 16]. The most important issue in the biotechnological engineering of implantable devices is the interface between living tissues and artificial man-made implantable devices. Cardiac defibrillators/pacemakers, deep brain neurostimulators, spinal cord stimulators, gastric stimulators, foot drop implants, cochlear implants, insulin pumps, retinal implants, implantable neural electrodes, muscle implants, and other implantable devices must perform their functions by directly interacting with the respective organs to improve their natural operation

Implantable Bioelectronics, First Edition. Edited by Evgeny Katz.
© 2014 Wiley-VCH Verlag GmbH & Co. KGaA. Published 2014 by Wiley-VCH Verlag GmbH & Co. KGaA.

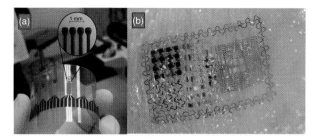

Figure 1.1 (a) Flexible bioelectronic devices allow interfacing with a biological tissue. (b) A new type of biosensor uses flat, flexible electronics ("tattoo"-bioelectronics) printed on a thin rubbery sheet, which can stick to human skin for at least 24 h. (Photos "a" and "b" were kindly provided by Prof. Joseph Wang, University California San Diego, USA, and Prof. John A. Rogers, University of Illinois at Urbana-Champaign, USA, respectively.)

Figure 1.2 Prof. Warwick had his nervous system wired to a robotic hand allowing its remote control. (a) A 100-electrode array surgically implanted into the median nerve fibers of the left arm allowed electrical reading of nerve signals. (b) The robotic hand was remotely controlled by signals from the researcher's nervous system. (Photos "a" and "b" were kindly provided by Prof. Kevin Warwick, University of Reading, UK.)

or substitute the missing function. Implantable medical devices can also restore function by integrating with nondamaged tissue within an organ. The artificially generated electrical and sometimes electromechanical activity in each of these cases must be engineered within the context of the physiological system and its biological characteristics. For example, in one of the recent research projects [17], a nervous system was wired to a robotic hand, allowing its remote control (Figure 1.2). Neural signals were transmitted to various technological devices to directly control them, in some cases via the Internet, and feedback to the brain was obtained from, for example, the fingertips of the robot hand [17].

Highly integrated systems also make possible the development of implantable devices that can sense their biological environment in real time and properly respond to the changing conditions. Integrated *"Sense-and-Act"* systems for intelligent drug delivery have emerged, contributing to the novel concept of personalized medicine and appear particularly important for advancing point-of-care and end-user applications [18]. Although very sophisticated digital electronics can provide perfect internal operation of the implantable devices, their interfacing with the biological environment requires further advancement. New materials and novel

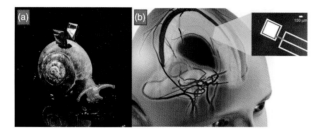

Figure 1.3 Implantation of biofuel cells in a living tissue can provide electrical power harvested from metabolic species for activating implantable bioelectronic devices. (a) A biofuel cell implanted in a free moving snail. (b) Conceptual schematic design for an implanted biofuel cell harvesting power from the cerebrospinal fluid, showing a plausible site of implantation within the subarachnoid space. The inset is a micrograph of one prototype, showing the metal layers of the anode (central electrode) and cathode contact (outer ring) patterned on a silicon wafer. (Photo "a" originates from Prof. E. Katz laboratory, Clarkson University, NY, USA. Part "b" is adapted from Ref. [23] with permission.)

concepts are needed for improved interfacing of the biological and electronic systems. Improving biocompatibility, via surface chemistry, is critical for enabling future implantable bioelectronic devices. Information processing by the integrated biological/electronic system requires novel computational approaches because natural information processing is conceptually different from the digital operation used in modern electronics. New methods for harvesting and managing energy to power implantable devices are required [19, 20]. They can be based on bio-inspired approaches using, for example, implantable biofuel cells harvesting energy from the internal physiological resources [21–23] (Figure 1.3). Revolutions in miniaturized electronic devices, cognitive science, bioelectronics, bio-inspired unconventional computing, nanotechnology, and applied neural control technologies are resulting in breakthroughs in the integration of humans and machines. The interactions of electronic computing elements, wireless information processing systems, advances in prosthetic devices, and artificial implants facilitate the merging of humans with machines. These exciting advancements lay the foundation for the development of bionic animal/human–machine hybrids [24] (Figures 1.4 and 1.5). Apart from biomedical applications, one can foresee bioelectronic self-powered "cyborgs" capable of autonomous operation using power from biological sources, utilized in environmental monitoring, homeland security, and military applications.

The present book summarizes the diverse subareas of implantable bioelectronics including the modification of biological cells, interfacing tissues, and particularly nervous systems with electronics, harvesting energy from biological sources using implantable biofuel cells and creating "cyborgs" where the function of biological organisms is highly integrated with electronic systems and machines. The variety of systems described in the book and their possible applications are really impressive! While some systems and their applications represent the present level of technology, others are at the interface with future advancements. Possible revolutionary changes in a human's life can be expected on the basis of the rapid progress in the

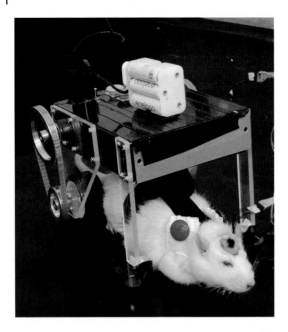

Figure 1.4 Brain–machine interface allowing control of moving robotic vehicles. Rat–robot hybrid involves implanted neural electrodes that allow the rat's brain signals to control a motorized vehicle. (Photo was kindly provided by Prof. Kunihiko Mabuchi and Dr. Osamu Fukayama, The University of Tokyo, Japan.)

Figure 1.5 "Cyborgs" with electronically integrated biomachine parts: (a) A giant flower beetle wears an electronic backpack that allows researchers to wirelessly control its flight. (b) A robotic hand controlled by brain signals can substitute for the missing hand of a disabled person. (Photos "a" and "b" were kindly provided by Prof. Michel M. Maharbiz, University of California, Berkeley, USA, and by the Rehabilitation Institute of Chicago, USA, respectively.)

technology integrating the human body with machines. This requires not only novel technological solutions but also careful ethical considerations. This book aims at summarizing the achievements in this rapidly developing multifaceted research area providing background for further progress and helping in understanding of various aspects in this complex scientific field.

References

1. Willner, I. and Katz, E. (eds) (2005) *Bioelectronics: From Theory to Applications*, Wiley-VCH Verlag GmbH, Weinheim.
2. Pethig, R.R. and Smith, S. (2012) *Introductory Bioelectronics: For Engineers and Physical Scientists*, John Wiley & Sons, Ltd, Chichester.
3. Sarpeshkar, R. (2010) *Ultra Low Power Bioelectronics: Fundamentals, Biomedical Applications, and Bio-Inspired Systems*, Cambridge University Press, Cambridge.
4. Someya, T. (ed) (2013) *Stretchable Electronics*, Wiley-VCH Verlag GmbH, Weinheim.
5. Katz, E. (ed) (2012) *Biomolecular Information Processing – From Logic Systems to Smart Sensors and Actuators*, Wiley-VCH Verlag GmbH, Weinheim.
6. Banica, F.-G. (2012) *Chemical Sensors and Biosensors: Fundamentals and Applications*, John Wiley & Sons, Ltd, Chichester.
7. Katz, E. and Willner, I. (2004) *Angew. Chem. Int. Ed.*, **43**, 6042–6108.
8. Katz, E. and Willner, I. (2004) *ChemPhysChem.*, **5**, 1084–1104.
9. Shipway, A.N., Katz, E., and Willner, I. (2000) *ChemPhysChem.*, **1**, 18–52.
10. Li, S., Singh, J., Li, H., and Banerjee, I.A. (eds) (2011) *Biosensor Nanomaterials*, Wiley-VCH Verlag GmbH, Weinheim.
11. Cai, J., Cizek, K., Long, B., McAferty, K., Campbell, C.G., Allee, D.R., Vogt, B.D., La Belle, J., and Wang, J. (2009) *Sens. Actuat. B*, **137**, 379–385.
12. DeMason, C., Choudhury, B., Ahmad, F., Fitzpatrick, D.C., Wang, J., Buchman, C.A., and Adunka, O.F. (2012) *Ear Hearing*, **33**, 534–542.
13. Hayes, D.L., Asirvatham, S.J., and Friedman, P.A. (eds) (2013) *Cardiac Pacing, Defibrillation and Resynchronization: A Clinical Approach*, Wiley-Blackwell, Chichester.
14. Barold, S.S., Stroobandt, R.X., and Sinnaeve, A.F. (2010) *Cardiac Pacemakers and Resynchronization Step-by-Step*, Wiley-Blackwell, Chichester.
15. Zhou, D. and Greenbaum, E. (eds) (2009) *Implantable Neural Prostheses 1: Devices and Applications*, Springer, Dordrecht.
16. Hakim, N.S. (ed) (2009) *Artificial Organs*, Springer, London.
17. Warwick, K. and Ruiz, V. (2008) *Neurocomputing*, **71**, 2619–2624.
18. Spekowius, G. and Wendler, T. (eds) (2006) *Advances in Healthcare Technology: Shaping the Future of Medical Care*, Springer, Dordrecht.
19. Sue, C.-Y. and Tsai, N.-C. (2012) *Appl. Energy*, **93**, 390–403.
20. Yun, J., Patel, S.N., Reynolds, M.S., and Abowd, G.D. (2011) *IEEE Trans. Mob. Comput.*, **10**, 669–683.
21. Halámková, L., Halámek, J., Bocharova, V., Szczupak, A., Alfonta, L., and Katz, E. (2012) *J. Am. Chem. Soc.*, **134**, 5040–5043.
22. Zebda, A., Cosnier, S., Alcaraz, J.-P., Holzinger, M., Le Goff, A., Gondran, C., Boucher, F., Giroud, F., Gorgy, K., Lamraoui, H., and Cinquin, P. (2013) *Sci. Rep.*, **3**, 1516, art. #1516.
23. Rapoport, B.I., Kedzierski, J.T., and Sarpeshkar, R. (2012) *PLoS ONE*, **7**, art. #e38436.
24. Johnson, F.E. and Virgo, K.S. (eds) (2006) *The Bionic Human: Health Promotion for People with Implanted Prosthetic Devices*, Humana Press, Totowa, NJ.

2
Magnetically Functionalized Cells: Fabrication, Characterization, and Biomedical Applications

Ekaterina A. Naumenko, Maria R. Dzamukova, and Rawil F. Fakhrullin

2.1
Introduction

The ability to control implantable devices using external stimuli is deemed to be of paramount importance. Currently, the use of magnetically facilitated control over spatial distribution and functional triggering is regarded as a versatile means to regulate the interaction of functional materials (i.e., drug-carrying containers, magnetically labeled polymer composites, hybrid organic–inorganic materials, etc.). The main reasons behind such enormous popularity of magnetic functional materials are the simplicity of preparation and the straightforward control of functional properties (such a spatial distribution, controllable release, etc.). Obviously, the use of magnetically responsive materials is not limited to biomedical applications; a number of recent reviews comprehensively describe the recent progress in fabrication of magnetically functionalized biomaterials and devices [1–3]. However, the most intriguing applications of magnetically responsive materials are associated with biomedicine [4]. Originally, the use of magnetic nanoparticles (MNPs) in biomedicine was mostly attributed to magnetic separation techniques [5], where magnetically labeled microparticles tailored with certain receptors were used to selectively target and then separate the receptor-specific molecules or even whole cells. The great potential is attributed to the successful application of magnetic nanomaterials (including MNPs and magnetic composites, such as magnetized carbon nanotubes [6]) as MRI contrast agents [7].

Currently, superparamagnetic MNPs (alternatively termed as *SPIONs – superparamagnetic iron oxide nanoparticles*) are either commercially available or can be easily synthesized following the published procedures. Various methods of chemical synthesis of MNPs result in nanoparticles having different size distribution, surface chemistry, and charge, which can be effectively utilized in biomedical applications. The reader is directed to more specific reviews for further information on chemical synthesis and characterization of MNPs [4, 8]. In this chapter, we provide an overview of the current methodology of labeling living cells with MNPs and their use in biomedical applications. Magnetically labeled cells, both microbial and mammal (particularly human cells), are gaining the attention of biomedical

Implantable Bioelectronics, First Edition. Edited by Evgeny Katz.
© 2014 Wiley-VCH Verlag GmbH & Co. KGaA. Published 2014 by Wiley-VCH Verlag GmbH & Co. KGaA.

scientists as the potential areas of their applications include such important directions of biomedicine as biosensors, cell delivery, and tissue engineering.

2.2
Magnetic Microbial Cells

Magnetically labeled microbial cells are regarded as an important tool in biotechnology and biomedicine. The introduction of magnetic functionality onto the microbial cells allows widening the areas of the biotechnological applications of the cells. Basically, the magnetically responsive "exterior" of microbial cells facilitates their use as miniature magnets combined with the enzymatic factory filled with functional cellular enzymes and genetic apparatus. Currently, two major approaches of magnetic functionalization of microbial cells exist; one is based on the direct deposition of MNPs, whereas the other employs polymer-mediated deposition of nanoparticles. Both approaches rely primarily on utilization of SPIONs, consisting mostly of mixtures of magnetite and maghemite. From the practical point of view, the most important feature of the MNPs deposited onto the cells is their ability to be effectively redispersed, thus avoiding irreversible aggregation of magnetic cells.

2.2.1
Direct Deposition of MNPs onto Microbial Cells

The pioneering reports by Ivo Safarik and coauthors [9] for the first time demonstrated the direct deposition of MNPs onto microbial cells. Magnetic ferrofluids were used to fabricate a thin layer of MNPs on yeast cells. The direct deposition of MNPs is based, mostly, on nonspecific adsorption of nanoparticles onto intact cells, or onto cells specifically modified prior to MNPs deposition.

One of the first, the relatively simple procedure of fabrication of magnetically modified fodder yeast cells (*Kluyveromyces fragilis*) using perchloric-acid-stabilized MNPs, was achieved after the repeated washing of cells with 0.1 M acetic acid in order to remove the extractable cell wall compounds, which facilitated the nonspecific precipitation of ferrofluid onto the cells. Typically, fodder yeast cells were washed six to eight times with excessive amounts of 0.1 M acetic acid. Next, the ferrofluid was added to the suspension of prewashed yeast cells and then mixed and incubated at room temperature for 1 h. To remove the unabsorbed ferrofluid, the cells were washed again, first with 0.1 M acetic acid and then with copious amounts of water. The magnetized cells were separated from the liquid phase using a magnetic separator [10]. The feasibility of this approach was further demonstrated for magnetization of microscopic algae *Chlorella vulgaris* cells [11]. Although this approach results in the fabrication of magnetically responsive cells, the treatment with acids severely reduces the number of viable cells after magnetization. Nonviable magnetic cells may find applications *per se*, that is, as adsorbents or catalysts; however, viable magnetic cells allow fully realizing the potential of the cellular enzymatic machinery combined with magnetic responsiveness.

An alternative strategy to render the magnetic functionality on living cells is based on adsorption of ferrofluid onto intact cells. Unlike with the use of the acetic acid wash cycles (as described above), in this case the cells are not introduced into harsh acidic environments, which helps to maintain the viability of the cells. Cell pellets were suspended in a carrier buffer (i.e., basic (pH 10.6) 0.1 M glycin–NaOH buffer) and then introduced into tetramethylammonium (TMA)-hydroxide-stabilized ferrofluid, citrate-stabilized ferrofluid, or perchloric-acid-stabilized ferrofluid at acidic pH. After mixing, the resulting suspensions were incubated for 1 h and then the magnetically functionalized yeast cells were washed with saline buffers and separated using a magnetic separator. This resulted in the successful deposition of MNPs arranged as both single nanoparticles and larger aggregates deposited onto the outer parts of the cell walls of yeast cells, as confirmed using SEM (scanning electron microscopy) and TEM (transmission electron microscopy) imaging of cells. It is noteworthy that there is no penetration of MNPs into the cytoplasm, indicating that the cell walls are not permeable to ferrofluids [12]. The surface topography of MNPs-coated cells was quite rough [9], which may facilitate the adhesive properties of cells or improve the absorption of metal ions, and so on.

2.2.2
Polymer-Mediated Deposition of MNPs onto Microbial Cells

2.2.2.1 Layer-by-Layer Magnetic Functionalization of Microbial Cells

Polymer-mediated assembly of functional nanomaterials is regarded as a versatile way to functionalize living microbial cells with a range of nanomaterials, including noble metal nanoparticles [13], carbon nanotubes [14], graphene nanosheets [15], and halloysite nanotubes [16]. In addition, deposition of layer-by-layer (LbL)-assembled polymers was applied to facilitate the fabrication of inorganic nano-organized shells on the living cells. The LbL-mediated assembly of nanomaterials is not limited to microbial cells [17] but was also applied for the functionalization of mammal cells [18] and viruses [19]. The strategy for the polymer-mediated deposition of functional nanomaterials (nanotubes and nanospheres) is shown schematically in Figure 2.1. Living cells, as well as the micron-sized colloid particles, possess a certain surface charge (typically, all microbial cells exhibit negative ζ-potentials in water), so that deposition begins with the introduction of the cells into the solution of a positively charged polymer. After the appropriate incubation period (followed by extensive washing), the cells are introduced into the negatively charged polymer solution, and the procedure can be repeated for virtually unlimited number of cycles. Once the desired number of polymer multilayers is obtained, the cells can also be modified with charged nanomaterials (i.e., carbon nanotubes or gold nanoparticles, as shown schematically in Figure 2.1). After the deposition of nanoparticles/nanotubes, additional layer(s) of polyelectrolytes are typically deposited to "seal" the nanoparticles on the cell walls and thus prevent disassembly during any mechanical manipulations with cells.

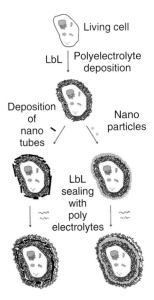

Figure 2.1 A sketch depicting the deposition of oppositely charged polycation (red) and polyanion (blue) multilayered films doped with nanomaterials (nanotubes and nanoparticles) onto isolated living cells. (Reproduced with permission from [20]. Copyright 2010 The Royal Society of Chemistry.)

As one may see, the procedure is extremely simple and tunable and allows the following:

- Adjusting the number of polyelectrolyte layers, thus controlling the thickness of the coatings with the nanometer precision.
- Varying the architecture and composition of the polyelectrolyte layers; virtually any charged polymer, both synthetic and biopolymer, can be utilized in this process.
- Controlling the density of the nanomaterials deposited, which can be achieved by modifying the concentration of nanoparticles and the number of nanoparticles layers.
- Facilitating the deposition of LbL-nanoparticles-coated cells onto other functional surfaces via the control of the outmost layer charge.

A number of polycation/polyanion pairs have been utilized so far in fabrication of LbL-coated cells [17]. Currently, the researchers are focusing on finding out the appropriate biogenic polyelectrolyte pairs as an alternative to the popular synthetic polyelectrolytes.

LbL-mediated deposition of MNPs was first demonstrated in the fabrication of surface-functionalized magnetic yeast cells [21]. Intact yeast cells were subjected to the LbL assembly of polyelectrolyte coating prior to the deposition of MNPs. Two different approaches were demonstrated, depending on the surface charge of the nanoparticles and polymer coatings. In the first approach, the negatively charged

TMA-hydroxide-stabilized magnetic nanorods were used (Figure 2.2); therefore, the negatively charged yeast cells were layer-wise coated with poly(allylamine hydrochloride) (PAH) and poly(sodium polystyrene sulfonate) (PSS) (likely to be the most popular polyelectrolytes used in LbL assembly on cells). The following polyelectrolyte architecture was obtained: PAH/PSS/PAH (in this case the outermost layer provided the cells with the positive charge); then a single layer of negatively charged TMA-stabilized magnetic nanorods was deposited and stabilized on the cells walls with the additional PAH/PSS bilayer. Alternatively, positively charged PAH-stabilized nanospheres were utilized. In this case, the yeast cells were LbL coated with polymers with the architecture PAH/PSS/PAH/PSS before the deposition of positively charged magnetic nanospheres (in this case, the outermost layer provided the cells with the negative charge). Here, to stabilize the MNPs on yeast cell walls, a single PSS outermost layer was deposited on the cellular surface of the cells. The tunability of this approach allowed fabricating the effectively magnetized yeast cells using both negatively and positively charged magnetic nanomaterials.

Figure 2.2 demonstrates the typical transmission electron microscopic images of the intact and LbL/MNPs-functionalized yeast (thin sections embedded into the epoxy resin). The sequential deposition of MNPs and polymer layers resulted in fabrication of the relatively uniform thick (80–130 nm) layer of MNPs (either nanorods (b) or nanospheres (b)) on the cells walls of the yeast. Remarkably, no nanoparticles were detected in the cytoplasm of the magnetized cells. For comparison, the image of the intact cells is shown in Figure 2.2a.

The fact that the nanoparticles were concentrated exclusively on the outer parts of cell walls explains why the viability in LbL-functionalized magnetic yeast cells was not jeopardized, when compared with intact cells. Moreover, the magnetic functionalization of genetically modified green fluorescent protein (GFP)-expressing yeast did not affect the fluorescence intensity of the expressed GFP, which further demonstrates the cellular viability and functionality [21].

However, the deposition of LbL polymer coatings implies the several cycles of polyelectrolyte deposition/washing. Currently, in surface functionalization applications, these cycles can be performed using robotized systems (i.e., automatic dipping robots, and washers); however, these systems are not always directly applicable to biological cells. Moreover, each cycle reduces the concentration of the cells owing to adsorption, and so on. Finally, prolonged expositions of cells to concentrated polymer solutions may also potentially reduce the viability and functionality of the cellular enzymes. Therefore, the alternative methods have been proposed, which are based on the single-step deposition of polymer-coated MNPs.

2.2.2.2 Single-step Polymer-mediated Magnetic Functionalization of Microbial Cells

To perform the magnetic functionalization using a rapid one-step approach, the nanoparticles must readily adhere to the negatively charged cells. This can be achieved if the positively charged polyelectrolytes are deposited on MNPs rather than on microbial cells. Positively charged polyelectrolyte-modified MNPs readily

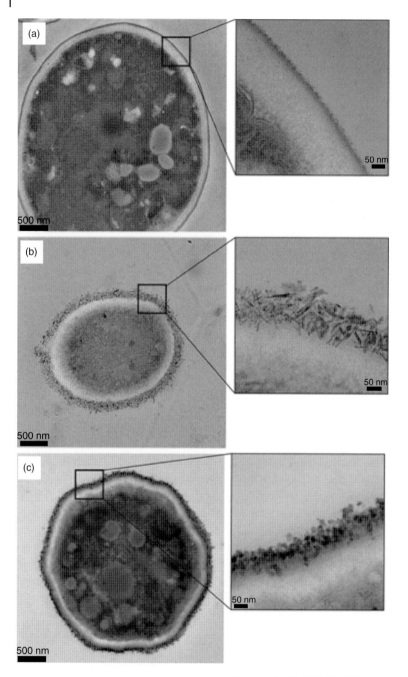

Figure 2.2 Transmission electron microscopy images of thin sections of (a) native yeast cells; (b) yeast cells coated with PAH/PSS/PAH/TMA-stabilized magnetic nanorods/PAH/PSS; and (c) yeast cells coated with PAH/PSS/PAH/PSS/PAH-stabilized magnetic nanospheres/PSS. (Reproduced with permission from [21]. Copyright 2010 The Royal Society of Chemistry.)

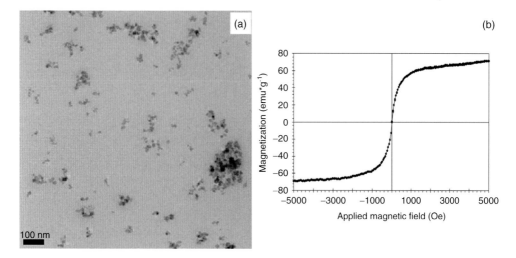

Figure 2.3 A typical TEM image of PAH-stabilized magnetic nanoparticles (a); and magnetization versus applied magnetic field curve demonstrating the superparamagnetic behavior of PAH-stabilized MNPs (at 300 K) (b). (Reproduced with permission from [22].)

adhere to negatively charged microorganisms because of multicenter adsorption, which is stable enough to hold the nanomaterials on the cell walls [20].

Living microalgae cells *Chlorella pyrenoidosa* were successfully coated with PAH-stabilized MNPs [22]. First, MNPs synthesized via the co-precipitation method were introduced into a concentrated aqueous solution of PAH, followed by intensive ultrasonication. As a result, a solution of spherical 15 nm MNPs was produced. The typical TEM image of PAH-stabilized MNPs is shown in Figure 2.3a. The nanoparticles exhibited superparamagnetic behavior (Figure 2.3b) and could be easily sterilized by filtering with a 220 nm syringe filter.

Superparamagnetic materials behave as ferromagnets if introduced into the external magnetic field, but demonstrate diamagnetic properties if removed from the magnetic field. This is very important in terms of the practical use of nano-coated magnetic cells because superparamagnetic nanoparticles help avoid unnecessary aggregation while filtration is a very simple and effective way of decontamination of MNPs. The single-step magnetization procedure is very simple and is based on co-incubation of PAH-stabilized MNPs with the living *C. pyrenoidosa* cells. The aqueous suspension of microalgae cells was introduced carefully into a vessel filled with excess of aqueous MNPs, which was followed by intensive shaking. After 10 min of incubation, the magnetized cells were separated using a strong permanent magnet and then washed to remove the unattached MNPs. The successful deposition of cationic PAH-stabilized MNPs onto the microalgae *C. pyrenoidosa* cells is facilitated by the electrostatic interactions between negatively charged cell walls (ζ-potential -18 ± 6 mV) and positively charged MNPs (ζ-potential $+48 \pm 5$ mV). Transmission electron microscopic images of thin-sectioned intact and MNPs-functionalized microalgae cells illustrate the deposition of MNPs (Figure 2.4a–c). Similarly, with

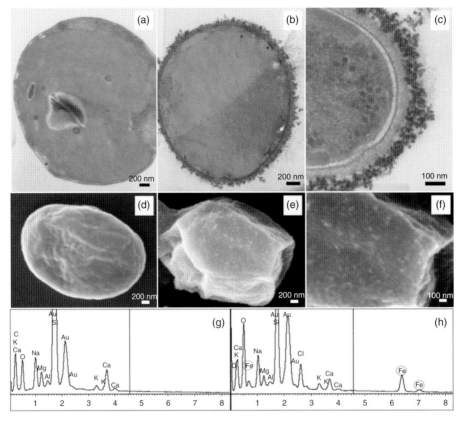

Figure 2.4 TEM images of the thin sections of (a) bare and (b, c) PAH-stabilized MNPs-coated *C. pyrenoidosa* cells. SEM images of (d) bare and (e, f) PAH-stabilized MNPs-coated *C. pyrenoidosa* cells; EDX spectra of (g) bare and (h) PAH-stabilized MNPs-coated *C. pyrenoidosa* cells. Note the circled peaks for Fe in (h). (Reproduced with permission from [22].)

LbL-magnetized cells, the uncoated cells possess a smooth cell wall surface, while the MNPs-coated microalgae are "decorated" with a 90 nm thick uniform layer of MNPs. These observations were further confirmed using SEM and energy-dispersive X-ray (EDX) spectroscopy. Figure 2.4d–f demonstrates the typical SEM images of intact and MNPs-coated microalgae. The corresponding EDX elemental analysis is shown in Figure 2.4g, h; one can see that the EDX spectra of the magnetically coated microalgae contain the characteristic Fe peaks, not present in control sample, which further supports the effective deposition of magnetic iron oxide nanoparticles [22].

Importantly, the cells were fully viable (which is demonstrated later, using the examples of the practical applications of the magnetic microbial cells). Later, a similar approach was applied for the successful magnetic functionalization of yeast cells [23], bacteria [24], and even microscopic nematodes (worms) *Caenorhabditis elegans* [25], which were effectively coated with 20 nm spherical iron oxide nanoparticles producing a uniform magnetic monolayer on intact cells. The major advantage of the one-step polymer-mediated deposition is that the procedure requires no more than 30 min, while the conventional LbL deposition may take up to 8 h to complete. Currently, the main challenge is to fabricate the MNPs functionalized with various cationic polymers to tailor such MNPs to cells without loss of viability.

2.2.3
Applications of Magnetically Modified Microbial Cells

2.2.3.1 Biosorbents and Biocatalysts

One of the most promising areas of application of magnetically functionalized cells is the fabrication of novel effective biological sorbents and catalysts. Here, the functionality of cellular enzymes and carbohydrates is supplemented with the functionality of MNPs deposited onto cells. This, for example, was utilized using magnetically functionalized yeast cells as adsorbent to effectively adsorb mercury ions from water samples [9] and several industrial dyes (such as acridine orange, congo red, safranin, etc.) [10]. The mentioned ligands are effectively absorbed by yeast cells, while the MNPs allow for the straightforward separation and removal of the absorbing cells from the purified samples. Another important area of application of magnetic microbial cells is their use in catalysis [26]. So far, hydrogen peroxide has been effectively decomposed using magnetically modified yeast cells in water solutions up to 2% concentration, leaving only negligible residual concentration of hydrogen peroxide after the treatment. It is noteworthy that MNPs-modified yeast cells were stable at 4 °C in buffers for at least 1 month [26], which simplifies the practical use of the magnetic cellular catalysts.

2.2.3.2 Whole-Cell Biosensors and Microfluidic Devices

Applications of magnetically modified microbial cells in the fabrication of whole-cell biosensors (biosensors based on viable cells immobilized on the transducer) are stimulated by the constant search for novel techniques of cell immobilization. Current techniques are typically based on chemical cross-linking of cells using appropriate linker molecules, physical adsorption, or incorporation of cells into a suitable matrix. Obviously, these immobilization methods suffer from reduced viability of the immobilized cell and irreversible immobilization. In addition, side effects caused by the cross-linking of molecules may cause false-negative or false-positive results. These drawbacks can be avoided if magnetic immobilization is used. Magnetic cells can be temporarily immobilized on biosensor surfaces using an external magnetic field, as shown in Figure 2.5. The miniature magnet positioned below the planar electrode secures the magnetically modified cells directly on the sensor surface.

So far, magnetically functionalized living *C. pyrenoidosa* microalgae [22] and Green Screen™ genetically modified GFP-expressing yeast cells [28] were utilized in the fabrication of whole-cell biosensors. MNPs (coated with PAH) were directly deposited onto the cell walls; then the MNPs-coated cells were accumulated and

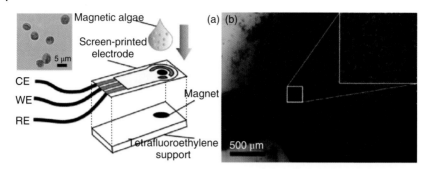

Figure 2.5 (a) A scheme illustrating the biosensor setup based on a screen-printed electrode, a tetrafluoroethylene support with the embedded permanent magnet and magnetically functionalized microalgae cells (CE – counter electrode, WE – working electrode, and RE – reference electrode). The inset shows a typical optical microscopy image of the magnetic cells; and (b) an epifluorescence microscopy image of the magnetic cells assembled on the screen-printed electrode (inset: a higher magnification image of the cells). (Reproduced with permission from [27]. Copyright 2010 The Royal Society of Chemistry.)

immobilized inside microfluidic devices (laminar-flow-based microchambers) [23] and on amperometric screen-printed electrodes [27], as shown in Figure 2.5. This immobilization approach is regarded as a biofriendly way to concentrate the cells on transducers because MNPs, unlike chemical cross-linking, do not affect the enzymatic [24] or genetic [28] apparatus of the cell.

Recently, a whole-cell biosensor based on magnetic retention of MNPs-functionalized microalgae cells on the surface of an electrochemical screen-printed electrode was reported, demonstrating high sensitivity toward triazine herbicides [27]. Next, the magnetically functionalized (using biocompatible PAH-stabilized 15 nm MNPs) GFP reporter yeast was used to detect genotoxicity via monitoring the exposure of the cells to a selected genotoxin (methyl methanesulfonate). A microfluidic device providing gradient mixing was utilized to simultaneously expose magnetic yeast to a range of concentrations of the genotoxin, and then the effective fluorescence emitted from the genotoxin-induced GFP was quantitatively measured. The magnetic retention of the yeast cells in the microchambers using an array of the miniature magnets, followed by removal (which can be achieved by the simple relocation of the magnets) and reloading, allows for the convenient and rapid toxicity screening as well as the fabrication of reusable sensors [23].

2.2.3.3 Remotely Controlled Organisms

Magnetic functionalization of microbial cells and even larger multicellular organisms can also be applied for remote spatial control of the MNPs-coated cells. Magnetically coated cells can be separated from the nonmagnetic, removed from the complex mixture, or spatially positioned using the magnetic field. Recently, *Acinetobacter baylyi* ADP1 bioreporter cells (genetically engineered to express bioluminescence induced by salicylate, toluene/xylene, and alkanes) were functionalized with 18 ± 3 nm MNPs. The magnetically functionalized bacteria can be remotely controlled and concentrated using an external magnetic field. As other magnetically modified species, A. baylyi cells were viable and functional in terms of sensitivity, specificity, and quantitative response to target toxins. It was shown that salicylate-detecting magnetic A. baylyi cells can be applied to detect the target molecules in sediments and soil. The semiquantitative detection of salicylate was performed after the discriminative recovery of magnetic bacteria using a permanent magnet. Obviously, the magnetically coated bioreporter cells are especially useful during analysis in complex environments, where the indigenous cells, microparticles, and other impurities may interfere with the direct measurement of bioreporter cells, and conventional purification methods (i.e., filtration, centrifugation, etc.) are not applicable [24]. In addition, magnetically modified microbial cells can be utilized in microfluidic systems for LbL polyelectrolyte coatings [29]. As was mentioned earlier, one of the main disadvantages of LbL assembly of polyelectrolytes on cells is the time-consuming procedure of polymer deposition/washing cycles. This can be avoided if a microfluidic device utilizing micropillars in a flow channel to continuously generate, encapsulate, and guide LbL polyelectrolyte microcapsules is used [30]. Microparticles traveling through the polymer solutions are coated in a fast and controllable manner; in addition, the miniaturization of coating devices makes them more accessible and minimizes the use of chemicals, and so on. A similar approach was proposed for the polyelectrolyte coating of magnetically modified microbial cells. The microfluidic device generated the laminar flow streams of polyelectrolytes and washing buffers across the chamber, and the magnetic cells were deflected sequentially through the co-flowing streams via an external permanent magnet positioned near the chambers, allowing polyelectrolytes deposition onto the magnetic cells immediately followed by the washing step. The procedure was rapid, not exceeding 90 s for one polymer layer deposition [29]. Obviously, the approach described can be further extended for use with other types of magnetically coated microbial cells and can also be extended to more complex microfluidic devices.

Not only microbial cells but also small multicellular organisms were surface-functionalized with biocompatible 15 nm iron oxide MNPs, which were attached to the cuticle of C. elegans nematodes. As long as the PAH-stabilized positively charged MNPs readily adhered to yeast, algae, and bacteria cell walls, the introduction and incubation (30 min) of the intact worms into the MNPs solution resulted in the uniform distribution of iron oxide MNPs on the cuticle of the nematodes. Such "ironclad" nematodes were viable and could be extracted from complex environments (i.e., sand samples or microbial suspensions) using a permanent magnet. Magnetic surface functionalization of nematodes allows for the effective extraction of magnetic microworms from the substrates containing nonmagnetic impurities or microbial cells. This can be utilized in the separation of magnetic worms from the multicomponent mixture and finds practical applications in toxicity assays of soil or water samples [25].

2.3
Magnetic Labeling of Mammal (Human) Cells

Unlike microbial species, isolated mammal cells lack the thick outer coating, such as cell walls, which seriously limits the approaches for modification of mammal cells with MNPs. However, the magnetic functionalization of mammal (and particularly human) cells is deemed to be a matter of paramount importance, since the magnetic modification (labeling) of human cells opens new avenues in biomedicine [31]. The following directions can be outlined – cell delivery [32], tissue engineering [33], and MRI imaging [34] of MNPs (SPION)-labeled cells. The fact that human cells are protected only by the fragile cellular membrane suggests two general strategies for magnetic labeling:

- intracellular labeling, where the MNPs are somehow introduced into the cellular cytoplasm and remain inside the magnetically labeled cells;
- extracellular (or surface) labeling, where the cellular membranes are coated with MNPs, without penetration into the cytoplasm.

Both approaches exhibit certain advantages and disadvantages, which are reviewed in this chapter.

2.3.1
Intracellular Labeling of Cells

2.3.1.1 Labeling with Anionic Magnetic Nanoparticles

Anionic MNPs (iron oxide cores stabilized with citrate [35], dextran [36], hyaluronic acid [34], etc.) were utilized in the intracellular magnetic labeling of isolated mammal and human cells. Typically, the MNPs are synthesized via the co-precipitation of Fe^{2+} and Fe^{3+} ions in aqueous media after the addition of alkali reagent, followed by washing and stabilization using the appropriately selected anion. Nanoparticles are then added to the culture media and the target cells are incubated in the presence of MNPs for extended periods of time (up to 72 h). Usually, the temperature is kept at 37 °C because reduced temperatures inhibit the MNPs internalization and trafficking process, reducing the overall magnetization of target cells [37]. During incubation, the dose-dependent uptake of MNPs occurs, which can be visualized using either the high-resolution microscopic techniques (i.e., TEM) [38] or SEM [39] flow cytometry [40] and Prussian Blue staining [35]. Figure 2.6 shows the typical TEM image of MNPs concentrated in cytoplasm of magnetically labeled mesenchymal stem cells (MSCs) (citrate-stabilized MNPs were used).

It is believed that the MNPs attach to the plasmatic membrane of cells, producing larger clusters; then, invagination of the membrane occurs, producing densely packed endosomes that further migrate into the cytoplasm and go through the typical endocytosis pathway (early endosome, late endosome, and lysosome, respectively [39]). Anionic MNPs are supposed to stimulate intracellular uptake via the endocytosis mechanism.

C 4h - citrate

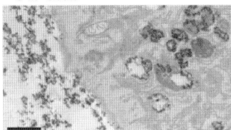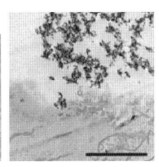

Figure 2.6 TEM images of non-aggregated MNPs inside the mesenchymal stem cells (scale bar − 1 μm). (Image reproduced with permission from [38]. Copyright WILEY-VCH.)

The intracellular concentration of MNPs directly affects the magnetic responsiveness of labeled cells; however, the high levels of nanoparticles in the cytoplasm may also affect cellular viability. So far, the low levels of acute toxicity of citrate-stabilized MNPs were demonstrated; for instance, MNPs-labeled human MSCs demonstrated unaffected morphology, growth rate, surface markers synthesis, and differentiation capacity [35]. However, the same study demonstrated the effect of high intracellular concentrations of MNPs – chondrogenesis and chemokine-induced migration (which was significantly decreased) in magnetically labeled cells [35].

2.3.1.2 Labeling with Cationic Magnetic Nanoparticles

Another approach for intracellular labeling of human cells is based on utilization of cationic MNPs coated with polycations [41] or organized into cationic magnetic liposomes (CML) [33]. Poly(L-lysine) (PLL)-coated MNPs were synthesized via the co-precipitation approach and stabilized using PLL (molecular weight 388 100 Da). It was demonstrated that the serum proteins, though adsorbing to the PLL-coated MNPs, do not remove PLL from the surface of MNPs; thus the MNPs preserve their positive charge and attach to the human MSC. After prolonged co-incubation of cells with PLL-stabilized MNPs, the cells were investigated using TEM and Prussian Blue stain. As in the case of anionic MNPs, MSC were labeled with nanoparticles concentrated predominantly in cellular endosomes [41].

CML were introduced by Ito *et al.* [42] to label human aortic endothelial cells and hepatocytes. Typically, the preparation of CML is based on the incorporation of nanosized Fe_3O_4 particles into the liposomes built from lipid mixtures (i.e., containing a mixture of *N*-(α-trimethylammonioacetyl)-didodecyl-D-glutamate, dilauroylphosphatidylcholine, etc. at certain ratios). The average particle hydrodynamic size of CML is around 150 nm, as measured using dynamic light scattering [43]. Labeling of cells is performed by co-incubation of substrate-attached cells with cell culture media containing CML; for instance, this approach was applied for the magnetic labeling of human cardiomyocytes, mouse fibroblasts, and canine urothelial cells [43]. The toxicity induced by CML toward the target cells was negligible, as studied *in vivo* using mice model.

2.3.2
Extracellular Labeling of Cells

Recently, another approach was proposed, which is based on the functionalization of isolated human cells via a single-step facile process using PAH-stabilized biocompatible MNPs, similarly to the functionalization of microbial cells. Here, the MNPs remain on the cellular membranes, without penetration into the cytoplasm. This approach is based on the electrostatic deposition of biocompatible polymer-stabilized MNPs on cellular membranes. HeLa human cells were used for surface functionalization with PAH-stabilized MNPs. High-magnification SEM and TEM images in Figure 2.7 demonstrate that the nanoparticles are arranged as a uniform monolayer on the membrane surface of the cell. EDX elemental analysis was further employed to demonstrate the distribution of MNPs on HeLa cells.

Interestingly, the cellular blebs (outstanding cytoplasm protrusions) were not coated with MNPs, which further indicates that SPIONs are concentrated around microvilli, are not taken into the cytoplasm, and can also be moved during the formation of blebs. Importantly, MNPs-coated cells preserve their viability after magnetization. Obviously, this approach can be extended to other human cell types, as demonstrated using cow embryo epithelial cells [44].

2.3.3
Applications of Magnetically Labeled Cells in Biomedicine

MNPs are widely employed in biomedicine, both *per se* and as components of more complex hybrid systems [45–47]. Here, we focus on just two, but very important, applications of magnetically labeled cells – magnetic resonance imaging and tissue engineering.

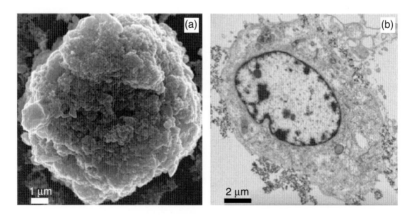

Figure 2.7 SEM (a) and TEM (b) images of MNPs-functionalized HeLa cells.

2.3.3.1 MRI Imaging of MNPs-Labeled Cells

SPIONs have been used as contrast-enhancing reagents for MRI imaging of tissues and organs [48]; however, the most intriguing application is the application of MRI imaging techniques in tracking the magnetically labeled cells within the body. MRI-mediated detection of magnetically labeled cells is a noninvasive approach allowing for the real-time monitoring of target cells both *in vitro* and *in vivo*. Typically, intracellular labeling with MNPs is employed for MRI imaging, using both commercially available and in-house synthesized MNPs [37]. Generally, size- and dose-dependent efficacy of MNPs-enhanced MRI contrast was observed [49]. Intramuscularly transplanted anionic-MNPs-labeled MSCs were monitored using MRI in rats, with a discrete dependence on surface coating of MNPs [35]. MRI imaging results were further confirmed using histological analysis combined with Prussian Blue stain. Cationic-MNPs-labeled stem cells were also prepared using PLL-stabilized MNPs. *In vitro* MR imaging of gelatin-embedded cells demonstrated visible contrast even at very low cell concentrations (200 cells μl^{-1}) when compared with control samples containing the same numbers of cells. After injection into rats, the MNPs-labeled implants were visible 3 days after transplantation, which suggests the use of MNPs-labeled cells in long-term MRI observations of implants [41].

Surface chemistry of MNPs also affects the functional properties of labeled cells; for instance, hyaluronic-acid-stabilized MNPs were used for *in vitro* (cells embedded into agarose gel samples) and *in vivo* (in mice models) MRI labeling of human ovarian cancer cells expressing hyaluronic acid receptors. MNPs-enhanced MRI images of high quality were obtained at low concentrations of MNPs (2.6 mg ml^{-1}), which suggests the high efficacy of MNPs-labeling [34]. Obviously, receptor-functionalized MNPs may find wider applications in areas that require the selective detection of certain areas/tissues in the human body (including implantable devices). Antibody-labeled MNPs [50] can also be utilized in the targeted labeling of cells *in vivo*. The applicability of MNPs-labeled cells for MRI monitoring was also demonstrated for proliferating cells [37]. Future work in this area is to be focused on signal enhancement and toxicity reduction, via surface-tailored molecules and MNPs size and shape [50].

2.3.3.2 MNPs-Mediated Cell Delivery and Tissue Engineering

Another important area where the magnetized cells may find important applications is tissue engineering – directed fabrication of tissues and potentially organs *de novo* using isolated cells as building blocks [51]. Tissue engineering aims at replacing the costly and inconvenient donor tissues/organ implantation approaches. Apart from overcoming ethical issues, the implantation of artificial tissues/organs allows overcoming immune incompatibility problems.

One of the major challenges in tissue engineering is the positioning and concentration of cells within a selected area on an extracellular scaffold or (in case of scaffold-free tissue engineering) without using any supporting scaffolds at all. Magnetically labeled cells are believed to be an important tool in tissue engineering [49]. The generalized scheme of magnetically assisted tissue engineering is shown

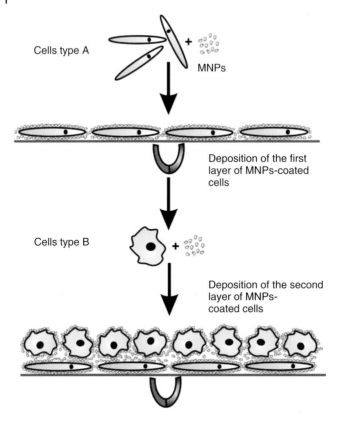

Figure 2.8 A scheme demonstrating the magnetically facilitated formation of multicellular layers.

in Figure 2.8. The deposition of uniform layers of magnetically labeled cells is facilitated by a permanent magnet, which can be positioned under a planar (as shown in Figure 2.8) or curved substrate.

Magnetically tagged cells were assessed for the magnetically facilitated cell retention in organs, which can be regarded as the simplest way to produce higher concentration of cells in a certain area of the body [32]. Furthermore, MNPs were applied to prepare artificial chain-like multicellular clusters *in vitro* consisting of a single type of cells. Protein (albumin)-coated MNPs were deposited onto mammal cells to facilitate the magnetically driven formation of ordered multicellular clusters. The biocompatible MNPs (~10–20 nm) were deposited on human endothelial cells; then, the application of the external magnetic field helped arrange the MNPs-labeled cells into long stable chains that preserved their morphology after the removal of the magnetic field. In addition, the feasibility of this approach was supported by the growth of the cellular chains if exposed to the nutrient medium [52].

More complex structures consisting of heterotypic layered cells were obtained using magnetically labeled rat hepatocytes and human aortic endothelial cells. The cells accumulated specifically at sites where magnets were positioned and formed tissue-mimicking clusters, where the heterotypic cells were layered and formed tight and closed contacts. Apart from the layered structure, the magnetically facilitated cell clusters were fully functional, secreting albumin at higher levels when compared with homotypic nonmagnetic constructs [42]. A similar approach based on CML-labeled cells was applied for fabrication of multilayered cell sheets that were transplanted subcutaneously into mice, and the histological examination of transplants retrieved after 14 days revealed that the sheet grafts produced well-vascularized tissues characterized with substantial mass, thickness, and cell density [33]. Moreover, a similar approach was applied to fabricate tubular structures fabricated using cylindrical magnets and magnetically labeled cells. Two types of artificial tissues were produced – urothelial cell layers and vascular tissue consisting of endothelial cells, smooth muscle cells, and fibroblasts [43]. Such multicellular tubular structures are a promising tool in the fabrication of hollow implants mimicking blood vessels, heart valves, and so on.

2.4
Conclusion

In this chapter, we introduced the reader to a vast and promising area of magnetically functionalized cells. The basic procedures for magnetic labeling of microbial and human cells demonstrated here are currently regarded as the beginning of a novel methodology of fabrication of magnetically responsive hybrid biomaterials. The major directions of the practical applications of magnetically labeled cells are focused on selective immobilization of cells (i.e., in biosensors or tissue engineering) or in remote manipulation with cells using the external magnetic field. Future work in this area will be focused on the utilization of multifunctional MNPs providing the cells not only with magnetic functionality but also with nutrients, protection, and better adhesive properties, among many others. Unquestionably, this area is one of the most interesting emerging biomedical topics with many things yet to be realized.

Acknowledgment

This study was supported by RFBR 12-04-33290 (Leading young scientists support) grant.

References

1. Safarik, I., Pospiskova, K., Horska, K., and Safarikova, M. (2012) Potential of magnetically responsive (nano) biocomposites. *Soft Matter*, **8**, 5407–5413.

2. Yuan, J., Xu, Y., and Müller, A.H.E. (2011) One-dimensional magnetic inorganic–organic hybrid nanomaterials. *Chem. Soc. Rev.*, **40**, 640–655.

3. Ray, P.Ch., Khan, S.A., Singh, A.K., Senapati, D., and Fan, Z. (2012) Nanomaterials for targeted detection and photothermal killing of bacteria. *Chem. Soc. Rev.*, **41**, 3193–3209.
4. Colombo, M., Carregal-Romero, S., Casula, M.F., Gutiérrez, L., Morales, M.P., Böhm, I.B., Heverhagen, J.T., Prosperi, D., and Parak, W.J. (2012) Biological applications of magnetic nanoparticles. *Chem. Soc. Rev.*, **41**, 4306–4334.
5. Durgadas, C.V., Sharma, C.P., and Sreenivasan, K. (2011) Fluorescent and superparamagnetic hybrid quantum clusters for magnetic separation and imaging of cancer cells from blood. *Nanoscale*, **3**, 4780–4787.
6. Korneva, G., Ye, H., Gogotsi, Y., Halverson, D., Friedman, G., Bradley, J.C., and Kornev, K.G. (2005) Carbon nanotubes loaded with magnetic particles. *Nano Lett.*, **5**, 879–884.
7. Terreno, E., Castelli, D.D., Viale, A., and Aime, S. (2010) Challenges for molecular magnetic resonance imaging. *Chem. Rev.*, **110**, 3019–3042.
8. Laurent, S., Forge, D., Port, M., Roch, A., Robic, C., Vander Elst, L., and Muller, R.N. (2008) Magnetic iron oxide nanoparticles: synthesis, stabilization, vectorization, physicochemical characterizations, and biological applications. *Chem. Rev.*, **108**, 2064–2110.
9. Yavuz, H., Denizli, A., Güngünes, H., Safarikova, M., and Safarik, I. (2006) Biosorption of mercury on magnetically modified yeast cells. *Sep. Purif. Technol.*, **52**, 253–260.
10. Safarik, I., Filipe, L., Rego, T., Borovska, M., Mosiniewicz-Szablewska, E., Weyda, F., and Safarikova, M. (2007) New magnetically responsive yeast-based biosorbent for the efficient removal of water-soluble dyes. *Enzyme Microb. Technol.*, **40**, 1551–1556.
11. Procházková, G., Šafařík, I., and Brányik, T. (2012) Surface modification of Chlorella vulgaris cells using magnetite particles. *Procedia Eng.*, **42**, 1778–1787.
12. Safarikova, M., Maderova, Z., and Safarik, I. (2009) Ferrofluid modified Saccharomyces cerevisiae cells for biocatalysis. *Food Res. Int.*, **42**, 521–524.
13. Fakhrullin, R.F., Zamaleeva, A.I., Morozov, M.V., Tazetdinova, D.I., Alimova, F.K., Hilmutdinov, A.K., Zhdanov, R.I., Kahraman, M., and Culha, M. (2009) Living fungi cells encapsulated in polyelectrolyte shells doped with metal nanoparticles. *Langmuir*, **25**, 4628–4634.
14. Zamaleeva, A.I., Sharipova, I.R., Porfireva, A.V., Evtugyn, G.A., and Fakhrullin, R.F. (2010) Polyelectrolyte-mediated assembly of multiwalled carbon nanotubes on living yeast cells. *Langmuir*, **26**, 2671–2679.
15. Yang, S.H., Lee, T., Seo, E., Ko, E.H., Choi, I.S., and Kim, B.S. (2012) Interfacing living yeast cells with graphene oxide nanosheaths. *Macromol. Biosci.*, **12**, 61–66.
16. Konnova, S.A., Sharipova, I.R., Demina, T.A., Osin, Y.N., Yarullina, D.R., Ilinskaya, O.N., Lvov, Y.M., and Fakhrullin, R.F. (2013) Biomimetic cell-mediated three-dimensional assembly of halloysite nanotubes. *Chem. Commun.*, **49**, 4208–4210.
17. Fakhrullin, R.F. and Lvov, Y.M. (2012) "Face-lifting" and "make-up" for microorganisms: layer-by-layer polyelectrolyte nanocoating. *ACS Nano*, **6**, 4557–4564.
18. Veerabadran, N.G., Goli, P.L., Stewart-Clark, S.S., Lvov, Y.M., and Mills, D.K. (2007) Nanoencapsulation of stem cells within polyelectrolyte multilayer shells. *Macromol. Biosci.*, **7**, 877–882.
19. Okada, T., Uto, K., Sasai, M., Lee, C.M., Ebara, M., and Aoyagi, T. (2013) Nanodecoration of the hemagglutinating virus of Japan envelope (HVJ-E) using a layer-by-layer assembly technique. *Langmuir*, **29**, 7384–7392.
20. Fakhrullin, R.F., Zamaleeva, A.I., Minullina, R.T., Konnova, S.A., and Paunov, V.N. (2012) Cyborg cells: functionalisation of living cells with polymers and nanomaterials. *Chem. Soc. Rev.*, **41**, 4189–4206.
21. Fakhrullin, R.F., García-Alonso, J., and Paunov, V.N. (2010) A direct technique

for preparation of magnetically functionalised living yeast cells. *Soft Matter*, **6**, 391–397.
22. Fakhrullin, R.F., Shlykova, L.V., Zamaleeva, A.I., Nurgaliev, D.K., Osin, Y.N., García-Alonso, J., and Paunov, V.N. (2010) Interfacing living unicellular algae cells with biocompatible polyelectrolyte-stabilised magnetic nanoparticles. *Macromol. Biosci.*, **10**, 1257–1264.
23. Garcia-Alonso, J., Fakhrullin, R.F., Paunov, V.N., Shen, Z., Hardege, J.D., Pamme, N., Haswell, S.J., and Greenway, G.M. (2011) Microscreening toxicity system based on living magnetic yeast and gradient chips. *Anal. Bioanal. Chem.*, **400**, 1009–1013.
24. Zhang, D., Fakhrullin, R.F., Özmen, M., Wang, H., Wang, J., Paunov, V.N., Li, G., and Huang, W.E. (2011) Functionalisation of whole cell bacterial reporters with magnetic nanoparticles. *Microb. Biotechnol.*, **4**, 89–97.
25. Minullina, R.T., Osin, Y.N., Ishmuchametova, D.G., and Fakhrullin, R.F. (2011) Interfacing multicellular organisms with polyelectrolyte shells and nanoparticles: a Caenorhabtidis elegans study. *Langmuir*, **27**, 7708–7713.
26. Dzamukova, M.R., Zamaleeva, A.I., Ishmuchametova, D.G., Osin, Y.N., Nurgaliev, D.N., Kiyasov, A.P., Ilinskaya, O.N., and Fakhrullin, R.F. (2011) A direct technique for magnetic functionalization of living human cells. *Langmuir*, **27**, 14386–14393.
27. Zamaleeva, A.I., Sharipova, I.R., Shamagsumova, R.V., Ivanov, A.N., Evtugyn, G.A., Ishmuchametova, D.G., and Fakhrullin, R.F.A. (2011) Whole-cell amperometric herbicide biosensor based on magnetically functionalised microalgae and screen-printed electrodes. *Anal. Methods*, **3**, 509–513.
28. García-Alonso, J., Fakhrullin, R.F., and Paunov, V.N. (2010) Rapid and direct magnetization of GFP-reporter yeast for micro-screening systems. *Biosens. Bioelectron.*, **25**, 1816–1819.
29. Tarn, M.D., Fakhrullin, R.F., Paunov, V.N., and Pamme, N. (2013) Microfluidic device for the rapid coating of magnetic cells with polyelectrolytes. *Mater. Lett.*, **95**, 182–185.
30. Kantak, C., Beyer, S., Yobas, L., Bansal, T., and Trau, D. (2011) A 'microfluidic pinball' for on-chip generation of Layer-by-Layer polyelectrolyte microcapsules. *Lab Chip*, **11**, 1030–1035.
31. Ito, A., Shinkai, M., Honda, H., and Kobayashi, T. (2005) Medical application of functionalized magnetic nanoparticles. *J. Biosci. Bioeng.*, **100**, 1–11.
32. Chaudeurge, A., Wilhelm, C., Chen-Tournoux, A., Farahmand, P., Bellamy, V., Autret, G., Ménager, C., Hagège, A., Larghéro, J., Gazeau, F., Clément, O., and Menasché, P. (2012) Can magnetic targeting of magnetically labeled circulating cells optimize intramyocardial cell retention? *Cell Transplant.*, **21**, 679–691.
33. Akiyama, H., Ito, A., Kawabe, Y., and Kamihira, M. (2010) Genetically engineered angiogenic cell sheets using magnetic force-based gene delivery and tissue fabrication techniques. *Biomaterials*, **31**, 1251–1259.
34. El-Dakdouki, M.H., El-Boubbou, K., Zhu, D.C., and Huang, X. (2011) A simple method for the synthesis of hyaluronic acid coated magnetic nanoparticles for highly efficient cell labelling and in vivo imaging. *RSC Adv.*, **1**, 1449–1452.
35. Andreas, K., Georgieva, R., Ladwig, M., Mueller, S., Notter, M., Sittinger, M., and Ringe, J. (2012) Highly efficient magnetic stem cell labeling with citrate-coated superparamagnetic iron oxide nanoparticles for MRI tracking. *Biomaterials*, **33**, 4515–4525.
36. Su, H., Liu, Y., Wang, D., Wu, C., Xia, C., Gong, Q., Song, B., and Ai, H. (2013) Amphiphilic starlike dextran wrapped superparamagnetic iron oxide nanoparticle clusters as effective magnetic resonance imaging probes. *Biomaterials*, **34**, 1193–1203.
37. Wilhelm, C. and Gazeau, F. (2008) Universal cell labelling with anionic magnetic nanoparticles. *Biomaterials*, **29**, 3161–3174.
38. Fayol, D., Luciani, N., Lartigue, L., Gazeau, F., and Wilhelm, C. (2013) Managing magnetic nanoparticle

aggregation and cellular uptake: a precondition for efficient stem-cell differentiation and MRI tracking. *Adv. Healthcare Mater.*, **2**, 313–325.

39. Berry, C.C., Wells, S., Charles, S., and Curtis, A.S. (2003) Dextran and albumin derivatised iron oxide nanoparticles: influence on fibroblasts in vitro. *Biomaterials*, **24**, 4551–4557.

40. Thorek, D.L. and Tsourkas, A. (2008) Size, charge and concentration dependent uptake of iron oxide particles by non-phagocytic cells. *Biomaterials*, **29**, 3583–3590.

41. Babic, M., Horák, D., Trchová, M., Jendelová, P., Glogarová, K., Lesný, P., Herynek, V., Hájek, M., and Syková, E. (2008) Poly(L-lysine)-modified iron oxide nanoparticles for stem cell labeling. *Bioconjug. Chem.*, **19**, 740–750.

42. Ito, A., Takizawa, Y., Honda, H., Hata, K., Kagami, H., Ueda, M., and Kobayashi, T. (2004) Tissue engineering using magnetite nanoparticles and magnetic force: heterotypic layers of cocultured hepatocytes and endothelial cells. *Tissue Eng.*, **10**, 833–840.

43. Ito, A., Ino, K., Hayashida, M., Kobayashi, T., Matsunuma, H., Kagami, H., Ueda, M., and Honda, H. (2005) Novel methodology for fabrication of tissue-engineered tubular constructs using magnetite nanoparticles and magnetic force. *Tissue Eng.*, **11**, 1553–1561.

44. Dzamukova, M.R., Naumenko, E.A., Zakirova, E.Y., Dzamukov, R.A., Shilyagin, P.A., Ilinskaya, O.N., and Fakhrullin, R.F. (2012) Polymer-stabilised magnetic nanoparticles do not affect the viability of magnetically-functionalised cells. *Cell. Transplant. Tissue Eng.*, **7**, 52–56.

45. Shubayev, V.I., Pisanic, T.R. II, and Jin, S. (2009) Magnetic nanoparticles for theragnostics. *Adv. Drug Deliv. Rev.*, **61**, 467–477.

46. Frey, N.A., Peng, S., Cheng, K., and Sun, S. (2009) Magnetic nanoparticles: synthesis, functionalization, and applications in bioimaging and magnetic energy storage. *Chem. Soc. Rev.*, **38**, 2532–2542.

47. Safarik, I. and Safarikova, M. (2002) Magnetic nanoparticles and biosciences. *Chem. Mon.*, **133**, 737–759.

48. Perea, H., Aigner, J., Heverhagen, J.T., Hopfner, U., and Wintermantel, E. (2007) Vascular tissue engineering with magnetic nanoparticles: seeing deeper. *J. Tissue Eng. Regen. Med.*, **1**, 318–321.

49. Reddy, L.H., Arias, J.L., Nicolas, J., and Couvreur, P. (2012) Magnetic nanoparticles: design and characterization, toxicity and biocompatibility, pharmaceutical and biomedical applications. *Chem. Rev.*, **112**, 5818–5878.

50. Gao, J., Gu, H., and Xu, B. (2009) Multifunctional magnetic nanoparticles: design, synthesis, and biomedical applications. *Acc. Chem. Res.*, **42**, 1097–1107.

51. Salinas, A.J., Esbrit, P., and Vallet-Regí, M. (2013) A tissue engineering approach based on the use of bioceramics for bone repair. *Biomater. Sci.*, **1**, 40–51.

52. Kelm, J.M. and Fussenegger, M. (2010) Scaffold-free cell delivery for use in regenerative medicine. *Adv. Drug Deliv. Rev.*, **62**, 753–764.

3
Untethered Insect Interfaces[1]

Amol Jadhav, Michel M. Maharbiz, and Hirotaka Sato

3.1
Introduction

The development of wireless telemetry systems for small, free-flying, and walking insects began in the early 1990s with applications ranging from neuromuscular recording [3–18] to using radio transmitters to study the long-range movements of insects [19–24]. Table 3.1 shows the wireless systems used for neuromuscular recording and stimulation of free-flying insects. All of the early devices employed custom-made radios hand-assembled from surface mount electronic components and lacked digital processing, onboard memory, or programmability. Detection of emitted signals was usually carried out using complimentary radio receivers or spectrum analyzers (and the relevant biological information was de-convolved from the analog radio signals). Kutsch *et al.* [3] developed a 0.42 g telemetry backpack (including battery) for a locust (*Schistocerca gregaria*, 3.0–3.5 g, payload capacity 0.5 g). The backpack had a single-channel transmitter to wirelessly acquire electromyograms (EMGs) of a single flight muscle of interest. Their modified backpack had a dual-channel transmitter and it allowed the researchers to tease out the function of the locust's proprioceptors [10], a feat difficult or impossible to do with tethered insects. Ando *et al.* developed a 0.23 g dual-channel telemetry backpack to measure and compare EMGs from a pair of flight muscles of a male hawkmoth, *Agrius convolvuli*, during pheromone-triggered zigzag flight [5, 11, 12, 18].

To the best of our knowledge, three different groups have recently developed wireless systems that can both transmit and receive data from free-flying insects' flight control (Figure 3.1 and Figures 2 and 3 in [1]). Flight control requiring multiple-channel stimulation during complex, long-duration controlled flight requires onboard digital processing, memory, and programmability in addition to efficient radio systems. Each group developed systems with different trade-off

---

[1] The material in this chapter is drawn from and expands on prior open access publications [1, 2]; the section on implantation draws from unpublished material.

Implantable Bioelectronics, First Edition. Edited by Evgeny Katz.
© 2014 Wiley-VCH Verlag GmbH & Co. KGaA. Published 2014 by Wiley-VCH Verlag GmbH & Co. KGaA.

Table 3.1 Representative systems used for neuromuscular recording and/or stimulation of free-flying insects.

Year	Purpose	Insect order	Species	Insect mass (g)	System mass (g)	Lifetime	Radio	Radio range (m)	References
1993	EMG	Locust orthoptera	*Schistocerca gregaria*	3.5	0.42	—	1 ch, T	25	Kutsch et al. [3]
1996	EMG	Locust orthoptera	*Schistocerca gregaria*	3.5	0.55	—	2 ch, T	20	Fischer et al. [7]
1999	EMG	Moth lepidoptera	*Agrius convolvuli*	—	0.4	30 min	2 ch, T	1	Kuwana et al. [5]
1999	EMG	Locust orthoptera	*Schistocerca gregaria*	3.5	0.3	—	1 ch, T	—	Fischer and Ebert (1999) [10]
1999	EMG	Locust orthoptera	*Schistocerca gregaria*	3.5	0.3	—	1 ch, T	—	Fischer and Kutsch [9]
2001	EMG	Moth lepidoptera	*Manduca sexta*	—	0.74	3 h	2 ch, T	2	Mohseni et al. [16]
2002	EMG	Moth lepidoptera	*Agrius convolvuli*	—	0.25	30 min	2 ch, T	5	Ando et al. [11]
2003	EMG	Locust orthoptera	*Schistocerca gregaria*	3.5	0.2	—	1 ch, T	—	Kutsch et al. (2003) [25]
2004	EMG	Moth lepidoptera	*Agrius convolvuli*	—	0.25	—	2 ch, T	—	Ando and Kanzaki [12]
2008	EMG	Moth lepidoptera	*Agrius convolvuli*	—	0.23	—	2 ch, T	3	Wang et al. [18]
2009	Flight control	Beetle coleoptera	*Mecynorhina torquata*	10	1.33	30 min active/24 h sleep mode	8 ch, T-R	20	Sato et al. [2, 26] Maharbiz and Sato [27]
2009[a]	Flight control	Moth lepidoptera	*Manduca sexta*	2.5	0.65	—	3 ch, R	—	Bozkurt et al. [28]
2009, 2010	Flight control	Moth lepidoptera	*Manduca sexta*	2.5	1	—	4 ch, R	—	Daly et al. (2009, 2010) [29, 30]

[a]The system used helium balloons to provide additional lift for the insect.
EMG, electromyogram; T, transmitter; R, receiver.
(Reproduced from [1] with permission.)

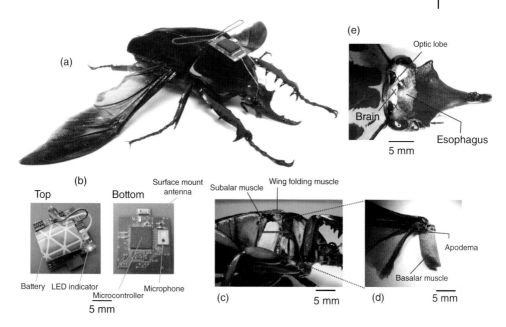

Figure 3.1 A beetle hybrid system. (a) Overview of the stimulator-mounted beetle (*Mecynorhina torquata*, 6 cm, 10 g, 3 g payload capacity). (b) The beetle stimulator has a surface mount ceramic antenna, and its total mass is 1.22 g (a lithium ion rechargeable battery included). (c–e) anatomical pictures of (c,d) muscular stimulation site (right basilar muscle), and (e) optic lobe and brain stimulation sites (the fat bodies and tracheae were removed to provide the clear images). The blue bar and letters of X indicate the electrode inserted length and positions, respectively. (Reproduced from [1] with permission.)

choices in terms of functionality, weight, and complexity. Sato et al. describe an eight-channel system built around the Texas Instruments CC2431 microcontroller with built-in transceiver; careful programming of the microcontroller allows for half hour of flight time and approximately 24 h of battery life in sleep mode (Figure 3.1; [2, 26, 27]). The use of surface mount ceramic antennae results in a very small package in terms of size, mass, and inertial effect on the flying insect (Figure 3.1b). Bozkurt et al. [28] developed a custom-built, two-channel AM receiver that used pulse-position modulation via a super-regenerative architecture, which was fed into a PIC12F615 microcontroller (see also Figure 2 in [1]). Daly et al. [29, 30] developed a custom silicon system-on-chip receiver operating at 3–5 GHz on the 802.15.4a wireless standard that interfaced with an onboard Texas Instruments MP430 microcontroller; the receiver was remarkable for its extremely low-power operation (2.5 mW, 1.4 nJ bit^{-1}) for a data rate of 16 Mb s^{-1} (see also Figure 3 in [1]). Driven primarily by technological developments in ultra-low-power-distributed sensor networks, low-power microcontrollers equipped with internal radios are now highly accessible.

3.2
Systems for Tetherless Insect Flight Control

3.2.1
Various Approaches to Tetherless Flight Control

In this chapter, we focus primarily on systems designed to perturb or control the free flight of insects; these systems ideally require the triggering of flight initiation and cessation as well as the free-flight adjustment of orientation with three degrees of freedom [33, 34]. It is important to note that all published attempts at free-flight control rely on the insect to "fly itself" while periodically introducing extraneous input to bias the free-flight trajectory. A sufficiently sophisticated system is, in effect, to wrap a synthetic control loop around the existing biological one; the idea of interfering with a biological control loop using an extraneous, synthetic loop has a long history. In insects, the motif has been used repeatedly in studies of motor control and biomechanics [35].

To date, wireless flight control of insects has relied on either neuromuscular or neuronal stimulation. In either case, the chosen interface and complimentary stimulation protocol (i.e., electrode geometry, electrode implantation method, and stimulus conditions) must generate reproducible, quantifiable alterations to insect flight in a way that is robust to the harsh conditions before, during, and after free flight. Free-flying insects routinely impact objects (shocks and hard impact are observed not just in flight or during accidents, but very often while landing); the vibrations of the center of mass can be substantial and at frequencies (50–200 Hz) that can resonantly couple to extraneous mechanical components (i.e., 3 cm dipole antennas); the legs or wings themselves can interfere with operation during normal flight. All of these conditions invariably lead to mechanical drift of the implanted electrodes over the lifetime of the insects. Successful, robust stimulation schemes in free flight have thus focused on combinations of the following three motifs: (i) the direct stimulation of a large, easily accessible muscle in the insect [2, 28, 31, 36, 37]; (ii) the direct stimulation of a relatively large ensemble of neurons in a ganglion [2, 31]; and (iii) the targeted stimulation of nerves in a nerve cord [29, 38–40].

3.2.2
Neurostimulation for Initiation of Wing Oscillations

In adult *Mecynorhina ugandensis* beetles, the abrupt darkening of the environment during untethered free flight led to the almost immediate cessation of flight. This led Sato and Maharbiz [1] to hypothesize that light levels and corresponding changes in neural activity at the optic lobes might strongly modulate flight initiation and cessation. In fact, potential pulses applied between two electrodes implanted near the base of the left and right optic lobes could elicit flight initiation and cessation with very high success rates. Implantation into the optic lobe yielded a much higher success rate and did not affect the beetle's ability to steer in free flight

[26, 27]. Ten insects initiated flight in response to stimulation, with the median number of stimulation waveforms required to initiate flight being 19 (range 1–59). One stimulation waveform was a pair of biphasic square pulses (1 ms per each pulse, 4 ms pitch), with the median response time from the first stimulation to flight initiation being 0.5 s (range 0.2–1.4 s). Median flight duration in response to stimulation was 46 s (range 33–2292 s). Stimulation voltage between 2 and 4 V did not affect the number of stimuli required to initiate flight, response time from stimulation to flight, or flight duration in *Mecynorhina torquata* (Mann–Whitney U-tests, $P = 0.13, 0.46, 0.35$, respectively). Data on stimulated flight bouts in individual beetles are summarized in [2]. Once flight was initiated by our stimulation, the flight tended to persist without additional stimulation irrespective of whether the beetle was in the tethered or free flight condition. During normal flight, the nervous system of the beetle produces a pulse train with approximately 50 ms period to the basalar muscles [41, 45]. Artificially induced flight lasted far longer than 50 ms: median flight durations were 2.5 s (range 0.2–1793.1 s) for *Cotinis texana* and 45.5 s (range 0.7–2292.1 s) for *M. torquata*. Among six given insects, flight bout duration correlated with neither beetle mass nor stimulus amplitude. Furthermore, the beetle adopted a normal flight posture and continued flying in the air after the stimulus was turned off, indicating that the tonic neural signals required for flight maintenance continued after stimulus.

A relatively long-duration pulse applied between optic lobes, which effectively clamps the voltage between the lobes, stopped flight for *M. torquata*. Ten insects were tested in tethered condition and each test was repeated 10 times, that is, 100 tests in total. Data on cessation of flight in individual insects are summarized in [2]. All 10 insects tested were forced to stop flying by an amplitude of 6.0 V or less. The majority (77%) stopped with 2–3 V amplitude. The median amplitude was 3.0 V (range 2–6 V). The majority (87%) showed quite short response time, < 100 ms. In *Manduca sexta*, Bozkurt et al. [28] showed that stimulation of the antennal lobes with 20 Hz, 3.5 V peak-to-peak pulses elicited flight while stimulation of the same site with 50 Hz, 3.5 V peak-to-peak pulses ceased flight.

During flight, the frequency and stroke amplitude of wing oscillation could be manipulated with the neural stimulator in beetles [27, 37, 46]. For *C. texana*, it was observed that progressively shortening the time between positive and negative potential pulses delivered to the area of the brain between the optic lobes led to the "throttling" of flight where the beetle's normal 76 Hz wing oscillation was strongly modulated by the 0.1–10 Hz applied stimulus [2, 46]. A repeating program of 3 s, 10 Hz, 3.0 V pulse trains followed by a 3 s pause (no stimulus) resulted in alternating periods of higher and lower pitch flight [26, 46]. In a similar manner, in *M. torquata*, brain stimulus at 100 Hz led to depression of flight (see [1]: Figure 4B, tethered flight: Movie S2 in supplementary material, free flight: Movie S3 in supplementary material). Set on a custom pitching gimbal, *M. torquata* could be repeatedly made to lower the angle to the horizon when stimulated. The change in the length of the envelope of the blurry region around the wing suggests that the wing stroke amplitude was clearly reduced (see [1]: the tethered flight: Movie S2 in supplementary material). Ten of eleven tested beetles showed the tendency.

3.2.3
Extracellular Stimulation of the Muscles to Elicit Turns

One of the classical methods for studying flight control in tethered animals is through the use of changing visual cues within the insect's field of view. Employing arrays of light-emitting diodes (LEDs) or digital projection on a screen, visual features such as scrolling stripes, moving shapes, and changing horizons elicit very strong maneuvering responses from many insects (e.g., [37, 47, 48]. Given that most insect compound eyes cannot move to track targets, visual cues that induce locomotion responses often also elicit strong motions of the head. In fact, the contraction of flight muscles is usually preceded by the rotation of the head toward the aimed direction [48]. Exploiting this, Bozkurt et al. [28] stimulated neck muscles to induce turning in flying moths. An electrical stimulus delivered to the neck muscles via thin wire electrodes implanted in the M. sexta elicited yawing in balloon-assisted flight. In contrast to attempts at direct stimulation of the wing muscles, neck-muscle stimulation avoids damage to the complex linkages and muscles of the wing. For insects whose small size or wing muscle complexity makes direct wing muscle stimulation prohibitive (i.e., bees and flies), this method has decided advantages.

Tsang et al. [38, 39] developed a microfabricated polyimide multisite flexible split-ring electrode (FSE), which could be implanted around the insect's ventral nerve cord just below the fourth abdominal segment during stage 16 of pupation (2 days prior to emergence). This approach was informed by the fact that changes in an insect's center of gravity can be used to adjust flight orientation and trajectory [49]. In moths, stimulation of the ventral cord with tungsten wires elicited abdominal motions, "presumably by activating motoneurons or interganglionic interneurons" [38–40]. Each FSE contained six independently addressable electrodes; potential pulse trains were applied between pairs of electrodes on the FSE (for a total of 15 possible stimulation pairs). The application of 1–500 ms, 1–5 V potential pulses with frequencies varying between 50 and 333 Hz elicited directional contraction of the abdomen depending on the electrode pair chosen. Interestingly, the direction of contraction for a given electrode pair varied not only from animal to animal but also between the pupal and adult stages of the same animal, implying not only "movement of the FSE but probably also ... developmental differences in the location and identity of axons in the nerve cord and changes in the mechanical articulation of the abdomen." By adjusting voltage levels and frequency, the abdominal response could be graded – an important consideration for future studies of free-flight control. Abdominal contractions in loosely tethered moths elicited by the FSE were shown to correspond to changes in flight path [1, 39].

Asymmetric stimulation of the muscles that actuate an insect's wings can be used to generate turns. In beetles, for instance, turns could be elicited by stimulus of the left and right basalar muscles with positive potential pulse trains. In *C. texana*, the basalar muscles normally contract and extend at 76 Hz when they are stimulated by approximately 8 Hz neural impulses from the beetle nervous system [41, 45]. It has been reported that the flight muscles in *Cotinis* produce maximum power when they are stimulated directly by electrical pulses at 100 Hz [41]. During the flight of tethered *C. texana*, a turn could be elicited by applying 2.0 V, 100 Hz positive potential pulse trains to the basalar muscle opposite to the intended turn direction. A right turn, for example, was triggered by stimulating the left basalar muscle. In free-flying *M. torquata*, turning was elicited when either of the left or right basalar muscles was stimulated in the same manner as *C. texana* but at 1.3 V (Movie S4 in supplementary material in [1, 2]). The success rates for left and right turn were 78% ($N=42$) and 66% ($N=68$), respectively. One second of left- and right-stimulation of free-flying beetles resulted in a 1.7 and −9.0 median roll to the ground and 20.0 and 32.4 median rotations parallel to the ground, respectively (see [2] for data on stimulus turns in free-flying *M. torquata*).

3.3
Implantable Bioelectronics in Insects

Beyond the issue of control, insects that undergo complete metamorphosis may present a unique system with which to study synthetic–organic interfaces, an idea recently posited in Paul *et al.* [42]. Several groups have begun to explore the advantages, in mechanical, electrical, or surgical contexts, of interfaces implanted in insects during pupation, that is, prior to emergence as adults (Paul *et al.*, 2006; Bozkurt *et al.*, 2008, 2009, [43]; Chung and Erickson, [44]) [28, 36, 38–40, 46]. Given the extensive reworking of the insect physiology during pupation, it is tempting to hypothesize that interfaces inserted during this period could somehow co-opt the developmental processes for an engineering advantage; this has not yet been conclusively shown. The process does provide engineering advantages, however. For example, both the authors and others have found that insertion of foreign objects transcutaneously in the pupal stages often results in mechanically robust implants as the shell hardens around the structure post-emergence (Paul *et al.*, 2006) [46]. Complex surgeries, as that required in [38–40], are easier in pupa than in adults. Moreover, for insects with large interstitial areas (such as horned beetles), a significant mass of material can be introduced into the pupa, which becomes incorporated into the insect (provided nerves, gut, or muscle was not severed in the insertion, Bozkurt *et al.*, 2008).

3.3.1
Example: Insertion of Flexible Substrates into the Developing Eye

We recently sought to determine whether the robustness and developmental plasticity of holometabolous insect pupation [50, 51] allowed for the incorporation of thin flexible substrates into the developing eye of the *Zophobas morio* beetle. The eventual goal of this work would be the introduction of substrates for high-density recording from complex sensory organs. The morphology of the beetle species *Z. morio* is not very well studied but these holometabolous insects are closely related to the smaller *Tenebrio molitor* (Tenebrionidae). Similarly, synaptogenesis and axon recruitment during eye development have been studied for a number of species but not for *Zophobas*. However, the development of the fly visual system is well studied [32, 52] and provides a guide for timing implantations during pupation. Specifically, fly studies have shown that the extension of retinal axons toward the brain is carried out in the early pupal phases, followed by a period of synaptogenesis in the mid- and late-pupal stages [53, 54]. Moreover, a number of studies reported that damage to afferent processes extending from sensory organs to the brain were robust to rewiring or rerouting upon injury [50, 55–57]. Given this, we hypothesized that the introduction of perforated, sufficiently thin substrates (Figure 3.2a) normal to the plane of axonal extension in the mid-pupa phase (between 20 and 50% pupation) would allow for process innervation through that substrate and be robust to the injury of processes already present (i.e., damaged afferent processes might re-extend). This period corresponded to the onset of pigmentation in the eye, when the morphogenetic furrow had reached the anterior edge of the eye and had begun to innervate the columns [32].

As in *Tenebrio* (and many other insects), changes in hormonal titer, notably a decrease in juvenile hormone [58], take place when the larvae attain a particular size (for *Zophobas* ~30 mm length or ~0.7–1 g mass). The juvenile hormone (ecdysone and its late product 20-hydroxyecdysone [59, 60]) leads to the activation of pathways that regulate the development of the eye and other organs [61, 62]. As with *Tenebrio*, *Zophobas* possesses external eye imaginal disks that appear de novo during late larval stage and whose cells proliferate during pupation. The transparent cuticle covering the pupa makes it is easy to study the development of the compound eye (Figure 3.3). The proximal photoreceptor or retinula cells (R-cells) accumulate pigment granules, which are clearly visible under white light microscopy (Figure 3.3b and inset). Structurally, these R-cells are surrounded by distal cone cells (C-cells), primary pigment cells, and a variable number of secondary as well as tertiary pigment cells [63]. Eye differentiation begins with a morphogenetic furrow that moves from posterior to anterior of the imaginal disk, laying evenly spaced neural cell clusters. A low-to-moderate level of ecdysteroid is necessary for the proliferation of the morphogenetic furrow; near the end of the eye development process, high levels trigger maturation of ommatidia (the functional unit of the insect eye that includes a fully developed cornea, crystalline cone, pigment cells, photoreceptor cells, and a single axon that innervates the ommatidia and extends to the insect brain).

Anatomical investigations of the structures of the eye during larval, pupal, and adult phases were carried out to determine the insertion location, angle, and depth. The interfaces were inserted 200 μm above the posterior-most eye edge so that the incision could be made above the retina, lamina, and medulla, and enter the rear end of the visual ganglion at the lobula plate (Figure 3.2b). The smaller axonal projections meet in the lobula plate; several hundred branches come together at the plate to form larger visual projection neurons that further extend to the protocerebrum. The antennal and visual tract neuronal pathways project around the same region, that is, posterior-median protocerebrum [64]. The base of the antennal lobe near the glomeruli is innervated with interneurons that extend to

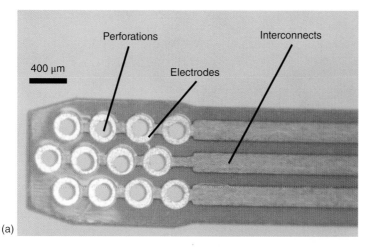

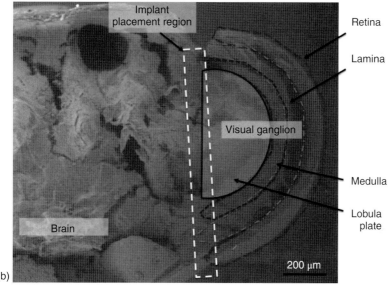

Figure 3.2 (a) Whitelight image of an interface prior to implantation. (b) Scanning electron micrograph of a *Zophobas morio* eye sectioned transverse to the anterior–posterior axis highlighted to show the implant location relative to the relevant gross morphological details of the eye.

the brain. Interfaces were inserted at the base of the antennae near the glomeruli, 200–300 μm posterior of the location on the head at which the neuronal bundles (running along the antenna) merged with the tissue. Owing to the transparency of the pupal cuticle, these structures are easy to find with the naked eye.

Beetle pupae were cold-anesthetized (−20 °C, 2–3 min) prior to implant surgery and then securely immobilized on a plastic dish. The specimen was placed upside

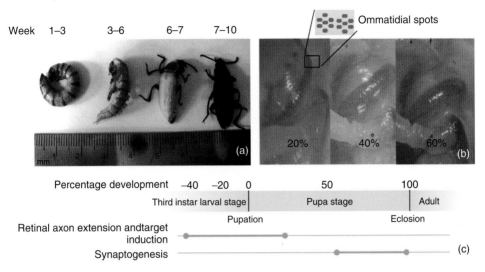

Figure 3.3 (a) Images of *Zophobas morio* at different stages of pupation by week (larva, pupa, early emerged adult, adult). (b) White-light images of the developing eye and antenna in *Zophobas morio* during pupation by percentage pupation completed; note how transparent the cuticle of the insect is. The developing eye is visible at the center of the image and is recognizable by the characteristic rows of dark ommatidial spots (running diagonally top right to bottom left in these images). As the eye develops, more ommatidial spots become evident. The developing antenna and its base just below the eye are visible in the lower right quadrant of each image. (c) A timeline of the relevant processes in eye and antenna development just before, during, and after pupation. (Adapted from anatomical observations and [32].)

down with its abdomen facing up and the head was held in position with damp tissue paper to avoid any movement during the procedure. After aligning the head under the microscope, a fine incision was made around the developing eye disc with a needle having a sharp tip (diameter $< 100\,\mu m$); the connective tissue fold above the eye served as a landmark. In order to limit the loss of hemolymph and any damage to internal structures during the surgical procedure, it was essential to perform the implantation in one attempt. The sterilized device was then inserted slowly through the cut along the anteroposterior axis at an angle of $400°-55°$ to the anterior–posterior head axis until the electrode pad (marking on the device) was inside the eye at the required site. After the procedure, the area around the implant was dried and a drop of saline was placed at the interface. The beetle was held in this position overnight until the tissue around the implant sealed properly and the device was fixed in the position. The pupa was then released and returned to an isolated chamber where it was kept for the rest of the metamorphosis period until recording was performed post-emergence. A similar implantation procedure was followed for the device placement in the antennal lobe. Prior to recording, the beetles were anesthetized and restrained in a plastic holder and the whole body was securely held, leaving the head and antennae fully exposed. In order to avoid head movement during the recording, the head was fixed with adhesive to the plastic holder, which was then mounted onto the recording table.

The insect survival rate (measured as the number of insects of an ensemble alive 2 h postimplant) was insensitive to the specific polymer used for the implant (parylene, polyimide, and bio-resorbable poly-L-lactide, PLLA). Independent of the material, implantation too early in the pupation period (<25% pupation) showed slightly lower survival rates but a significant increase in malformed, nonfunctional organs, or sluggish behavior post-emergence (Figure 3.4). In contrast, implants performed after eclosion usually resulted in premature death within a few days and obviously damaged any nonfunctional organs. Implantation in mid-pupation resulted in very high survival rates, normal motility and behavior (Movie M1), and life spans equivalent to normal unimplanted adult life spans (2–3 weeks). Moreover, in all cases, the adult organs were fully formed (Figure 3.5) and, in most but not all cases, apparently functional. Eye functionality was measured with behavioral assays where either light or odor stimulus was presented. All unimplanted beetles (seven of seven), all explanted beetles (three of three), and most beetles implanted during mid-pupation (five of eight) repeatedly and robustly

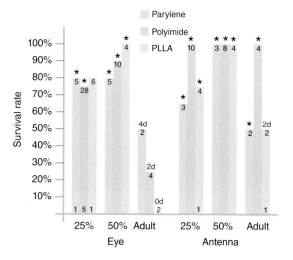

Figure 3.4 Survival rates and malformations for implants in either eye or antenna are not strongly dependent on the materials of the implant but do strongly depend on the time of implant. The survival rate as a function of material and time of implant (as percent completion of pupation) are plotted. The black numbers in each bar represent the number of insects implanted for a given condition. The number above the bar represents the average number of days that the animals survived postimplant (e.g., 2d for 2 days); for simplicity, an * indicates that all animals survived, on average, as long as unimplanted animals (2–3 weeks). The bold numbers at the bottom of each bar indicates the number of animals that emerged with malformations; malformations typically included nonmoving or absent antennae, malformed, or absent elytra and/or wings or weak to no motility. Note that very early implants (25% pupation) are more likely to produce malformed organs.

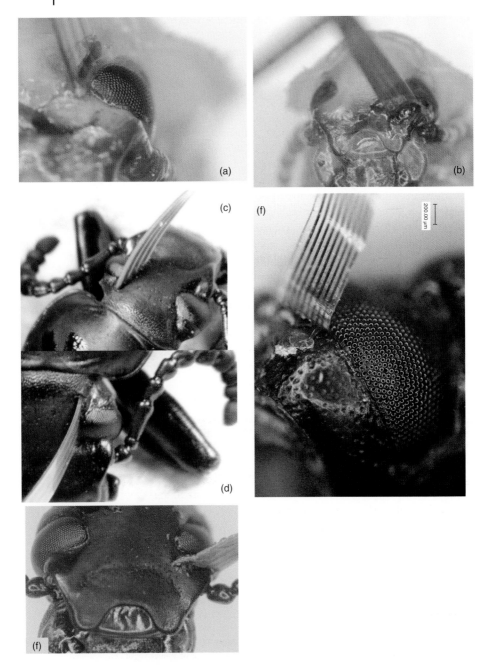

Figure 3.5 (a) An implanted polyimide recording interface in the pupal eye; (b) an identical interface implanted into the base of the pupal antenna; (c,d) two images of the same implant in the live adult viewed from behind and above the eye; (e) the implant in (b) in the adult insect; and (f) a 10 μm-thick parylene implant in the adult eye.

moved toward a light source when presented. These beetles also displayed a startle light response, indicating that the phototransduction mechanism was active and at least some signals were delivered to the brain. (To ensure that the stimulus did not come from the unimplanted eye, it was coated with two layers of black paint prior to the experiment.) Similarly, all unimplanted (10 of 10), all explanted (4 of 4), and most mid-pupal implanted (6 of 7) beetles responded to a slice of potato presented as an odor stimulus within 5 mm of the antenna with rapid antenna movement toward the source of the odor. Ongoing work explores the quality of multisite electrophysiological recordings taken with implants inserted with the above method.

3.4
Conclusions

The ability to control the flight of insects and receive information from onboard sensors would have many applications. In biology, the ability to control insect flight would be useful for studies of insect communication, pollination and mating behavior, and flight energetics, and for studying the foraging behavior of insect predators such as birds, as has been done with terrestrial robots [65]. The technology may also enable new types of experiments relevant to neuroscience as it relates to insect flight. Remote stimulation and recording systems, coupled with flight arenas equipped with real-time motion-capture systems, can trigger motor responses decoupled from the insect's sensory inputs while tracking the resultant changes in flight behavior and recovery. This can be done while simultaneously recording neural or neuromuscular signals from the insect. This is an area of interest in our laboratory, specifically as it informs or improves the ability to elicit controlled reactions from flying insects. Moreover, the ability to take real-time motion data allows for the timing of control signals referenced to a specific state of the insect in flight (e.g., stimulation is applied only at specific orientations, velocities, rotations, etc.), which may help tease out the control circuits at work in the insect.

In engineering, electronically controllable insects could be useful models for insect-mimicking M/NAVs (micro/nano air vehicles) [66–70]. Furthermore, tetherless, electrically controllable insects themselves could be used as M/NAVs and serve as couriers to locations not easily accessible to humans or terrestrial robots.

References

1. Sato, H. and Maharbiz, M.M. (2010) Recent developments in the remote radio control of insect flight. *Front. Neurosci.*, **4**, art # 199.
2. Sato, H., Berry, C.W., Peeri, Y., Baghoomian, E., Casey, B.E., Lavella, G., VandenBrooks, J.M., Harrison, J.F. and Maharbiz, M.M. (2009) Remote radio control of insect flight. *Front. Integr. Neurosci.*, **3**, art # 24.
3. Kutsch, W., Schwarz, G., Fischer, H., and Kautz, H. (1993) Wireless transmission of muscle potentials during free

flight of a locust. *J. Exp. Biol.*, **185** (1), 367–373.

4. Kutsch, W. (2002) Transmission of muscle potentials during free flight of locusts. *Comput. Electron. Agric.*, **35** (2), 181–199.

5. Kuwana, Y., Ando, N., Kanzaki, R., and Shimoyama, I. (1999) A radiotelemetry system for muscle potential recordings from freely flying insects. Engineering in Medicine and Biology 21st Annual Conferences and the 1999 Annual Fall Meeting of the Biomedical Engineering Society. Proceedings of the First Joint BMES/EMBS Conference, 1999, Vol. 2, p. 846.

6. Kuwana, Y., Shimoyama, I., and Miura, H. (1995) Steering control of a mobile robot using insect antennae. Proceedings of the 1995 IEEE/RSJ International Conference on Intelligent Robots and Systems 95.'Human Robot Interaction and Cooperative Robots', Vol. 2, pp. 530–535.

7. Fischer, H., Kautz, H., and Kutsch, W. (1996) A radiotelemetric 2-channel unit for transmission of muscle potentials during free flight of the desert locust, *Schistocerca gregaria*. *J. Neurosci. Methods*, **64** (1), 39–45.

8. Holzer, R. and Shimoyama, I. (1997) Locomotion control of a bio-robotic system via electric stimulation. Proceedings of the 1997 IEEE/RSJ International Conference on Intelligent Robots and Systems, 1997. IROS'97, Vol. 3, pp. 1514–1519.

9. Fischer, H. and Kutsch, W. (1999) Timing of elevator muscle activity during climbing in free locust flight. *J. Exp. Biol.*, **202** (24), 3575–3586.

10. Fischer, H. and Ebert, E. (1999) Tegula function during free locust flight in relation to motor pattern, flight speed and aerodynamic output. *J. Exp. Biol.*, **202** (6), 711–721.

11. Ando, N., Shimoyama, I., and Kanzaki, R. (2002) A dual-channel FM transmitter for acquisition of flight muscle activities from the freely flying hawkmoth, Agrius convolvuli. *J. Neurosci. Methods*, **115** (2), 181–187.

12. Ando, N. and Kanzaki, R. (2004) Changing motor patterns of the 3rd auxiliary muscle activities associated with longitudinal control in freely flying hawkmoths. *Zool. Sci.*, **21** (2), 123–130.

13. Colot, A., Caprari, G., and Siegwart, R. (2004) Insbot: design of an autonomous mini mobile robot able to interact with cockroaches. Proceedings of ICRA'04, 2004 IEEE International Conference on Robotics and Automation, 2004, Vol. 3, pp. 2418–2423.

14. Cooke, S.J., Hinch, S.G., Wikelski, M., Andrews, R.D., Kuchel, L.J., Wolcott, T.G., and Butler, P.J. (2004) Biotelemetry a mechanistic approach to ecology. *Trends Ecol. Evol.*, **19** (6), 334–343.

15. Takeuchi, S. and Shimoyama, I. (2004) A radio-telemetry system with a shape memory alloy microelectrode for neural recording of freely moving insects. *IEEE Trans. Biomed. Eng.*, **51** (1), 133–137.

16. Mohseni, P., Najafi, K., Eliades, S.J., and Wang, X. (2005) Wireless multichannel biopotential recording using an integrated FM telemetry circuit. *IEEE Trans. Neural Syst. Rehabilit. Eng.*, **13** (3), 263–271.

17. Lemmerhirt, D.F., Staudacher, E.M., and Wise, K.D. (2006) A multitransducer microsystem for insect monitoring and control. *IEEE Trans. Biomed. Eng.*, **53** (10), 2084–2091.

18. Wang, H., Ando, N., and Kanzaki, R. (2008) Active control of free flight manoeuvres in a hawkmoth, Agrius convolvuli. *J. Exp. Biol.*, **211** (3), 423–432.

19. Hedin, J. and Ranius, T. (2002) Using radio telemetry to study dispersal of the beetle *Osmoderma eremita*, an inhabitant of tree hollows. *Comput. Electron. Agric.*, **35** (2), 171–180.

20. Sword, G.A., Lorch, P.D., and Gwynne, D.T. (2005) Insect behaviour: migratory bands give crickets protection. *Nature*, **433** (7027), 703.

21. Holland, R.A., Wikelski, M., and Wilcove, D.S. (2006) How and why do insects migrate? *Science*, **313** (5788), 794–796.

22. Wikelski, M., Moxley, J., Eaton-Mordas, A., Lopez-Uribe, M.M., Holland, R., Moskowitz, D., Roubik, D.W., and Kays, R. (2010) Large-range movements of neotropical orchid bees observed via

radio telemetry. *PLoS ONE*, **5** (5), art. # e10738.
23. Wikelski, M., Moskowitz, D., Adelman, J.S., Cochran, J., Wilcove, D.S., and May, M.L. (2006) Simple rules guide dragonfly migration. *Biol. Lett.*, **2** (3), 325–329.
24. Pasquet, R.S., Peltier, A., Hufford, M.B., Oudin, E., Saulnier, J., Paul, L., Knudsen, J.T., Herren, H.R., and Gepts, P. (2008) Long-distance pollen flow assessment through evaluation of pollinator foraging range suggests transgene escape distances. *Proc. Natl. Acad. Sci. U.S.A.*, **105** (36), 13456–13461.
25. Kutsch, W., Berger, S. and Kautz, H. (2003) Turning manoeuvres in free-flying locusts: two-channel radio-telemetric transmission of muscle activity. *J Exp Zool A Comp Exp Biol.* 1, **299** (2), 139–150.
26. Sato, H., Peeri, Y., Baghoomian, E., Berry, C.W., and Maharbiz, M.M. (2009) Radio-controlled cyborg beetles a radio-frequency system for insect neural flight control. IEEE 22nd International Conference on Micro Electro Mechanical Systems, MEMS 2009, pp. 216–219.
27. Maharbiz, M.M. and Sato, H. (2010) Cyborg beetles. *Sci. Am.*, **303** (6), 94–99.
28. Bozkurt, A., Gilmour, R.F., and Lal, A. (2009) Balloon-assisted flight of radio-controlled insect biobots. *IEEE Trans. Biomed. Eng.*, **56** (9), 2304–2307.
29. Daly, D.C., Mercier, P.P., Bhardwaj, M., Stone, A.L., Aldworth, Z.N., Daniel, T.L., Voldman, J., Hildebrand, J.G., and Chandrakasan, A.P. (2010) A pulsed UWB receiver SoC for insect motion control. *IEEE J. Solid-State Circuits*, **45** (1), 153–166.
30. Daly, D.C., Mercier, P.P., Bhardwaj, M., Stone, A.L., Voldman, J., Levine, R.B., Hildebrand, J., Levine, R.B., Hildebrand, J.G. and Chandrakasan, A.P. (2010). A pulsed UWB receiver SoC for insect motion control. *IEEE J. Solid-State Circuits*, **45**, 153–166. doi: 10.1109/JSSC.2009.2034433
31. Bozkurt, A., Gilmour, R.F., Sinha, A., Stern, D., and Lal, A. (2009) Insect machine interface based neurocybernetics. *IEEE Trans. Biomed. Eng.*, **56** (6), 1727–1733.
32. Ting, C.Y. and Lee, C.-H. (2007) Visual circuit development in drosophila. *Curr. Opin. Neurobiol.*, **17**, 65–72.
33. Dudley, R. (2002) *The Biomechanics of Insect Flight: Form, Function, Evolution*, Princeton University Press.
34. Taylor, G.K. (2001) Mechanics and aerodynamics of insect flight control. *Biol. Rev.*, **76** (4), 449–471.
35. Nishikawa, K., Biewener, A.A., Aerts, P., Ahn, A.N., Chiel, H.J., Daley, M.A., Daniel, T.L., Full, R.J., Hale, M.E., Hedrick, T.L. et al (2007) Neuromechanics: an integrative approach for understanding motor control. *Integr. Comp. Biol.*, **47** (1), 16–54.
36. Bozkurt, A., Lal, A., and Gilmour, R. (2008) Electrical endogenous heating of insect muscles for flight control. 30th Annual International Conference of the IEEE Engineering in Medicine and Biology Society, EMBS 2008, pp. 5786–5789.
37. Sato, H., Berry, C.W., Casey, B.E., Lavella, G., Yao, Y., VandenBrooks, J.M., and Maharbiz, M.M. (2008) A cyborg beetle: insect flight control through an implantable, tetherless microsystem. IEEE 21st International Conference on Micro Electro Mechanical Systems, 2008, MEMS 2008, pp. 164–167.
38. Tsang, W.M., Aldworth, Z., Stone, A., Permar, A., Levine, R., Hildebrand, J.G., Daniel, T., Akinwande, A.I., and Voldman, J. (2008) Insect flight control by neural stimulation of pupae-implanted flexible multisite electrodes. The Proceeding of μ TAS, pp. 1922–1924.
39. Tsang, W.M., Stone, A.L., Aldworth, Z.N., Hildebrand, J.G., Daniel, T.L., Akinwande, A.I., and Voldman, J. (2010) Flexible split-ring electrode for insect flight biasing using multisite neural stimulation. *IEEE Trans. Biomed. Eng.*, **57** (7), 1757–1764.
40. Tsang, W.M., Stone, A., Aldworth, Z., Otten, D., Akinwande, A.I., Daniel, T., Hildebrand, J.G., Levine, R.B., and Voldman, J. (2010) Remote control of

a cyborg moth using carbon nanotube-enhanced flexible neuroprosthetic probe. 2010 IEEE 23rd International Conference on Micro Electro Mechanical Systems (MEMS), pp. 39–42.
41. Josephson, R.K., Malamud, J.G., and Stokes, D.R. (2000) Power output by an asynchronous flight muscle from a beetle. *J. Exp. Biol.*, **203** (17), 2667–2689.
42. Paul, A., Bozkurt, A., Ewer, J., Blossey, B. and Lal, A. (2006) *Surgically implanted micro-platforms in manduca-sexta, in Solid State Sensor Actuator Workshop*, Hilton Head Island, 209–211.
43. Bozkurt, A., Paul, A., Pulla, S., Ramkumar, R., Blossey, B., Ewer, J., Gilmour, R. and Lal, A. (2007) Microprobe microsystem platform inserted during early metamorphosis to actuate insect flight muscle, in Proceedings of IEEE International Conference on Proceedings of IEEE International Conference on Micro Electro Mechanical Systems (MEMS 2007), Kobe, 405–408.
44. Chung, A.J. and Erickson, D. (2009). Engineering insect flight metabolics using immature stage implanted microfluidics. Lab. Chip 9, 669.
45. Josephson, R.K., Malamud, J.G., and Stokes, D.R. (2000) Asynchronous muscle a primer. *J. Exp. Biol.*, **203** (18), 2713–2722.
46. Sato, H., Berry, C.W., and Maharbiz, M.M. (2008) Flight control of 10 gram insects by implanted neural stimulators. Solid State Sensor Actuator Workshop, pp. 90–91.
47. Tu, M. and Dickinson, M. (1994) Modulation of negative work output from a steering muscle of the blowfly Calliphora vicina. *J. Exp. Biol.*, **192** (1), 207–224.
48. Berthoz, A., Graf, W., and Vidal, P.P. (1992) *The Head-Neck Sensory Motor System*, Oxford University Press, New York.
49. Ellington, C.P. (1984) The aerodynamics of hovering insect flight. I. The quasi-steady analysis. *Philos. Trans. R. Soc. London, Ser. B: Biol. Sci.*, **305** (1122), 1–15.
50. Meinertzhagen, I.A. (2001) Plasticity in the insect nervous system. *Adv. Insect Physiol.*, **28**, 84–167.
51. Horridge, G. (1968) Affinity of neurons in regeneration. *Nature*, **219**, 737–740.
52. Clandinin, R.T. and Zipursky, S. (2002) Making connections in the fly visual system. *Neuron*, **35**, 827–841.
53. Frolich, A. and Meinertzhagen, I.A. (1983) Quantitative features of synapse formation in the fly's visual system. I. The presynaptic photoreceptor terminal. *J. Neurosci.*, **3** (11), 2336–2349.
54. Meinertzhagen, I.A. (1989) Fly photoreceptor synapses: their development; evolution; and plasticity. *J. Neurobiol.*, **20** (5), 276–294.
55. Jan, Y.N., Christoph, I., Barbel, S., and Jan, L.Y. (1985) Formation of neuronal pathways in the lmaginal discs of drosophila melanogaster. *J. Neurosci.*, **5** (9), 2453–2464.
56. Nowel, M.S. (1981) Formation of the retina–lamina projection of the cockroach: no evidence for neuronal specificity. *J. Embryol. Exp. Morphol.*, **62**, 241–258.
57. Stern, M., Scheiblich, H., Eickhoff, R., Didwischus, N., and Bicker, G. (2010) Regeneration of olfactory afferent axons in the locust brain. *J. Comp. Neurol.*, **520** (4), 679–693.
58. Nadia, A., Quennedey, A., Pitoizet, N., and Delbecque, J.-P. (1997) Ecdysteroid titres in a tenebrionid beetle; Zophobas atratus: effects of grouping and isolation. *J. Insect Physiol.*, **43** (9), 815–821.
59. Tissota, M. and Stocker, R.F. (2000) Metamorphosis in Drosophila and other insects: the fate of neurons throughout the stages. *Prog. Neurobiol.*, **62**, 89–111.
60. Truman, J.W., Hiruma, K., Allee, J.P., MacWhinnie, S.G.B., and Champlin, D.T. (2006) Juvenile hormone is required to couple imaginal disc formation with nutrition in insects. *Science*, **312** (5778), 1385–1388.
61. Champlin, D.T. and Truman, J.W. (1998) Ecdysteroids govern two phases of eye development during metamorphosis of the moth Manduca sexta. *Development*, **125**, 2009–2018.
62. Markus, F., Rambold, I., and Melzer, R.R. (1996) The early stages of ommatidial development in the flour beetle

Tribolium casteneum. *Dev. Genes Evol.*, **206**, 136–146.
63. Markus, F. (2003) Evolution of insect eye development: first insights from fruit fly; grasshopper and flour beetle. *Integr. Comp. Biol.*, **43**, 508–521.
64. Maronde, U. (1991) Common projection areas of antennal and visual pathways in the honey bee brain; Apis Mellifera. *J. Comp. Neurol.*, **309**, 328–340.
65. Michelsen, A., Andersen, B.B., Kirchner, W.H., and Lindauer, M. (1989) Honeybees can be recruited by a mechanical model of a dancing bee. *Naturwissenschaften*, **76** (6), 277–280.
66. Wu, W.-C., Schenato, L., Wood, R.J., and Fearing, R.S. (2003) Biomimetic sensor suite for flight control of a micromechanical flying insect: design and experimental results. Proceedings of ICRA'03, IEEE International Conference on Robotics and Automation, 2003, Vol. 1, pp. 1146–1151.
67. Schenato, L., Wu, W.C., and Sastry, S. (2004) Attitude control for a micromechanical flying insect via sensor output feedback. *IEEE Trans. Robot. Autom.*, **20** (1), 93–106.
68. Deng, X., Schenato, L., Wu, W.C., and Sastry, S.S. (2006) Flapping flight for biomimetic robotic insects: part I-system modeling. *IEEE Trans. Robot.*, **22** (4), 776–788.
69. Deng, X., Schenato, L., and Sastry, S.S. (2006) Flapping flight for biomimetic robotic insects: part II-flight control design. *IEEE Trans. Robot.*, **22** (4), 789–803.
70. Wood, R.J. (2008) The first takeoff of a biologically inspired at-scale robotic insect. *IEEE Trans. Robot.*, **24** (2), 341–347.

4
Miniaturized Biomedical Implantable Devices[1]

Ada S.Y. Poon

4.1
Introduction

The current state of medical care is primarily aligned toward "fixing" the patients when their disease has evolved significantly with detrimental consequences. As a matter of fact, most people take better care of their cars than of their health primarily because of the fact that the cars have sophisticated diagnostics and monitoring tools to assist the user and remind them of when and what kind of maintenance is required. This results in preventative care that is much cheaper and significantly prolongs the life of the vehicle and improves its quality. It is common knowledge that without proper maintenance, vehicles do not last nearly as long as they do when they are properly taken care of. Why, then, do people not maintain their health in much the same manner as they maintain their cars? Well, the simple fact is that people do not have similar gauges that tell them when they should get more sleep, reduce their stress, eat healthier, or even see a doctor. Instead, people have to rely on the way they feel for feedback about their health conditions. Thus, if they have a fever and start coughing, they might visit a doctor who makes diagnoses based on a few unreliable data points provided by the patients who are generally unfamiliar with the field of pathology. This often results in inaccurate diagnoses and therefore wrong treatment. In addition, because patients usually wait to see doctors when the disease progresses and causes complications, it becomes much more difficult, expensive, and longer to treat the disease. Instead, if people were continuously aware of the state of their well-being, they not only get the necessary treatment sooner but also potentially prevent the disease altogether. Therefore, accurate and timely information about the state of one's well-being and vital signs can be used to guide her/his lifestyle choices such as diet, quality of sleep, and stress levels. This timely information can be provided from a tiny implantable device that can monitor one's well-being from within.

People who are genetically predisposed to or those who have already developed certain medical conditions can also benefit from diagnostic and therapeutic implantable devices. Certain forms of diseases, especially of neural origin, currently do not have any chemical or drug therapies. The only known existing solution is

---

[1] A part of the text is adapted from [4, 9] with copyright permission.

Implantable Bioelectronics, First Edition. Edited by Evgeny Katz.
© 2014 Wiley-VCH Verlag GmbH & Co. KGaA. Published 2014 by Wiley-VCH Verlag GmbH & Co. KGaA.

through neurostimulation. For instance, certain regions of brain respond remarkably well to electrical stimulation to treat the debilitating effects of disorders such as chronic pain, essential tremor, Parkinson's disease, dystonia, major depression, and Tourette syndrome without causing permanent damage to physiological or anatomical structures [1–3]. Cardiac pacemaker is another example of a widely used implantable therapeutic device that has a tremendous impact on prolonging lives of people with chronic heart diseases.

However, current state-of-the-art commercially available implantable devices rely on batteries to power them, resulting in bulky form factors. The size of the devices makes them difficult to implant, requiring expensive invasive surgery. Moreover, the batteries only last for 3–5 years, requiring the patient to undergo a surgery to replace the batteries with the recovery period lasting up to several days. To make these devices accessible for battery replacement, the pacemaker is placed under the patient's skin somewhere on the chest with long leads running subcutaneously to the region where the actual stimulation has to be performed. In the case of deep brain stimulation (DBS) device, the lead runs to the top of a patient's head and is inserted deep inside the brain through an opening in the skull. The leads, their placement, and implantation surgeries can cause complications and significantly increase the risk of infection. Therefore, replacing batteries with alternative energy sources can help dramatically reduce the device sizes and thus alleviate these serious problems, and will be the future direction for implantable devices.

In addition to solving the battery problem, future implantable devices that are small and versatile enough to support distributed biosensing and localized operations would revolutionize modern medicine. Implantable systems might soon be an integral part of minimally invasive diagnostic, therapeutic, and surgical treatments to attain more accurate diagnosis and enhance the success rate of complex procedures. In fact, the movie *Fantastic Voyage* well describes the future vision for these implants. In the movie, a submarine with doctors and necessary navigation and therapeutic equipment is miniaturized and inserted inside the body of a patient to treat the problem area from the inside. Even though this still appears to be something far from reality, much advancement has been done technologically toward realization of this grand vision. Many of the essential components have already been developed and demonstrated such as locomotion in fluid medium [4, 5], energy harvesting for miniaturized implants [6–8], efficient communication [7, 9], actuation and drug delivery [5, 10, 11], and low-power diagnostics [12–14].

The following sections explain in more detail system consideration for implantable devices and provide references to the implementation of essential components for most active battery-less implantable devices. First, energy harvesting, and specifically radio frequency (RF) power transfer, as a pathway to miniaturization of the implantable devices is discussed; a typical implantable device architecture and the essential blocks necessary for its implementation are then presented; energy-efficient and robust communication for implantable devices including the forward and reverse data links are then described; two examples of millimeter-sized, fully wireless implantable devices are provided that demonstrate the feasibility of therapeutic and diagnostic devices.

4.2
Energy Harvesting as a Pathway to Miniaturization

Miniaturization of implantable devices enables their proliferation to formerly nonexisting applications and implantation areas, providing better insight into pathologies and tools to treat them with the added benefit of comfort to the patient. In addition, doctors can implant these devices using minimally invasive techniques such as delivering them through a needle. Moreover, removing batteries or replacing them with rechargeable ones eliminates the need for repeat surgeries to replace them and significantly reduces the risk of infection. Miniaturization of implants necessitates either replacing the batteries as an energy source with a much smaller alternative source or eliminating the batteries altogether. Therefore, energy harvesting is one of the most attractive pathways to miniaturization. It can help move away from the current bulky devices toward tiny devices.

Although many forms of energy harvesting exist, depending on the application some are more appropriate than others. For instance, thermoelectric gradient (TEG)-based energy harvesting can potentially provide enough energy to power up very low-power sensing circuitry at a temperature gradient of several degrees celsius. Such a gradient is difficult to attain, however, when the implant is completely inside the body and would only be practical for applications such as smart patches on the skin surface for runners or athletes. For most applications, however, RF energy harvesting is the most appropriate choice because it provides the highest energy density among the common harvesting techniques. Some of the most common energy-harvesting techniques and their corresponding power densities are summarized in Table 4.1 [15]. From the table, it can be seen that electromagnetic (EM) or RF power transfer delivers the highest power density per unit area, and thus has the highest potential for miniaturization.

Until recently, many devices focused on using inductive coupling for wireless powering. For practical implant depths, this leads to operation at low frequencies on the order of a few tens of megahertz, such as the popular 13.56 MHz industrial, scientific, and medical (ISM) band. The main reasoning behind this is that tissue losses increase with increasing frequency, resulting in degraded power transfer efficiency. Recent research has shown, however, that the optimal frequency of operation for transcutaneous power transfer is in the low-gigahertz range [16]. Intuitively, as the devices scale down in size, so does the power-harvesting

Table 4.1 A summary of a few popular energy-harvesting techniques and the corresponding power densities.

Principle and constraints	Power density ($\mu W\,mm^{-2}$)
Glucose biofuel cell utilizing glucose from blood (5 mM)	2.8
Thermoelectric, $\Delta T = 5\,°C$	0.6
Piezoelectric micro bender, $f \sim 800\,Hz$, $2.25\,m\,s^{-2}$	< 0.2
Electromagnetic power transfer	10–1000

antenna. As the antenna size is reduced, its efficiency increases with frequency and so do the tissue losses, resulting in the optimal operating frequency in the low-gigahertz range for millimeter-sized implantable devices. Moreover, higher frequency operation naturally provides higher bandwidth for the data link. Most existing implantable devices that require higher data rates, such as cochlear implants, and retinal and other prosthetic devices, have to trade off bandwidth for a higher quality factor. Higher quality factor results in more efficient power-harvesting operation but limits the maximum data rate that can be achieved. Thus, at a reasonable quality factor of 10, the bandwidth is only 1.56 MHz at 15.6 MHz center frequency but is as much as 200 MHz at 2 GHz. Another added benefit of higher frequency operation is desensitization of transmit and receive antennas to relative alignment and orientation because the operation happens to be in the midfield and not the near field in inductive coupling. This means that implants operate more robustly even with the uncertainty in the implant's position and orientation with respect to the external source. In addition, it is possible to optimize the external transmit antenna to focus the power delivered to the device while minimizing the tissue absorption and thus adhering to the specific absorption rate (SAR) regulations and further improving power transfer efficiency [9, 17, 18].

4.3
Implementation of Implantable Devices

Figure 4.1 shows a typical active battery-less implantable system. It consists of an external device that provides power and control signals to the implantable devices and receives the sensed signals from those devices. The coupling between the

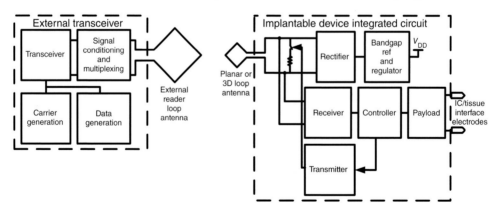

Figure 4.1 Typical implantable system block diagram. The coupling between the external device and the implantable device is via a pair of loosely coupled antenna structures. The implantable device consists of power- harvesting and management block, receiver and transmitter, controller and some auxiliary circuits, and a payload defined by an application.

external and the implantable devices is via a pair of loosely coupled transmit and receive antennas, which can be modeled by a two-port network. Although implants are restricted in size and power consumption, we have more freedom in designing the external transmit antenna.

4.3.1
RF Power Harvesting

Battery-less implantable devices that rely on RF power harvesting need a power management unit that typically consists of a rectifier, bandgap reference circuit to provide a stable and invariable voltage or current reference, and a regulator to maintain a constant supply voltage for the rest of the active circuits. In addition, the antenna has to be matched to the impedance seen, looking into the RF input of the integrated circuit (IC) to maximize the harvested power. In the following, we briefly describe these components and their design trade-offs.

4.3.1.1 Matching Network

A matching network is necessary to maximize the delivered power to the IC load by conjugate matching the antenna impedance at the frequency of operation. The matching network topology can be L, Π, transformer, or a higher order matching network. The trade-offs are among bandwidth, complexity, and chip area. In [19], the authors provide theoretical performance bounds in terms of Bode–Fano limit for quality factor and bandwidth trade-off for matching networks of various orders. Since on-chip capacitors often have a much higher quality factor compared to inductors or bond-wires in the low-gigahertz regime, a matching network using only capacitors will have a higher voltage gain and may occupy a smaller area than a matching network consisting of inductors or a transformer. In addition, the coupling between the transmit and receive antenna structures varies with the surrounding biological tissue composition. The matching network can achieve improved performance if it is adaptive in real time. For example, a tunable matching is realized using a fixed inductance and a digitally controlled nine-element binary weighted capacitor array in [6]. The capacitor array is programmed on the basis of a gradient search algorithm and improves the delivered power by 3–6 dB.

4.3.1.2 Rectifier

The rectifier converts the AC signal to a DC voltage that can be used to supply the active circuitry in the implantable IC. Depending on the induced EMF at the antenna and the quality factor of the matching network, several rectifier topologies can be employed in implantable devices: diode-connected metal oxide semiconductor (MOS), native MOS, and self-driven synchronous rectifier (SDSR) topologies. The critical trade-offs for these rectifier topologies are the on-resistance and reverse conduction current, which limit the efficiency of the rectifier. Typically, the RF input voltage amplitude is too low for one rectifier stage to generate sufficient voltage at the output to power up the on-chip circuitry. In this case, several charge pump-connected rectifier stages can be used to increase the output voltage to an applicable level. SDSR topology provides high efficiency at low RF amplitudes and is therefore suitable for many implantable applications. Moreover, the rectifier efficiency can be optimized according to its application by adjusting the number of stages in the voltage multiplier and the capacitance per stage, and the size of transistors in each stage for a given load [19] provides a good reference that summarizes SDSR-based rectifier optimization.

4.3.1.3 Regulator and Bandgap Reference

The key function of the regulator is to provide a stable supply voltage for on-chip circuitry even as the input voltage or load current experiences variations. Existing regulators can be grouped into two main categories: linear regulators and switching regulators. Linear regulators are simple in design but inefficient for high drop-out voltage with large load currents because the excess voltage is dropped across a regulation resistor dissipating the power as heat. Switching regulators, on the other hand, can achieve very high efficiency, commonly over 80%, and can regulate to lower, higher, or inverted output voltage as compared to its input. The trade-off, however, is the complexity in the design of the controller, and the need for an on-chip clock and large passive components that often need to be off-chip.

Both linear and switching regulators require a stable reference voltage, which necessitates a bandgap reference circuit. Such circuits reject supply noise and are insensitive to ambient temperature by employing proportional to absolute temperature (PTAT) and complementary to absolute temperature (CTAT) components. Ueno provides a thorough review of sub-microwatt reference circuits that avoid the use of bipolar junction transistors or off-chip resistors in [20]. These circuits rely on MOS transistors biased in subthreshold and linear regimes, which are particularly suitable for application in biomedical devices because of their low power consumption and the lack of external passive components.

4.3.1.4 Low-Power Controller and Auxiliary Circuits in the Implant Functional Block

Low-power digital circuits can be designed by operating transistors at low supply voltages near subthreshold or even in deep subthreshold region. This limits the maximum operating frequency of logic gates, which, however, is adequate for many biomedical devices designed to capture low frequency physiological signals. The optimal operating condition to achieve best power efficiency for digital circuits is when leakage power dissipation equals the dynamic power dissipation. The leakage power can be further reduced by implementing Schmitt trigger logic as shown in [21], which allows the reduction of the supply voltage while maintaining robust operation. IN addition, power-gating parts of the controller and even some of the analog blocks can further lead to the reduction in power consumption.

Start-up circuitry for the initial power-on is necessary to ensure that the antenna impedance maintains a match and that the chip enters a known state. Isolating and delaying the turn-on of the digital supply can help minimize the crowbar current during the chip turn-on. Furthermore, a power-on-reset (POR) signal can be generated shortly after the digital supply is enabled and used to reset the controller to a known state. A more detailed description of the start-up circuit implementation can be found in [4].

4.3.2 Wireless Communication Link

Along with wireless powering, wireless communication plays a key role in reducing the size of implants because ultimately, it is necessary to either instruct the implant to do some form of therapy or in the case of diagnostic implants, it is necessary to retrieve the measurements. Existing implementations of communication links in implantable devices are not suitable for many applications because of their poor energy efficiency. For example, most existing implants suffer from significant power consumption in maintaining an environmentally robust, high data rate communications link, which limits their size and performance. In [22], the authors report that increasing temporal resolution and the number of channels in cochlear implants would improve perceived sound quality in hearing-impaired patients, which requires an increase in the data rate, and consequently results in high power consumption. The number of electrodes and encoding resolution in intraocular retinal prostheses are also constrained by the achievable data rate, as stated by Liu *et al.* [23]. Applications, such as cochlear implants or retinal visual prostheses, would benefit from high-efficiency wireless powering and communication, and lead to miniature systems. It is critical to maximize energy efficiency because only limited amount of power can be delivered to these implants, limiting the power budget for their operation. Therefore, the entire system has to be designed to operate very efficiently. Moreover, since the primary function of implantable devices is to diagnose or perform a treatment, the communication power budget has to be minimized to be able to allocate sufficient power to the payload.

Taking advantage of this higher frequency simplifies the design of the data link as well as the antenna and the matching network. In addition, it allows decoupled optimization of the power harvesting and telemetry circuitry, and also desensitizes the link gain to antenna orientation. The power performance of these devices can be further enhanced by implementing them in newer technology processes. The power efficiency of the implantable devices can also be improved from the communications point of view. For instance, moving away from synchronous modulation schemes could reduce power consumption of data transceivers from tens of picojoules per bit in today's demodulators [24] to sub-picojoules per bit.

4.3.2.1 Forward Data Link

Many communication schemes for implantable devices have been demonstrated in recent literature spanning amplitude, frequency, and phase modulation and ranging from binary to higher order encoding in complexity. One common justification for constant amplitude modulation is the need for constant power flow since on-chip energy storage may be infeasible. This is the reason why many designers choose phase or frequency shift keying modulation over amplitude modulation. The disadvantage, however, is the need for a precise on-chip clock or carrier synchronization in order to receive error-free data. Integrating synchronization circuitry in the implant receiver significantly increases power consumption, which can easily offset its advantages. In addition, having a precise clock may require the use of external components such as crystal reference, which would in turn increase the device size. Therefore, a reasonable compromise is achieved by taking advantage of an asynchronous data link as in the case of amplitude shift keying (ASK) with data encoded in the pulse width (PW) that avoids the need for clock and carrier synchronization. However, in order to maintain uninterrupted power flow to the device, the modulation depth for the ASK modulation must be low enough to provide the required average power. A high-level block diagram of an efficient ASK-PW demodulator is shown in Figure 4.2a.

As mentioned earlier, for efficient data transfer on top of the power carrier, modulation depth has to be minimized. Reducing modulation depth requires a sensitive comparator for the envelope and its average. The threshold voltage is ideally set to $V_{th} = \frac{V_{env,max} + V_{env,min}}{2}$, where $V_{env,max/min}$ are the maximum and the minimum voltages of the RF envelope. However, maximum and minimum envelope voltages are not known *a priori*. Therefore, the front end of the demodulator consists of not only an envelope detector but also an envelope-averaging circuit. For devices that are forward link limited, as in the case of most therapeutic and many diagnostic devices, the envelope-detection circuitry can be passive and thus save additional power. The envelope detector can be implemented utilizing a cross-coupled PMOS rectifier topology with a small time constant for the envelope detector and the envelope can be averaged using a passive RC filter or a large time constant replica cross-coupled PMOS rectifier to provide dynamic reference for ASK demodulation. Passive envelope detectors and envelope-averaging circuits are shown in Figure 4.2b. The envelope and its zero-crossings are extracted by a comparator that compares the envelope to its average and also acts as an amplifier that outputs a full-swing digital waveform. The resulting digital waveform can be used as a clock and fed to a resettable integrator, which integrates the duration of the high envelope amplitude and is reset by the falling edge of the envelope. The output of the integrator is compared to a fixed threshold voltage, which is set on the basis of the desired integrator time constant and the data rate requirements. Long pulses provide enough time for the integrator to cross the threshold voltage and will cause the second comparator to output a digital "1," whereas short pulses will result in a digital "0." The digital data stream can be delayed and latched into memory registers using the falling edge of the clock waveform from the first comparator. A more detailed demodulator block diagram is shown in Figure 4.2c, while the output waveform of each stage is shown in Figure 4.2d.

The key benefit of this topology is its configurability. It can operate with an adjustable modulation depth or data rate, and RF power levels. The modulation depth can be reduced by increasing the input device size of the differential pairs in the comparator to overcome the associated offset or even using an active envelope detector or a preamplifier stage. The data rate can be adjusted by controlling the integrator time constant or the threshold voltage of the second comparator. Various RF power levels can be accommodated by controlling the resistive divider ratio

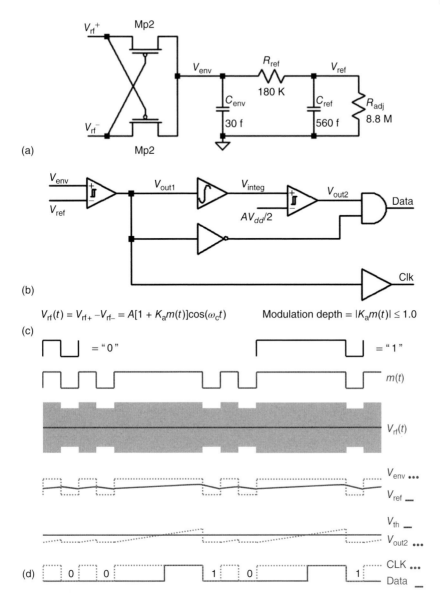

Figure 4.2 (a) Passive envelope detector and envelope-averaging circuit; (b) high-level block diagram of ASK-PWM demodulator; (c) ASK modulation equation, where $m(t)$ represents data encoded in pulses of different duration; and (d) sample waveforms of modulation waveform $m(t)$, received RF envelope $V_{rf}(t)$, output of envelope detector V_{env}, and envelope-averaging circuit V_{ref}, output of the integrator V_{int} compared to threshold voltage V_{th}, and data and clock waveforms. (Adapted from [9] with copyright permission.)

$\frac{R_2}{R_1+R_2}$ of the envelope-averaging circuit. All these parameters can be designed if the operating conditions are known *a priori* or can be tuned before or during the operation by incorporating some feedback mechanism or configured through feed forward data stream.

4.3.2.2 Reverse Data Link

Reverse data link from the implantable device to the external reader can be implemented in many different ways ranging from very complex, power-hungry transmitters to simple, low-power solutions. For the first type, a dedicated transmitter with a local oscillator can be used to transmit data to the external reader. This approach allows for full-duplex communication at the cost of high complexity and power consumption, such as the one used in [25, 26]. The advantage of such an approach is that a longer range can be achieved with increased size and energy efficiency as high as 450 pJ bit^{-1}. For implantable devices that tend to be forward link limited, however, this is definitely not an optimal approach. For forward link limited devices load modulation – modulation of load impedance as seen by the antenna – yields a much more power efficient solution. Load can be modulated in several different ways – resistive, reactive, or a combination of the two. Depending on the link transfer function and how the load is varied, the phase and/or amplitude of the carrier will be modulated. In addition, load modulation can be easily combined with continuous time sensing and processing that was recently described by Schell and Tsividis [27]. They demonstrated a system that converts an analog waveform into a digital representation without a clock or sampling. Not only does this eliminate the need for precise clock generation but it also saves power during idle time intervals because the signal is not sampled when there is no activity. This is perfectly suitable for a variety of biological waveforms as they tend to have long periods of low or no activity. The modulator that is driven by the event-driven sampling will only transmit data to the external reader during physiological activity and will save energy during the idle state. Because the forward link is asynchronous and does not require a clock, the event-driven sampling with load modulation minimizes power for the reverse data link by enabling the circuitry only when necessary.

Although backscatter link is a great choice because of its inherent efficiency it can suffer from high bit error rate (BER) due to varying tissue composition, which can occur rapidly as with breathing or more slowly as with changes in body mass. The compensation for these variations and unpredictability in operating environment can be accomplished with a reconfigurable modulating load, data rate, and backscattered PW.

Analysis in [28] provides a method for determining the optimal load to minimize BER for binary backscattering modulation. Two key criteria must be satisfied: maximizing the average backscattered power and the Euclidean distance between the reflection coefficients on the Smith chart. To satisfy these criteria, the IC can be designed to have different loads to accommodate a variety of conditions as shown in Figure 4.5. The pass gates and the modulation transistor are sized to balance the loading on the RF path while minimizing the parasitic effective series resistance

(ESR). Figure 4.3 shows the Smith chart for the device in air and implanted in a porcine heart with the structural antenna mode A_s, reflection coefficient Γ_1 corresponding to antenna impedances in both air and tissue (matching was designed for tissue), and reflection coefficients $\Gamma_{2,k}$ corresponding to the three loads for backscatter modulation. It is clear from the chart that the capacitive load and the inductive load should be used for the device in tissue and in air, respectively.

To accommodate a variety of applications that may require high and low data rates, data rate configurability is important. Thus, the device can be configured to have high data rate for applications such as an array of implantable cardiac probes described in the next section, cochlear, retinal, or neural prosthetic devices. Alternatively, some applications such as intraocular pressure sensors or pH sensors require much slower data rates and the device can be reconfigured to do so, thus saving power and operating on a lower power budget. This configurability can be accomplished by adjusting the clock frequency. The decoder is unaffected by this frequency change because each data packet can incorporate a preamble that can be used by the external decoder to synchronize to the proper data rate. The overhead of transmitting a preamble is justified by not having to have a precise clock on the implantable device, saving both power and area. In addition, the backscattered pulses can be made adjustable. As explained in [29], shorter pulses minimize the time that the RF path is mismatched and thus not harvesting power. However, very short pulses are spectrally inefficient and result in reduced signal quality as the higher frequencies are filtered out by the antennas, matching network, or other elements in the reverse link. Therefore, adjustability of the PW allows for optimization of the data rate while maintaining an adequate BER.

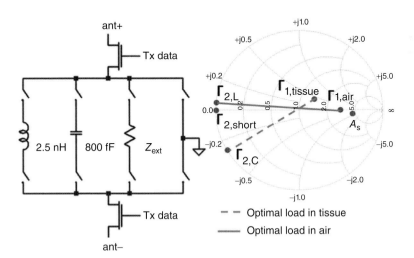

Figure 4.3 Configurable load for backscatter link and Smith chart representation.

4.3.3
Payload and Applications: Locomotive Implant and Implantable Cardiac Probe

As was mentioned previously, application determines the actual payload for the implantable device. Although each application is unique and requires specialized implementation of the device and therefore design specifications and its corresponding trade-offs, there are many common features that the implantable devices share. Therefore, it is possible to group them into actuation for therapeutics and sensing for diagnostics. Two applications that demonstrate the feasibility of both the miniaturized battery-less therapeutics and diagnostics devices are millimeter-sized locomotive implantable device and implantable cardiac probe.

4.3.3.1 Actuation for Therapeutics: Millimeter-Sized Wirelessly Powered and Remotely Controlled Locomotive Implant

Implantable devices capable of *in vivo* controlled motion can serve a variety of existing applications, and they also open up new possibilities for emerging noninvasive medical technologies. Drug delivery is an especially attractive application, as drugs can be precisely targeted to problematic regions with minimal disturbance to the rest of the body. In addition, guidance through fluid cavities could enhance both endoscopic and cardiac procedures that currently rely on catheter systems, such as angioplasty and coronary stent treatments, cardiac arrhythmia ablation surgeries, and diagnostic techniques such as endomyocardial biopsies. Heart disease is the leading cause of death in the US and worldwide, and this technology could improve the effectiveness of these procedures as well as reduce the time and therefore the cost. With these enhancements, it is possible to develop new noninvasive procedures for cardiac care or endoscopy that can aid in prevention, early detection, and treatment of a variety of conditions.

The key challenge in designing an effective locomotive implant is the limited power budget, as existing fluid propulsion methods have significant power requirements, and batteries hinder the potential for miniaturization. Existing wireless devices rely on batteries and have very limited or no motion control. These devices are not small enough to travel through the circulatory system, and are most widely used in the gastrointestinal (GI) tract for endoscopy [29]. There are many proposed systems being researched that attempt to enhance motion either with mechanical actuation or by manipulating passive magnetic structures with external fields [30, 31]. Mechanical techniques tend to require significant power, have complex designs and moving parts, and have very low thrust efficiency as they are scaled down. Devices relying on passive magnetic fields require complex external equipment, have limited controllability in small regions (typically a few cubic centimeters), and move very slowly as they are scaled down [32]. Because of the challenges in developing a highly efficient and scalable propulsion method, wireless locomotive devices have not been possible. An alternative to both the mechanical and passive magnetic propulsion techniques can be accomplished with the manipulation of Lorentz forces on current-carrying wires. These forces require a simple static magnetic field, and are power efficient and controllable. In our prior work, we evaluated two mechanisms for generating propulsion with these forces [33, 34]. The first is based on magnetohydrodynamics (MHD), in which current is driven directly through a conductive fluid. This current flowing through the fluid experiences a force in the magnetic field, and the device experiences an equal and opposite force that propels it forward. The second method relies on asymmetries in fluid drag forces experienced by an oscillating structure. The oscillations are achieved with alternating currents flowing through a loop of wire, which experiences torques in the magnetic field. With loops in different orientations, the structure can be oscillated to create "swimming" motions similar to fins on a fish. Both of these methods have significant potential for operating efficiently at very small sizes, and can enable fully wireless locomotive implants. In this paper, we present a fully wireless device capable of controlled motion in water. An external transmitter powers the device wirelessly with data modulated on the power carrier.

Propulsion is achieved through the described methods based on the manipulation of Lorentz forces, and the prototype operates with either method. Figure 4.4a shows the conceptual operation of the prototype traveling through the bloodstream with MHD propulsion. The constructed device is comprised of a 2 mm × 2 mm receive antenna and an IC fabricated in 65 nm CMOS that includes a matching network, a rectifier, a regulator, a demodulator, a digital controller, and high-current drivers that interface with the propulsion system. This prototype can navigate through fluids, and with further refinement it may be able to navigate through the circulatory system, enabling a variety of new medical procedures.

MHD propulsion drives electric currents through fluids, so the efficiency of this method depends on the fluid conductivity. An electrode configuration allowing for thrust and steering is described in Figure 4.4b. The conductivity of human blood varies approximately from 0.2 to 1.5 S m^{-1} depending on the concentration of blood cells [35]. This translates to a load of less than 300 Ω at the device, which varies with the size, shape, and distance between the electrodes as well as the temperature and applied voltage. Stomach acids tend to have higher conductivities but also vary significantly with normal biological processes. In the analysis provided in [36], the required current for a given speed can be estimated as a function of the size of the device and the background magnetic field.

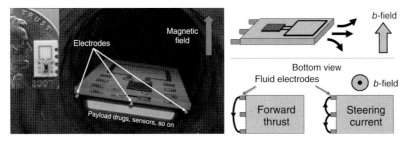

Figure 4.4 Conceptual operation of the device in the bloodstream; operation of magneto-hydrodynamic (MHD) propulsion. (Adapted from [4] with copyright permission.)

The thrust force for MHD propulsion is the Lorentz force on the current flowing through the fluid. These forces are given in the equation below, where I is the current in the wire, $\mathbf{l}$ is a vector denoting the length and direction of the wire, and $\mathbf{b}$ is the background magnetic field

$$F = I\mathbf{l} \times \mathbf{b}$$

These forces scale linearly with the length of the wire $\mathbf{l}$, which allows for the operation of very small devices. It scales more slowly than high Reynolds drag forces, which means that for smaller devices constant current scaling results in higher speeds; and it scales evenly with low Reynolds drag forces, which means that constant current scaling results in a constant speed. In addition, the amount of force is linearly proportional to the background magnetic field, so the performance of this method improves with stronger magnetic fields. To estimate the speed accurately, numerical simulations of the fluid mechanics are performed. Fluid simulations based on incompressible Navier–Stokes flows predict the fluid drag forces, and from these forces the steady-state velocity can be extracted. In Figure 4.5, the required current is estimated for a given speed as a function of the size of the device with a background magnetic field of 0.1 T, which can be generated with permanent magnets. This analysis shows that millimeter-sized devices should be able to achieve speeds on the order of centimeters per second with approximately 1 mA of current.

The current that can be driven depends on fluid conductivity, and has significant nonlinear variations with electrode area, electrode materials, applied voltage, ions in the fluid, and fluid temperature. To drive 1 mA through blood (which has the lowest conductivity of the targeted fluids), roughly 300 mV is required, resulting in a power consumption of around 300 μW. As the fluid conductivity increases, the required power decreases. These power requirements are within the

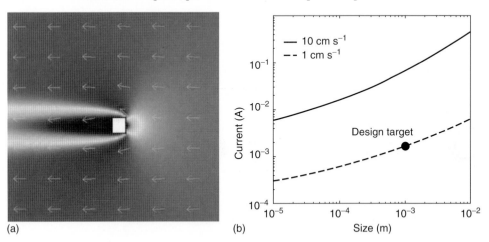

Figure 4.5 Simulated performance of MHD propulsion. (a) An example of the device moving through a fluid and (b) simulation results for the scaling of the device. (Adapted from [4] with copyright permission.)

bounds of optimized wireless powering techniques through tissue, so miniaturized locomotive implantable devices are possible with this method.

The experimental setup for MHD propulsion is shown in Figure 4.6. During propulsion testing, the external antenna tracked the device at a distance ranging from 2 to 5 cm. Data is continuously transmitted with commands to control the motion. The device achieves speeds of up to 0.53 cm s^{-1} in a 0.06 T field with approximately 1 mA, and can be navigated successfully along the surface of the water. Performance improves as the magnetic field is increased, so MRI systems will generate approximately 100 times as much propulsion force. Flows are strong in the arteries and heart, so this device would be most useful in peripheral veins where the flow is under 10 cm s^{-1}, or in the GI tract where normal peristalsis occurs at 2–4 cm min^{-1}.

Although this application demonstrates the feasibility of propulsion of implantable devices through fluid-filled cavities in the body, the concept can be extended to therapeutic methods, such as neuromodulation or pacemakers. In fact, some of the recent research has shown that it is possible to achieve similar results from much lower power stimulation as conventional pacemakers, if the electrodes and stimulation pattern are optimized [37]. Therefore, many of the building blocks used for the locomotive implant could be used to design novel miniaturized implantable therapeutic devices.

4.3.3.2 Low-Power Sensing for Diagnostics: Implantable Intracardiac Probe

Cardiac arrhythmia is abnormal electrical activity in the heart that causes the heartbeat to be either too fast or too slow. Cardiac arrhythmias affect more than 5 million people nationwide, and result in more than 1.2 million hospitalizations and 400 000 deaths each year in the United States [38]. The most common and the most lethal electrical disturbances of the heart are both caused by altered electrical

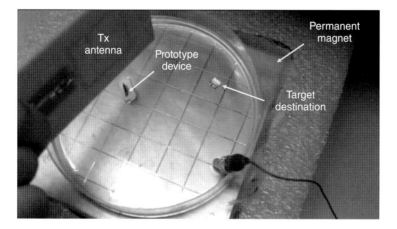

Figure 4.6 MHD propulsion test setup in which the device is navigated to the target destination. The prototype was encapsulated in epoxy and positioned atop a Styrofoam piece for flotation. The fluid conductivity was adjusted to simulate conductivity of blood. (Adapted from [4] with copyright permission.)

conduction patterns [39]. In order to find these disturbances, precise mapping of the local electrical depolarization pattern is required [40, 41]. The conventional mapping method is a catheter-based point-by-point mapping procedure that provides the propagation electrical wave front, but is invasive and a time-intensive procedure. Moreover, it does not provide the real-time information of the pattern.

Given these difficulties, miniaturized implantable low-power sensors can greatly enhance the detection and treatment of cardiac arrhythmia. Miniaturization of implantable sensors with antenna and electrodes enables their placement in a large number of locations on the heart with a simple method of minimally invasive implantation. Then, using wireless power transfer and wireless communication, we can noninvasively reconstruct the heart electrical pathway and deduce the source of rhythm disturbance continuously in real time and as needed throughout the patient's life-span because of the low-power operation of the sensor.

In order to describe the implementation of the miniaturized implantable low-power sensor in detail, we introduce individual blocks as shown in Figure 4.7 from the general design considerations to specific details unique to this application. First, a brief introduction to biopotential electrodes is presented. Later, the readout front end is presented, which defines the quality of the extracted signals, including three major blocks: high pass filter, amplifier, and low pass filter. Finally, low-power analog to digital converters (ADCs) are presented.

Electrode design is often treated as the least important component of biopotential sensing system design. However, when it comes to achieving the optimal signal quality with minimal system power dissipation, designers should understand the principles of operation and trade-offs associated with biopotential electrode design. First of all, electrode can be thought of as transducer between the body and the readout circuit that converts the ionic current into electronic current [13]. It can be placed on the skin with the configuration of the unipolar electrodes with ground electrode or implanted inside the body with bipolar electrodes. In this implantable cardiac probe example, as there is no way to wire the ground electrode with the system, a bipolar electrode is used. The main design consideration is to find the optimum size and material of the metal plate. Depending on these parameters, the equivalent circuit model of the electrode can be obtained, which determines the impedance of electrode. It is important to minimize electrode

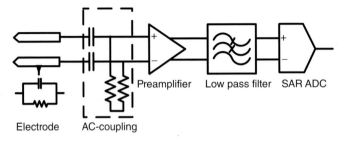

Figure 4.7 Architecture of the analog front end for the acquisition of extracellular potential with electrode equivalent circuit model.

impedance to attain high signal-to-noise ratio because it decreases the intrinsic thermal noise associated with the metal–electrolyte interface and increases the input signal amplitude by reducing the attenuation ratio of electrode impedance to the finite readout front end input impedance. Reduced electrode impedance can be accomplished by either increasing the geometric size of the electrode or by increasing the surface area through control of electrode roughness. Here, we emphasize the design trade-off that reduced electrode noise reduces total power consumption by relieving the input referred noise requirement of the readout front end. However, since the electrode size is often limited to the dimensions of the sensor for the application, the electrode impedance has to be optimized to reduce the electrode noise. For example, if the cardiac probe has to fit into the coronary artery, the sensor should be smaller than $1\,mm^2$ on the side and then each electrode and space between them together should be optimized within $1\,mm^2$. In addition, the electrode material should be chosen from inert metals that do not react with surrounding tissues, such as platinum or gold to make them biocompatible. Therefore, careful consideration of all the design trade-offs significantly relaxes other design requirements and enables designers to reduce power consumption of the overall system. Electrodes interface with the readout front end in which the preamplifier is the most important block to determine the input referred noise because the noise of the following blocks is attenuated by the preamplifier gain. In addition, common mode noise interference can be significant at the input of the amplifier requiring the preamplifier to have high common mode rejection ratio (CMRR) and power supply rejection ratio (PSRR) with minimal power consumption. Moreover, flicker noise ($1/f$ noise) in CMOS technology further complicates the design of low-noise circuits in the sub-hertz to few tens of hertz frequency band because of the very low frequency content of the biological signals and extremely small amplitudes on the order of hundreds of microvolts. To meet these specifications, instrumentation amplifier architectures and chopper modulation techniques are often used in the preamplifier stage [14]. A high pass filter or AC-coupling block interfaces with the bipolar electrode and rejects slow DC variations and electrode offset, which come from localized field potentials and electrode impedance mismatch. Because of the very low cutoff frequency, large capacitor and resistor are necessary to produce a large time constant. A large capacitor can occupy more area, which is not desirable with respect to the miniaturization of the implantable sensor. For this reason, pseudo-PMOS resistors are often employed for large resistance in AC-coupling blocks [42]. Moreover, in order to reject high-frequency noise and 50/60 Hz interference, a low pass filter is necessary. Switched capacitor filter and Gm-C filter are possible candidates for the low pass filter implementation [43]. However, switched capacitor filter requires precise on-chip clock to accurately set the cutoff frequency, which is difficult to implement in miniaturized implantable devices because of power and size constraints. In the example of cardiac probe, tunable third order elliptic Gm-C low pass filter is implemented after the preamplifier.

The last stage of the readout front end consists of the ADC, which provides the digitized output of the signal to the transmitter for wireless transmission to the external reader, which will do further signal processing in the digital domain. Charge-sharing successive approximation ADCs and comparator-based asynchronous binary search ADCs are often selected for biomedical applications because of their low power consumption, small area, and simple topology compared to the other architectures [44, 45]. The sampling frequency and resolution are typically determined on the basis of the desired application. For the implantable heart sensor, an 8-bit comparator-based asynchronous binary search ADC was designed with asynchronous signal sampling due to the intermittent operation of the converter.

In this section, we have presented not only the specific sensing application, intracardiac mapping, but also design considerations of the critical building blocks for many sensing applications. Electrodes, amplification and signal conditioning, and ADC are essential for any biopotential acquisition system and the design consideration can be easily extended to other applications such as neural implantable sensor or implantable pressure sensor. As a payload for the implantable device, the implantable sensor offers excellent potential and opportunity for new treatments and diagnostics in medicine.

4.4
Conclusion

This chapter has provided some insight into the current state-of-the-art commercially available implantable devices and how they fall short of the ultimate goal of preventative care rather than acute care. Then, we provided a grand vision of a future implantable device that will enable preventative care and empower both doctors with novel tools and allow people to take better care of their own well-being. We have also discussed that energy harvesting and RF power transfer in particular are key to miniaturization of implantable devices. System design considerations and trade-offs were then discussed in some detail followed by typical implant architecture and an overview of the essential blocks including power management and data transceiver. Two examples of implants were provided that demonstrate feasibility of novel miniaturized wirelessly powered implantable therapeutic and diagnostic devices.

References

1. WebMD http://www.webmd.com/parkinsons-disease/deep-brain-stimulation (accessed 06 September 2013).
2. University of Pittsburgh Neurological Surgery, http://www.neurosurgery.pitt.edu/imageguided/movement/stimulation.html (accessed 06 September 2013).
3. Wikipedia Deep Brain Stimulation http://en.wikipedia.org/wiki/Deep_brain_stimulation (accessed 06 September 2013).
4. Pivonka, D., Yakovlev, A., Poon, A., and Meng, T. (2012) A mm-sized wireless powered and remotely controlled locomotive implant. *IEEE Trans. Biomed. Circ. Syst.*, **6** (6), 523–532 (invited).
5. Tabatabaei, S.N., Duchemin, S., Girouard, H., and Martel, S. (2012) Towards MR-navigable nanorobotic carriers for drug delivery into the brain. IEEE International Conference on Robotics and Automation.

6. O'Driscoll, S., Poon, A.S.Y., and Meng, T.H. (2009) A mm-sized implantable power receiver with adaptive link compensation. *Proceedings of the IEEE International Solid-State Circuits Conference (ISSCC)*, February 2009.
7. Yakovlev, A., Pivonka, D., Meng, T.H., and Poon, A.S.Y. (2012) A mm-sized wirelessly powered and remotely controlled locomotive implantable device. *Proceedings of the IEEE International Solid-State Circuits Conference (ISSCC)*, San Francisco, CA, February 2012.
8. Chen, G. et al. (2011) A cubic-millimeter energy-autonomous wireless intraocular pressure monitor. *Proceedings of the IEEE International Solid-State Circuits Conference (ISSCC)*, San Francisco, CA, February 2011.
9. Yakovlev, A., Kim, S., and Poon, A.S.Y. (2012) Implantable biomedical devices: wireless powering and communication. *IEEE Commun. Mag.*, **50** (4), 152–159 (invited).
10. Gultepe, E., Randhawa, J.S., Kadam, S., Yamanaka, S., Selaru, F.M., Shin, E.J., Kalloo, A.N., and Gracias, D.H. (2013) Biopsy with thermally-responsive untethered microtools. *Adv. Mater.*, **25**, 514–519.
11. Stevenson, C.L., Santini, J.T. Jr., and Langer, R. (2012) Reservoir-based drug delivery systems utilizing microtechnology. *Adv. Drug Deliv. Rev.*, **64** (14), 1590–1602.
12. Liao, Yu-Te. et al. (2012) A 3-μW CMOS glucose sensor for wireless contact-lens tear glucose monitoring. *J. Solid-State Circuits*, **47** (1), 335–344.
13. Webster, J.G. (1992) *Medical Instrumentation: Application and Design*, 2nd edn, Houghton Mifflin, Boston, MA.
14. Yoo, H.-J. and van Hoof, C. (eds) (2011) *Bio-Medical CMOS ICs*, Springer, New York.
15. Rabaey, J.M., Mark, M., Chen, D., Sutardja, C., Chongxuan, T., Gowda, S., Wagner, M., and Werthimer, D. (2011) Powering and communicating with mm-size implants. DATE – Design, Automation and Test in Europe, 14–18 March 2011, pp. 1–6.
16. Poon, A.S.Y., O'Driscoll, S., and Meng, T.H. (2007) Optimal operating frequency in wireless power transmission for implantable devices. *Proceedings of the IEEE Engineering in Medicine and Biology Society Annual International Conference (EMBC)*, Lyon, France, August 2007, pp. 5673–5678.
17. Kim, S., Ho, J.S., Chen, L.Y., and Poon, A.S.Y. (2012) Wireless power transfer to a cardiac implant. *Appl. Phys. Lett.*, **101**, 073701.
18. Kim, S., Ho, J.S., and Poon, A.S.Y. (2012) Wireless power transfer to miniature implants: transmitter optimization. *IEEE Trans. Antenn. Propag.*, **60** (10), 4838–4845.
19. Mandal, S. and Sarpeshkar, R. (2007) Low power CMOS rectifier design for RFID applications. *IEEE Trans. Circuits Syst. I*, **54**, 1177–1188.
20. Ueno, K. et al (2009) A 300 nW, 15 ppm/C, 20 ppm/V CMOS voltage reference circuit consisting of sub-threshold MOSFETs. *IEEE J. Solid-State Circuits*, **44** (7), 2047–2054.
21. Lotze, N. and Manoli, Y. (2011) A 62 mV 0.13 μm CMOS standard-cell-based design technique using Schmitt-trigger logic. Technical Digest of 2011 IEEE International Solid-State Circuits Conference (ISSCC), Paper 19.5, San Francisco, CA, February 2011.
22. Rubinstein, J.T. (2004) How cochlear implants encode speech. *Curr. Opin. Otolaryngol. Head Neck Surg.*, **12**, 444–448.
23. Liu, W., Fink, W., Tarbell, M., and Sivaprakasam, M. (2005) Image processing and interface for retinal visual prostheses. IEEE International Symposium on Circuits and Systems, ISCAS, May 23–26, 2005, Vol. 3, pp. 2927–2930.
24. Ghenim, A. et al. (2010) A full digital low power DPSK demodulator and clock recovery circuit for high data rate neural implants. IEEE International Conference on Electronics, Circuits, and Systems (ICECS), December 2010, pp. 571–574.
25. Zhang, Y., Zhang, F., Shakhsheer, Y., Silver, J.D., Klinefelter, A., Nagaraju, M., Boley, J., Pandey, J., Shrivastava, A., Carlson, E.J., Wood, A., Calhoun, B.H., and Otis, B. (2013) A batteryless 19 uW MICS/ISM-band energy harvesting

body sensor node SoC for ExG applications. *IEEE J. Solid-State Circuits*, **48** (1), 199–213.

26. Pandey, J. and Otis, B.P. (2011) A Sub-100 μW MICS/ISM band transmitter based on injection-locking and frequency multiplication. *Solid-State Circuits, IEEE Journal of*, **46** (5), 1049–1058.

27. Schell, B. and Tsividis, Y. (2008) A continuous-time ADC/DSP/DAC system with no clock and with activity-dependent power dissipation. *IEEE J. Solid-State Circuits*, **43** (11), 2472–2481.

28. Bletsas, A., Dimitriou, A.G., and Sahalos, J.N. (2010) Improving backscatter radio tag efficiency. *IEEE Trans. Microw. Theory Tech.*, **58** (6), 1502–1509.

29. Delvaux, M. and Gay, G. (2006) Capsule endoscopy in 2005: facts and perspectives. *Best Pract. Res. Clin. Gastroenterol.*, **20** (1), 23–39.

30. Abbott, J. (2007) Robotics in the small part I: microrobotics. *IEEE Robot. Auto. Mag.*, **14** (2), 92–103.

31. Tamaz, S., Gourdeau, R., Chanu, A., Mathieu, J.-B., and Martel, S. (2008) Realtime MRI-based control of a ferromagnetic core for endovascular navigation. *IEEE Trans. Biomed. Eng.*, **55** (7), 1854–1863.

32. Nelson, B. (2009) Towards nanorobots. Solid-State Sensors, Actuators and Microsystems Conference, Transducers 2009, June 2009, pp. 2155–2159.

33. Pivonka, D., Poon, A.S.Y., and Meng, T.H. (2009) Locomotive micro-implant with active electromagnetic propulsion. *Proceedings of the IEEE Engineering in Medicine and Biology Society Annual Internationl Conference (EMBC)*, September 2009, pp. 6404–6407.

34. Pivonka, D., Meng, T., and Poon, A. (2010) Translating electromagnetic torque into controlled motion for use in medical implants. *Proceedings of the IEEE Engineering in Medicine and Biology Society Annual International Conference (EMBC)*, September 2010, pp. 6433–6436.

35. Visser, K. (1989) Electric conductivity of stationary and flowing human blood at low frequencies. *Proceedings of the IEEE Engineering in Medicine and Biology Society Annual International Conference (EMBC)*, November 1989, pp. 1540–1542.

36. Le, H.-P., Seeman, M., Sanders, S., Sathe, V., Naffziger, S., and Alon, E. (2010) A 32 nm fully integrated reconfigurable switched-capacitor dc-dc converter delivering 0.55 W/square mm at 81 percent efficiency. Solid-State Circuits Conference – Digest of Technical Papers (ISSCC), February 2010, pp. 210–211.

37. Lin, Z., Sarosiek, I., and McCallum, R.W. (2007) Gastrointestinal electrical stimulation for treatment of gastrointestinal disorders: gastroparesis, obesity, fecal incontinence, and constipation. *Gastroenterol. Clin. North Am.*, **36** (3), 713–734.

38. Medtech Insight http://www.medtechinsight.com/ReportA230.html (accessed 06 September 2013).

39. Kneeland, P.P. and Fang, M.C. (2009) Trends in catheter ablation for atrial fibrillation in the United States. *J. Hosp. Med.*, **4**, E1–E5. doi: 10.1002/jhm.445

40. Ben-Haim, S.A., Osadchy, D., Scnuster, I., Gepstein, L., Hayam, G., and Josephson, M.E. (1996) Nonfluoroscopic, in vivo navigation and mapping technology. *Nat. Med.*, **2**, 1393–1395.

41. Gepstein, L., Hayam, G., and Ben-Haim, S.A. (1997) A novel method for nonfluoroscopic catheter-based electroanatomical mapping of the heart: in vitro and in vivo accuracy results. *Circulation*, **21**, 1268–1278.

42. Harrison, R.R. and Charles, C. (2003) A low-power low-noise CMOS amplifier for neural recording applications. *IEEE J. Solid-State Circuits*, **38** (6), 958–965.

43. Brodersen, R.W., Gray, P.R., and Hodges, D. (1979) MOS switched-capacitor filters. *Proc. IEEE*, **67** (1), 61–75.

44. Tsividis, Y. (1994) Integrated continuous-time filter design—an overview. *Solid-State Circuits, IEEE Journal of*, **29** (3), 166–176.

45. McCreary, J.L. and Gray, P.R. (1975) All-MOS charge redistribution analog-to-digital conversion techniques. I. *IEEE J. Solid-State Circuits*, **10** (6), 371–379.

5
Cross-Hierarchy Design Exploration for Implantable Electronics
Mrigank Sharad and Kaushik Roy

5.1
Introduction

A multitude of neurological and cardiovascular illnesses that cannot be mitigated by medication alone have resulted in a significant growth in the number of patients that require implantable electronic devices. These range from sensors, gastric and cardiac pacemakers, and cardioverter defibrillators to deep-brain, nerve, and bone stimulators [1]. The advancement of implantable electronic devices has benefited from the parallel growth in the knowledge of various aspects of the human biologic system and the development of technologies capable of interfacing with living tissues and organs at micro and nanoscales [2].

Long-term implants for different biomedical applications present specific engineering challenges related to minimization of energy consumption, miniaturization, and stable performance. As a result, design of implantable bioelectronic devices has emerged as an evolving field of research, with ever-increasing application space. Development in device technology and circuit design methods, clubbed with advances in the field of sensors, signal-processing algorithms, wireless communication methodologies, and power delivery techniques, have opened new avenues for this exciting multidisciplinary research domain [3].

Energy efficiency and other design metrics of each of the different components of implantable bioelectronic systems may depend on several factors and design decisions along the IC design hierarchy. This may include the choice of technology, circuit architecture design, trade-off between different performance characteristics, and, algorithm- and system-level optimizations. In this chapter, we present an overview of the basic concepts related to the design of low-power implantable electronics, with such a cross-hierarchy approach.

5.2
System Overview of a Generic Bioelectronic Implant

A conceptual diagram for a generic biomedical implant is depicted in Figure 5.1. The overall architecture can be broadly divided into parts for signal acquisition,

Implantable Bioelectronics, First Edition. Edited by Evgeny Katz.
© 2014 Wiley-VCH Verlag GmbH & Co. KGaA. Published 2014 by Wiley-VCH Verlag GmbH & Co. KGaA.

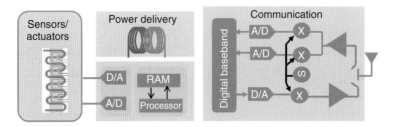

Figure 5.1 A generic biomedical implant system.

analog preprocessing, data conversion, digital signal processing (DSP), memory, data communication, and power delivery [4].

Technology scaling has enabled small-form factors for area-constrained bioelectronic devices, facilitating the implementation of complex information-processing functionality, both in analog and digital domains. Novel circuit design techniques, power-aware signal-processing algorithms, along with exploration of emerging device technologies (for computation and memory) may hold the key to the ultralow-energy budgets for implantable ICs [5]. RF signaling and noninvasive power delivery methods have shown promise for the development of complex, interlinked system of bioelectronic devices that can communicate and cooperate among themselves to realize superior and reliable heath monitoring systems [6]. Emerging techniques for energy harvesting can enable scavenging from widely varying and disperse energy sources, ranging from heat and light to physical vibrations and the chemistry of body fluids [7]. Such techniques may help elongate the life span of implantable systems by minimizing the dependence on battery life.

In this chapter, we review the basic cross-hierarchy low-power design concepts, ranging from the application of suitable device technologies and circuits schemes to algorithms and architecture-level optimizations that aim to achieve impactful improvements in the key design metrics of bioelectronic implants. Multiple design concepts can be applicable to the optimization of a given system sub-block. For instance, reduction in the power consumption for DSP blocks may be achieved by circuit-level design techniques, such as the application of subthreshold circuits. A higher level perspective, on the other hand, may involve exploring smart and adaptive trade-offs between design metrics such as precision and performance in order to achieve low power budgets [3]. Research may also justify climbing down the ladder of IC design hierarchy and exploring the application of novel transistor models, if their advantages over the prevalent bulk CMOS technology can be clearly indentified. In the present scenario, such cross-hierarchy design options may be available for all the different aforementioned sub-blocks of implantable bioelectronic systems that call for multidisciplinary and cross-hierarchy research and exploration.

5.3
Circuit Design for Low-Power Signal Processing

Generally, two processing domains exist: analog and digital. Analog signal processing is directly mapped using the input–output characteristics of transistors. Certain complex computations can be performed very efficiently by exploiting these. However, sensitivities to environment, biasing, noise, and variation limit their dynamic range. DSP, on the other hand, quantizes the signal and achieves enhanced robustness against these sensitivities.

The critical data-processing components of biomedical electronic devices can be accordingly classified into analog blocks such as sensor interface, analog-to-digital converter (ADC), and communication transceivers and digital blocks that include the DSP and memory units. Stringent energy constraints dominate circuit design choices for implantable biomedical systems. Circuit design choices for these blocks are discussed in the following subsections. The basic design challenges for the fundamental circuits at scaled technology nodes are analyzed, which set the prologue for the discussion on the exploration of emerging device technologies suitable for ultralow power biomedical circuit design in the following section.

5.3.1
Design Challenges for Low-Power Bioelectronic Sensor Interface

The interface between man-made electronic circuits and living organisms is one of the focal points of bioelectronics implant design. Biointerfaces have evolved tremendously during the last two decades by integrating new technologies for sensing and fabrication, by exploiting MEMS (micro-electromechanical systems)-based and, more recently, nanotechnology-based implementation [8]. Still, at present, we are far from a complete system where a seamless two-way interface exists between the electronic world and the biological system. Such an interface will be instrumental in allowing novel systems to be designed. Figure 5.2a provides a conceptual view of one such integrated bioelectronic system, which includes on-chip microfluidics, sensing elements, and analog front end [8].

The main challenges related to bioelectronic neural interfaces include

- low-noise amplification of signals that have a very low amplitude, typically in the range of several tens of microvolts; proper handling of $1/f$ noise, since biosignals are typically below 10 kHz;
- control over the thermal dissipation into surrounding cortical or body tissues; advances in fabrication technologies, circuit design, as well as system-level control are required;
- energy scavenging, enabling the harvesting of power from human activity or metabolism;
- increasing spatial resolution of the sensors, and solving consequent issues related to sensor, addressing high data rates to be routed, stored, and transmitted.

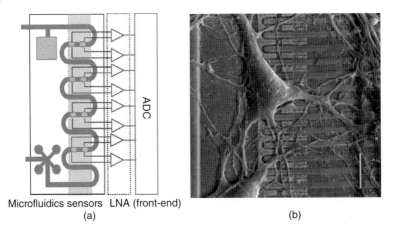

Figure 5.2 (a) Diagram of a typical integrated microfluidic biosensor interface and (b) neuro biosensors. (From [9].)

The main analog blocks in the acquisition and processing of neural signals include low-noise amplifiers that typically provide usable output signals in the millivolt range, filters that limit the range of the in-band signals, optional spike detection circuits, and the ADC(s). Typical interelectrode pitches in the range of 50–100 µm in most systems presented so far has enabled most analog blocks to be embedded in each acquisition/stimulation channel. As the bioelectronic sensor interface circuits are designed for low-frequency operation, the main design challenge for these analog blocks is the $1/f$ noise that mandates the use of gate lengths greater than 180 nm. Especially, the input stage of the first amplifier in the chain needs to use fairly large transistors. This poses scalability challenges and limits the spatial resolution of sensing channels and electrodes [10].

Nanotechnologies and nanometer-scale biosensor structures are recently emerging to provide applicable solutions in the domain of biomedical engineering. Physical phenomena that occur in the nanometer range or are related to quantum transports are exploited to create novel transistor and sensor devices. Protein, enzyme, and DNA sensors, as well as neural cell sensors, have already been reported [11]. If nanotechnologies can be perceived as the ultimate integration goal, interfacing nanodevices with standard microelectronics remains an open issue.

5.3.2
Analog Signal Processing Using Subthreshold Circuits

There is often an inevitable degree of analog preprocessing required before the digitalization of a biological signal [3]. This may include, for instance, the use of band-limiting filters that discard the less informative frequency content of the input data, thereby reducing the power consumption in the DSP blocks [5]. Analog circuits exploit transistors' input–output characteristics in appropriate operating regions to compute complex functions directly upon the analog signal. Realization of such computation in the digital domain may take a lot many transistors and lower dynamic switching power.

However, reducing the static power consumption in such analog processing blocks, arising from constant biasing currents, is critical for maximizing their overall benefits. This is generally achieved through subthreshold operation of analog circuits. The threshold voltage V_t separates the operating conditions of a MOSFET (with drain current I_D, gate-to-source voltage V_{GS}, and drain-to-source voltage V_{DS}) into two regimes: super-V_t and sub-V_t, given by Equations 5.1 and 5.2, respectively:

$$I_D = I_0 e^{\frac{V_{GS}-V_t}{nth}} \left(1 - e^{\frac{-V_{DS}}{nth}}\right) : V_{GS} < V_t \quad (5.1)$$

$$I_D = K' \frac{W}{L}((V_{GS} - V_t)V_{DS} - 0.5 V_{DS}^2) : V_{GS} > V_t \quad (5.2)$$

Here, I_0, k', and $\frac{W}{L}$ are physical and geometric parameters. Equation (5.2), the sub-threshold (sub-V_t) regime, has an exponential dependence on both V_{GS} and V_t; as a result, I_D reduces by more than an order of magnitude for every 100 mV (assuming subthreshold swing to be less than 100mv/decade) decrease in ($V_{GS} - V_t$), and the relative currents in sub-V_t can be very low for a given device width. Accordingly, the relative transconductance, g_m (which is defined as $\frac{\partial I_D}{\partial V_{GS}}$), is also low. Importantly, however, the transconductance efficiency, $\frac{g_m}{I_D}$, is highest in sub-V_t [5]. The biasing current, I_D, sets the operating point of the transconductor and typically determines its power consumption because it must be held static to avoid output distortion. Therefore, sub-V_t has superior power efficiency. The associated cost is reduced circuit speed due to the lower absolute output current; however, for biomedical applications, where the processing bandwidths are typically less than 1 MHz, the speed is sufficient, particularly where V_{GS} is close to V_t.

The key challenges in subthreshold analog design involve high sensitivity to transistor mismatch, power supply noise, and temperature that mandates sophisticated biasing circuits [12]. Linearity achievable in sub-V_t analog circuits is also significantly lower as compared to other regimes, which may prohibit cascading of multiple sub-V_t stages to prevent the loss of information in the data due to signal distortion. Figure 5.3 shows the diagram for a sub-V_t circuit design for a gammatone filter employed in cochlear implants [13].

5.3.3
Analog-to-Digital Conversion

Post amplification and analog preprocessing, the analog-mode biomedical signals need to be quantized using an ADC before they can be processed digitally using a sophisticated DSP. ADC requirements depend on system characteristics, namely, bandwidth and dynamic range; therefore, system optimization must consider ADC power, which can be a significant portion of the total power. The energy per conversion, which is an important metric for ADCs, increases as the dynamic range and sampling rate requirements increase. An empirical figure of merit (FOM) for

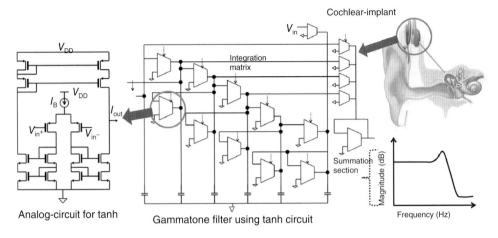

Figure 5.3 Sub-V_t analog circuit for gammatone filters employed in cochlear implants.

ADCs normalizes their power consumption to the Nyquist sampling rate, F_S, and the dynamic range, expressed as 2^{ENOB} (where ENOB is the effective number of bits output) [5]:

$$\text{FOM} = \frac{\text{Power}}{(2^{ENOB} F_S)} \quad (5.3)$$

State-of-the-art converters today achieve an FOM as low as 4 fJ per conversion step. However, generally, dynamic ranges beyond those yielded by 8-bit converters have a steeper power increase due to device noise limitations in the ADC circuits; the same is true when sampling rates exceed tens of megahertz because devices must be biased in upper super-threshold regime. Successive approximation register (SAR) ADC, shown in Figure 5.4, is the most common topology used for low-power biomedical implants [14].

Here, a digital-to-analog converter (DAC) is implemented using capacitor array. During input acquisition, the negative of the input voltage is sampled on the top

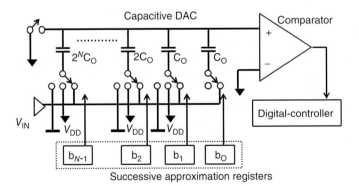

Figure 5.4 Schematic diagram of an SAR-ADC.

plates, and, subsequently during the conversion, the binary weighted capacitors are successively switched between ground and V_{DD} by a digital controller, performing a binary search that eventually converges to the digital output code. Importantly, all components, except the comparator, are either passive or digital. Accordingly, the voltage scaling and clock-gating can enable low-energy SAR-ADC implementations. The comparator, particularly in high-resolution SAR converters, whose gain requirements are immense, often dominates power consumption.

5.3.4
Low-Power Digital Signal Processing

Because devices in a digital circuit have a large noise margin, they are robust to the distortion and noise effects plaguing analog circuits in advanced technologies. The ability to benefit accordingly from technology scaling trends has enabled elaborate digital architectures that are highly reconfigurable and energy efficient, offering great possibilities for biomedical devices [5]. Using the low-power techniques described in this section, the high level of agility afforded by digital circuits can be aggressively leveraged for general biomedical applications.

The dominant source of power in digital circuits is called *active power*, P_{ACT}, and is consumed during logic transitions when charge must be transferred from the supply voltage, V_{DD}, to the physical capacitance of a signal-carrying circuit node, C_L. P_{ACT} for each node is given by $\alpha_{0 \to 1} C_L V_{DD}^2 f_{clk}$. Digital logic gates conventionally perform a new computation every clock cycle; however, depending on the circuit functionality and implementation, transitions might not occur at the clock frequency, f_{clk}. Therefore, $\alpha_{0 \to 1}$ reflects the average number of $0 \to 1$ transitions per clock cycle and can be greater than 1 due to glitches. The actual energy consumed for an entire function, which is a relevant metric with regard to the energy constraint, can be determined by summing the total capacitance that transitions during the course of the operation, $C_{TOT} = N \sum_{i=\text{all nodes}} \alpha_{0 \to 1} C_{L,i}$, where N represents the number of clock cycles required to complete the function. Then, the total active energy is given by

$$E_{ACT} = C_{TOT} V_{DD}^2 \tag{5.4}$$

5.3.4.1 V_{DD} Scaling and Parallel Processing

From Equation 5.4, reducing the supply voltage yields significant energy savings but at the cost of reduced performance due to lower output currents available to switch the circuit node capacitances [15]. In biomedical applications, the reduced performance that comes with voltage scaling is often acceptable because the accompanying energy savings are so significant [16]. More generally, however, in some real-time applications, system-level throughput constraints, modest as they may be, must be met. In such cases, parallelism, combined with voltage scaling, can be employed to simultaneously achieve the required performance and energy efficiency [17]. Here, a hardware unit is replicated M times, so that each unit can operate at a frequency reduced by a factor of M while maintaining the same overall performance. The reduced frequency per unit, however, enables voltage scaling and improved energy efficiency. This is because in parallel digital designs, $Nf_{lo}CV_{DDlo}^2$ is less than $Cf_{hi} V_{DDhi}^2$. Of course, the necessity of interface combiners imposes additional overhead; nonetheless, system partitioning and algorithm redesign often allow for highly parallelized implementations with very low overhead.

It should be noted that architectural and algorithmic techniques have also been developed and widely used to reduce energy by minimizing $\alpha_{0 \to 1}$ [3]. For example, balancing logic delays from timing path inputs can avoid glitching, and optimizing logic function implementations, while considering the input signal statistics, can result in a reduced amount of total switching.

5.3.4.2 Dynamic Voltage and Frequency Scaling

A low-power design model for many biomedical implantable and wearable devices, including biosensor networks [18], consists of reactive nodes that perform only minimal monitoring operations for the vast majority of the time. On detection of an event, their operational state is elevated to execute the appropriate signal processing along with actuation or data transmission. In either case, the performance demand, or workload, increases instantaneously. While circuits operating statically at a reduced voltage cannot accommodate the performance increase, a widely used technique is to dynamically increase their voltage [19]. As shown in Figure 5.5, this approach utilizes a voltage-controlled oscillator (VCO) to set the circuit's operating frequency and a voltage regulator to optimally set V_{DD}. The VCO replicates the critical speed-limiting signal path in the digital circuit, ensuring that its period is sufficient for the circuit delays at any given V_{DD}. The loop works by comparing an input reference clock, which sets the desired frequency based on the system performance constraint, with the VCO frequency. If the VCO frequency is lower, the controller issues a command to the voltage regulator to increase V_{DD}, which subsequently increases the VCO and circuit frequency; if the VCO frequency is higher, the opposite command is issued until the circuit frequency eventually matches the reference. An alternative approach uses a lookup table to map the desired circuit performance to a predetermined V_{DD}; here, the regulator V_{DD} can be immediately set in a feed-forward manner.

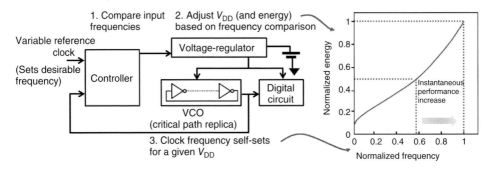

Figure 5.5 Dynamic voltage frequency scheme.

5.3.4.3 Standby Mode Power Reduction

Minimizing circuit node activity, $\alpha_{0\to 1}$, is also an effective method for reducing active power. A simple architectural solution is to partition designs into fine-grained functional blocks and implement independent standby modes for each, as shown in Figure 5.6. Then, during standby, the local clock (L_{CLK}) can be gated, prohibiting logic transitions (i.e., $\alpha_{0\to 1} = 0$) and eliminating active power consumption.

The power reduction techniques described thus far only address active power. In standby, however, an additional source of power – leakage power, P_{LKG} – becomes significant. P_{LKG} refers to idle-mode current that flows while logic circuits are in a high or low state, and it is particularly prominent in advanced technologies, where, by design, V_t is low in order to maximize performance. As shown in Figure 5.6, even when V_{GS} is reduced to zero, some finite leakage current continues to flow. To eliminate it, the supply voltage must also be gated using high-threshold header devices that have much lower leakage current [20]. Of course, header device control, as well as power supply voltage (VV_{DD}) recovery after standby, has some associated energy overhead, $E_{Overhead}$. A simulation of $E_{Overhead}$, as well as the circuit's leakage energy, E_{LKG}, is shown in Figure 5.6 for a finite impulse response (FIR) filter implementation after entering standby (at $t=0$). Power gating only saves energy in this example if the duration of the standby mode exceeds the break-even time of 50 μs, and it should not be applied for shorter periods of inactivity.

5.3.4.4 Minimum Energy Subthreshold Operation

When aggressive voltage scaling is employed, such that the supply voltage is near or below the MOSFET threshold voltage, the resulting leakage energy becomes significant even during normal circuit operation because the circuit delay increases exponentially. In this scenario, the power supply cannot be gated until the operation is complete, and the leakage power integrates over the operating time. Consequently, E_{LKG} is given by Equation (5.5):

$$E_{LKG} = \int_0^{P_{OPT}} P_{LKG} dt \tag{5.5}$$

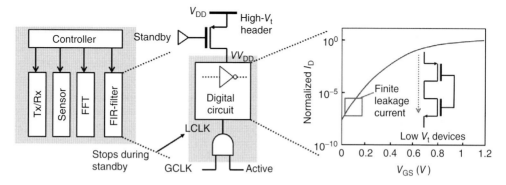

Figure 5.6 Fine-grained power gating for reducing leakage power.

Although scaling V_{DD} is highly effective for reducing E_{ACT}, the circuit delay increases rapidly at very low voltages and E_{LKG} increases correspondingly. The opposing energy trends are shown in Figure 5.7 and they give rise to a minimum total energy voltage [20]. Importantly, this minimum energy voltage occurs in the sub-V_t regime (V) for most practical digital circuits, and the energy savings exceed by one order of magnitude compared with the nominal supply voltage.

An example of circuit-level optimization using the foregoing techniques can be found in [21a]. This work achieved more than 90% improvement in energy efficiency over a previous design [21b] by the use of minimum energy subthreshold operation and data-dependent fine-grained clock and supply gating.

In the next subsection, we briefly discuss the advantages of FinFETs as a device technology for low-energy subthreshold circuit design.

5.3.5
FinFETs for Ultralow Voltage Subthreshold Circuits

In this section, we discuss the advantages offered by FinFETs for low-voltage, low-power signal-processing circuit schemes discussed in the foregoing sections.

As discussed earlier, the subthreshold region operation of conventional bulk MOSFETs offers relatively lower energy dissipation for both analog and digital circuits. For digital circuits, the energy savings result from lower dynamic energy and leakage current resulting from V_{DD} scaling. However, at the same time, circuit delay increases, resulting in higher leakage energy per cycle. This limits the maximum achievable voltage scaling for ultralow-power applications. This effect becomes even more prominent for scaled technology nodes because of increased short-channel effects (SCEs) [22]. SCE stems from increasing influence of drain-to-source voltage, V_{DS}, upon the channel current which, in turn, translates to weaker control of gate voltage upon the device characteristics. As a result, ultralow-power subthreshold circuit design may not reap the benefits of technology scaling, as opposed to high-performance designs for which leakage makes relatively minor

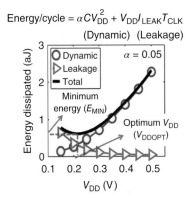

Figure 5.7 Determination of energy-optimal operating point.

contribution to the total energy budget. The analog designs mainly benefit from the increased $\frac{g_m}{I_D}$ ratio in the subthreshold regime; however, increased SCE adversely affects several important characteristics of analog designs such as linearity, dynamic range, and lower intrinsic cutoff frequency f_T [23].

Several device optimization techniques have been proposed to counter SCE in scaled technology nodes. For instance, the use of halo implant and graded channel doping has been advocated for enhancing gate control. However, such methods offer limited scalability due to enhanced random dopant fluctuation in scaled devices [24]. In the following paragraphs, we briefly discuss two such new and emerging device options that can be promising for ultralow-energy subthreshold circuit design, suitable for biomedical implants.

FinFET, a multigate architecture, has received considerable attention in recent years owing to the suppression of SCEs and excellent scalability, and has thus been regarded as a possible candidate for device scaling at the end of International Technology Roadmap for Semiconductors [22]. The device structure for FinFET is shown in Figure 5.8a. It consists of thin silicon channels or "fins" surrounded by gate oxide from three sides. Larger MOS transistors employed in a circuit use multiple such fins with a common gate contact.

Dual-gate structure of a FinFET provides superior gate control over channel current leading to improved subthreshold slope [23], as shown in Figure 5.8b.

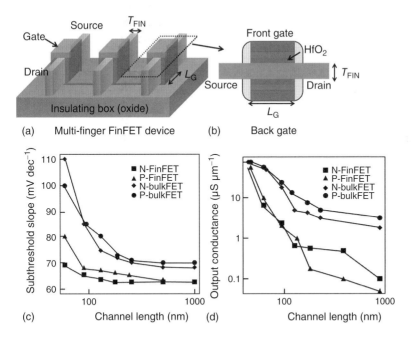

Figure 5.8 (a) Device structure for FinFET, (b) cross-section view of a single Fin, comparison (data from [23]) between FinFET and bulk CMOS devices with respect to (c) subthreshold slope, and (d) output conductance.

A higher subthreshold slope can facilitate the use of lower supply voltage for a given performance constraint. The narrow silicon fins can be fully depleted and can be intrinsic (threshold voltage being modulated by gate workfunction), leading to mitigation of RDF effects. Mitigation of RDF can also facilitate improved device scaling. Fully depleted FinFET also achieves significantly lower output transconductance as compared to bulk MOSFET [24]. This translates to higher output small-signal impedance and hence higher gain for analog amplifier circuits. The plot given in Figure 5.8d compares the transconductance of FinFET with bulk CMOS and shows more than an order of magnitude improvement for longer channel lengths (> 60 nm, normally used for analog circuits). Owing to the use of confined thin-film channel, a FinFET offers significantly lower parasitic capacitance values, leading to high performance for both digital and analog circuits [24].

In summary, FinFETs possess key advantages over bulk FETs: better scalability, reduced leakage, excellent subthreshold slope, better voltage gain, and higher performance. This makes them attractive for ultralow power-frequency digital and analog circuit designs for biomedical implants.

5.4
Architecture-Level Optimizations for Low-Power Data Processing

Architecture- and system-level design decisions can be equally useful in enhancing the energy efficiency of a bioelectronic hardware. Such higher level optimizations invariably break links with lower level design choices. For instance, the use of subthreshold circuits for energy-optimal operation may be best exploited by parallel processing architectures. As discussed earlier, in such designs low-voltage and low-performance computing blocks are replicated to meet the throughput requirement. In this section, we briefly touch upon two different system-level design optimizations that can be useful for energy-optimal operation of a complex implantable system. The first method deals with the optimal partition of computational task between analog and digital blocks, whereas the second example treats the dynamic trade-offs between computation accuracy and energy efficiency.

5.4.1
Optimal Apportioning of Computation Task to Analog and Digital Blocks

A default encoding for many sensory computations is a high-speed high-precision analog-to-digital conversion followed by extensive DSP [3]. This solution is highly flexible and robust but is rarely the most power-efficient encoding: It does not exploit the fact that the actual meaningful output information is often orders of magnitude less than the raw information in the numbers of the signal, and the technology's basis functions are more powerful than just switches. It may be better to preprocess the information in an analog manner before digitization and then quantize and sample higher level information at significantly lower speed and/or precision. There is an optimal amount of analog preprocessing before a signal-restoring or

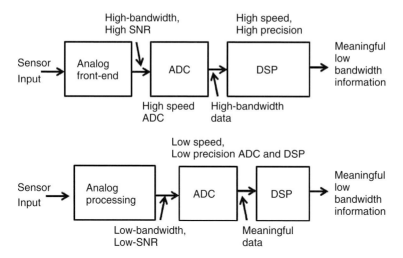

Figure 5.9 Optimum amount of analog preprocessing before digitization minimizes power consumption.

quantizing digitization is performed: For too little analog preprocessing, efficiency is degraded because of an increase in power consumption in the ADC and digital processing, and for too much analog preprocessing, efficiency is degraded because the costs of maintaining precision in the analog preprocessor become too high. The optimum depends on the task, technology, and topology. Figure 5.9 illustrates the idea visually.

In a practical design, other considerations of flexibility and modularity that are complementary to efficiency will also affect the location of this optimum. Examples of systems where analog preprocessing has been used to reduce power by more than an order of magnitude include the highly digitally programmable cochlear implant processor and electroencephalogram processor described in Ref. [25]. The retina in the eye compresses nearly 36 Gb s^{-1} of incoming visual information into nearly 20 Mb s^{-1} of output information before sampling and quantization. The field of "compressed sensing" attempts to exploit similar bandwidth-reduction schemes before sampling and digitization to improve efficiency [26].

The design for implantable seizure prosthesis presented in [27] employed a simple 1-bit ADC to capture the amplitude surge in the seizure signal. This was followed by time-windowed counting of "high" outputs of the ADC to detect the seizure. This design exploited the inherent analog amplitude and timing characteristics of seizure spikes to avoid the use of complex DSP.

By symmetry, digital output information from a DAC should be post-processed with an optimal amount of analog processing before it interfaces with an actuator or communication link. In general, this principle illustrates that energy-efficient systems are optimally partitioned between the analog and digital domains.

5.4.2
Approximate Computing for Low Power

A recent trend in low-power design has been the use of reduced precision "approximate processing" methods for reducing arithmetic activity and average chip power dissipation. Such designs treat power and arithmetic precision as system parameters that can be traded off versus each other on an *ad hoc* basis. For instance, an approximate filtering technique dynamically reduces the filter order on the basis of the input data characteristics [28]. More specifically, the number of taps of a frequency-selective FIR filter is dynamically varied on the basis of the estimated stop-band energy of the input signal. The resulting stop-band energy of the output signal is always kept under a predefined threshold. This technique results in power savings by a factor of 6 for speech inputs, and can also be implemented using dedicated programmable processors.

Another interesting example of dynamic precision adaptation for analog processing is presented in [29]. This work exploited the fact that the outputs of many types of sensors are sparse in the time domain. Data compression was achieved by performing signal-activity-dependent deterministic sampling by way of an asynchronous ADC. The authors further improved the performance of the asynchronous ADC by implementing an adaptive resolution (AR) algorithm, which varied the quantizer resolution on the basis of the rate of change of the input. This

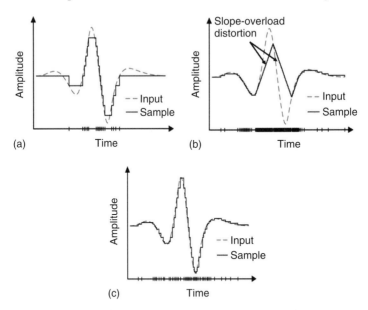

Figure 5.10 Example waveforms of a signal asynchronously sampled with the quantizer resolution: (a) set to the maximum value that meets the input bandwidth requirement of the signal; (b) set to the minimum value that meets the dynamic range requirement of the signal; and (c) varied with the input slope, allowing both the input bandwidth and dynamic range requirements of the signal to be met.

overcomes the trade-off between dynamic range and input bandwidth typically seen in asynchronous ADC designs, and increases the amount of data compression that can be achieved. This scheme is pictorially depicted in Figure 5.10.

Combination of adaptive precision modulation in analog and digital domains may achieve enhanced energy benefits for bioelectronic hardware [6]. However, the cost of control generation must be considered for determining the optimal degree of adaptation in such designs.

5.5
Design of Energy-Efficient Memory

The design of robust, high-density subthreshold memory can be critical in determining the viability of sophisticated biomedical implants. As in the case of analog and digital computing blocks, subthreshold region operation has been the most general approach toward lowering the power consumption of the conventional static random access memory (SRAM) for energy-constrained systems. In this section, we analyze the design challenges with the design of conventional SRAM for such low-voltage operation. Following this, the prospects of emerging non-volatile magnetic random access memory (MRAM) is discussed as a possible technology alternative.

5.5.1
Design Challenges with Subthreshold SRAM

The design of robust low-voltage SRAM is extremely challenging and has been the subject of intense study recently [30]. The key challenges in low-voltage SRAM design are discussed in the following subsections.

5.5.1.1 On-Current to Off-Current Ratio
The dramatic reduction in the on-current to off-current ratio (I_{on}/I_{off}) observed in the near-threshold and subthreshold regions is one of the most fundamental challenges facing low-voltage memory designers. This determines the theoretical upper bound of the number of cells sharing one bitline [31]. For a small I_{on}/I_{off} ratio, it becomes difficult to distinguish between the read current of the accessed cell and the cumulative leakage current of unaccessed cells (Figure 5.11).

5.5.1.2 Sizing Constraints
The second challenge facing low-voltage memory designers is a change in gate-sizing requirements. Since subthreshold current is exponentially dependent on V_t, any skew between the nominal threshold voltages of PMOS and NMOS devices can lead to dramatic shifts in the ratio at low voltages. The read stability and write stability of a conventional 6-T SRAM cell are heavily dependent on the relative strengths of the pull-up, pull-down, and pass transistor devices [32]. The skewed ratios observed at low voltage may therefore lead to an unstable memory cell.

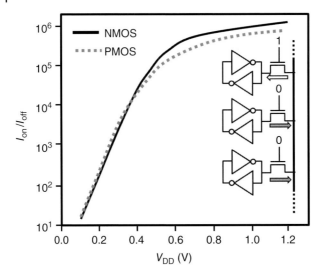

Figure 5.11 On-/off-current ratio as a function of supply voltage. The inset shows contention between read current and bitline leakage during a read operation of 6-T SRAM cell.

5.5.1.3 Variability

The final, and arguably most important, challenge in low-voltage SRAM design is the heightened sensitivity to process, voltage, and temperature variations. Owing to the exponential dependence of drive current on terminal voltages, and temperature, even small variations lead to large fluctuations in transistor drive current. Mismatch between the cross-coupled inverters is particularly concerning since it can lead to widespread functional failure.

All the aforementioned design constraints become increasingly more stringent with technology scaling. Next, we discuss MRAM as a future candidate for energy-efficient on-chip memory technology that can be suitable for the design of biomedical implants.

5.5.2
Spin Transfer Torque MRAM (STT-MRAM) for Energy-Efficient Memory Design

Among different classes of emerging memory technologies, spin transfer torque (STT) MRAM is attractive because of its several desirable features such as nonvolatility, high density, endurance, and energy efficiency resulting from zero leakage [33]. The conventional STT-MRAM (STT-C) device, shown in Figure 5.12a, is a simple magnetic-tunnel junction (MTJ) that consists of two nanoscale ferromagnetic layers, a "fixed" layer and a "free" layer, separated by a tunneling barrier in the form of an oxide [34]. Parallel or antiparallel spin polarities in the two ferromagnetic layers determine the low- and high-resistance states of the device, respectively, and hence the stored data bit. In a standard, STT-C bit-cell, data is written into the free layer of the MTJ by injecting charge current into the high-resistance tunneling device. This

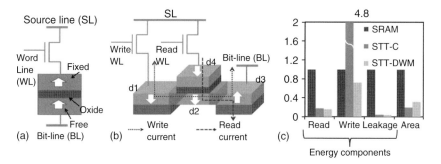

Figure 5.12 (a) Standard two-terminal STT-MRAM, (b) three-terminal STT-MRAM using domain wall shift based write, and (c) energy and area comparison of bit-cells in (a,b) with SRAM.

write scheme can consume four times higher write energy as compared to SRAM. Although write energy is generally a minor fraction of the total cache energy, this can be a significant overhead for ultralow power applications.

Recently proposed STT-MRAM devices with decoupled read–write paths [35] can overcome the aforementioned limitation. Such a bit-cell (shown in Figure 5.12b) employs a separate write path that consists of a free magnetic domain (d2) placed between two complementary fixed domains (d1 and d3). The write mechanism involves current-induced magnetic domain-wall shift (DWS) across the two ends of the free domain. This results in the flipping of its spin polarity, depending upon the direction of current flow. Owing to the low-resistance magnetometallic write path, such a device can achieve low-voltage, low-energy write operations [36]. Figure 5.12c compares the different energy components of the STT-MRAM cells with 6-T SRAM. Notably, DWS-based MRAM achieves large reductions in the two dominant components, namely, read (being more frequent than write) and leakage (negligibly small for nonvolatile MRAM) [37].

Recent spin torque experiments have shown the possibility of large reduction in the critical write currents for such DWS-based memory cells. With appropriate device circuit co-optimization, STT-MRAM devices, with energy-efficient write schemes, can emerge as promising candidates for ultralow power on-chip memory for biomedical implants.

5.6
Wireless Communication Power Delivery

The discussion on low-power bioelectronic implants would be incomplete without an overview of power delivery and communication systems for such devices. Wireless communication is the dominant method of communication as well as power delivery for biomedical devices. The primary method of wireless linking is either via low-frequency, near-field inductive coupling or higher frequency, far-field wireless transmission [38]. A generic block diagram for such a link is depicted in

82 | *5 Cross-Hierarchy Design Exploration for Implantable Electronics*

Figure 5.13. Application of these two methods for power data communication and power delivery is discussed in the following subsections [39].

5.6.1
Near-Field Electromagnetic Wireless Communication

Near-field communication operates on the principle of electromagnetic induction between two nearby coils. Data transfer is often bidirectional, consisting of a forward and a reverse link. Owing to the volume and energy constraints faced by implanted systems, near-field communication links are typically limited in range to a few centimeters. One example application of near-field communication is for data and power transfer from an external cochlear speech processor to its associated implanted stimulator [5]. This application requires data rates on the order of 1 Mbps and a distance of only a few centimeters, operating at carrier frequencies such as 49 MHz. For emerging biomedical applications, such as neurostimulators and artificial retina, higher data rates are required. This can be achieved by far-field electromagnetic wireless communication.

5.6.2
Far-Field Electromagnetic Wireless Communication

To realize high data rates at communication distances longer than a few centimeters, far-field electromagnetic communication is preferable to near-field communication. Far-field communication links for biomedical devices operate at carrier frequencies of hundreds of megahertz and above. The majority of low-power wireless transceivers are half-duplex where data transmission and reception do not occur simultaneously, and thus there is a transmit/receive switch connecting either the transmitter or the receiver to the antenna.

A frequent trade-off in many portable systems is the balance between computation and communication power costs. If one computes too little, there is too much information to transmit, which increases communication costs. If one computes too much, there is little information to transmit, which reduces communication costs but increases computation costs. The optimum depends on the relative costs of each. For instance, such trade-offs affect digitization of bits that need to be

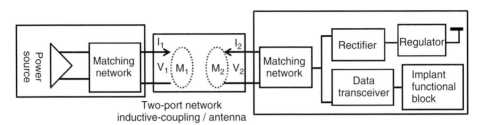

Figure 5.13 Typical implantable system block diagram. The coupling between the external device and the implantable device is via a pair of antenna structures.

transmitted wirelessly in brain–machine interfaces and in neural prosthetics and the general design of complex systems composed of parts that need to interact with each other.

5.6.3
Wireless Energy Transfer

For implanted systems, batteries cannot be easily removed without surgery; hence, alternate wireless energy-transfer approaches must be considered. To supply energy to implanted devices, wireless electromagnetic energy transfer is an effective technique [5]. It involves transmitting electromagnetic energy into the body and collecting it via a coil or antenna. The use of low-frequency electromagnetic energy transfer has been demonstrated for cochlear implants, radio-frequency identification (RFID) implants, retinal prosthetics [40], and neurostimulators [41]. The physical process underlying low-frequency wireless energy transfer is near-field electromagnetic induction. Near-field wireless energy transfer results in electromagnetic fields that can heat tissue and generate electromagnetic interference on nearby electronics. For small implanted applications, where volume constraints limit the size of the receive coil, the practical range of energy transfer is limited to a few centimeters. To increase the transfer distance and increase efficiency, both transmitter and receiver can be tuned to induce resonant coupling [38].

5.7
Conclusion

Long-term implants for different biomedical applications present specific engineering challenges related to minimization of energy consumption, miniaturization, and stable performance. As a result, design of implantable bioelectronic devices has emerged as an evolving field of research, with an ever-increasing application space. Development in device technology and circuit design methods, clubbed with advances in the field of sensors, signal-processing algorithms, wireless communication methodologies, and power delivery techniques have opened new avenues for this exciting multidisciplinary research domain.

The energy efficiency and other design metrics of each of the different components of implantable bioelectronic systems may depend upon several factors and design decisions along the IC design hierarchy. This may include the choice of technology, circuit architecture design, trade-off between different performance characteristics, and, algorithm- and system-level optimizations. In this chapter, we presented an overview of recent developments and proposals for the hardware implementation of low-power implantable electronics with such a cross-hierarchy approach.

References

1. Manivasagam, G., Dhinasekaran, D., and Rajamanickam, A. (2010) *Recent Patents Corros. Sci.*, **2**, 40–54.
2. Nel, A.E., Mädler, L., Velegol, D., Xia, T., Hoek, E.M., Somasundaran, P., and Thompson, M. (2009) *Nat. Mater.*, **8** (7), 543–557.
3. Sarpeshkar, R. (2010) *Ultra Low Power Bioelectronics*, Vol. 1, Cambridge University Press, Cambridge.
4. Yu, H. and Najafi, K. (2003) *Digest of Technical Papers, ISSCC*, IEEE, pp. 194–487.
5. Chandrakasan, A.P., Verma, N., and Daly, D.C. (2008) *Annu. Rev. Biomed. Eng.*, **10**, 247–274.
6. Otto, C., Milenkovic, A., Sanders, C., and Jovanov, E. (2006) *J. Mobile Multimedia*, **1** (4), 307–326.
7. Olivo, J., Carrara, S., and De Micheli, G. (2011) *IEEE Sens. J.*, **11** (7), 1573–1586.
8. Guiducci, C., Schmid, A., Gürkaynak, F.K., and Leblebici, Y. (2008) *Design Automation and Test in Europe*, ACM Press, pp. 1328–1333.
9. Fromherz, P. (2003) in *Nanoelectronics and Information Technology* (ed. R. Wase), Wiley-VCH Verlag GmbH, Berlin, pp. 781–810.
10. Zou, X., Xu, X., Yao, L., and Lian, Y. (2009) *IEEE J. Solid-State Circuits*, **44** (4), 1067–1077.
11. Jokerst, J.V., Jacobson, J.W., Bhagwandin, B.D., Floriano, P.N., Christodoulides, N., and McDevitt, J.T. (2010) *Anal. Chem.*, **82** (5), 1571–1579.
12. Mandal, S. and Sarpeshkar, R. (2008) *IEEE Trans. Biomed. Circuits Syst.*, **2** (4), 301–315.
13. Ngamkham, W., Sawigun, C., Hiseni, S., and Serdijn, W.A. (2010) *Proceedings of 2010 IEEE International Symposium on May, 2010*, IEEE, pp. 969–972.
14. Zhang, D., Bhide, A., and Alvandpour, A. (2011) *Proceedings of the ESSCIRC*, IEEE, pp. 467–470.
15. Soeleman, H., Roy, K., and Paul, B.C. (2001) *IEEE Trans. VLSI Syst.*, **9** (1), 90–99.
16. Kim, C.I., Soeleman, H., and Roy, K. (2003) *IEEE Trans. VLSI Syst.*, **11** (6), 1058–1067.
17. Wang, A., Chandrakasan, A.P., and Kosonocky, S.V. (2002) *Proceedings of the IEEE Computer Society Annual Symposium on VLSI, 2002*, IEEE, pp. 5–9.
18. Raskovic, D. and Giessel, D. (2009) *IEEE Trans. Inf. Technol. Biomed.*, **13** (6), 903–909.
19. Powell, H.C., Barth, A.T., and Lach, J. (2009) Proceedings of the 4th International Conference on Body Area Networks, ICST, 2002, art# 15.
20. (a) Roy, K., Mukhopadhyay, S., and Mahmoodi-Meimand, H. (2003) *Proc. IEEE*, **91** (2), 305–327. (b) Raychowdhury, A. (2005) *IEEE Trans. Very Large Scale Integr. (VLSI) Syst.*, **13** (11), 1213–1224.
21. (a) Sharad, M., Gupta, S.K., Raghunathan, S., Irazoqui, P.P., and Roy, K. (2012) *ACM J. Emerg. Technol. Comput. Syst. (JETC)*, **8** (2), 10. (b) Markandeya, H., Karakonstantis, G., Raghunathan, S., Irazoqui, P., and Roy, K. (2010) *ISLPED*, ACM Press, pp. 301–306.
22. Kim, J.J. and Roy, K. (2004) *IEEE Trans. Electron Dev.*, **51** (9), 1468–1474.
23. Bansal, A., Mukhopadhyay, S., and Roy, K. (2007) *IEEE Trans. Electron Dev.*, **54** (6), 1409–1419.
24. Subramanian, V., Parvais, B., Borremans, J., Mercha, A., Linten, D., Wambacq, P., and Decoutere, S. (2005) *Proceedings of IEDM*, IEEE, pp. 898–901.
25. (a) Sarpeshkar, R., Salthouse, C., Sit, J.J., Baker, M.W., Zhak, S.M., Lu, T.T., and Balster, S. (2005) *IEEE Trans. Biomed. Eng.*, **52** (4), 711–727. (b) Avestruz, A.T., Santa, W., Carlson, D., Jensen, R., Stanslaski, S., Helfenstine, A., and Denison, T. (2008) *IEEE J. Solid-State Circuits*, **43** (12), 3006–3024.
26. Aviyente, S. (2007) *Workshop on Statistical Signal Processing, August, 2007*, IEEE, pp. 181–184.
27. Raghunathan, S., Gupta, S.K., Ward, M.P., Worth, R.M., Roy, K., and Irazoqui, P.P. (2009) *J. Neural Eng.*, **6** (5), art. # 056005.

28. Amirtharajah, R. and Chandrakasan, A.P. (2004) *IEEE J. Solid-State Circuits*, **39** (2), 337–347.
29. Trakimas, M. and Sonkusale, S.R. (2011) *IEEE Trans. Circuits Syst.*, **58** (5), 921–934.
30. Chang, I.J., Kim, J.J., Park, S.P., and Roy, K. (2009) *IEEE J. Solid-State Circuits*, **44** (2), 650–658.
31. Raychowdhury, A., Mukhopadhyay, S., and Roy, K. (2005) *Proceedings of ICCAD*, IEEE, pp. 417–422.
32. Zhai, B., Hanson, S., Blaauw, D., and Sylvester, D. (2008) *IEEE J. Solid-State Circuits*, **43**, 2338–2348.
33. Zhu, J.G. (2008) *Proc. IEEE*, **96** (11), 1786–1798.
34. Li, J., Ndai, P., Goel, A., Salahuddin, S., and Roy, K. (2010) *TVLSI*, **18**, 1710–1723.
35. Fukami, S., Suzuki, T., Nagahara, K., Ohshima, N., Ozaki, Y., Saito, S., and Sugibayashi, T. (2009) *Symposium on VLSI Technology*, IEEE, pp. 230–231.
36. Venkatesan, R., Sharad, M., Roy, K., and Raghunathan, A. (2013) DWM-TAPESTRI - an energy efficient all-spin cache using domain wall shift based writesDATE'13 Proceedings of the Conference on Design, Automation and Test in Europe, pp. 1825–1830.
37. Sharad, M., Venkatesan, R., Raghunathan, A., and Roy, K. (2013) Proceedings of International Symposium on Low Power Electronics and Design (ISLPED), *IEEE*, pp. 64–69.
38. RamRakhyani, A.K., Mirabbasi, S., and Chiao, M. (2011) *IEEE Trans. Biomed. Circuits Syst.*, **5** (1), 48–63.
39. Yakovlev, A., Kim, S., and Poon, A. (2012) *IEEE Commun. Mag.*, **50** (4), 152–159.
40. Wang, G., Liu, W., Sivaprakasam, M., Zhou, M., Weiland, J.D., and Humayun, M.S. (2006) A dual band wireless power and data telemetry for retinal prosthesis. Annual International Conference of the IEEE EMBS, pp. 4392–4395.
41. Karthaus, U. and Fischer, M. (2003) *IEEE J. Solid-State Circuits*, **38** (10), 1602–1608.

6
Neural Interfaces: from Human Nerves to Electronics

Jessica D. Falcone, Joav Birjiniuk, Robert Kretschmar, and Ravi V. Bellamkonda

6.1
Introduction

Preliminary clinical work on the cochlear implant, the first neural interface, began in the 1960s. Designed to revive hearing for the clinically deaf [1], the cochlear implant represents the first successful integration of a stimulating electrode within the peripheral nervous system (PNS). Since Food and Drug Administration (FDA) approval in 1984, over 219 000 people have been implanted worldwide [2]. As demonstrated by the cochlear implant, neural interfacing can restore sensory information. Interfacing with the central nervous system (CNS) and PNS will aid in the rehabilitation of amputees, paralysis survivors, and patients with neurodegenerative diseases. Within the course of this chapter, electrode designs for interfacing with the CNS and PNS are discussed. The spatial selectivity and biological hurdles of these electrodes are addressed, as well as current research for improved sensitivity and chronic integration. Finally, current clinical applications for electrical stimulation of the nervous system are presented and future research directions are proposed.

6.2
Fusing Robotics with the Human Body: Interfacing with the Peripheral Nervous System

Electrode designs for the PNS are greatly influenced by the nerve structure. While less invasive electrodes sample a larger selection of neural signals, the best signal resolution is obtained by situating the electrode at the axon. By taking advantage of the natural regenerative capabilities of the PNS, integrative electrodes can be developed for improved signal recording. The PNS is a closed-feedback loop, and seamless access to functional pathways (i.e., motor and sensory) will provide the patient with a more intuitively controlled prosthetic.

Implantable Bioelectronics, First Edition. Edited by Evgeny Katz.
© 2014 Wiley-VCH Verlag GmbH & Co. KGaA. Published 2014 by Wiley-VCH Verlag GmbH & Co. KGaA.

6.2.1
The Anatomy of Peripheral Nerves

Examination of the PNS anatomy provides key interfacing information. The PNS connects the brain and spinal cord to peripheral sensors and muscles. Exiting from the brain and spinal cord are 43 pairs of nerves, which are designated as the cranial, cervical, thoracic, lumbar, sacral, and coccygeal nerves [3]. These axonal extensions compose the PNS and extend throughout the entire body. The majority of axons within the periphery are myelinated, which provides an insulating sheath around the axon and increases the conduction speed of the action potentials. To form a single nerve, axons are grouped into fascicles and surrounded by connective tissue [4]. The connective tissue contains three layers: the epineurium, perineurium, and endoneurium. The epineurium is the outermost layer, which contains blood vessels that support the nerve. The perineurium surrounds each fascicle and provides tensile support to the nerve and acts as a diffusion barrier. The endoneurium is found within the fascicles and provides the supporting extracellular matrix and cells for the axons, such as fibroblasts and Schwann cells [4].

6.2.1.1 Glial Cells of the Peripheral Nervous System

It is important to note that when peripheral nerves are stimulated, several other cell types are signaled. Principal among the peripheral glia are Schwann cells, which are responsible for myelinating the axons, and are the first responders during neural injury [5]. Schwann cells can either be myelinating or ensheathing. Myelinating Schwann cells wrap around the sensory and motor axons, providing the myelin sheath and subsequent signal insulation for action potential conduction. Ensheathing (or unmyelinated) Schwann cells loosely envelop smaller unmyelinated axons. During injury, myelinating Schwann cells immediately respond by upregulating neurotrophic factors and halting the production of myelin. Both myelinating and ensheathing Schwann cells are recruited to the neural injury site. These cells continue to proliferate and provide a physical scaffold (bands of Büngner) for the regenerating nerve. Schwann cells also aid neurite outgrowth by secreting neurotrophic factors, such as nerve growth factor (NGF), and extracellular matrix components, such as laminin [6].

6.2.1.2 Functional Afferent and Efferent Pathways

To convey information from the sensory and motor systems, the entire nervous system is divided into three functional parts: afferent neurons (peripheral), efferent neurons (peripheral), and interneurons (central). Afferent neurons propagate electrical signals from peripheral organs and tissues to the CNS. These sensory fibers can be myelinated or unmyelinated. The receptors are located within the skin, muscle, and deep tissue, and convey a range of external information to the brain, including mechanical, noxious, and thermal stimuli [4]. Located within the periphery, the afferent cell body contains two axonal processes, a long process (peripheral) that interfaces with the sensory receptors and another short process (central) that connects to the spinal cord or brain. Afferent neurons are unique in that they lack dendrites and signaling is strictly carried out along the axon branches. The somatic sensory, visceral sensory, and special sensory systems all provide input to the afferent neurons [3].

Efferent neurons project electrical signals from the CNS to the peripheral muscles and glands. The cell body of the efferent neurons is located within the CNS, along with the dendrites and part of the axon. The remaining axon extends throughout the PNS. The efferent neurons are divided into the somatic motor and the autonomic motor nervous system [3]. The somatic system contains efferent motor neurons, which are responsible for conscious movement. These motor neurons include alpha-motor fibers, which excite skeletal muscle fibers, and gamma-motor fibers, which excite muscle fibers [4]. The efferent neurons in the autonomic system stimulate glands, smooth muscles, cardiac muscles, and gastrointestinal neurons (aka unconscious, automatic biological functions) [3].

6.2.2
Interfacing with the Periphery for Recording and Stimulation

A variety of noninvasive and invasive technologies are available for stimulating and recording within the PNS. Noninvasive methods include surface electrodes such as electromyography (EMG) and magnetoneurography. Invasive peripheral electrodes include extraneural, intrafascicular, and regeneration-based electrodes [7]. The main trade-off between these designs is resolution versus sampling size. EMG can receive signals from several different muscle groups at a low cost to the patient. However, these signals are not truly neural. Invasive electrodes interact more specifically with axons, but most electrodes are limited to a small sampling range, which usually resides within the implanted fascicle.

6.2.2.1 Noninvasive Electrodes

Noninvasive electrodes are designed for external application to collect internally generated electrical signals. One common modality is EMG, which can be placed on the skin and can record muscle activation. This design is quite adaptable and can be changed to record or stimulate various muscle groups [4]. Kuiken *et al.* have developed a novel neurosurgical procedure that redirects the amputee's remaining nerves in the amputated limb to the surface of the pectoral muscles. Following surgery, the patient was able to "feel" her phantom hand when the skin was stimulated on the pectoral. The patient was then outfitted with a robotic prosthetic arm with EMG recording electrodes. By flexing the pectoral muscle at the different "hand" areas, recorded surface signals were then used to direct the prosthetic arm [8, 9]. While the main advantage of EMG is the noninvasive electrodes, the lack of neural localization inhibits signal clarity. In addition, the electrodes require frequent repositioning, usually on a daily basis, which requires regular recalibration. If the positioning is not correct, weakened signals or improper stimulation may occur [10].

Another method of noninvasively gathering signals from the PNS employs magnetoneurography. In this method, an electrode is applied to the skin, but instead of detecting currents, it detects the small magnetic fields generated by the motion of ions during neural action potentials. Using this method, high temporal and spatial resolution can be achieved, but the equipment is not easy to use. The technology must be highly sensitive as the magnetic fields are relatively weak, and for stronger signals, a room blocking external magnetic fields should be utilized. Magnetoneurography is still in development and has not yet become a commonly used signal-collecting method [10].

6.2.2.2 Extraneural Electrodes

Cuff Electrodes Cuff electrodes completely surround the nerve with an insulator that contains an interior electrode placed in contact with the nerve (Figure 6.1a). This insulating material, which is usually made of silicone, is designed to increase the sensitivity of the electrode and better isolate the signals from the nerve [4, 10]. Cuff electrodes cause minimal damage to the nerve as the nerve trunk is not punctured during the insertion. In more mobile implantation sites, the design must be flexible to prevent damage to the nerve. The size of cuff electrodes is adjustable, allowing for implantation on a variety of nerves [10]. However, the same design elements that make the cuff electrode more favorable also interfere with the recording capability. Separating the cuff from the fascicles is the connective tissue, which interferes with the sensitivity of the electrodes. Cuff electrodes are only able to detect strong nerve signals from fascicles located at the edges of the nerve, greatly reducing neural selectivity [4].

Flat Interface Nerve Electrode (FINE) The flat interface nerve electrode (FINE) is a variation of the cuff electrode with a structural design change to increase the electrode interaction with the fascicles [11]. Similar to the cuff electrode, the FINE surrounds the nerve, but unlike the cuff electrode, the FINE mechanically flattens and reconfigures the nerve (Figure 6.1b). Altering the shape of the nerve increases the surface area for electrode connection and allows access to single units generated within the inner fascicles [10]. Anatomical limits do exist for neural reshaping; if the nerve is compressed too severely by the FINE, permanent damage may occur. When the changes are kept relatively small, there is no severe or long-lasting damage [7].

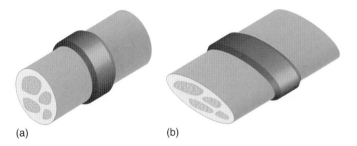

Figure 6.1 (a) Cuff electrode and (b) LIFE electrode.

6.2.2.3 Intrafascicular Electrodes

Unlike extraneural electrodes, intrafascicular electrodes penetrate the connective tissue of the nerve to record from interior fascicles. The location of the intrafascicular electrode increases selectivity of single-unit action potentials and increases detection of weaker signals [12]. However, the invasiveness of the intrafascicular electrodes increases the likelihood of infection, damage to the nerve, and chronic inflammation. This tissue response can prevent the successful chronic implantation of electrodes [13].

Longitudinal Implanted Intrafascicular Electrode (LIFE) Longitudinal implanted intrafascicular electrodes (LIFEs) are teflon-insulated, platinum–iridium wires with the recording site located at the end of the wire [14]. These wires are inserted parallel to the fascicles inside the nerve, and then, by puncturing the endoneurium, the electrode site is guided inside a single fascicle (Figure 6.2a) [4, 10]. The LIFE is flexible to adjust for shifts in the nerve placement due to the motion of the body. Designs lacking flexibility have been shown to cause increased inflammation, resulting in encapsulation of the electrode and decreased functionality [15]. The chronic compatibility of the LIFE has been shown to be 6 months with minimal nerve damage [10]. The main limitation of the LIFE is the recording restriction of axons only located within the implanted fascicle [4, 13]. However, a clinical trial has been conducted in which the LIFE was implanted into the peripheral nerve of eight amputees. The electrodes were able to both record from and stimulate the peripheral nerve, allowing patients to control a virtual hand and receive tactile responses from a prosthetic arm [16, 17].

Transverse Intrafascicular Multichannel Electrode (TIME) Unlike the LIFE, the transverse intrafascicular multichannel electrode (TIME) records from multiple fascicles along the nerve. Using semiconductor processes, a single polyimide shank is fabricated with multiple platinum electrodes along the shank [13]. The TIME is then inserted into the nerve, perpendicularly across multiple fascicles, for significantly greater spatial selectivity (Figure 6.2b) [4]. An array of TIMEs can also be used in parallel to innervate multiple nerves, significantly increasing

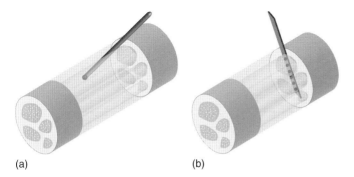

Figure 6.2 (a) LIFE electrode and (b) TIME electrode.

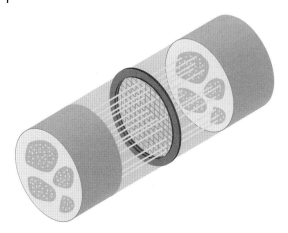

Figure 6.3 Regeneration-based electrode.

simultaneous axonal stimulation. This design is relatively new and has been shown to work in small animal models and is currently being tested in humans [13].

6.2.2.4 Regeneration-Based Electrodes

Regeneration-based electrodes are designed to regrow axons into the electrode. A 10 μm thick polyimide scaffold has recording electrodes situated within each channel [18]. When the regeneration-based electrode is implanted, the nerve is first severed, and the electrode is implanted at the nerve stump (Figure 6.3). Individual axons grow through the scaffolding, and action potentials can then be measured and generated [19]. A major advantage of this design is the ability to record from individual axons across multiple fascicles. Current designs are biocompatible and remain stable for months [4].

6.2.2.5 Research Designs and Challenges

Regeneration-Based Electrode Designs For the regeneration-based electrodes, a significant trade-off has been observed between the channel diameter and axonal regeneration. An ideal channel diameter would be between 2 and 10 μm, for which there are two reasons. Such a small diameter would allow only a single axon to grow through the hole, and the density of available recording sites would be greatly increased. However, axons do not regenerate through scaffolds with such small diameters [4, 20]. Holes of a larger diameter (between 50 and 100 μm) have successfully regenerated axons in the periphery, but selectivity is sacrificed [19].

Another challenge with the regeneration-based electrodes is the recorded signal strength. The recording sites in the regeneration-based electrodes are located outside of the myelinated axon, leading to a significant reduction in neural signal amplitude. A biophysical solution remedies this problem. Increasing the resistance of the extracellular fluid proportionally increases the voltage of the action potential generated by the axon. As the resistivity of the extracellular fluid cannot be altered *in vivo*, an alternate solution would be to decrease the volume of the extracellular space. By creating a regeneration-based electrode with longer, insulated channels, the axonal voltage is naturally amplified for feasible neural recordings [21, 22].

Distinguishing Sensory and Motor Axons The goal of peripheral interfacing is to create a fully functional closed-feedback loop. The person using the prosthetic arm should not only be able to move the arm but also receive sensory feedback from the hand interacting with the environment. However, this requires differentiation of sensory (afferent) and motor (efferent) nerves at the interface. Several methods are being researched for sensory feedback. The first is to develop a substitute sensory feedback loop by applying pressure, vibration, and/or thermo-feedback to the residual limb [23]. Although this method has had some success, the natural circuitry is still bypassed, making the control of a prosthetic less intuitive. Another strategy is to reinnervate the sensory nerves to a portion of the residual limb or torso. By applying pressure to these reinnervation points, the patient feels as if his phantom limb is being touched [24]. The final solution is direct electrical stimulation of the nerve in response to environmental stimulation. Patients implanted with the LIFE were able to feel graded responses to touch and movement of a prosthetic hand [16], and this method shows great promise.

6.3
Listening to the Brain: Interfacing with the Central Nervous System

Unlike the PNS, the CNS is immunologically isolated. The introduction of a foreign body (i.e., an electrode) activates a complex molecular and cellular cascade that often interferes with chronic implantation. An understanding of the glial cells in the brain and the response to foreign materials introduced into the microenvironment is critical. Although less invasive electrodes for neural recording are available, spatial selectivity is sacrificed, resulting in less resolved signals for computational control. Biochemically augmented electrode interfaces as well as neurologically integrative biomaterials are currently being explored.

6.3.1
Glial Cells of the Central Nervous System

6.3.1.1 Microglia – Sentinels of the Brain
Microglia, which are related to macrophages, represent the wound healing cellular component of the brain [25]. When in the resting state, microglia constantly survey the surrounding neural environment. These cells are highly mobile and extend and retract cellular projections at a high rate [26]. As such, microglia are highly sensitive to changes within the brain microenvironment, and the detection of altered homeostasis immediately prompts the microglia into an activated state [27]. When activated, microglial morphology changes from a highly branched to an amoeboid state. At this stage, phagocytosis of foreign bodies and lytic enzyme

secretion become the microglia's primary role [28]. Activated microglia serve a multifunctional, and often contradictory, role. Activated microglia can promote neural regeneration by secreting neurotrophic factors, as well as by removing synapses of damaged neurons [29, 30]. At the same time, activated microglia are highly neurotoxic. A host of damaging pro-inflammatory cytokines (such as TNF-α) and neurotoxic factors (including reactive oxygen species and nitric oxide) are secreted during activation [27, 28].

6.3.1.2 Astrocytes – Cellular Support for Neurons

The primary functions of astrocytes are to support neurons by providing structural integrity, regulating the chemical environment, and maintaining ionic homeostasis. Astrocytes also play a key role in the maintenance of neural vasculature. Perivascular astrocytic endfeet interact with blood–brain barrier capillaries, forming tight junctions [5]. In the nonreactive form, astrocytes possess star-shaped appendages composed of glial fibrillary acid protein (GFAP). Injury transforms astrocytes into the reactive phenotype, which is characterized by an increase in GFAP filaments, hypertrophy, proliferation, and an increased production of inflammatory and neurotrophic factors [28].

Astrocytes maintain ionic homeostasis in the brain. During normal brain activity, neurons produce K^+, which are recycled by the K^+ spatial buffer mechanism. An increase in extracellular K^+ surrounding astrocytes leads to diffusion of K^+ into the astrocyte, which prevents depolarization of the neurons. Cytoplasmic bridges connect astrocytes throughout the brain. These connections allow the astrocyte to efflux K^+ away from the source neurons to a less active part of the brain. The astrocytic perivascular endfeet then shuttle the K^+ into the perivascular space, located within the blood–brain barrier. When neural activity ceases, K^+ then reenters the brain through reverse K^+ spatial buffering, thereby preserving K^+ within the brain [5, 31].

6.3.2
Interfacing with the Brain for Recordings

There are currently three categories of recording electrodes for the brain: (i) noninvasive electroencephalography (EEG), (ii) extracortical electrocorticography (ECoG), and (iii) invasive intracortical electrodes. Across all of these modalities, the main difference is the proximity of the electrode interface with the neural tissue. The more invasive the electrode, the better the signal resolution; however, the stability of the electrode decreases and the health risk to the patient increases. A balance must be achieved between functionality and resolution.

6.3.2.1 Noninvasive Electrodes

EEGs are completely noninvasive surface electrodes that are attached to the scalp of the patient. While EEG recordings do not require surgery, the biggest caveat with this modality is low signal and lack of neural spatial resolution. The skull, which is about 2 cm thick, can attenuate the neural voltages by 3 orders of magnitude.

The physical separation of the skull also requires more sampling over a larger surface area for more accurate signals [32, 33]. Another compounding factor to EEG recordings is the accidental recording of muscle movements of the scalp [34]. In order to accommodate for these limitations, a variety of signal processing algorithms have been developed for EEG-controlled computers and prosthetics; however, these algorithms are beyond the scope of this chapter.

6.3.2.2 Extracortical Electrodes

ECoGs consist of a planar array of electrodes arranged in a flexible grid that is placed directly on the surface of the brain. While a craniotomy and removal of the dura is required to situate these electrodes, the ECoG resides within the subdural space, and not within the brain. This lack of neural penetration allows ECoGs to be used in temporary recording settings, where more detailed neural activity is required. For example, patients with refractory epilepsy, who remain unresponsive to drugs, can elect to undergo brain surgery to remove the epileptic loci. Presurgical evaluation of neural activity during a seizure is necessary. By temporarily implanting the patient with an ECoG, the recorded neural activity aids the surgeon in localizing the epileptogenic focus [35]. Using the same surgical principle, epileptic patients outfitted with ECoG arrays over the cortex were tested for prosthetic control. Patients were able to modulate cortical activity, producing distinctive neural signals for different parts of the cortex (particularly differentiation of the hand and elbow). Through training, patients were also able to control a computer mouse in two dimensions [34, 36]. Recently, a paralyzed patient was implanted with a 32-electrode high-density ECoG, and over the course of 28 days, successfully differentiated upper limbs within the sensorimotor cortex and controlled a computer mouse within a three-dimensional space [37]. As yet, the electrode viability of ECoGs has not been evaluated on a chronic timeline in humans.

6.3.2.3 Invasive Intracortical Electrodes

As opposed to EEG and ECoG, intracortical electrodes directly penetrate the brain. The physical proximity of intracortical electrodes to neurons makes recording of single-unit activity feasible. There are a variety of intracortical electrodes, and the most prominent are Michigan electrodes, microwires, and Utah electrodes. Each electrode type has different design elements with varying list of advantages and disadvantages. The major hurdle to overcome with intracortical electrodes is chronic recording *in vivo*. However, the signal recording of intracortical electrodes is vastly superior to their less invasive counterparts.

Michigan Electrodes The Michigan electrode array is composed of silicon and is fabricated using semiconductor processing techniques (Figure 6.4c). The silicon is etched to form planar shank electrodes, with multiple recording sites along the z-direction of the shank [38]. These recording sites allow for multiple simultaneous recordings at different layers of the cortex. The length of the shanks can be varied from 3 to 5 mm, with a thickness of 15 μm. The width of the probes ranges from

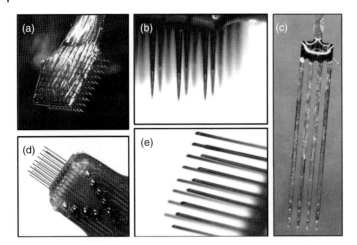

Figure 6.4 (a) 100-channel Utah electrode array (Cyberkinetics); (b) magnified view of the Utah electrode recording sites; (c) 1×4 Michigan electrode array (NeuroNexus) – note the four recording sites at the end of each shank; (d) microwire electrode array (Tucker–Davis technologies); and (e) magnified view of the microwire recording sites. (Adapted from Ref. [39] with permission.)

15 to 50 μm, and either single shank or four shank arrays, with 150 μm spacing, can be selected. The recording site areas of the Michigan electrodes are ∼1250 μm^2, and the precharacterization impedance is very high at 270 kΩ at 1 kHz frequency [39]. Michigan electrodes are commonly used in animal work, particularly murine models. The fabrication techniques of these electrodes allow for precise control over thickness and cable tethering, as well as the addition of features such as fluid-release channels and perforated wells [32, 40].

Microwire Electrodes Microwires are the oldest form of recording electrodes for neurons [28]. Manufactured with a conductive stainless steel or tungsten core, the wire is then insulated with cytocompatible polyimide or Teflon (Figure 6.4d). The insulation is removed from the tip of the wire to reveal the recording site (Figure 6.4e). The overall diameter of the wire is 50 μm, and the length ranges from 2 to 8 mm. The configuration of the microwires varies from a single shank to 100 shank arrays, with 200–300 μm spacing. Microwires have the largest recording site area at 2300 μm^2, and the baseline electrode impedance at 1 kHz frequency is 20 kΩ [39]. Microwire electrodes are most commonly used in nonhuman primate experiments. Miguel Nicolelis and his research team first validated the efficacy of chronic microwire electrodes in a monkey. At 1 month, 50% of the electrodes were recording single units with a signal-to-noise ratio of 5 : 1, and at 18 months, recordings were still possible from one monkey [41]. Nicolelis et al. then demonstrated the feasibility of a cortical electrode-controlled robotic arm in a monkey model [42]. This work was further extended by Andrew Schwartz and his laboratory by implanting microwires into the motor cortex of a monkey.

The animal was then trained to control a robotic arm with the electrodes to feed itself [43].

Utah Electrodes The only electrode array with FDA approval, which has been implanted into humans, is the Utah electrode [44]. Utah electrodes are three-dimensional needles that are fabricated from a block of boron-doped silicon (Figure 6.4a). A microsaw and etchant form the needles from the silicon. Next, the shanks are insulated with parylene-C and the recording tips are exposed and coated with platinum (Figure 6.4b) [45]. The shanks of the Utah electrodes are 1–1.5 mm in length, and the arrays contain anywhere from 1 to 100 shanks. The recording tips have a diameter of 5 μm and a total surface area of 760 μm^2. At 1 kHz frequency, the electrode impedance is 195 kΩ [39]. The Utah electrode array was first implanted into the cortex of a cat, enabling recordings of neural signals for up to 13 months [46]. As a result of the international collaboration by the Applied Physics Laboratory, the first mind-controlled prosthetic for tetraplegia has been developed using the Utah electrode [47]. In these clinical trials, viable neural signals were recorded in patients past 1000 days [44]. Most notably, a paralyzed patient who was implanted with a Utah electrode array was able to feed herself from a robotic arm that she controlled with her mind [47].

6.3.2.4 Research Designs and Challenges

Improving the recording fidelity of cortical electrodes is the driving force in the field of neural interfacing. Electrode viability is extremely variable, with recordings lasting anywhere from a few days to a few years. Out of the above-listed electrodes, each with very different geometries, fabrications, and biological footprints, a superior array has yet to emerge. The primary hypothesis for eliminating electrode failure requires the reduction of the brain's foreign body response. When an electrode is implanted into the brain, an intense astroglial response is observed, as well as neurodegeneration (Figure 6.5). Electrode sites must be within 100 μm of the neurons to record any activity, and the physical blockage of the astroglial scar, as well as neuronal death, greatly interferes. The multiphasic and mechanistic nature of the astroglial scar makes identification of solutions challenging. The immune response surrounding the electrodes shifts through several different phases, and the biochemical and physical relationship between the glia and neurons is complex. Several research strategies are currently being implemented to improve neural recording longevity.

The Astroglial Scar Within 100 μm of the electrode–tissue interface, astrocytes and microglia play key roles. During the acute timeframe (between implantation and 2 weeks post implantation), reactive astrocytes and reactive microglia are heavily recruited to the implant sight, and significant neuronal loss occurs as well. Reactive microglia levels peak at 2 weeks. At the post-acute window (between 2 and 8 weeks postimplantation), reactive astrocyte recruitment peaks and then plateaus, while reactive microglia decrease. It is important to note that while there is a reduction in activated microglia, astrocytes still remain in a more elevated reactive state.

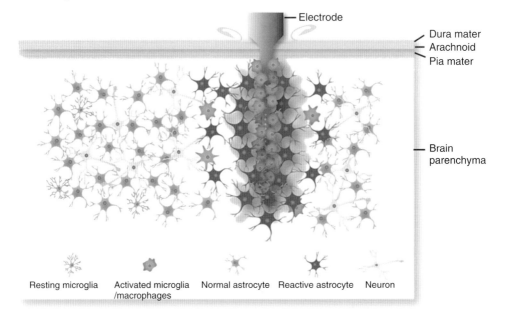

Figure 6.5 Cartoon representation of CNS cellular response to a foreign implant. Activated microglia (represented by the amoeboid morphological change) and activated astrocytes (represented by the darkened color for increased GFAP production) congregate at the tissue–electrode interface, while neurons recede. (Reproduced from Ref. [48] with permission.)

During the post-acute phase, the neural population remains stable. At chronic time points (from 8 to 16 weeks postimplantation), however, while reactive astrocyte and microglia levels remain stable, a significant neural degeneration is observed [28, 49, 50, 48].

Reducing Inflammation due to Brain Micromotion The brain is not firmly fixed within the skull, and in fact floats in the cerebral spinal fluid [5]. This micromotion has been hypothesized to instigate the chronic inflammatory response at the electrode–tissue interface [51]. Different engineering approaches have been proposed to minimize this physical disturbance. The foremost has been to produce thinner and more flexible electrodes. Both Michigan and Utah electrodes are fabricated from silicon, which has a Young's modulus of 165 GPa. The Young's modulus of the brain is 15 kPa, 7 orders of magnitude less [52]. Therefore, more flexible electrodes have been fabricated to combat the tissue–electrode mismatch. Popular materials include polydimethlysiloxane (PDMS) and parylene-C, which have Young's modulus of ~1 GPa [53]. However, increased flexibility poses a new problem: insertion. To remedy this issue, insertion shuttles [53], degradable dip-coatings for temporary hardening [54], and a combination of the two have all been proposed [55] with some success. However, none of these designs have been validated in a chronic *in vivo* animal model.

The electrode connectors have also been identified as a source of irritation due to micromotion. Traditional electrodes have the external connector directly attached to electrode. After insertion, the connector is then cemented to the skull, forming a rigid, fixed point. By installing a tether between the connector and the actual electrode, the inserted electrodes are able to "float" within the brain. Both the manufacturers of Michigan electrodes (NeuroNexus) and Utah electrodes (Cyberkinetics) offer electrodes with tethered cables. However, further research is still required to determine whether tethered or cemented electrodes actually prevent astroglial scar formation.

Biochemical–Electrode Integration for Improved Neural Growth An alternative approach to ensuring neural survival at the electrode–tissue interface is to introduce biochemicals at the interface. Bioactive coatings [56–59], controlled release systems [60, 61], and seeded neural cells [62–65] have all been incorporated into conventional electrodes. Subsequently, a variety of biochemicals have also been explored for release. One such example is neurotrophic factors, which are peptides that promote growth and survival of neural tissue. Endogenous neurotrophic factors are responsible for the building of neural circuitry during development and for the survival of the neurons. Within the CNS, neurotrophic factors such as BDNF, NT-3, GDNF, and CNTF play integral roles [66]. Coating electrodes with neurotrophic factors or developing a sustained release of the molecules could improve neural survival at the electrode interface. Another approach to positively altering the neural microenvironment around the electrode involves neural stem cells. Neural stem cells secrete a host of beneficial molecules for neural development and survival. By seeding the neural stem cells onto the electrodes, neurons are attracted to the interface, thereby improving recordings [63, 65].

Philip Kennedy and his research team developed a novel electrode design for an integrative neural interface, termed the *neurotrophic conical electrode*. The design consisted of a hollow cone, with a recording electrode situated inside. The cone was then filled with a piece of sciatic nerve from the PNS. When implanted *in vivo*, molecules secreted by the sciatic nerve attracted neurites from the surrounding tissue. The neurites eventually grew through the cone, fully integrating the device into the cortex. The neurotrophic conical electrode has since been implanted into humans with some success. However, the lack of multiple recording sites greatly limits the signal processing capabilities of the electrodes [67–71].

6.4
Electrical Modulation of the Human Nervous System: Stimulation and Clinical Applications

The previous sections provided an in-depth view of the field of neural interfacing for prosthetic integration. However, several novel interfacing discoveries have been made with surprisingly profound impacts on different diseases and functionalities within the human body. Installing a pacemaker with a stimulating electrode deep within a patient's brain can stop muscle tremors. Nerve gaps incapable of natural healing can be breached by applying electrical stimulation. Patients suffering unmanageable chronic pain can experience relief of symptoms through spinal cord stimulation. Most interestingly, the nervous system is now known to be connected to the immune system, with neural stimulation leading to immunomodulation. The impact of neural interfacing extends beyond sensory and motor restoration.

6.4.1
Deep Brain Stimulation

Since the late twentieth-century, deep brain stimulation (DBS) has become a standard means of treating a number of conditions remaining refractory to medical therapy. These include Parkinson's disease, essential tremor, major depression, and Tourette's syndrome. Several electrodes are implanted under stereotactic guidance into deep nuclei of the brain relevant to the patient's condition, and are connected to a stimulator implanted subcutaneously in the patient [72]. Intraoperative recording allows for functional localization within stereotactic range.

6.4.1.1 Biological Mechanisms
Though much of the mechanism of DBS remains unknown, there exists certain prevailing theories as to its effect. It is believed that high-frequency stimulation interferes with neuronal activity [73]. Investigators and clinicians have observed that DBS can simultaneously suppress and promote activity in certain nuclei. This contradictory evidence has revealed that DBS is able to reduce intrinsic neuronal firing at the soma while increasing synaptic output [74]. In essence, the high-frequency stimulation patterns are believed to mask the intrinsic, lower frequency circuit activity associated with pathology, thereby confusing the circuitry. Simultaneous orthodromic and antidromic signal propagation may play a role in this mechanism [75]. Current research is ongoing to understand the exact changes in neuronal firing and synaptic changes that occur during DBS, as well as the involvement of astrocytes in mediating the effects of DBS. Astrocytes are also implicated as potential mediators of DBS, as they sense neural transmission and release transmitters that can potentiate or depress certain circuits [76].

6.4.1.2 Electrode Design and Stimulation
Electrode and pulse design is an important area of research in current DBS progress. Electrode design for DBS constitutes a balance between sufficient activation of neuronal targets without activation of nontarget elements [77]. Similarly, pulse design tries to balance adequate stimulation without delivering damaging levels of current that may ablate the surrounding tissue. Modeling studies have revealed that increased electrode impedance [78], capacitance [79], and relative diameter [80] all reduce the volume of activated tissue surrounding the electrode. This can be used to understand the effects of electrode failure (i.e., increased electrode impedance)

on stimulation as well as determine methods to target certain neural populations selectively by location. Furthermore, simulations of segmented electrodes have been found to achieve greater activating functions, requiring lower stimulation intensity to obtain sufficient neuronal recruitment [81]. In addition, irregular intervals of decreased stimulus can minimize the tremor-suppressing effects of DBS [82].

Biocompatibility and failure rates are critical in assessing the viability of DBS as a long-term interface. Reports have shown lead fracture, migration, infection, and skin erosion as possible complications following implantation of DBS electrodes [83]. These issues can lead to functional failure of the electrodes and a need for reimplantation. More recently, cases have been reported of hematoma induced by DBS procedure, eventually leading to electrode shift [84, 85]. Tissue damage may also be possible as a result of overheating of electrodes exposed to magnetic fields during MR imaging [86]. These types of complications may affect a significant number of patients, but are not indicative of overall effectiveness of this therapy.

6.4.1.3 Research Designs and Challenges

Active research in DBS includes stimulation functionality, and applications in movement, affective, and cognitive disorders beyond Parkinson's disease. There is also interest in developing closed-loop systems that can both record and stimulate from regions of interest in the brain [77]. This setup would allow for precise tuning and feedback control of specific patterns of activity and is being tested in simulations [87] and *in vivo* [88]. Research has also elucidated some understanding of how the body's reaction, via chronic inflammation, affects the survival of implanted electrodes and surrounding tissue [50]. Such efforts may help in the design of more biocompatible electrodes. Clinically, DBS is being applied to treat refractory Tourette's syndrome [89] as well as to relieve anhedonia in patients with clinical depression [90]. As more neural circuits involved in pathology are discovered, DBS will continue to be a useful clinical tool for treating a wide range of diseases.

6.4.2
Electrical Modulation of Nerve Regeneration

Peripheral nerve injury and repair remain a significant problem for clinicians. Nerve injury occurs as a result of trauma, neoplastic conditions, neuropathic processes (such as in diabetes mellitus), or surgical manipulation [72]. Nerve regeneration is a complex, temporal process involving inflammatory response, neurotrophic and chemotactic factors, and neural activity. Electrical stimulation of nerve tissue may provide a method for accelerated nerve regrowth and repair following injury. Original experiments showed that minor crush injury preconditions nerves toward post-transection regeneration [91]. Investigative efforts have evaluated electrical stimulation for this preconditioning signal [92]. In fact, if appropriately timed, this type of stimulation increases the speed and effectiveness of axonal recovery following injury [93, 94].

6.4.2.1 Biological Mechanisms

Further work has uncovered some of the mechanisms behind this improved regeneration. On an electrophysiological level, calcium currents associated with injury induce a cAMP pathway that redifferentiates the site proximal to injury into an axonal growth cone [95]. These effects are mediated by retrograde communication with the neuronal soma, leading to increased expression of the neurotrophic factor, BDNF, and one of its receptors, TrkB [96]. It is important to note that these factors are upregulated in the regenerating neurons themselves but not in surrounding supporting cells [97]. Furthermore, the regenerative process is believed to involve axonal sampling of distal Schwann cell tubes and the staggered timing of regeneration [98], both of which may be affected by electrical stimulation. In addition, electrical stimulation of the nerve may provide greater specificity of axonal outgrowth to motor or sensory targets [99].

6.4.2.2 Electrode Stimulation

The exact stimulation parameters for optimizing regeneration are also an active area of study. Experiments have shown that short-term stimulation regimens are more effective than prolonged patterns in increasing the speed and extent of repair [94, 100]. These findings have been confirmed with functional testing of affected nerves [101]. The route of delivery of these stimulation regimens is also an important aspect of their efficiency. Specialized biomaterials with conductive elements may help induce neurite sprouting at the site of injury [102]. These types of platforms, however, may act as a physical barrier and may prevent certain fiber types from achieving full regenerative potential [103]. Despite these limitations, electrical stimulation has been shown in pilot studies to improve axonal recovery time in humans [104].

6.4.3
Pain Modulation

Though much is still unknown about the neural circuitry of pain, electrical stimulation of both CNS and PNS components is being investigated as a method of pain modulation. Part of the difficulty in assessing the effectiveness of such methods is the lack of objective measures of pain on an absolute scale. Despite this, significant research has provided evidence that pain modulation is possible via neural interfaces.

6.4.3.1 Biological Mechanisms

PNS Stimulation Peripheral nervous stimulation can modulate pain via the afferent nociceptive pathways. Evidence of such ascending pathways has been observed in patients undergoing vagus nerve stimulation (VNS) for conditions such as epilepsy [105] and depression [106]. The underlying theory is that vagus projections to central nuclei, such as the nucleus tractus solitarius and locus coeruleus, are responsible for inhibitory pathways mediating pain perception. Different stimulation parameters have opposing activating and conduction blocking effects. Conflicting results led to further investigation in animals, providing evidence that vagus stimulation provides inhibitory input via central mechanisms [107]. This mechanism is also behind pain relief via transcutaneous electrical nerve stimulation (TENS), which can increase heat pain threshold [108].

Stimulation of spinal elements may be responsible for modulating local circuits as well as central pathways [109]. Conditioning stimuli applied for short durations can lead to long-term circuit modulation and ultimately pain modulation. In particular, these stimuli may be able to "break" long-term potentiation of established circuits, thus modifying expectant pain. These results could be used to achieve chronic pain inhibition.

CNS Stimulation   Methods of CNS stimulation address pain by directly activating descending inhibitory pathways. Pioneering studies in the 1960s and 1970s involved implantation of deep brain electrodes in animals and achieving analgesia sufficient to perform laparotomy [110]. These stimulation patterns were able to produce specific analgesia without affecting nearby sensory pathways traversing other diencephalon nuclei [111]. Both periventricular gray (PVG) [112] and periaqueductal gray (PAG) [113] are locations specifically implicated in such pathways. PAG stimulation is traditionally believed to release endocannabinoids such as anandamide, which affect central pain processing [114]. However, more recent evidence reveals modulation by such compounds on dorsal horn neurons in addition to central sensory neurons [115, 116]. Thus, both central processing [117] and peripheral reflexes can be modulated by a single stimulus location. These different levels of control may allow specific manipulations at both spinal and central locations.

6.4.3.2   Clinical Outcomes

Human studies have been performed in patients receiving ablative surgeries [118] or with chronic, refractory, neuropathic pain [119]. Though serious complications and permanent sequelae were not reported, the implantation of such devices is still a somewhat contentious topic. The long-term effectiveness of such therapy is still being studied in patients with chronic stimulators. New research is also investigating the use of more noninvasive methods, such as transcranial magnetic stimulation (TMS) as well as transcranial direct current stimulation (tDCS), to activate central pain-suppressing centers [120].

6.4.4
Electrical Modulation of Inflammation

6.4.4.1   The Vagus Nerve and Stimulation

The vagus nerve is the tenth cranial nerve and extends throughout the body, innervating the heart, lungs, and visceral organs [121, 122]. Belonging to the autonomic neural system, the vagus nerve is a two-way feedback system [122]. The afferent extensions relay information from the pharynx, larynx, thorax, and abdomen to the brain, and the efferent extensions innervate muscles throughout the thorax and abdomen [3]. Electrical stimulation of the vagus nerve has proven to be a clinically relevant tool. Most prominently, for patients with drug-resistant epilepsy, VNS of afferent fibers at high frequencies can chronically prevent seizures [123–125]. A pacemaker is subcutaneously connected to a stimulating electrode coil that is wrapped around the left cervical vagus nerve in the neck [125]. By pulse stimulating the vagus nerve at frequencies between 20 and 30 Hz, this therapy has been shown to reduce seizures by 50% after 2–3 years [123, 125].

The biological mechanisms behind VNS seizure prevention are not fully understood. The vagus nerve feeds into the solitary nucleus, which then projects into other cortical regions. Researchers hypothesize that during VNS, there is increased activation in the thalamus (as well other components of the autonomic nervous system) and decreased activation in the limbic system, all of which connect to the solitary nucleus and may prevent seizures [125]. VNS clinical trials have also shown mood improvements in epileptic patients, and VNS is now FDA-approved to treat depression [125, 126]. Other research benefits of VNS include improvement of memory [127], control of ischemia-induced arrhythmias [128], increased lifespan of patients with chronic heart failure [129], and most interestingly, regulation of the inflammatory response [130].

6.4.4.2 Cholinergic Anti-Inflammatory Pathway

Research from Kevin Tracey and his group revealed a distinct connection between the nervous and immune systems. When vagal efferents are stimulated at low frequencies, the inflammatory response induced by infection is reduced. The autonomic nervous system represents an elaborate feedback loop responsible for controlling such necessary functions as heart rate, temperature, blood pressure, and glucose levels [131]. Therefore, the vagus nerve's control of the anti-inflammatory response, termed the *inflammatory reflex*, is completely plausible.

The classic inflammation paradigm is a diffusible pathway, which is controlled by cytokines transported through the circulatory system. During inflammation, the stimulus (damaged tissue) activates macrophages that release pro-inflammatory cytokines. Inflammatory cells, including neutrophils and monocytes, are recruited to the inflammation site from the blood. The inflammatory cells in turn release anti-inflammatory cytokines to repair the damaged tissue. As opposed to the diffusible pathway (which functions on the order of hours), the inflammatory reflex provides an immediate localized response to the injury site [122, 131].

By stimulating the vagus nerve, the pro-inflammatory cytokine, TNF-α, which is normally upregulated in response to infection, is suppressed. This effect is part of the cholinergic anti-inflammatory pathway (Figure 6.6). When the vagus nerve is stimulated, vagus synapses that terminate in the celiac ganglion innervate fibers that connect to the spleen. The splenic nerve is adrenergic, and when stimulated, releases norepinephrine. Norepinephrine activates memory T-cells that secrete acetylcholine into the spleen. Cytokine-producing macrophages in the spleen possess the $\alpha 7$ subunit of the nicotinic acetylcholine receptor, which, when activated by acetylcholine, significantly inhibits cytokine production, particularly

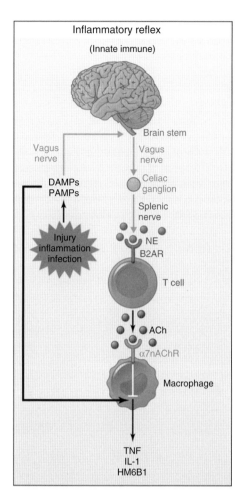

Figure 6.6 Cholinergic anti-inflammatory pathway activated by inflammation. (Adapted from Ref. [133] with permission.)

TNF-α [122, 130–132]. The discovery of the cholinergic anti-inflammatory pathway sheds much needed light on the maintenance of *in vivo* immune homeostasis.

6.5
Future Directions for Neural Interfacing

The successful integration of machines with biological systems requires the production of minimally inflammatory, chronic neural interfaces with closed-feedback loops. The intersection of biology, chemistry, and engineering has propelled the field of neural interfacing forward, and further collaborations will continue to do so. An improved interface for the CNS is required for chronic implantation. A more thorough understanding of immunology and molecular signaling during acute and chronic electrode implantation is required. Disruption of the blood–brain barrier acutely introduces foreign myeloid cells from the blood into the brain. These cells could have chronic implications for a sustained inflammatory state at the electrode interface. Identifying the elements in this signal cascade could elucidate improvements for electrode design and methodology. Researching the various material properties at the electrode–tissue interface is another avenue of exploration. A range of synthetic and natural materials with unique properties are currently under investigation [134]. The mechanical environment at the neural interface changes as the implantation duration progresses. Mechanical optimization of electrode size and flexibility to minimize acute insertion damage and chronic inflammation is needed. For the PNS, the research aims are application driven. Selective recording and stimulation of motor and sensory axons are required to close the prosthetic loop. Identification and precise interaction with sensory and motor fibers are a necessity. Further engineering to develop more integrative and sensitive electrodes will aid in the differentiation of axons. Recruitment of sensory and motor axons to the electrode is another possibility and may be achieved through tailored biochemical signaling to both axons and Schwann cells.

As seen throughout this chapter, neural interfacing has the potential to open many interesting doors. The most recent work of Nicolelis *et al.* demonstrates the plasticity of the rodent brain, and the ability of the cortex to re-form and process new (and unnatural) sensorial stimulations. By affixing an infrared sensor to an implanted stimulating electrode, rats' cortices were stimulated when infrared light was detected. The rats soon began to use the infrared sensor as a "sixth sense" [135]. This research drives home a major point; with neural interfaces, we have the opportunity to not only replace lost function but also to potentially augment neural function. Advances in neural electrode design and computing, coupled with an ever-increasing understanding of physiological pathways and responses, can not only improve one's quality of life but also lead to a future with profound implications for what it means to be human.

References

1. House, W.F. (1976) Cochlear implants. *Ann. Otol. Rhinol. Laryngol.*, **85** (Suppl. 2), 1–93.
2. National Institutes of Health (2010) Cochlear Implants – Fact Sheet. http://report.nih.gov/nihfactsheets/Pdfs/CochlearImplants(NIDCD).pdf
3. Strang, K.T., Raff, H., and Widmaier, H.R.E.P. (2006) *By Eric P. Widmaier – Vander's Human Physiology*, 10th edn, McGraw-Hill Higher Education, New York.
4. Navarro, X., Krueger, T.B., Lago, N., Micera, S., Stieglitz, T., and Dario, P. (2005) A critical review of interfaces with the peripheral nervous system for the control of neuroprostheses and hybrid bionic systems. *J. Peripher. Nerv. Syst.*, **10**, 229–258.
5. Kandel, E.R., Schwartz, J.H., and Jessel, T.M. (1991) *Principles of Neural Science*, McGraw-Hill Professional Publishing, New York.
6. Gaudet, A.D., Popovich, P.G., and Ramer, M.S. (2011) Wallerian degeneration: Gaining perspective on

inflammatory events after peripheral nerve injury. *J. Neuroinflammation*, **8**, art. # 110.

7. Hess, A. and Dunning, J. (2007) Development of a microfabricated flat interface nerve electrode based on liquid crystal polymer and polynorbornene multilayered structures. IEEE EMBS Conference on Neural Engineering, pp. 32–35.

8. Kuiken, T.A., Miller, L.A., Lipschutz, R.D., Lock, B.A., Stubblefield, K., Marasco, P.D., Zhou, P., and Dumanian, G.A. (2007) Targeted reinnervation for enhanced prosthetic arm function in a woman with a proximal amputation: a case study. *Lancet*, **369**, 371–380.

9. Ohnishi, K., Weir, R.F., and Kuiken, T.A. (2007) Neural machine interfaces for controlling multifunctional powered upper-limb prostheses. *Expert Rev. Med. Devices*, **4**, 43–53.

10. Badia, J., Boretius, T., Andreu, D., Azevedo-Coste, C., Stieglitz, T., and Navarro, X. (2011) Comparative analysis of transverse intrafascicular multichannel, longitudinal intrafascicular and multipolar cuff electrodes for the selective stimulation of nerve fascicles. *J. Neural Eng.*, **8**, art # 036023.

11. Tyler, D.J. and Durand, D.M. (2002) Functionally selective peripheral nerve stimulation with a flat interface nerve electrode. *IEEE Trans. Neural Syst. Rehabil. Eng.*, **10**, 294–303.

12. Rutten, W.L.C. (2002) Selective electrical interfaces with the nervous system. *Annu. Rev. Biomed. Eng.*, **4**, 407–452.

13. Boretius, T., Yoshida, K., Badia, J., Harreby, K., Kundu, A., Navarro, X., Jensen, W., and Stieglitz, T. (2012) A transverse intrafascicular multichannel electrode (TIME) to treat phantom limb pain — towards human clinical trials. 2012 4th IEEE RAS EMBS International Conference on Biomedical Robotics and Biomechatronics, pp. 282–287.

14. Lawrence, S.M., Dhillon, G.S., and Horch, K.W. (2003) Fabrication and characteristics of an implantable, polymer-based, intrafascicular electrode. *J. Neurosci. Methods*, **131**, 9–26.

15. Lefurge, T., Goodall, E., Horch, K., Stensaas, L., and Schoenberg, A. (1991) Chronically implanted intrafascicular recording electrodes. *Ann. Biomed. Eng.*, **19**, 197–207.

16. Dhillon, G.S. and Horch, K.W. (2005) Direct neural sensory feedback and control of a prosthetic arm. *IEEE Trans. Neural Syst. Rehabil. Eng.*, **13**, 468–472.

17. Dhillon, G.S., Lawrence, S.M., Hutchinson, D.T., and Horch, K.W. (2004) Residual function in peripheral nerve stumps of amputees: implications for neural control of artificial limbs. *J. Hand Surg. Am.*, **29**, 605–615.

18. Riso, R.R. (1999) Strategies for providing upper extremity amputees with tactile and hand position feedback-moving closer to the bionic arm. *Technol. Health Care*, **7**, 401–409.

19. Kawada, T., Zheng, C., Tanabe, S., Uemura, T., Sunagawa, K., and Sugimachi, M. (2004) A sieve electrode as a potential autonomic neural interface for bionic medicine. Conference Proceeding of the IEEE Engineering in Medicine and Biology Society, Vol. 6, pp. 4318–4321.

20. Laurell, T., Drott, J., Wallman, L., Montelius, L., Bjursten, L.M., Lundborg, G., and Danielsen, N. (1996) Perforated silicon membranes as an artificial neural contact pad-chip design and tissue implant response. Proceeding of the 18th Annual International Conference Engineering in Medicine and Biology Society, Vol. 1, pp. 243–244.

21. FitzGerald, J.J., Lago, N., Benmerah, S., Serra, J., Watling, C.P., Cameron, R.E., Tarte, E., Lacour, S.P., McMahon, S.B., and Fawcett, J.W. (2012) A regenerative microchannel neural interface for recording from and stimulating peripheral axons in vivo. *J. Neural Eng.*, **9**, art # 016010.

22. FitzGerald, J.J., Lacour, S.P.S., McMahon, S.B., and Fawcett, J.W. (2008) Microchannels as axonal amplifiers. *IEEE Trans. Biomed. Eng.*, **55**, 1136–1146.

23. Davalli, A.S. (2000) Biofeedback for upper limb myoelectric prostheses. *Technol. Disability*, **13**, 161–172.

24. Sensinger, J.W., Kuiken, T., Farrell, T.R., and Weir, R.F. (2005) Phantom limb sensory feedback through nerve transfer surgery. Proceeding of the 2005 MyoElectric Control. Prosthetics Symposium, pp. 1–5.
25. Kim, S.U. and de Vellis, J. (2005) Microglia in health and disease. *J. Neurosci. Res.*, **81**, 302–313.
26. Nimmerjahn, A., Kirchhoff, F., and Helmchen, F. (2005) Resting microglial cells are highly dynamic surveillants of brain parenchyma in vivo. *Science*, **308**, 1314–1318.
27. Hanisch, U.-K. and Kettenmann, H. (2007) Microglia: active sensor and versatile effector cells in the normal and pathologic brain. *Nat. Neurosci.*, **10**, 1387–1394.
28. Polikov, V.S., Tresco, P.A., and Reichert, W.M. (2005) Response of brain tissue to chronically implanted neural electrodes. *J. Neurosci. Methods*, **148**, 1–18.
29. Trapp, B.D., Wujek, J.R., Criste, G.A., Jalabi, W., Yin, X., Kidd, G.J., Stohlman, S., and Ransohoff, R. (2007) Evidence for synaptic stripping by cortical microglia. *Glia*, **55**, 360–368.
30. Cullheim, S. and Thams, S. (2007) The microglial networks of the brain and their role in neuronal network plasticity after lesion. *Brain Res. Rev.*, **55**, 89–96.
31. Abbott, N.J., Rönnbäck, L., and Hansson, E. (2006) Astrocyte-endothelial interactions at the blood-brain barrier. *Nat. Rev. Neurosci.*, **7**, 41–53.
32. Schwartz, A.B., Cui, X.T., Weber, D.J., and Moran, D.W. (2006) Brain-controlled interfaces: movement restoration with neural prosthetics. *Neuron*, **52**, 205–220.
33. Cooper, R., Winter, A.L., Crow, H.J., and Walter, W.G. (1965) Comparison of subcortical, cortical and scalp activity using chronically indwelling electrodes in man. *Electroencephalogr. Clin. Neurophysiol.*, **18**, 217–228.
34. Leuthardt, E.C., Schalk, G., Wolpaw, J.R., Ojemann, J.G., and Moran, D.W. (2004) A brain-computer interface using electrocorticographic signals in humans. *J. Neural Eng.*, **1**, 63–71.
35. Behrens, E., Zentner, J., and Roost, D.V. (1994) Subdural and depth electrodes in the presurgical evaluation of epilepsy. *Acta Neurochir. (Wien)*, **128**, 84–87.
36. Schalk, G., Miller, K., Anderson, N., Wilson, J., Smyth, M., Ojemann, J., Moran, D., Wolpaw, J.R., and Leuthardt, E. (2008) Two-dimensional movement control using electrocorticographic signals in humans. *J. Neural Eng.*, **5**, 75–84.
37. Wang, W., Collinger, J.L., Degenhart, A.D., Tyler-Kabara, E.C., Schwartz, A.B., Moran, D.W., Weber, D.J., Wodlinger, B., Vinjamuri, R.K., Ashmore, R.C., Kelly, J.W., and Boninger, M.L. (2013) An electrocorticographic brain interface in an individual with tetraplegia. *PLoS One*, **8**, art # e55344.
38. Najafi, K., Wise, K., and Mochizuki, T. (1985) A high-yield IC-compatible multichannel recording array. *IEEE Trans. Electron Devices*, **32**, 1206–1211.
39. Ward, M., Rajdev, P., and Ellison, C. (2009) Toward a comparison of microelectrodes for acute and chronic recordings. *Brain Res.*, **2**, 183–200.
40. Polikov, V.S., Block, M.L., Fellous, J.-M., Hong, J.-S., and Reichert, W.M. (2006) In vitro model of glial scarring around neuroelectrodes chronically implanted in the CNS. *Biomaterials*, **27**, 5368–5376.
41. Nicolelis, M.A.L., Dimitrov, D., Carmena, J.M., Crist, R., Lehew, G., Kralik, J.D., and Wise, S.P. (2003) Chronic, multisite, multielectrode recordings in macaque monkeys. *Proc. Natl. Acad. Sci. U.S.A.*, **100**, 11041–11046.
42. Carmena, J.M., Lebedev, M.A., Crist, R.E., O'Doherty, J.E., Santucci, D.M., Dimitrov, D.F., Patil, P.G., Henriquez, C.S., and Nicolelis, M.A.L. (2003) Learning to control a brain-machine interface for reaching and grasping by primates. *PLoS Biol.*, **1**, art # e42.
43. Velliste, M., Perel, S., Spalding, M.C., Whitford, A.S., and Schwartz, A.B. (2008) Cortical control of a prosthetic arm for self-feeding. *Nature*, **453**, 1098–1101.

44. Simeral, J.D., Kim, S.-P., Black, M.J., Donoghue, J.P., and Hochberg, L.R. (2011) Neural control of cursor trajectory and click by a human with tetraplegia 1000 days after implant of an intracortical microelectrode array. *J. Neural Eng.*, **8**, art # 025027.
45. Campbell, P.K., Jones, K.E., Huber, R.J., Horch, K.W., and Normann, R.A. (1991) A silicon-based, three-dimensional neural interface: manufacturing processes for an intracortical electrode array. *IEEE Trans. Biomed. Eng.*, **38**, 758–768.
46. Rousche, P.J. and Normann, R.A. (1998) Chronic recording capability of the Utah Intracortical Electrode Array in cat sensory cortex. *J. Neurosci. Methods*, **82**, 1–15.
47. Hochberg, L.R., Bacher, D., Jarosiewicz, B., Masse, N.Y., Simeral, J.D., Vogel, J., Haddadin, S., Liu, J., Cash, S.S., van der Smagt, P., and Donoghue, J.P. (2012) Reach and grasp by people with tetraplegia using a neurally controlled robotic arm. *Nature*, **485**, 372–375.
48. Potter, K.A., Buck, A.C., Self, W.K., and Capadona, J.R. (2012) Stab injury and device implantation within the brain results in inversely multiphasic neuroinflammatory and neurodegenerative responses. *J. Neural Eng.*, **9**, art # 046020.
49. McConnell, G.C., Rees, H.D., Levey, A.I., Gutekunst, C.-A., Gross, R.E., and Bellamkonda, R.V. (2009) Implanted neural electrodes cause chronic, local inflammation that is correlated with local neurodegeneration. *J. Neural Eng.*, **6**, art # 046020.
50. He, W. and Bellamkonda, R. (2008) in *Indwelling Neural Implants: Strategies for Contending with the In Vivo Environment (Frontiers in Neuroscience)* (ed W. Reichert), CRC Press.
51. Maynard, E.M., Fernandez, E., and Normann, R.A. (2000) A technique to prevent dural adhesions to chronically implanted microelectrode arrays. *J. Neurosci. Methods*, **97**, 93–101.
52. Lee, H., Bellamkonda, R.V., Sun, W., and Levenston, M.E. (2005) Biomechanical analysis of silicon microelectrode-induced strain in the brain. *J. Neural Eng.*, **2**, 81–89.
53. Gilgunn, P. and Khilwani, R. (2012) An ultra-compliant, scalable neural probe with molded biodissolvable delivery vehicle. IEEE 25th International Conference on Micro Electro Mechanical Systems (MEMS), pp. 56–59.
54. Lewitus, D., Smith, K.L., Shain, W., and Kohn, J. (2011) Ultrafast resorbing polymers for use as carriers for cortical neural probes. *Acta Biomater.*, **7**, 2483–2491.
55. Kozai, T. and Kipke, D. (2009) Insertion shuttle with carboxyl terminated self-assembled monolayer coatings for implanting flexible polymer neural probes in the brain. *J. Neurosci. Methods*, **184**, 199–205.
56. He, W., McConnell, G.C., and Bellamkonda, R.V. (2006) Nanoscale laminin coating modulates cortical scarring response around implanted silicon microelectrode arrays. *J. Neural Eng.*, **3**, 316–326.
57. Winter, J.O., Cogan, S.F., and Rizzo, J.F. III, (2007) Neurotrophin-eluting hydrogel coatings for neural stimulating electrodes. *J. Biomed. Mater. Res. Part B Appl. Biomater.*, **81**, 551–563.
58. Zhong, Y. and Bellamkonda, R.V. (2007) Dexamethasone-coated neural probes elicit attenuated inflammatory response and neuronal loss compared to uncoated neural probes. *Brain Res.*, **1148**, 15–27.
59. Azemi, E., Lagenaur, C., and Cui, X. (2011) The surface immobilization of the neural adhesion molecule L1 on neural probes and its effect on neuronal density and gliosis at the probe/tissue interface. *Biomaterials*, **32**, 681–692.
60. Luo, Z., Cai, K., Hu, Y., Zhao, L., Liu, P., Duan, L., and Yang, W. (2011) Mesoporous silica nanoparticles end-capped with collagen: redox-responsive nanoreservoirs for targeted drug delivery. *Angew. Chem. Int. Ed.*, **50**, 640–643.
61. Rohatgi, P., Langhals, N.B., Kipke, D.R., and Patil, P.G. (2009) In vivo performance of a microelectrode neural

probe with integrated drug delivery. *Neurosurg. Focus*, **27**, art # e8.
62. Pine, J., Maher, M., Potter, S., Ta, Y., and Tatic-lucic, S. (1996) A cultured neuron probe. 18th Annual International Conference of the IEEE Engineering in Medicine and Biology Society, pp. 2133–2135.
63. Purcell, E.K., Singh, A., and Kipke, D.R. (2009) Alginate composition effects on a neural stem cell-seeded scaffold. *Tissue Eng. Part C Methods*, **15**, 541–550.
64. Purcell, E.K., Seymour, J.P., Yandamuri, S., and Kipke, D.R. (2009) In vivo evaluation of a neural stem cell-seeded prosthesis. *J. Neural Eng.*, **6**, art # 049801.
65. Azemi, E., Gobbel, G.T., and Cui, X.T. (2010) Seeding neural progenitor cells on silicon-based neural probes. *J. Neurosurg.*, **113**, 673–681.
66. Lanni, C., Stanga, S., Racchi, M., and Govoni, S. (2010) The expanding universe of neurotrophic factors: therapeutic potential in aging and age-associated disorders. *Curr. Pharm. Des.*, **16**, 698–717.
67. Kennedy, P.R. (1989) The cone electrode: a long-term electrode that records from neurites grown onto its recording surface. *J. Neurosci. Methods*, **29**, 181–193.
68. Kennedy, P.R. and Bakay, R.A. (1998) Restoration of neural output from a paralyzed patient by a direct brain connection. *Neuroreport*, **9**, 1707–1711.
69. Guenther, F.H., Brumberg, J.S., Wright, E.J., Nieto-Castanon, A., Tourville, J.A., Panko, M., Law, R., Siebert, S.A., Bartels, J.L., Andreasen, D.S., Ehirim, P., Mao, H., and Kennedy, P.R. (2009) A wireless brain-machine interface for real-time speech synthesis. *PLoS One*, **4**, art # e8218.
70. Brumberg, J.S., Nieto-Castanon, A., Kennedy, P.R., and Guenther, F.H. (2010) Brain-computer interfaces for speech communication. *Speech Commun.*, **52**, 367–379.
71. Kennedy, P.R., Mirra, S.S., and Bakay, R.a. (1992) The cone electrode: ultrastructural studies following long-term recording in rat and monkey cortex. *Neurosci. Lett.*, **142**, 89–94.
72. Brunicardi, F., Andersen, D., Billiar, T., Dunn, D., Hunter, J., Matthews, J., and Pollock, R.E. (2009) *Schwartz's Principles of Surgery*, McGraw-Hill Professional, New York.
73. Hashimoto, T., Elder, C.M., Okun, M.S., Patrick, S.K., and Vitek, J.L. (2003) Stimulation of the subthalamic nucleus changes the firing pattern of pallidal neurons. *J. Neurosci.*, **23**, 1916–1923.
74. McIntyre, C.C., Grill, W.M., Sherman, D.L., and Thakor, N.V. (2004) Cellular effects of deep brain stimulation: model-based analysis of activation and inhibition. *J. Neurophysiol.*, **91**, 1457–1469.
75. Hammond, C., Ammari, R., Bioulac, B., and Garcia, L. (2008) Latest view on the mechanism of action of deep brain stimulation. *Mov. Disord.*, **23**, 2111–2121.
76. Vedam-Mai, V., van Battum, E.Y., Kamphuis, W., Feenstra, M.G.P., Denys, D., Reynolds, B.A., Okun, M.S., and Hol, E.M. (2012) Deep brain stimulation and the role of astrocytes. *Mol. Psychiatry*, **17**, 124–131, 115.
77. Kringelbach, M.L., Jenkinson, N., Owen, S.L.F., and Aziz, T.Z. (2007) Translational principles of deep brain stimulation. *Nat. Rev. Neurosci.*, **8**, 623–635.
78. Butson, C.R., Maks, C.B., and McIntyre, C.C. (2006) Sources and effects of electrode impedance during deep brain stimulation. *Clin. Neurophysiol.*, **117**, 447–454.
79. Butson, C.R. and McIntyre, C.C. (2005) Tissue and electrode capacitance reduce neural activation volumes during deep brain stimulation. *Clin. Neurophysiol.*, **116**, 2490–2500.
80. Butson, C.R. and McIntyre, C.C. (2006) Role of electrode design on the volume of tissue activated during deep brain stimulation. *J. Neural Eng.*, **3** (1), 1–8.
81. Wei, X.F. and Grill, W.M. (2005) Current density distributions, field distributions and impedance analysis of segmented deep brain stimulation electrodes. *J. Neural Eng.*, **2**, 139–147.

82. Birdno, M.J., Kuncel, A.M., Dorval, A.D., Turner, D.A., Gross, R.E., and Grill, W.M. (2012) Stimulus features underlying reduced tremor suppression with temporally patterned deep brain stimulation. *J. Neurophysiol.*, **107**, 364–383.
83. Blomstedt, P. and Hariz, M.I. (2005) Hardware-related complications of deep brain stimulation: a ten year experience. *Acta Neurochir. (Wien)*, **147**, 1061–1064.
84. Oyama, G., Okun, M.S., Zesiewicz, T.A., Tamse, T., Romrell, J., Zeilman, P., and Foote, K.D. (2011) Delayed clinical improvement after deep brain stimulation-related subdural hematoma. Report of 4 cases. *J. Neurosurg.*, **115**, 289–294.
85. Servello, D., Sassi, M., Bastianello, S., Poloni, G.U., Mancini, F., and Pacchetti, C. (2009) Electrode displacement after intracerebral hematoma as a complication of a deep brain stimulation procedure. *Neuropsychiatr. Dis. Treat.*, **5**, 183–187.
86. Finelli, D.A., Rezai, A.R., Ruggieri, P.M., Tkach, J.A., Nyenhuis, J.A., Hrdlicka, G., Sharan, A., Gonzalez-Martinez, J., Stypulkowski, P.H., and Shellock, F.G. (2002) MR imaging-related heating of deep brain stimulation electrodes: in vitro study. *AJNR Am. J. Neuroradiol.*, **23**, 1795–1802.
87. Feng, X.-J., Greenwald, B., Rabitz, H., Shea-Brown, E., and Kosut, R. (2007) Toward closed-loop optimization of deep brain stimulation for Parkinson's disease: concepts and lessons from a computational model. *J. Neural Eng.*, **4**, L14–L21.
88. Rosin, B., Slovik, M., Mitelman, R., Rivlin-Etzion, M., Haber, S.N., Israel, Z., Vaadia, E., and Bergman, H. (2011) Closed-loop deep brain stimulation is superior in ameliorating parkinsonism. *Neuron*, **72**, 370–384.
89. Mink, J.W., Walkup, J., Frey, K.A., Como, P., Cath, D., Delong, M.R., Erenberg, G., Jankovic, J., Juncos, J., Leckman, J.F., Swerdlow, N., Visser-Vandewalle, V., and Vitek, J.L. (2006) Patient selection and assessment recommendations for deep brain stimulation in Tourette syndrome. *Mov. Disord.*, **21**, 1831–1838.
90. Schlaepfer, T.E., Cohen, M.X., Frick, C., Kosel, M., Brodesser, D., Axmacher, N., Joe, A.Y., Kreft, M., Lenartz, D., and Sturm, V. (2008) Deep brain stimulation to reward circuitry alleviates anhedonia in refractory major depression. *Neuropsychopharmacology*, **33**, 368–377.
91. Brushart, T.M., Gerber, J., Kessens, P., Chen, Y.-G., and Royall, R.M. (1998) Contributions of pathway and neuron to preferential motor reinnervation. *J. Neurosci.*, **18**, 8674–8681.
92. Al-Majed, A.A., Neumann, C.M., Brushart, T.M., and Gordon, T. (2000) Brief electrical stimulation promotes the speed and accuracy of motor axonal regeneration. *J. Neurosci.*, **20**, 2602–2608.
93. Mendonça, A.C., Barbieri, C.H., and Mazzer, N. (2003) Directly applied low intensity direct electric current enhances peripheral nerve regeneration in rats. *J. Neurosci. Methods*, **129**, 183–190.
94. Ahlborn, P., Schachner, M., and Irintchev, A. (2007) One hour electrical stimulation accelerates functional recovery after femoral nerve repair. *Exp. Neurol.*, **208**, 137–144.
95. Ziv, N.E. and Spira, M.E. (1997) Localized and transient elevations of intracellular Ca^{2+} induce the dedifferentiation of axonal segments into growth cones. *J. Neurosci.*, **17**, 3568–3579.
96. Al-Majed, A.A., Brushart, T.M., and Gordon, T. (2000) Electrical stimulation accelerates and increases expression of BDNF and trkB mRNA in regenerating rat femoral motoneurons. *Eur. J. Neurosci.*, **12**, 4381–4390.
97. English, A.W., Schwartz, G., Meador, W., Sabatier, M.J., and Mulligan, A. (2007) Electrical stimulation promotes peripheral axon regeneration by enhanced neuronal neurotrophin signaling. *Dev. Neurobiol.*, **67**, 158–172.
98. Witzel, C., Rohde, C., and Brushart, T.M. (2005) Pathway sampling by

regenerating peripheral axons. *J. Comp. Neurol.*, **485**, 183–190.
99. Brushart, T.M., Jari, R., Verge, V., Rohde, C., and Gordon, T. (2005) Electrical stimulation restores the specificity of sensory axon regeneration. *Exp. Neurol.*, **194**, 221–229.
100. Geremia, N.M., Gordon, T., Brushart, T.M., Al-Majed, A.a., and Verge, V.M.K. (2007) Electrical stimulation promotes sensory neuron regeneration and growth-associated gene expression. *Exp. Neurol.*, **205**, 347–359.
101. Haastert-Talini, K., Schmitte, R., Korte, N., Klode, D., Ratzka, A., and Grothe, C. (2011) Electrical stimulation accelerates axonal and functional peripheral nerve regeneration across long gaps. *J. Neurotrauma*, **28**, 661–674.
102. Schmidt, C.E., Shastri, V.R., Vacanti, J.P., and Langer, R. (1997) Stimulation of neurite outgrowth using an electrically conducting polymer. *Proc. Natl. Acad. Sci. U.S.A.*, **94**, 8948–8953.
103. Castro, J., Negredo, P., and Avendano, C. (2008) Fiber composition of the rat sciatic nerve and its modification during regeneration through a sieve electrode. *Brain Res.*, **1190**, 65–77.
104. Gordon, T., Chan, K.M., Sulaiman, O.A.R., Udina, E., Amirjani, N., and Brushart, T.M. (2009) Accelerating axon growth to overcome limitations in functional recovery after peripheral nerve injury. *Neurosurgery*, **65**, A132–A144.
105. Kirchner, A., Birklein, F., Stefan, H., and Handwerker, H.O. (2000) Left vagus nerve stimulation suppresses experimentally induced pain. *Neurology*, **55**, 1167–1171.
106. Borckardt, J.J., Kozel, F.A., Anderson, B., Walker, A., and George, M.S. (2005) Vagus nerve stimulation affects pain perception in depressed adults. *Pain Res. Manag.*, **10**, 9–14.
107. Multon, S. and Schoenen, J. (2005) Pain control by vagus nerve stimulation: from animal to man...and back. *Acta Neurol. Belg.*, **105**, 62–67.
108. Marchand, S., Bushnell, M.C., and Duncan, G.H. (1991) Modulation of heat pain perception by high frequency transcutaneous electrical nerve stimulation (TENS). *Clin. J. Pain*, **7**, 122–129.
109. Guan, Y., Wacnik, P.W., Yang, F., Carteret, A.F., Chung, C.-Y., Meyer, R.A., and Raja, S.N. (2010) Spinal cord stimulation-induced analgesia: electrical stimulation of dorsal column and dorsal roots attenuates dorsal horn neuronal excitability in neuropathic rats. *Anesthesiology*, **113**, 1392–1405.
110. Reynolds, D.V. (1969) Surgery in the rat during electrical analgesia induced by focal brain stimulation. *Science (New York)*, **164**, 444–445.
111. Mayer, D., Wolfle, T., Akil, H., Carder, B., and Liebeskind, J. (1971) Analgesia from electrical stimulation in the brainstem of the rat. *Science*, **174**, 1351–1354.
112. Nandi, D., Aziz, T., Carter, H., and Stein, J. (2003) Thalamic field potentials in chronic central pain treated by periventricular gray stimulation – a series of eight cases. *Pain*, **101**, 97–107.
113. Walker, J.M. (1999) Pain modulation by release of the endogenous cannabinoid anandamide. *Proc. Natl. Acad. Sci. U.S.A.*, **96**, 12198–12203.
114. Hohmann, A.G. and Suplita, R.L. (2006) Endocannabinoid mechanisms of pain modulation. *AAPS J.*, **8**, E693–708.
115. Duggan, A.W. and Morton, C.R. (1983) Periaqueductal grey stimulation: an association between selective inhibition of dorsal horn neurones and changes in peripheral circulation. *Pain*, **15**, 237–248.
116. Morgan, M.M., Sohn, J.-H., and Liebeskind, J.C. (1989) Stimulation of the periaqueductal gray matter inhibits nociception at the supraspinal as well as spinal level. *Brain Res.*, **502**, 61–66.
117. Porreca, F. (2002) Chronic pain and medullary descending facilitation. *Trends Neurosci.*, **25**, 319–325.
118. Richardson, D.E. and Akil, H. (1977) Pain reduction by electrical brain stimulation in man. Part 2: chronic self-administration in the periventricular gray matter. *J. Neurosurg.*, **47**, 184–194.

119. Young, R.F., Kroening, R., Fulton, W., Feldman, R.A., and Chambi, I. (1985) Electrical stimulation of the brain in treatment of chronic pain. Experience over 5 years. *J. Neurosurg.*, **62**, 389–396.
120. Borckardt, J.J., Reeves, S., and George, M.S. (2009) The potential role of brain stimulation in the management of postoperative pain. *J. Pain Manag.*, **2**, 295–300.
121. Gray, H., Pick, T.P., and Howden, R. (1974) *Gray's Anatomy: The Unabridged Running Press Edition of the American Classic*, Running Press, New York.
122. Tracey, K.J. (2002) The inflammatory reflex. *Nature*, **420**, 853–859.
123. Morris, G. and Mueller, W. (1999) Long-term treatment with vagus nerve stimulation in patients with refractory epilepsy. *Neurology*, **53**, 1731–1735.
124. Ben-Menachem, E., Manon-Espaillat, R., Ristanovic, R., Wilder, B.J., Stefan, H., Mirza, W., Tarver, W.B., and Wernicke, J.F. (1994) Vagus nerve stimulation for treatment of partial seizures: 1. A controlled study of effect on seizures. *Epilepsia*, **35**, 616–626.
125. Bonaz, B., Picq, C., Sinniger, V., Mayol, J.F., and Clarençon, D. (2013) Vagus nerve stimulation: from epilepsy to the cholinergic anti-inflammatory pathway. *Neurogastroenterol. Motil.*, **25**, 208–221.
126. Elger, G., Hoppe, C., Falkai, P., Rush, A.J., and Elger, C.E. (2000) Vagus nerve stimulation is associated with mood improvements in epilepsy patients. *Epilepsy Res.*, **42**, 203–210.
127. Clark, K.B., Naritoku, D.K., Smith, D.C., Browning, R.A., and Jensen, R.a. (1999) Enhanced recognition memory following vagus nerve stimulation in human subjects. *Nat. Neurosci.*, **2**, 94–98.
128. Ando, M., Katare, R.G., Kakinuma, Y., Zhang, D., Yamasaki, F., Muramoto, K., and Sato, T. (2005) Efferent vagal nerve stimulation protects heart against ischemia-induced arrhythmias by preserving connexin43 protein. *Circulation*, **112**, 164–170.
129. Schwartz, P.J., De Ferrari, G.M., Sanzo, A., Landolina, M., Rordorf, R., Raineri, C., Campana, C., Revera, M., Ajmone-Marsan, N., Tavazzi, L., and Odero, A. (2008) Long term vagal stimulation in patients with advanced heart failure: first experience in man. *Eur. J. Heart Fail.*, **10**, 884–891.
130. Borovikova, L.V., Ivanova, S., Zhang, M., Yang, H., Botchkina, G.I., Watkins, L.R., Wang, H., Abumrad, N., Eaton, J.W., and Tracey, K.J. (2000) Vagus nerve stimulation attenuates the systemic inflammatory response to endotoxin. *Nature*, **405**, 458–462.
131. Tracey, K. (2007) Physiology and immunology of the cholinergic anti-inflammatory pathway. *J. Clin. Invest.*, **117**, 289–296.
132. Rosas-Ballina, M., Olofsson, P.S.P., Ochani, M., Valdés-Ferrer, S.I., Levine, Y.A., Reardon, C., Tusche, M.W., Pavlov, V.A., Andersson, U., Chavan, S., Mak, T.W., and Tracey, K.J. (2011) Acetylcholine-synthesizing T cells relay neural signals in a vagus nerve circuit. *Science*, **334**, 98–101.
133. Tracey, K.J. (2012) Immune cells exploit a neural circuit to enter the CNS. *Cell*, **148**, 392–394.
134. Bellamkonda, R.V., Pai, S.B., and Renaud, P. (2012) Materials for neural interfaces. *MRS Bull.*, **37**, 557–561.
135. Thomson, E.E., Carra, R., and Nicolelis, M.A.L. (2013) Perceiving invisible light through a somatosensory cortical prosthesis. *Nat. Commun.*, **4**, art # 1482.

7
Cyborgs – the Neuro-Tech Version
Kevin Warwick

7.1
Introduction

Science fiction has, for many years, looked to a future in which robots are intelligent and Cyborgs – human/machine amalgams – are commonplace. *The Terminator*, *The Matrix*, *Blade Runner* and I, Robot are all good examples of this. However, until the last decade any consideration of what this might actually mean in the future real world was not necessary because it was all science fiction and not scientific reality. Now, however, science has not only done a catching up exercise but, in bringing about some of the ideas thrown up by science fiction, has introduced practicalities that the original story lines did not appear to extend to (and in some cases, still have not extended to).

A cursory glance through books or encyclopedias indicates a wide variety of views as to what actually constitutes a Cyborg. Some say any biotechnological link is sufficient and hence by riding a bicycle or wearing glasses you can be so classified. Others meanwhile point to us all being Cyborgs as our brains adapt to the technological world in which we live. The view taken here is perhaps more in line with the picture painted by science fiction. We are interested in humans who are enhanced in some way through the use of implant technology that becomes integral – either that or literally a new creature is realized, in Frankenstein fashion, from both biological and technological components.

In this chapter, a series of different experiments in linking biology and technology together in a cybernetic manner, essentially ultimately combining humans and machines in a relatively permanent merger, are considered. Key to this is that it is the overall final Cyborg system that is important. Where a brain is involved, which surely it is, it must not be seen as a stand alone entity but rather as part of an overall system – adapting to the system's needs – the combined cybernetic creature is the system of importance. Each of the experiments discussed is something that has been practically realized, that is, we are looking here at actual real world experiments as opposed to mere philosophical speculations.

Each of the experiments is described in its own section. While there is distinct overlap between the sections, they each throw up individual considerations.

Implantable Bioelectronics, First Edition. Edited by Evgeny Katz.
© 2014 Wiley-VCH Verlag GmbH & Co. KGaA. Published 2014 by Wiley-VCH Verlag GmbH & Co. KGaA.

Following a description of each investigation some pertinent issues on the topic are therefore discussed. Points have been raised with a view to near term future technical advances and what these might mean in a practical scenario. It has not been the case of an attempt here to present a fully packaged conclusive document, rather the aim has been to open up the range of research being carried out, see what's actually involved and look at some of its implications.

In the next two sections, we first consider growing a brain and embodying it in a robot frame and then using deep brain stimulation (DBS) for medical treatment of neurological problems. Then we look at a more general purpose neurological implant before briefly considering the more commonly encountered external electroencephalography (EEG) electrode route. The final two applications involve what has become endearingly referred to as *BioHacking*, which consists of implanting permanent magnets into finger tips for sensory substitution and radio frequency identification device (RFID) implants for person monitoring and recognition.

7.2
Biological Brains in a Robot Body

The first area considered might not immediately be at all familiar to the reader. When one first thinks of linking a brain with technology then it is probably in terms of a brain already functioning and settled within its own body – could there possibly be any other way? Well, in fact there can be! Here, we consider the possibility of a fresh merger where a brain is first grown and then given its own body in which to operate.

When one initially thinks of a robot it may be a little wheeled device that springs to mind [1] or perhaps a metallic head that looks roughly human-like [2]. Whatever the physical appearance, our thoughts tend to be that the robot might be operated remotely by a human, as in the case of a bomb disposal robot, or it may be controlled by a simple computer program, or even may be able to learn with a microprocessor as its technological brain. In all these cases, we regard the robot simply as a machine. But, what if the robot has a biological brain made up of brain cells (neurons), possibly even human neurons?

Growing biological neurons under laboratory conditions on an array of non-invasive electrodes provide an attractive alternative with which to realize a new form of robot controller. An experimental control platform, a robot body, can move around in a defined area purely under the control of such a network and the effects of the brain controlling the body can be witnessed. While this is extremely interesting from a robotics perspective, it also opens up a new approach to the study of the development of the brain itself because of the sensory-motor embodiment. Investigations can, in this way, be carried out into memory formation and reward/punishment scenarios – elements that underpin the basic functioning and growth mechanisms of a brain.

7.2 Biological Brains in a Robot Body

Growing networks of brain cells (around 100 000 – 150 000 at present) *in vitro* typically commences by using enzymes to separate neurons obtained from fetal rodent cortical tissue. They are then grown (cultured) in a specialized chamber, in which they can be provided with suitable environmental conditions (e.g., appropriate temperature) and nutrients. An array of electrodes embedded in the base of the chamber (a multielectrode array; MEA) acts as a bidirectional electrical interface with which to provide signals to the culture and to monitor signals from the culture. This enables electrical signals to be supplied both for input stimulation and for recordings to be taken as outputs from the culture. The neurons in such cultures spontaneously connect, communicate, and develop within a few weeks, giving useful responses for typically 3 months at present. To all intents and purposes, it is rather like a brain in a jar!

The brain is grown in a glass specimen chamber lined with a flat "8 × 8" MEA that can be used for real-time recordings (see Figure 7.1). In this way, it is possible to separate the firings of small groups of neurons by monitoring the output signals on the electrodes. Thereby, a picture of the global activity of the entire network

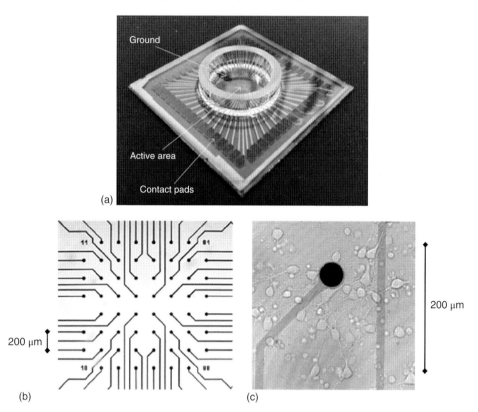

Figure 7.1 (a) A multielectrode array (MEA) showing the electrodes, (b) electrodes in the center of the MEA seen under an optical microscope, and (c) an MEA at ×40 magnification, showing neuronal cells in close proximity to an electrode.

can be formed. It is also possible to electrically stimulate the culture via any of the electrodes to induce neural activity. The MEA therefore forms a bidirectional interface with the cultured neurons [3, 4].

The brain is then coupled to its physical robot body [5]. Sensory data fed back from the robot is delivered to the culture, thereby closing the robot–culture loop. Thus, the processing of signals can be broken down into two discrete sections (i) "culture to robot," in which live neuronal activity is used as the decision-making mechanism for robot control and (ii) "robot to culture," which involves an input mapping process from the robot sensor to stimulate the culture.

The actual number of neurons in a brain depends on natural density variations in seeding the culture in the first place. The electrochemical activity of the culture is sampled and this is used as input to the robot's wheels. Meanwhile, the robot's (ultrasonic) sensor readings are converted into stimulation signals received by the culture, closing the feedback loop.

Once the brain has grown for several days, which involves the formation of some elementary neural connections, an existing neuronal pathway through the culture is identified by searching for strong relationships between (input–output) pairs of electrodes. Such pairs are defined as those electrode combinations in which neurons close to one electrode respond to stimulation from the other electrode at which the stimulus was applied more than 60% of the time and respond no more than 20% of the time to stimulation on any other electrode.

A rough input–output response map of the culture can be created by cycling through all of the electrodes in turn. In this way, a suitable input–output electrode pair can be chosen in order to provide an initial decision-making pathway for the robot. This is then employed to control the robot body – for example, if the ultrasonic sensor is active and we wish the response to cause the robot to turn away from the object being located ultrasonically (possibly a wall) in order to keep moving.

For simple experimentation purposes at this time, the intention is for the robot, as seen in Figure 7.2, to follow a forward path until it nears a wall, at which point the front sonar value decreases below a threshold, triggering a stimulating pulse. If the responding/output electrode registers activity, the robot turns to avoid the wall. During experiments, the robot turns spontaneously whenever activity is registered on the response electrode. The most relevant result is the occurrence of the chain of events: wall detection–stimulation–response. From a neurological perspective, it is of course also interesting to speculate why there is activity on the response electrode when no stimulating pulse has been applied.

As an overall control element for direction and wall avoidance the cultured brain acts as the sole decision-making entity within the overall feedback loop. Clearly, one important aspect then involves neural pathway changes, with respect to time, in the culture between the stimulating and recording electrodes.

In terms of research, learning, and memory, investigations are generally at an early stage. However, the robot can be witnessed to improve its performance over time in terms of its wall avoidance ability in the sense that neural pathways that

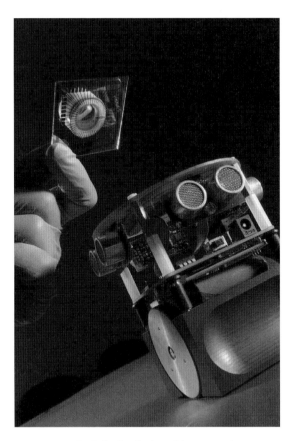

Figure 7.2 Wheeled robot body and brain together.

bring about a satisfactory action tend to strengthen purely through the process of being habitually performed – learning from habit.

The number of variables involved is considerable and the plasticity process, which occurs over quite a period of time, is dependent on such factors as initial seeding and growth near electrodes as well as environmental transients such as temperature and humidity. Learning by reinforcement – rewarding good actions and punishing bad is merely investigative research at this time.

On many occasions the culture responds as expected, on other occasions it does not, and in some cases it provides a motor signal when it is not expected to do so. Nevertheless, does it "intentionally" make a decision different to the one we would have expected? We can only guess.

In terms of robotics, it has been shown by this research that a robot can successfully have a biological brain with which to make its "decisions." The culture size is merely due to the present-day limitations of the experimentation described. Indeed three-dimensional structures are presently being investigated. Increasing the complexity from two dimensions to three dimensions realizes a figure of over

30 million neurons for the three-dimensional case – not yet reaching the 100 billion neurons of a perfect human brain, but well in tune with the brain size of many other animals.

This area of research is however expanding rapidly. Not only is the number of cultured neurons increasing, but the range of sensory input is also being expanded to include audio and visual inputs. Such richness of stimulation will no doubt have a dramatic effect on culture development. The potential of such systems, including the range of tasks they can deal with, also means that the physical body can take on different forms. There is no reason, for example, that the body could not be a two-legged walking robot, with rotating head and the ability to walk around in a building.

It is certainly the case that understanding neural activity becomes more difficult as the culture size increases. With a three-dimensional structure, monitoring activity deep within the central area, as with a human brain, becomes extremely complex, even with needle-like electrodes. In fact the present 100 000 neuron cultures are already far too complex at present for us to gain an overall insight. When they are grown to sizes such as 30 million neurons and beyond, clearly the problem is significantly magnified. Looking a few years ahead, it seems quite realistic to assume that such cultures will become larger, potentially growing into sizes of billions of neurons. On top of this, the nature of the neurons may be diversified. At present, rat neurons are generally employed in studies. However, human neurons are also now being cultured, allowing for the possibility of a robot with a human neuron brain. If this brain then consists of billions of neurons, many social and ethical questions will need to be asked [6].

For example – if the robot brain has roughly the same number of human neurons as a typical human brain then could/should it have rights similar to humans? Also – what if such creatures had far more human neurons than in a typical human brain – for example, a million times more – would they make all future decisions rather than regular humans?

7.3
Deep Brain Stimulation

In this section, we consider a technique that is initially therapeutic but, as shown later, also has implications as far as enhancement is concerned.

Different types of brain–computer interfaces (BCIs) are employed either for research purposes or for standard medical procedures. The number actually in position and operating at any one time is steadily growing, a trend that is likely to increase in the years ahead.

As an example, the number of Parkinson's disease (PD) patients is estimated to be 120–180 out of every 100 000 people, although the percentage is increasing rapidly as life expectancies increase. For decades, researchers have exerted considerable effort to understand more about the disease and to find methods to successfully limit its symptoms [7], which are most commonly periodic (and frequently acute) muscle tremor and/or rigidity. Many other symptoms such as haunched stooping may however occur in later stages of PD.

Several approaches exist to treat this disease. In its early stages, the drug levodopa (L-dopa) has been the most common one since 1970. However, it is found that the effectiveness of L-dopa decreases as the disease worsens and severity of the side effects increases, something that is far more apparent when PD is contracted by a younger person.

Surgical treatment, such as lesioning, is an alternative when drug treatments have become ineffective. Lesioning can alleviate symptoms, thus reducing the need for drug therapy altogether. A further alternative treatment of PD by means of DBS only became possible when the relevant electrode technology became available from the late-1980s onwards. From then on, many neurosurgeons have moved to implanting neurostimulators connected to deep brain electrodes positioned in the thalamus, sub-thalamus, or globus pallidus for the treatment of tremor, dystonia, and pain.

A typical DBS device contains an electrode lead with four or six cylindrical electrodes at equally spaced depths attached to an implanted pulse generator (IPG), which is surgically positioned below the collar bone. DBS has many advantages such as being reversible. It is also potentially much less dangerous than lesioning and is, in many cases, highly effective. However, it presently utilizes a continuous current simulation at high frequency, resulting in the need for regular battery replacement every 24 months or so. The cost of battery replacement, the time-consuming surgery involved, and the trauma of repetitive surgery for battery replacement severely limits the patients who can benefit, particularly those who are frail, or have problems with their immune system or are not particularly wealthy.

The obvious solution, namely, remote inductive battery recharging is fraught with problems such as the size of the passive coil that needs to be implanted and nasty chemical discharges that occur within the body – even then, the mean time between replacements is only marginally improved. Another solution to prolong the battery life is simply to improve battery technology. However, the link between the price of battery and battery life is clear. If we are considering here a battery that could potentially supply the stimulation currents required over a 10 or 20 year period, then the technology to achieve this in a low-cost, implantable, durable form is not on the horizon.

Ongoing research involving the author is aimed at developing an "intelligent" stimulator [8, 9]. The idea of the stimulator is to produce warning signals before the tremor starts so that the stimulator only needs to generate signals occasionally instead of continuously – in this sense operating in a similar manner as a heart pacemaker.

Artificial intelligence (AI) tools have been shown to successfully provide tremor onset prediction. In either case, data input to the network is provided by the measured electrical local field potentials obtained by means of the deep brain electrodes, that is, the network is trained to recognize the nature of electrical activity deep in the human brain and to predict (several seconds ahead) the subsequent tremor onset outcome. In this way, the DBS device is "intelligent" when the stimulation is only triggered by the AI system.

Many issues exist with the AI system as much preprocessing of the brain data is necessary along with frequency filtering to minimize the difficulty of prediction. Comparative studies are now ongoing to ascertain which AI method appears to be the most reliable and accurate in a practical situation.

It is worth pointing out here that false-positive predictions (that is, the AI system indicating that a tremor is going to occur when in fact this is not the case) are not so much of a critical problem. The end result with such a false positive is that the stimulating current may be applied when it is not strictly necessary. In any event no actual tremor would occur, which is a good outcome for the patient in any case; however, unnecessary energy would have been used – in fact, if numerous false predictions occurred the intelligent stimulator would tend to operate in the same way as the present "blind" stimulator. The good news is that results show that the network can be readily tuned to avoid the occurrence of most false positives anyway.

Missing the prediction of a tremor onset is extremely critical and is simply not acceptable. Such an event would mean that the stimulating current would not come into effect and the patient would actually suffer from tremors occurring.

While deep brain implants are, as described, aimed primarily to provide current stimulation for therapeutic purposes, they can also have a broader portfolio in terms of the effects they can have within the human brain. It is worth stressing that in all cases further implantations are at this time forging ahead with little consideration being given to the more general technical, biological, and ethical issues that pervade.

The same physical stimulator that is used for the treatment of PD is also employed, albeit in fewer instances at present, for cases of Tourette's syndrome, epilepsy, and even clinical depression. In many peoples' eyes it is probable that the use of deep brain stimulators for the treatment of PD, epilepsy, or Tourette's syndrome is perfectly acceptable because of the standard of living it can effect for the individual recipient. However, long-term modifications of brain organization can occur in each case, causing the brain to operate in a completely different manner; for example, there can be considerable long-term mental side effects in the use of such technology. The stimulators, when positioned in central areas of the brain, can cause other direct results, including distinct emotional changes. The picture is therefore not one of merely overcoming a medical problem – it is far more complex.

As described here, "intelligent" deep brain stimulators are starting to be designed [8]. In such a case, a computer (artificial brain) is used to understand the workings of specific aspects of the human brain. The job of the artificial brain is to monitor the normal functioning of the human brain so that it can accurately predict a spurious event, such as a Parkinson tremor, several seconds before it actually occurs. In other words, the artificial brain's job is to outthink the human brain and to stop it from doing what it "normally" wants to do. Clearly, the potential for this system to be applied for a broad spectrum of different uses is enormous.

7.4
General Purpose Brain Implants

In the previous section, a description has been given of a BCI, which is used for therapeutic purposes, to overcome a medical/neurological problem. However, even there it is possible to consider employing such technology to give individuals abilities not normally possessed by humans. Human Enhancement!!

With more general BCIs, the therapy enhancement situation is more complex. In some cases, it is possible for those who have suffered an amputation or have received a spinal injury due to an accident to regain control of devices via their (still functioning) neural signals [10]. Meanwhile, stroke patients can be given limited control of their surroundings, as indeed can those who have conditions such as motor neuron disease.

Even with these cases the situation is not exactly simple, as each individual is given abilities that no normal human has – for example, the ability to move a cursor around on a computer screen from neural signals alone [11]. The same quandary exists for blind individuals who are allowed extra sensory input, such as sonar (a bat-like sense) – it does not repair their blindness but rather allows them to make use of an alternative sense.

Some of the most impressive human research to date has been carried out using the microelectrode array, shown in Figure 7.3. The individual electrodes are 1.5 mm long and taper to a tip diameter of less than 90 µm. Although a number of trials not using humans as a test subject have occurred, human tests are at present limited to two groups of studies. In the second of these, the array has been employed in a recording only role, most notably recently as part of (what was called) the "Braingate" system. Essentially, electrical activity from a few neurons monitored by the array electrodes was decoded into a signal to direct cursor movement. This enabled an individual to position a cursor on a computer screen, using neural

Figure 7.3 A 100 electrode, 4×4 mm microelectrode array, shown on a UK 1 pence piece for scale.

signals for control combined with visual feedback. The same technique was later employed to allow the individual recipient, who was paralyzed, to operate a robot arm [12]. The first use of the microelectrode array (shown in Figure 7.3) has, however, considerably broader implications that extend the capabilities of the human recipient.

Actually deriving a reliable command signal from a collection of captured neural signals is not necessarily a simple task, partly because of the complexity of signals recorded and partly because of time constraints in dealing with the data. In some cases, however, it can be relatively easy to look for and obtain a system response to certain anticipated neural signals – especially when an individual has trained extensively with the system. In fact, neural signal shape, magnitude, and waveform with respect to time are considerably different to the other signals that it is possible to measure in this situation.

The interface through which a user interacts with technology provides a layer of separation between what the user wants the machine to do and what it actually does. This separation imposes a cognitive load that is proportional to the difficulties experienced. The main issue is interfacing the human motor and sensory channels with the technology in a reliable, durable, effective, and bidirectional way. One solution is to avoid this sensorimotor bottleneck altogether by interfacing directly with the human nervous system.

An individual human connected in this way can potentially benefit from some of the advantages of machine/AI, for example, rapid and highly accurate mathematical abilities in terms of "number crunching," a high speed, almost infinite, internet knowledge base, and accurate long-term memory. In addition, it is widely acknowledged that humans have only five senses that we know of, whereas machines offer a view of the world that includes infrared, ultraviolet, and ultrasonic signals, to name but a few.

Humans are also limited in that presently they can only visualize and understand the world around them in terms of a limited three-dimensional perception, whereas computers are quite capable of dealing with hundreds of dimensions. Most importantly, the human means of communication, essentially transferring a complex electrochemical signal from one brain to another via an intermediate, often mechanically slow and error-prone medium (e.g., speech), is extremely poor, particularly in terms of speed, power, and precision. It is clear that connecting a human brain by means of an implant with a computer network could in the long-term open up the distinct advantages of machine intelligence, communication, and sensing abilities to the implanted individual.

As a step toward a more broader concept of brain–computer interaction, in the first study of its kind, the microelectrode array (as shown in Figure 7.3) was implanted into the median nerve fibers of a healthy human individual (the author) during 2 h of neurosurgery in order to test *bidirectional* functionality in a series of experiments. A stimulation current directly into the nervous system allowed information to be sent to the user, while control signals were decoded from neural activity in the region of the electrodes [13]. A number of experimental trials were successfully concluded [14]:

Figure 7.4 Experimentation with an ultrasonic sensory input.

1) Extrasensory (ultrasonic) input was successfully implemented (see Figure 7.4).
2) Extended control of a robotic hand across the internet was achieved, with feedback from the robotic fingertips being sent back as neural stimulation to give a sense of force being applied to an object (this was achieved between Columbia University, New York (USA), and Reading University, England).
3) A primitive form of telegraphic communication directly between the nervous systems of two humans (the author's wife assisted) was performed [14].
4) A wheelchair was successfully driven around by means of neural signals.
5) The color of jewelry was changed as a result of neural signals – also the behavior of a collection of small robots.

In all of the above cases, it can be regarded that the trial proved useful for purely therapeutic reasons, for example, the ultrasonic sense could be useful for an individual who is blind or the telegraphic communication could be very useful for those with certain forms of motor neuron disease. However, each trial can also be seen as a potential form of enhancement beyond the human norm for an individual. Indeed, the author did not need to have the implant for medical purposes to overcome a problem but rather for scientific exploration. The question then arises as to how far things should be taken. Clearly, enhancement by means of BCIs opens up all sorts of new technological and intellectual opportunities; however, it also throws up a raft of different ethical considerations that need to be addressed.

When ongoing experiments of the type just described involve healthy individuals where there is no reparative element in the use of a BCI, but the main purpose of the implant is to enhance an individual's abilities, it is difficult to regard the operation as being for therapeutic purposes. Indeed the author, in carrying out such experimentation, specifically wished to investigate actual, practical enhancement possibilities [13, 14]. From the trials it is clear that extrasensory input is one practical possibility that has been successfully attempted (shown in Figure 7.4); however, improving memory, thinking in many dimensions, and communication by thought alone are other distinct potential, yet realistic, benefits, with the latter of these also having been investigated to an extent. To be clear – all these things appear to be possible (from a technical viewpoint at least) for humans in general.

As we presently stand, to get the go ahead for an implantation in each case (in the United Kingdom) requires ethical approval from the local authority at the hospital in which the procedure is carried out, and, if it is appropriate for a research procedure, also approval from the research and ethics committee of the establishment involved. This is quite apart from Devices Agency approval if a piece of equipment, such as an implant, is to be used on many individuals. Interestingly, no general ethical clearance is needed from any societal body – yet, the issues are complex.

7.5
Noninvasive Brain-Computer Interfaces

Perhaps the most studied BCI is that involving EEG and this is due to several factors. Firstly, there is no need for surgery with potential infection and/or side effects. As a result ethical approval requirements are significantly less and, because of the ease of electrode availability, costs are significantly lower than with other methods.

It is also a portable procedure, involving electrodes which are merely stuck on to the outside of a person's head and can be set up in a laboratory with relatively little training and little background knowledge and taking little time – it can be done then and there, on the spot. As a consequence, some researchers not so well versed in the field sometimes often encounter the feeling that BCI = EEG.

The number of electrodes employed for experimental purposes can vary from a small number, 4–6, to the most commonly encountered 26–30, to well over 100 for those attempting to achieve better resolution. As a result, it may be that individual electrodes are attached at specific locations or a cap is worn in which the electrodes are prepositioned. The care and management of the electrodes also varies considerably between experiments – from those in which the electrodes are positioned dry and external to hair to those in which hair is shaved off and gels are used to improve the contact made.

Some studies are employed more in the medical domain, for example, to study the onset of epileptic seizures in patients; however, the range of applications is widespread. A few of the most typical and/or interesting are included here as much to give an idea of possibilities and ongoing work rather than for a complete overview of the present state of play.

Typical are those in which subjects learn to operate a computer cursor in this manner [15]. It must be pointed out here, however, that even after significant periods of training (many months), the process is slow and usually requires several attempts before success is achieved. Along much the same lines, numerous research groups have used EEG recordings to switch on lights, control a small robotic vehicle, and control other analog signals [16, 17]. A similar method was employed, with a 64-electrode skull cap, to enable a quadriplegic to learn to carry out simple hand movement tasks by means of stimulation through embedded nerve controllers [18].

It is possible also to consider the uniqueness of specific EEG signals, particularly in response to associated stimuli, potentially as an identification tool [19]. Meanwhile, interesting results have been achieved using EEG for the identification of intended finger taps, whether the taps occurred or not, with high accuracy. This is useful as a fast interface method as well as a possible prosthetic method [20].

While EEG experimentation is relatively cheap, portable, and easy to set up, in a completely different light, yet still within the category of noninvasive techniques, both functional magnetic resonance imaging (fMRI) and magnetoencephalography (MEG) have also been successfully employed. fMRI brain scans use a strong, permanent magnetic field to align nuclei in the brain region being studied to ascertain blood flow at specific times in response to specific stimuli. As was reported earlier, they can therefore be used as a marker to figure out where there is activity in the brain when an individual thinks about moving his or her hand.

The equipment is, however, necessarily cumbersome and relatively expensive. As a result of the cost and equipment availability, experimentation in this area is by no means as widespread as that for EEG. Results have nevertheless been obtained in reconstructing images from such scans [21] and matching visual patterns from watching videos with those obtained in a time-stamped manner from the fMRI scans being recorded [22].

In terms of the concept of Cyborgs, there is a big question mark here. Certainly the technology enables the individual to learn how to operate technology through their brain signals; however, at the end of the experiment the electrodes can be merely detached – they are certainly not "part" of the person.

7.6
Subdermal Magnetic Implants

The next area to be considered is that of subdermal magnetic implants [23]. This involves the controlled stimulation of mechanoreceptors by an implanted permanent magnet manipulated through an external electromagnet. A suitable magnet and implant site are required for this along with an external interface for manipulating the implant. Clearly, issues such as magnetic field strength sensitivity and frequency sensitivity are important.

Implantation is an invasive procedure and hence implant durability is an important requirement. Only permanent magnets retain their magnetic strength over a very long period of time and are robust to various conditions. This restricts the type of magnet that can be considered for implantation to permanent magnets. Hard ferrite, neodymium, and alnico are easily available, low-cost permanent magnets deemed suitable.

The magnetic strength of the implant magnet contributes to the amount of agitation the implant magnet undergoes in response to an external magnetic field and also determines the strength of the field that is present around the implant location.

The skin on the human hand contains a large number of low-threshold mechanoreceptors that allow humans to experience in great detail the shape, size, and texture of objects in the physical world through touch. The highest density of mechanoreceptors is found in the fingertips, especially of the index and middle fingers. They are responsive to relatively high frequencies and are most sensitive to frequencies in the range 200–300 Hz.

For reported experiments [23], the pads of the middle and ring fingers were the preferred sites for magnet implantation. A simple interface containing a coil mounted on a wire frame and wrapped around each finger was designed for generating the magnetic fields to stimulate movement in the magnet within the finger. The general idea was that the output from an external sensor be used to control the current in the wrapped coil. As the signals detected by the external sensor change, these in turn are reflected in the amount of vibration experienced through the implanted magnet.

A number of application areas have already been experimented on, as reported in [23], the first being ultrasonic range information. This scenario connects the magnetic interface to an ultrasonic ranger for navigation assistance. Distance information from the ranger was encoded via the ultrasonic sensor as variations in frequency of current pulses, which in turn were passed on to the electromagnetic interface.

It was found that this mechanism allowed a practical means of providing reasonably accurate information about the individual's surrounding toward navigational assistance. The distances were intuitively understood within a few minutes of use and were enhanced by distance "calibration" through touch and sight.

A further application involves reading Morse signals. This application scenario applies the magnetic interface toward communicating text messages to humans using an encoding mechanism suitable for the interface. Morse code was chosen for encoding because of its relative simplicity and ease of implementation.

In this way, text input can be encoded as Morse code and the dots and dashes transmitted to the interface. The dots and dashes can be represented as either frequency or magnetic field strength variations.

7.7
RFID Implants

The final Cyborg experiment to be considered is the implantation of an RFID as a token of identity. In its simplest form, such a device transmits by radio a sequence of pulses that represent a unique number. The number can be preprogrammed to act rather like a PIN number on a credit card. So, with an implant of this type in place, when activated, the code can be checked by computer and the identity of the carrier specified.

Such implants have been used as a sort of fashion item to gain access to night clubs in Barcelona and Rotterdam (The Baja Beach Club), as a high security device by the Mexican Government or as a medical information source (having been approved in 2004 by the US Food and Drug Administration, which regulates medical devices in the United States). In the latter case, information on an individual's medication, for conditions such as diabetes, is stored in the implant. As it is implanted, details cannot be forgotten, the record cannot be lost, and it will not be easily stolen.

An RFID implant does not have its own battery. It has a tiny antenna and microchip enclosed in a silicon or glass capsule. The antenna picks up power remotely when passed near a larger coil of wire that carries an electric current. The power picked up by the antenna in the implant is employed to transmit by radio the particular signal encoded in the microchip. Because there is no battery or any moving parts, the implant requires no maintenance; so once it has been implanted it can stay there.

The first such RFID implant to be put in place in a human occurred on 24 August 1998 in Reading, England. It measured 22 mm by 4 mm diameter. The body selected was that of the author of this article. The doctor involved (George Boulos) burrowed a hole in the upper left arm, pushed the implant into the hole, and closed the incision with a couple of stitches.

The main reason for selecting the upper left arm for the implant was that we were not sure how well it would work. It was reasoned that, if the implant was not working, it could be waved around until a stronger signal was transmitted. It is interesting that most present-day RFID implants in humans are located in a roughly similar place (the left arm or hand), even though they do not have to be. Even in the James Bond film, *Casino Royale* (the new version), Bond himself has an implant in his left arm.

The RFID implant allowed the author to control lights, open doors, and be welcomed "Hello" when he entered the front door at Reading University. Such an implant could be used in humans for a variety of identity purposes – for example, as a credit card, as a car key or (as is already the case with some other animals) a passport, or at least a passport supplement.

The use of implant technology as an extra identity device has been with us now for some time. As yet, however, no credit card company has offered a major incentive in terms of extra security or lower costs. It is suspected that if a company did so, the take up might well be considerable. However, the broad discussion on security and privacy issues regarding mass RFID deployment has started.

To this time, there have been many recipients of RFID implants and all echo the sentiments of many cochlear implant and heart pacemaker users – the implant quickly becomes perceived as being part of the body and what the user understands to be their body includes the technological enhancement. In essence, the boundaries between man and machine become merely theoretical.

This development in our traditional notion of what constitutes our body and its boundaries leads to two notable repercussions. Firstly, it becomes possible to talk in terms of an enhanced human virus. Secondly, the development of our concept of the body impacts on certain human rights, in particular the right to bodily integrity. Bodily integrity constitutes a right to do with one's body whatever one wants. When technology is an integral part of a human body it becomes subject to the rights of that body.

7.8
Conclusions

In this chapter a look has been taken at several different types of Cyborgs. Experimental cases have been reported on in order to indicate how humans, and/or animals for that matter, can merge with technology in this way – thereby throwing up a range of social and ethical considerations as well as technical issues. In each case, reports on actual practical experimentation results have been given, rather than merely some theoretical concept. Further details for each of these can be found in a variety of publications, for example, Refs. [24–26].

When considering robots with biological brains, this could ultimately mean human brains operating in a robot body. Therefore, should such a robot be given rights of some kind? If one was switched off would this be deemed as cruelty to robots? More importantly at this time – should such research forge ahead regardless? Before too long, we may well have robots with brains made up of human neurons that have the same sort of capabilities as those of the human brain – is this acceptable?

In DBS we looked at some of the issues raised by seemingly therapeutic-only implants such as those used for the treatment of PD – a relatively standard procedure. However, the present implant does throw up possible problems concerning responsibility if a malfunction occurs but when an intelligent, predictive implant is employed should this be acceptable, even for therapeutic reasons, when a computer brain is outwitting a human brain and stopping it from doing what it naturally wants to do? If you cannot do what your own brain wants you to do, then what?

Meanwhile, in the section on a more general purpose invasive brain implant as well as implant employment for therapy, a look was taken at the potential for human enhancement. Already extrasensory input has been scientifically achieved, extending the nervous system over the internet and a basic form of thought communication. So, if many humans upgrade and become part machine (Cyborgs) themselves, what would be wrong with that? If ordinary (non-implanted) humans are left behind as a result, then what is the problem? If you could be enhanced, would you have any problem with it?

Then came a section on the much more standard EEG electrodes, which are positioned externally and which therefore are encountered much more frequently. Unfortunately, the resolution of such electrodes is relatively poor and they are indeed only useful for monitoring and not for stimulation. Hence, issues surrounding them are somewhat limited. We may well be able to use them to learn a little more about how the brain operates but it is difficult to see them ever being used for highly sensitive control operations when several million electrodes feed into the information transmitted by each electrode.

A quick look was then taken at subdermal magnetic implants. This type of connection has, until recently, been investigated more by body modification artists rather than by scientists and hence application areas are still relatively few. While involving an invasive procedure it still is relatively straightforward in comparison with areas such as DBS or MEAs fired into the nervous system. It is expected therefore that this will become an area of considerable interest over the next few years with many more potential application areas being revealed.

Finally, RFID implants have been shown to enhance an individual by allowing technology to be brought into action depending on how the user employs the implant. In my own case, my body and the building were integrated as a system. As a Cyborg your body is not simply the physical entity as we now know it. When you are inextricably linked with technology your body includes wherever the technology takes you. Your brain and body do not have to be in the same place.

Apart from taking a look at the procedures involved, the aim in this chapter has been to have a look at some of the ethical and social issues as well. Some technological issues have also been pondered on in order to open a window on the direction that developments are heading. In each case a firm footing has been planted on actual practical technology rather than on speculative ideas. In a sense the overall idea is to open up a sense of reflection such that further experimentation can be guided by the informed feedback that results.

References

1. Bekey, G. (2005) *Autonomous Robots*, MIT Press.
2. Brooks, R. (2002) *Robot*, Penguin.
3. Chiappalone, M., Vato, A., Berdondini, L., Koudelka-Hep, M., and Martinoia, S. (2007) Network dynamics and synchronous activity in cultured cortical neurons. *Int. J. Neural Syst.*, **17**, 87–103.
4. DeMarse, T., Wagenaar, D., Blau, A., and Potter, S. (2001) The neurally controlled animat: biological brains acting with simulated bodies. *Auton. Robot.*, **11**, 305–310.
5. Warwick, K., Nasuto, S., Becerra, V., and Whalley, B. (2010) Experiments with an in-vitro robot brain, in *Instinctive Computing*, Lecture Notes in Artificial Intelligence, Springer, Vol. 5987 (ed. Y. Cai), pp. 1–15.
6. Warwick, K. (2010) Implications and consequences of robots with biological brains. *Ethics Inf. Technol.*, **12**(3), 223–234.
7. Pinter, M., Murg, M., Alesch, F., Freundl, B., Helscher, R., and Binder, H. (1999) Does deep brain stimulation of the nucleus ventralis intermedius affect postural control and locomotion in Parkinson's disease? *Mov. Disord.*, **14**(6), 958–963.
8. Pan, S., Warwick, K., Gasson, M., Burgess, J., Wang, S., Aziz, T., and Stein, J. (2007) Prediction of Parkinson's Disease tremor onset with artificial neural networks. Proceedings of the IASTED Conference BioMed 2007, Innsbruck, Austria, February 14–16, 2007, pp. 341–345.
9. Wu, D., Warwick, K., Ma, Z., Burgess, J., Pan, S., and Aziz, T. (2010) Prediction of Parkinson's disease tremor onset using radial basis function neural networks. *Expert Syst. Appl.*, **37**(4), 2923–2928.
10. Donoghue, J., Nurmikko, A., Friehs, G., and Black, M. (2004) Development of

a neuromotor prosthesis for humans, in *Advances in Clinical Neurophysiology, Supplements to Clinical Neurophysiology*, Elsevier, Vol. 57, pp. 588–602.

11. Kennedy, P., Andreasen, D., Ehirim, P., King, B., Kirby, T., Mao, H., and Moore, M. (2004) Using human extra-cortical local field potentials to control a switch. *J. Neural Eng.*, **1**(2), 72–77.

12. Hochberg, L., Serruya, M., Friehs, G., Mukand, J., Saleh, M., Caplan, A., Branner, A., Chen, D., Penn, R., and Donoghue, J. (2006) Neuronal ensemble control of prosthetic devices by a human with tetraplegia. *Nature*, **442**, 164–171.

13. Warwick, K., Gasson, M., Hutt, B., Goodhew, I., Kyberd, P., Andrews, B., Teddy, P., and Shad, A. (2003) The application of implant technology for cybernetic systems. *Arch. Neurol.*, **60**(10), 1369–1373.

14. Warwick, K., Gasson, M., Hutt, B., Goodhew, I., Kyberd, P., Schulzrinne, H., and Wu, X. (2004) Thought communication and control: a first step using radiotelegraphy. *IEE Proc. Commun.*, **151**(3), 185–189.

15. Trejo, L., Rosipal, R., and Matthews, B. (2006) Brain-computer interfaces for 1-D and 2-D cursor control: designs using volitional control of the EEG spectrum or steady-state visual evoked potentials. *IEEE Trans. Neural Syst. Rehabil. Eng.*, **14**(2), 225–229.

16. Millan, J., Renkens, F., Mourino, J., and Gerstner, W. (2004) Non-invasive brain-actuated control of a mobile robot by human EEG. *IEEE Trans. Biomed. Eng.*, **51**(6), 1026–1033.

17. Tanaka, K., Matsunaga, K., and Wang, H. (2005) Electroencephalogram-based control of an electric wheelchair. *IEEE Trans. Robot.*, **21**(4), 762–766.

18. Kumar, N. (2008) Brain Computer Interface, Cochin University of Science & Technology Report, August 2008.

19. Palaniappan, R. (2008) Two-stage biometric authentication method using thought activity brain waves. *Int. J. Neural Syst.*, **18**(1), 59–66.

20. Daly, I., Nasuto, S., and Warwick, K. (2011) Single tap identification for fast BCI control. *Cogn. Neurodyn.*, **5**(1), 21–30.

21. Rainer, G., Augath, M., Trinath, T., and Logothetis, N. (2001) Nonmonotonic noise tuning of BOLD fMRI signal to natural images in the visual cortex of the anesthetized monkey. *Curr. Biol.*, **11**(11), 846–854.

22. Beauchamp, M., Lee, K., Haxby, J., and Martin, A. (2003) fMRI responses to video and point-light displays of moving humans and manipulable objects. *J. Cogn. Neurosci.*, **15**(7), 991–1001.

23. Hameed, J., Harrison, I., Gasson, M., and Warwick, K. (2010) A novel Human-machine interface using subdermal magnetic implants. Proceedings of the IEEE International Conference on Cybernetic Intelligent Systems, Reading, UK, September 2010, pp. 106–110.

24. Warwick, K. (2002) *I, Cyborg*, Century.

25. Warwick, K., Xydas, D., Nasuto, S., Becerra, V., Hammond, M., Downes, J., Marshall, S., and Whalley, B. (2010) Controlling a mobile robot with a biological brain. *Def. Sci. J.*, **60**(1), 5–14.

26. Bakstein, E., Burgess, J., Warwick, K., Ruiz, V., Aziz, T., and Stein, J. (2012) Parkinsonian tremor identification with multiple local field potential feature classification. *J. Neurosci. Methods*, **209**, 320–330.

8
Interaction with Implanted Devices through Implanted User Interfaces[1]

Christian Holz, Tovi Grossman, George Fitzmaurice, and Anne Agur

In his seminal article, Mark Weiser wrote, "the most profound technologies are those that disappear. They weave themselves into the fabric of everyday life until they are indistinguishable from it" [1]. Weiser's seminal vision is close to becoming today's reality. We now use mobile devices to place calls and send emails on the go, maintain our calendars and setup reminders, and quickly access information – anywhere, anytime. While these devices have not yet disappeared, the transition to ultra-small mobile devices is undeniable [2], and mobile devices have become an integral part of our lives.

In the medical domain, some devices have indeed reached the state of virtual invisibility. People have started to receive *implanted devices* for medical purposes, such as pacemakers and hearing aids. Active medical implants typically maintain life-crucial functionality (e.g., pacemakers), improve body functionality to restore normal living (e.g., hearing aids), or monitor the user's health [3]. Passive implants are also commonly used for medical purposes, such as for artificial joints. Active implanted devices provide a large number of benefits – benefits that are fundamental from a medical point of view and at the same time benefits the mobile community is striving toward: implanted devices along with the information they store always travel with the user; users can never lose or forget them, nor is there a need for manually attaching them. Implanted devices are available to the user *at all times*. Although they are invisible to other people, over 3 million people have implanted pacemakers alone [4].

However, current implanted devices involve a number of downsides. Although they are part of the user, users have currently no way to interact with them. To check on the status of their pacemaker, for example, a user needs to see a physician. If a pacemaker is running low on battery, the user will need to undergo replacement surgery, along with the risks and costs that such an operation entails.

[1] The material presented in this chapter is based on the following publication: Christian Holz, Tovi Grossman, George Fitzmaurice, and Anne Agur. 2012. Implanted user interfaces. In *Proceedings of the SIGCHI Conference on Human Factors in Computing Systems* (CHI '12). ACM, New York, NY, USA, pages 503–512.

Implantable Bioelectronics, First Edition. Edited by Evgeny Katz.
© 2014 Wiley-VCH Verlag GmbH & Co. KGaA. Published 2014 by Wiley-VCH Verlag GmbH & Co. KGaA.

Since it is unclear how a user might interact with an implanted device directly, we investigated the *implanted user interfaces* that such small devices provide when implanted underneath humans [5]. Although implanted devices have existed for a long time in the medical domain, they currently support only limited interaction and cannot support personal tasks. Unlike other types of mobile devices, such as wearables [6] or interactive clothing [7], implanted devices are with the user *at all times*. Implanting thus truly allows always-available interaction [8]. Before implanted user interfaces can become a reality for *interactive tasks*, numerous questions must be considered.

We discuss four core challenges of implanted user interfaces: how to sense input through the skin, how to produce output, how to communicate among one another and with external infrastructure, and how to remain powered. In the rest of this chapter, we first discuss these four challenges and then perform a technical evaluation, where we surgically implant seven devices into a specimen arm. We evaluate and quantify the extent to which traditional interface components, such as LEDs, speakers, and input controls work through skin (Figure 8.1b,c). Our main finding is that traditional interface components do work when implanted underneath human skin, which provides an initial validation of the feasibility of implanted user interfaces.

Motivated by these results, we demonstrate how to deploy a prototype implant on participants. We report the results of a qualitative evaluation using the prototype device (Figure 8.2a), in which we collected user feedback. As a substitute for actually implanting this device, we place it under a layer of artificial skin made from silicon, which affixes on the user's skin (Figure 8.2b). We conclude our exploration of implanted user interfaces with a comprehensive discussion of medical assessment, limitations, and projection into the future.

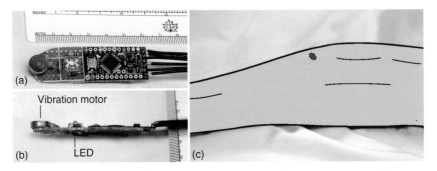

Figure 8.1 *Implanted user interfaces* allow users to interact with small devices through human skin. (a,b) This output device is implanted (c) underneath the skin of a specimen arm. The output is shown in Figure 8.16a. *Note: Throughout this chapter, illustrations have been used in place of actual photographs of the specimen to ensure ethical and professional standards are maintained.* (Adapted from [26] with permission.)

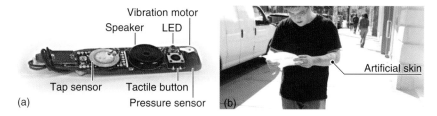

Figure 8.2 We covered a prototype device (a) with a layer of artificial skin (b) to collect qualitative feedback from use in an outdoor scenario. Participants received output triggers *through* the artificial skin and responded with input. (Adapted from [26] with permission.)

8.1
Implanted User Interfaces

We consider *implanted devices* as devices that are surgically and permanently inserted under the human skin. Implanting devices that possess user interfaces would allow users to *directly* interact with them, allowing them to support a wide range of applications and tasks, beyond the medical usages prevalent today.

Implanted devices have several advantages over mobile and wearable devices. First, implanted devices do not need to be manually attached to the user's body. They stay out of the way of everyday or recreational activities (e.g., swimming or showering). Second, implanted devices have the potential to be completely *invisible*. This would avoid any social stigma of having such devices. Third, implanted devices, along with the information they store and provide, always travel with the user; the user can never lose or forget to take them. The devices and applications become *part* of the user.

Humans have experimented with adding new abilities to their bodies, such as implanting a small magnet to their finger [9] or an RFID chip into their body. Masters and Michael discuss issues surrounding human-centric applications of RFID implants, such as automatically opening doors and turning on lights [10]. Warwick's Project Cyborg investigates user interaction through an implanted RFID chip with devices in the proximity, as well as the interaction of implants with user's nervous system [11]. Ullah *et al.* discuss in- and on-body wireless communication [12] in the context of *body area networks* [13]. Relevant work can also be found in the art community. For example, Stelarc attached an ear-replica to his arm, which used a miniature microphone to transmit recorded sounds wirelessly [14].

Despite these potential benefits, there has been little or no investigation of implanted user interfaces from a human–computer interaction (HCI) perspective. Given the continuous miniaturization of technology [2], we believe implanted user interfaces could become a reality in the future. Below, we outline some of the core design considerations, with the hope of bringing these issues to the attention of the HCI community.

8.1.1
Design Considerations

We see four core challenges associated with implanted user interfaces and their use through human skin: (i) providing input to and sensing input on implanted devices, (ii) perceiving output from and producing output from implanted devices, (iii) communication among implanted devices and with external devices, and (iv) power supply to implanted devices.

8.1.1.1 Input through Implanted Interfaces
Since implanted devices remain under the skin, they are not directly accessible through their interfaces. This makes providing input to them an interesting challenge.

One option is to use contact-based input through the skin, such as a button, which would additionally offer tactile and audible feedback to the user. Tap and pressure sensors allow devices to sense how strongly touches protrude the skin, whereas brightness and capacitive sensors detect a limited range of hover. Strategic placement of touch-based sensors could form an input surface on the skin that allows for tapping and dragging. Audio is an alternative implanted user interface. A microphone could capture speech input for voice activation.

Fully implanted and thus fully concealed controls require users to learn their locations, either by feeling them through skin or by indicating their location through small marks. Natural features such as moles could serve as such marks. Partial exposure, in contrast, would restore visual discoverability and allow for direct input. Exposing a small camera, for example, would allow for spatial swiping input above the sensor (e.g., optical flow of the fingerprint [2, 15]). All such input components, whether implanted or exposed, are subject to accidental activation, much like all wearable input components. Systems have addressed this, for example, by using a global on/off switch or requiring a certain device posture [16].

8.1.1.2 Output through Implanted Interfaces
Device output typically depends on the senses of sight (i.e., visual signals), hearing (i.e., audio signals), and touch (e.g., vibration and moving parts). Stimulation of other senses, such as taste and smell, is still only experimental (e.g., taste interfaces [17]).

The size constraints of small devices require sacrificing spatial resolution and leave room for only individual visual signals, such as LEDs. Furthermore, visual output may go unnoticed if the user is not looking directly at the source. Although audio output is not subject to such size constraints, its bandwidth is similar to the visual output of a single signal: varying intensities, pitches, and sound patterns [2]. Tactile output of single elements is limited to the bandwidth of pressure to the body and intensity patterns. Tactile feedback may be particularly suited toward implanted user interfaces, since it could provide output noticeable only to the host user and no one else.

8.1.1.3 Communication and Synchronization

To access and exchange data among each other or with external devices, implanted devices need to communicate.

If devices are fully implanted under the skin, communication will need to be wireless. Bluetooth is already being used to replace wired short-range point-to-point communication, such as for health applications (e.g., in body area networks [13]). Wi-Fi, as an alternative, transmits across longer distances at higher speeds, but comes at the cost of increased power usage and processing efforts. For interactive purposes, electrodes have been implanted for interactive purposes in the context of brain–computer interaction [18] and speech production [19].

Equipping implanted devices with an *exposed port* would enable tethered communication. Medical ports are already used to permit frequent injections to the circulatory system [20]. Ports and tethered connections are suitable for communication with external devices, but not between two devices implanted at different locations in a user's body. Such devices would still require wireless communication.

8.1.1.4 Power Supply through Implanted Interfaces

A substantial challenge for implanted devices is how they source energy. As power is at a premium, implanted devices should employ sleep states and become fully active only after triggering them.

A simple way to power an active implanted device is to use a replaceable battery. This is common with pacemakers, which typically need surgical battery replacement every 6–10 years. Rechargeable batteries would avoid the need for surgery and recharging could be wireless, through technology known as *inductive charging* [21]. If the implanted device is close to the skin surface, inductive charging may work through the skin [22]. Alternatively, an exposed port could provide tethered recharging to an implanted device. Finally, an implanted device could harvest energy from using the device or from body functions (e.g., heartbeats [23] or body heat [24]). We direct the reader to Starner's overview for more information [24].

8.1.2
Summary

We have described some of the key challenges and discussed possible components that could support the interface between the human and the implantable. However, there is little understanding of how well these basic interface components actually function underneath human skin.

8.2
Evaluating Basic Implanted User Interfaces

The purpose of this evaluation was to examine to what extent input, output, communication, and charging components remain useful when implanted underneath human skin. In addition, we provide a proof of concept that these devices can in fact be implanted, both fully under the skin and with exposed parts.

We performed this evaluation in collaboration with the Department of Surgery in the Division of Anatomy at the University of Toronto, Canada. The procedure of the study underwent full ethics review prior to the evaluation and received approval from the Research Ethics Board.

8.2.1
Devices

We evaluated 7 devices featuring 12 controls in total, which were traditional input and output components as well as components for synchronization and powering common in conventional mobile devices. As shown in Figure 8.3, we tested four basic sensors for direct touch input: button, pressure sensor, and tap sensor. In addition, we tested two devices that could potentially detect hover above the skin: capacitive and brightness sensor. We also tested a microphone for auditory input. For output, we tested an LED (visual), vibration motor (tactile), and speaker (audio). For charging, we evaluated an inductive charging mat, and for communication, we tested Bluetooth data transfer. These devices do not exhaust all possible implanted interface components, but we chose them as some of the more likely components that could be used.

Cables connected each of the devices to a laptop computer to ensure reliable connectivity and communication with the devices throughout the study (Figure 8.4). The laptop logged all signals sent from the input components on the devices, including device ID, sensor ID, sensed intensity, and timestamp. The laptop also logged time-stamped output triggers, including output component ID, intensity, and frequency.

All active devices used an ATmega328 microcontroller with a 10-bit precision analog-to-digital (AD) converter. The chip forwarded all measurements to the laptop and also computed length of impact as well as average and maximum intensities. We also recorded all background intensities separately.

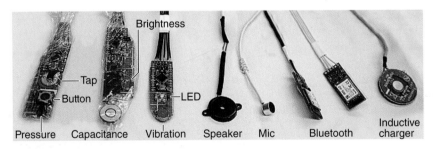

Figure 8.3 Devices implanted during the study. Plastic bags around devices prevent contact with tissue fluid. (Adapted from [26] with permission.)

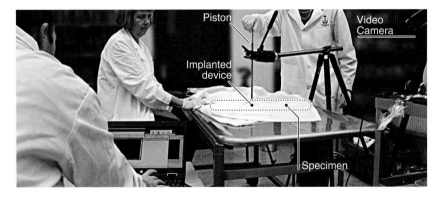

Figure 8.4 Study setup with input apparatus setup. A piston repeatedly dropped from controlled heights onto the sensors. (Adapted from [26] with permission.)

8.2.2
Experimenters

The study was administered by an experimenter and an experimenter assistant, both with HCI backgrounds, and an anatomy professor, who carried out all of the surgical procedures (Figure 8.4). Because the focus of this study was on the technical capabilities of the devices themselves, external participants were not necessary.

8.2.3
Procedure

We conducted the evaluation in two sessions. In the *baseline session*, the devices lay on the table shown in Figure 8.4. In the *implant session*, each of the seven devices was implanted into a cadaveric specimen, one at a time. An external video camera documented the entire implant session, and parts of the baseline session. The experimenter configured and initialized the devices through the laptop and monitored the incoming data, while the assistant performed the necessary interactions with the devices.

8.2.4
Medical Procedure

One lightly embalmed cadaveric upper limb specimen (dark-skinned male, 89 years old) was used for this study. With light embalming, the tissues remained pliable and soft, similar to fresh and unembalmed tissue [25]. The skin and subcutaneous tissues remained mobile.

Each of the seven devices was enclosed by two thin transparent plastic bags to prevent malfunction due to penetration by tissue fluid (as shown by the two left-most devices in Figure 8.3). To insert devices, the skin was incised and separated along the tissue plane between the skin and underlying subcutaneous tissue at the cut end of the limb, about 7.5 cm proximal to the elbow joint, which was 20 cm from the insertion point. Once the plane was established, a long metal probe was used to open the plane as far distally as the proximal forearm, creating a pocket for the devices. Each of the devices was inserted, one at a time, into the tissue plane and the wires attached to the devices were used to guide the device into the pocket between the skin and subcutaneous tissue of the proximal forearm (Figure 8.5). Distal to the insertion site of the device, the skin remained intact. All devices were fully encompassed by skin, with no space between device and skin or tissue, or any opening.

8.2.5
Study Procedure and Results

We now report the study procedure along with results separately for each of the seven devices.

8.2.5.1 Touch Input Device (Pressure Sensor, Tap Sensor, Button)

To produce input at controlled intensities, we built a stress test device as shown in Figure 8.4. The assistant dropped a piston from controlled heights onto each input sensor to produce predictable input events.

For the pressure and tap sensors, the piston was dropped from six controlled heights (2–10 cm in steps of 2 cm), repeated five times each, and the intensities from the sensors were measured. For the button, the piston was dropped from seven heights (3 mm, 7 mm, 1 cm, 2–10 cm in 2 cm steps), also repeated five times each, and we recorded if the button was activated. Subjectively, the piston dropping from 10 cm roughly compared to the impact of a hard tap on a tabletop system. Dropping from 1 cm produces a noticeable but very light tap.

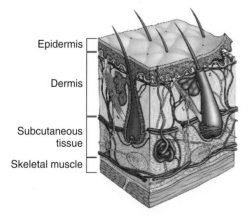

Figure 8.5 Illustration of skin layers. All devices were implanted between the skin and the subcutaneous fatty tissue. (Adapted from [26] with permission.)

8.2 Evaluating Basic Implanted User Interfaces

Apparatus Details The pressure sensor used a voltage divider with a circular 0.2" Interlink Electronics force sense resistor (100 g to 10 kg) and a 10 kΩ resistor. The button was a 12 mm (H4.3 mm) round PTS125 hardware button. The touch sensor was a Murata 20 mm piezoelectric disc. The microcontroller captured events at 30 kHz. The piston was a 60 g metal rod.>

Results

Force sensor: Skin softened the peak pressure of the dropping piston, whereas the softening effect shrunk with increasing impact force (Figure 8.6). We analyzed the measured voltages and, by relating them back to the force–resistance mapping in the datasheet, obtained an average of 3N in differences of sensing impact between conditions.

Button: Figure 8.7 illustrates the effect of skin dampening on the impact of the dropping piston. In the baseline condition, the piston always activated the button, whereas only dropping from a height of 1 cm and higher achieved enough force to activate the button through the skin at all times.

Tap sensor: In both conditions, the piezo tap sensor produced the maximum voltage our device could measure in response to the impact of the piston from all tested heights. The piston therefore activated the tap sensor reliably with all forces shown in Figure 8.6.

8.2.5.2 Hover Input Device (Capacitive and Brightness Sensor)

To produce hover input, the assistant used his index finger and slowly approached the sensor from above over the course of 3 s, rested his finger on it for 3 s, and then slowly moved his finger away. The assistant repeated this procedure five times for each of the two sensors.

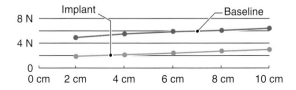

Figure 8.6 On average, skin accounts for 3 N overhead for impact forces on pressure and touch sensors. (Adapted from [26] with permission.)

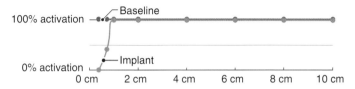

Figure 8.7 The piston activated the button from all tested heights in the baseline condition, but activated the button reliably only from a height of 1 cm and above when implanted. (Adapted from [26] with permission.)

8 Interaction with Implanted Devices through Implanted User Interfaces

Apparatus Details The capacitive sensor was a 24-bit, two-channel capacitance to digital converter (AD7746). The brightness sensor used a voltage divider with a 12 mm cadmium sulfide 10 MΩ photoresistor and a 10 kΩ resistor. Both sensors captured hover intensities at 250 Hz. Three rows of fluorescent overhead lighting illuminated the study room.

Results For both sensors, we averaged the five curves of measured signal intensities to account for noisy measurements.

> *Brightness sensor*: Without the finger present, the skin diffused incoming light, and resulted in reduced brightness (Figure 8.8a). The environmental light explains the differences in slopes between baseline and implant conditions; as the finger approaches the sensor, light reflected from surfaces can still fall in at extreme angles in the baseline condition. The skin, in contrast, diffuses light and thus objects approaching the sensor result in a less pronounced response.
>
> *Capacitive sensor*: Similar to the brightness sensor, the capacitive levels were offset when sensing through the skin (Figure 8.8b). The signal of a touching finger was comparably strong in the baseline condition, but caused only a milder difference in sensed capacitance through the skin.

8.2.5.3 Output Device (Red LED, Vibration Motor)

To evaluate the LED and motor, we used a descending staircase design to determine minimum perceivable intensities [27, 28]. For each trial, the experimenter triggered components to emit output at a controlled intensity level for a duration of 5 s. The assistant, a 32 year old male, served as the participant for the staircase study to determine absolute perception thresholds. The method started with full output intensity, which the participant could clearly perceive. The experimenter then decreased the intensity in discrete steps, and the participant reported if he could perceive it. If he did not, the experimenter increased output intensities in smaller steps until the participant could perceive it. We continued this procedure until

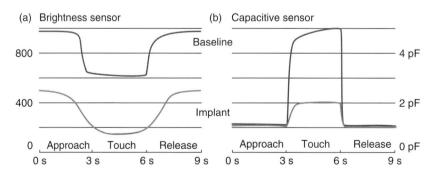

Figure 8.8 (a) Impact on sensed brightness and on sensed capacitance (b). Curves average the values of all five trials. (Adapted from [26] with permission.)

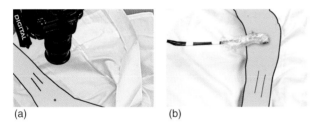

Figure 8.9 A camera captured the intensity of produced light (a) and an accelerometer measured vibration intensities (b). (Adapted from [26] with permission.)

the direction had reversed eight times [28, 29]. The last four reversal values then determined the absolute perception threshold [29].

At each step, we also measured the actual output intensities. We captured the LED output with a camera focusing onto the LED at a fixed distance, aperture, and exposure time (Figure 8.9a). An accelerometer placed directly above the vibration motor captured output intensities (Figure 8.9b).

Apparatus Details The LED was a red 3000 mcd square light. The vibration motor was a 10 mm (H3.4 mm) Precision Microdrives shaftless 310-101 vibration motor. The external camera was a Canon DSLR EOS5D and captured 16 bit RAW images.

Results

  LED: The staircase methodology yielded the absolute threshold for perceiving LED output at 8.1% intensity required in the baseline condition and 48.9% intensity required through the skin. Figure 8.10a shows the actually produced intensities determined by the external camera.

  Vibration motor: The accelerometer captured a signal through the skin only when the motor was powered at 40% intensity and higher; lower intensities were indistinguishable from background noise (Figure 8.10b). The baseline condition with the accelerometer resting on the motor directly shows an

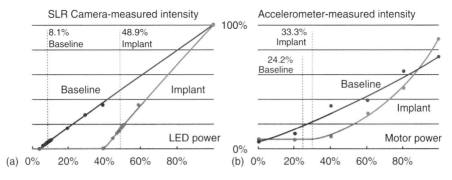

Figure 8.10 (a) Minimum perceivable LED intensity. (b) The accelerometer did not pick up a signal through skin at motor intensities of 40% and lower. Dotted lines indicate the participant's absolute perception thresholds. (Adapted from [26] with permission.)

expected linear decay. The shown values represent the mean standard deviation of the three values read by the accelerometer. The difference in personal perception of the vibration was small (24.2% vs 33.3%).

8.2.5.4 Audio Device (Speaker and Microphone)

To evaluate the speaker, we again used a descending staircase design to determine minimum perceivable audio levels. We conducted the evaluation from two distances: 25 cm (*close*) and 60 cm (*far*). These distances simulated holding the arm to one's ear to listen to a signal (*close*) and hearing the signal from a resting state with the arms beside one's body (*far*). The stimulus was a 1 kHz sine wave signal [30]. During each step, an external desktop microphone measured actual output signals from 5 cm away in the *close* condition, and 60 cm away in the *far* condition.

To evaluate the implanted microphone, we produced audio as input from two distances (25, 60 cm). Two desktop speakers pointed at the microphone and played 5 prerecorded sounds at 10 volume levels (100, 80, 60, 40, 20, 10, 8, 4, 2, 1%). Three of the sound playbacks were voice ("one," "two," "three," "user"), one was a chime sound.

Apparatus Details   The implanted microphone was a regular electret condenser microphone. The external microphone was an audio-technica AT2020 USB. The speaker was a Murata Piezo 25 mm piezoelectric buzzer. The laptop recorded from both microphones at 44.1 kHz with the microphone gain set to 1.

Results

Speaker: We first applied a bandpass filter of 100 Hz to the recorded signal around the stimuli frequency to discard background noise. The assistant could perceive the stimuli sound at a level of 5.2 dB at only 0.3% output intensity in the *baseline* session, and at 7% in the *implant* session (Figure 8.11). The perceivable decibel levels compare to other results [30]. Figure 8.11 illustrates the additional output intensity needed to achieve comparable sound pressures.

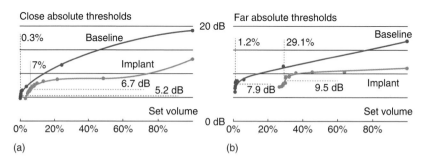

Figure 8.11   Sound perception through skin is possible, but the skin substantially takes away from the output intensity (a). This effect grows with the distance between the listener the and speaker (b). Dotted lines indicate absolute perception thresholds. (Adapted from [26] with permission.)

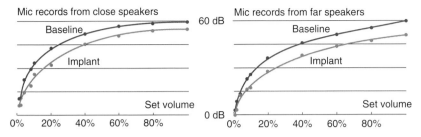

Figure 8.12 The differences in perceived sound intensities were nearly constant between the implant and the baseline session. (Adapted from [26] with permission.)

Microphone: The skin accounted for a difference in recorded sound intensities of 6.5 dB (±3 dB) for the close-speaker condition and 6.24 dB (±2.5 dB) in the far-speaker condition. At full output volume, the skin dampened the volume of the incoming sound by less than 2% when close by 25 cm away, but almost 10% with speakers 60 cm away (Figure 8.12).

8.2.5.5 Powering Device (Powermat Wireless Charger)

To evaluate the powering device, we docked the receiver to the powering mat (Figure 8.13). In the *baseline* session, the two devices docked directly. In the *implant* session, the receiver was implanted, and the powering mat was placed on the surface of the skin directly above the implant.

Once docked, we separately measured the voltages and currents the receiver supplied with a voltmeter and an ampere meter. We took five probes for each measurement, each time capturing values for 5 s for the meters to stabilize. We measured the voltages provided and the current drawn with four resistors: 2 kΩ, 1 kΩ, 100 Ω, and 56 Ω.

Apparatus Details   The powering device was a PMR-PPC2 Universal Powercube Receiver with a PMM-1PA Powermat 1X. The voltmeter and ampere meter was a VC830L digital multimeter.

Results   The Powermat receiver output a nominal voltage of 5.12 V in the baseline condition. Through skin, the provided voltage was unsubstantially smaller (5.11 V).

Figure 8.13 The wireless charging mat docks to the receiver, which is implanted inside the specimen. (Adapted from [26] with permission.)

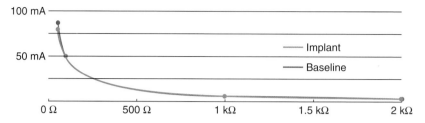

Figure 8.14 Skin affected the current provided through the wireless connection only at higher current values. (Adapted from [26] with permission.)

As shown in Figure 8.14, the skin did not impact the current drawn by the device for low resistances. For the 56 Ω resistor, the difference was 7 mA, still providing 80 mA, which should easily power an Arduino microcontroller.

8.2.5.6 Wireless Communication Device (Bluetooth Chip)

To test the performance of the wireless connection between two chips, one was external and one implanted with no space between chip and encompassing skin in the *implant* session. The *baseline* session tested both devices placed outside. We evaluated the connection at two speed levels (slow: 9600 bps, fast: 115 200 bps), sending arrays of data in bursts between the devices (16, 32, 128 kB) and calculating checksums for the sent packages. The receiving device output time-stamped logs of the number of received packages and its calculated checksum. The test was fully bidirectional, repeated five times and then averaged.

Apparatus Details The Bluetooth modules were Roving Networks RN-42 (3.3 V/26 µA sleep, 3 mA connected, 30 mA transmitting) connected to an ATmega328 controller. The RN-42 featured an on-board chip antenna with no external antenna.

Results For the slow transmission speed, no packet loss occurred in either condition. The effective speed rate was 874 B s^{-1} in both conditions (Figure 8.15a).

For the fast transmission speed, the devices received 74% of the sent packages on average in the baseline and 71% when sent through the skin (Figure 8.15b). The effective speed differed by 200 B s^{-1} (4.4 kB baseline vs 4.2 kB skin). We found no differences in direction.

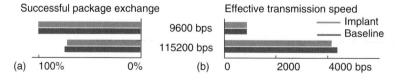

Figure 8.15 (a) Bluetooth exchanges data reliably when running slow, but comes with data loss when running fast. (b) Implanting affected fast transmission rates negatively. (Adapted from [26] with permission.)

8.2.6
Discussion

Overall, all traditional user interface components that were implanted worked under the skin, sensing input through the skin, emitting output that could be perceived through the skin, and charging and communicating wirelessly.

Regarding input, the skin required, as expected, user input to increase in intensity to activate sensor controls. Despite this required intensity overhead, all tested input sensors did perceive input through the skin, even at the lower levels of intensity we tested. This leaves enough dynamic range for the sensors' additional degrees of freedom, such as detecting varying pressure. As for hover detection, the skin incurs an offset of brightness and diminishes capacitive signals, but both sensors responded to the approaching finger.

While output appears diminished through the skin, detection is possible at low-enough intensity levels, such that output components, too, can leverage a range of intensities for producing output.

Powering the device through the skin yielded enough voltage to have powered any of the implanted devices. It is also enough to power our *3in3out* prototype device, which we describe in the next section. More measurements with lower resistances remain necessary to determine the maximum throughput of the tested inductive power supply beyond the 100 mA levels.

While skin affected the throughput of the fast wireless communication and accounted for a 3% higher loss of packages and a $0.2\,\text{kB}\,\text{s}^{-1}$ drop in speed, it did not affect the slow condition. The flawless wireless communication in 9600 bps enables reliable data exchange. Results found in the related area of body area networks differ, as transmission goes through the body or arm, not just the skin [22].

8.2.7
Exploring Exposed Components

In addition to quantitatively evaluating input components, we wanted to prototype an exposed implanted interface component. To do so, we mounted a Blackberry trackball control on the back of an Arduino Pro Mini 8 MHz board and soldered batteries to it. The trackball was a fully autonomous standalone device. We programmed the trackball to emit a different light color when the user swiped the ball horizontally or vertically.

To expose the roller ball, the skin and plastic cover over the roller ball were carefully incised using a scalpel. The incision was about 3 mm in length, so that only the roller ball was exposed. Once implanted into the specimen, the experimenters took turns interacting with the device, which worked as expected. Figure 8.16 illustrates the exposed trackball. Note that this exploration took place *after* the quantitative evaluation had fully finished. The incision made for this exploration had no effect on our earlier evaluation.

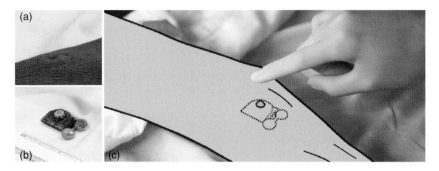

Figure 8.16 (a) Actual photograph of the LED output through the skin. (b) This standalone prototype senses input from an *exposed* trackball and (c) illuminates it in response. (Adapted from [26] with permission.)

8.3
Qualitative Evaluation

To explore initial user feedback on implanted user interfaces, we built and deployed a prototype device covered with a layer of artificial skin on users. Our goal was to gain initial insights on how users may feel about walking around with an interactive implanted device and to demonstrate how such devices can be prototyped and tested outside controlled laboratory conditions.

Study Device: We built the *3in3out* device specifically for the qualitative evaluation (Figure 8.2a). It features three input controls (button, tap sensor, pressure sensor) and three output components (LED, vibration motor, piezo buzzer). A Li-Po battery powers the standalone *3in3out* device.

The device implemented a game as an abstract task that involved receiving output and responding with input. At 30–90 s intervals, a randomly chosen output component triggered the user, who had to respond using the correct input: pressure sensor for the LED, tap sensor for the motor, and button for the speaker. While the LED kept blinking, the speaker and vibration motor repeated their output trigger every 10 s. Without a response, the trigger timed out after 1 min. Participants received points on the basis of the speed and accuracy of their responses.

8.3.1
Simulating Implants: Artificial Skin

We created artificial skin to cover our prototype and simulate actual implantation with the aid of a professional prosthetics shop, which had years of experience modeling body parts. The artificial skin is diffuse and diffused light, and dampened sound and vibration in roughly the same manner as the real skin in our evaluation. Participants in the study confirmed that the artificial skin qualitatively felt similar to real skin. As the focus of this study was on obtaining qualitative feedback, we did not calibrate the characteristics of the artificial skin to match the absolute quantitative properties of skin we measured in our evaluation. We did not need the artificial skin to match the Bluetooth properties of skin either, because our qualitative study did not include communication devices.

To create the artificial skin, we mixed Polytek Platsil-Gel 10 with Polytek Smith's Theatrical Prosthetic Deadener, which is known to produce silicone with skin-like feel and consistency. We added skin-color liquid foundation and enhanced the skin look with red, blue, and beige flocking. We then poured the silicone mixture into a mold customized to fit a human arm, added the device wrapped in Seran foil, and positioned a clay arm, so that the silicone assumed the correct shape. We then affixed the artificial skin to users' arms using ADM Tronics Pros-Aide medical grade adhesive. The final layer of artificial skin measured 4.5″ × 2″ (Figure 8.17) and was 1–2 mm thick above the device (i.e., similar to anterior surface skin [31], which we studied).

8.3.2
Task and Procedure

We designed a set of six primary tasks to distract from wearing the prototype device, which interrupted participants while carrying out those tasks: (i) ask a person for the time, (ii) board public transport and exit after two stops, (iii) ask a person for directions to the post office, (iv) pick up a free newspaper, (v) buy a coffee, and, finally, (vi) sit in a park, finish the coffee, and read the newspaper. Participants' secondary task was to respond to the triggers that the *3in3out* device emitted, and try to achieve a high score. The device recorded response times, errors, and point totals. The study took place in downtown Toronto, Canada, on a summer day, which represented a realistic worst-case scenario; both, direct sunlight and noise levels were very intense.

Participants first received a demonstration of the device and practiced its use. The participant then left the building to perform all primary tasks, and returned after approximately 60 min. Participants filled out a questionnaire after the study, sharing their impression when using the device in public environments and the reactions they received.

Figure 8.17 Artificial skin, created from silicon, covered the *3in3out* device to simulate implantation and allow for testing. (Adapted from [26] with permission.)

8.3.3
Participants

We recruited four participants (one female) from our institution. Participants were between 28 and 36 years old and wore the prototype device on their left arm. We reimbursed participants for using public transport and buying coffee.

8.3.4
Results

Overall, participants found the device easy to use. All liked the tap sensor ("easy to use") and button ("easy to find," "haptic feedback"), but none enjoyed the pressure sensor. For output components, all ranked the LED lowest for perception relative to the other output components, the speaker medium, and the vibration motor best ("really easy to notice"). While these results suggest that the device might work better in environments quieter and/or darker than the noisy city setting in direct sunlight, participants were able to see the LED blinking when looking at it.

While participants mentioned receiving curious looks from others when interacting with their arm, no external person approached a participant, even though they spent time in casual settings (e.g., coffee place and public transport).

Most importantly, the results of our study demonstrate that implanted user interfaces can be used to support interactive tasks. This evaluation also provides a methodology to pave the way for future evaluations and mockups of more elaborate devices and applications of implanted user interfaces.

8.4
Medical Considerations

While the goal of this paper is to consider implanted user interfaces from an HCI perspective, it is also important to discuss some of the medical considerations. Below we discuss some of the issues surrounding the feasibility of implanted user interfaces.

8.4.1
Location

In our study, the devices were implanted under the skin on the front of the forearm, just distal to the elbow joint. This location was chosen as the devices could be easily activated by an individual with the other hand and would not be in an area where damage by impact is likely. For the most part, these devices could be implanted deep into the skin in the subcutaneous tissue anywhere in the body where the devices are accessible and can transmit signals. This includes the upper and lower limbs, the chest wall, abdomen, and so on. Areas covered by thick skin, such as the palms and soles of the feet, would not be suitable for implantables, as the skin is too thick and tough to interact. The thickness of human skin ranges between 0.5 mm on the eyelids to 4+ mm on the palms and soles of the feet [31].

The superficial placement of the devices, directly under the skin, facilitates device activation and signal transmission. The devices can be inserted between the skin and subcutaneous tissue, providing a minimally invasive approach. The deep underlying tissues, for example, muscle, would not be disrupted. Similarly, pacemakers are placed under the skin in the chest or abdominal regions and the wires that are extending from the heart are connected to the pacemaker. Only a small skin incision that is later closed with sutures is needed to insert the pacemaker. The device remains stationary in its implanted location because of the fibrous nature of subcutaneous tissue.

The tracking ball was the only device we implanted that required surface exposure. The device worked very well under the experimental conditions, but much work needs to be done to assess the medical implications of a long-term insertion of an exposed device.

8.4.2
Device Parameters

Tissue fluid will penetrate a device that is not encased in a protective hull and affect its function. The hull's material must be carefully chosen to be pharmacologically inert and nontoxic to body tissues. For examples, pacemakers are typically made from titanium or titanium alloys, and the leads from polyether polyurethanes. *In vivo* testing would need to be carried out to determine what materials are most suitable.

The device should be as small as possible, so that it is easily implantable and cosmetically acceptable to the recipient. Functionality and minimal disruption of the contour of the skin are important considerations.

8.4.3
Risks

The main medical risk of implanting devices is infection. Infection can be caused by the procedure of implanting the devices. There are also possible risks to muscles if the device is implanted any deeper than the subcutaneous tissue. The material used for the casing could also possibly cause infections, so it will be important that the material being used passes proper testing. It is very difficult to hypothesize about other types of risks without performing testing. The wear of skin depends on the pressure applied to it; while paraplegics get sore skin from body weight resting on single spots through bones, the skin is unlikely to wear from manual pressure. The proposed input with implanted devices is short and low in force and intensity, making the skin unlikely to wear. One risk that is relatively low is that of the skin actually tearing. Skin is very strong and it is unlikely that the small devices would cause any damage. However, determining the long-term effects of interactions with implanted devices on skin requires further studies.

8.4.4
Implications and Future Studies

All of the input and output devices were functional under the experimental conditions of this study. Further cadaveric study is needed to determine if gender, skin color, and site of implantation affect device function. In the next phase, testing would also be carried out on unembalmed tissue, although the skin of lightly embalmed and unembalmed specimens is similar, loose, and pliable in both cases. Finally, the medical implications of long-term insertion of devices of this nature require detailed study.

8.5
Discussion and Limitations

The results of our study shows that traditional user interfaces for input, output, wireless communication, and powering function when embedded in the subcutaneous tissue of the forearm. Obtaining an evaluation of common components establishes the foundation for future investigations into more complex devices to explore the many other aspects of implanted user interfaces.

For example, we disregarded security concerns in our exploration. Wireless implanted devices need to prevent malicious activities and interactions from users other than the host user, such as stealing or altering stored information and manipulating the devices' operating system [32].

The processing capabilities of the devices that were implanted during the technical evaluation, as well as the *3in3out* device, require only simple processing on the microchip. More work is necessary to investigate if and how implanted devices can perform more computationally intensive operations (e.g., classification tasks using machine learning [8]) and how this affects the needs for power supply.

Social perception of implanted interfaces, both by host users as well as public perception, requires more studying. Although this has been studied with implanted *medical* devices [33], social perception of invisible and implanted *user interfaces* and devices remain to be examined.

We conducted our qualitative evaluation with participants in the summer, which is why all participants wore short-sleeve shirts. In the winter, cloth will additionally cover implanted input and output components [34] and interfere with interaction, which raises new challenges.

8.5.1
Study Limitations

Our technical evaluation comprised a single specimen. In addition, we carried out the staircase evaluations with a single participant. As such, the metrics we have collected can serve as baselines for future experimentations, but should be generalized with caution. Furthermore, our evaluation captured technical metrics from the devices, and not human factor results. In the future, it may be interesting to have external participants interact with the implanted devices and study task performance levels.

8.6 Conclusions

The technological transition society has made in the past 30 years is astounding. Technology, and the way we use it, continues to evolve and no one can tell what the future holds. Several experts have predicted that cyborgs are coming [35, 36], and devices will become indistinguishable from the very fabric of our lives [1]. If we look at how much has changed, it should not be hard to believe that we will one day interact with electronic devices that are permanent components of our body. Our work takes a first step toward understanding how exactly this might be accomplished, and begins to ask and answer some of the important technical and human factors, and medical questions.

References

1. Warwick, K. Project Cyborg, http://www.kevinwarwick.com/ (accessed 7 September 2013).
2. Moore, M.M. and Kennedy, P.R. (2000) *Proceedings of the Fourth International ACM Conference on Assistive Technologies (Assets '00)*, ACM Press, New York, pp. 114–120.
3. Drew, T. and Gini, M. (2006) *Proceedings of the 5th International Joint Conference Autonomous Agents and Multiagent Systems (AAMAS06)*, ACM Press, pp. 1534–1541.
4. Wolpaw, J.R., Birbaumer, N., McFarland, D.J., Pfurtscheller, G., and Vaughan, T.M. (2002) Brain-computer interfaces for communication and control. *Clin. Neurophysiol.*, **113** (6), 767–791.
5. Moore, K.L., Agur, A.M.R., and Dalley, A.F. (2010) *Essential Clinical Anatomy*, 4th edn, Lippincott, Williams & Wilkins, Philadelphia, PA, http//lww.com.
6. Starner, T. (1996) Human-powered wearable computing. *IBM Syst. J.*, **35** (3), 618–629.
7. Norman, D. (2001) Cyborgs. *Commun. ACM*, **44** (2), 36–37.
8. Rekimoto, J. (2001) GestureWrist and GesturePad: unobtrusive wearable interaction devices. Proceedings of ISWC '01, pp. 21–27.
9. Weiser, M. (1991) The computer for the 21st century. *Sci. Am.*, Special Issue on Communications, Computers, and Networks, **265** (3), 94–104.
10. Li, Z. and Wang, Z.L. (2011) Air/liquid pressure and heartbeat driven flexible fiber nanogenerators as micro-nano-power source or diagnostic sensors. *Adv. Mater.*, **23**, 84–89.
11. Ullah, S., Higgin, H., Siddiqui, M., and Kwak, K. (2008) *Proceedings of the 2nd KES International Conference on Agent and Multi-Agent Systems: Technologies and Applications (KES-AMSTA '08)*, Springer-Verlag, Berlin, Heidelberg, pp. 464–473.
12. Stelarc. Official website. http://www.stelarc.org (accessed 7 September 2013).
13. Huo, X., Wang, J., and Ghovanloo, M. (2008) A magneto-inductive sensor based wireless tongue-computer interface. *IEEE Trans. Neural Syst. Rehabil. Eng.*, **16** (5), 497–504.

14. Starner, T. (1995) The Cyborgs are Coming. Technical Report 318, Perceptual Computing, MIT Media Laboratory.
15. Holz, C. and Baudisch, P. (2010) *Proceedings of the SIGCHI Conference on Human Factors in Computing Systems (CHI '10)*, ACM Press, New York, pp. 581–590.
16. Hinckley, K., Pierce, J., Sinclair, M., and Horvitz, E. (2000) *Proceedings of the 13th Annual ACM Symposium on User Interface Software and Technology (UIST '00)*, ACM Press, New York, pp. 91–100.
17. Saponas, T.S., Tan, D.S., Morris, D., Balakrishnan, R., Turner, J., and Landay, J.A. (2009) *Proceedings of the 22nd Annual ACM Symposium on User Interface Software and Technology (UIST '09)*, ACM Press, New York, pp. 167–176.
18. Wishnitzer, R., Laiteerapong, T., and Hecht, O. (1983) Subcutaneous implantation of magnets in fingertips of professional gamblers-case report. *Plast. Reconstr. Surg.*, **71** (3), 473–474.
19. Brumberg, J.S., Nieto-Castanon, A., Kennedy, P.R., and Guenther, F.H. (2010) Brain-computer interfaces for speech communication. *Speech Commun.*, **52** (4, Silent Speech Interfaces), 367–379.
20. Kaiser, M. and Proffitt, D. (1987) Observers' sensitivity to dynamic anomalies in collisions. *Percept. Psychophys.*, **42** (3), 275–280.
21. Orth, M., Post, R., and Cooper, E. (1998) *CHI 98 Conference Summary on Human Factors in Computing Systems*, ACM Press, New York, pp. 331–332.
22. Masters, A. and Michael, K. (2005) *Proceedings of the Second IEEE International Workshop on Mobile Commerce and Services (WMCS '05)*, IEEE Computer Society, Washington, DC, pp. 32–41.
23. Levitt, H. (1971) Transformed up-down methods in psychoacoustics. *J. Acoust. Soc. Am.*, **49** (2), 467–477.
24. Smith, D.V. and Margolskee, R.F. (2001) Making sense of taste. *Sci. Am.*, **284** (3), 32–39.
25. Anderson, S.D. (2006) Practical light embalming technique for use in the surgical fresh tissue dissection laboratory. *Clin. Anat.*, **19** (1), 8–11.
26. Holz, C., Grossman, T., Fitzmaurice, G., and Agur, A. (2012) *Proceedings of the SIGCHI Conference on Human Factors in Computing Systems (CHI '12)*, ACM Press, New York, pp. 503–512.
27. Cornsweet, T.N. (1962) The staircase-method in psychophysics. *Am. J. Psychol.*, **75** (3), 485–491.
28. Lambert, M., Chadwick, G., McMahon, A., and Scarffe, H. (1988) Experience with the portacath. *Hematol. Oncol.*, **6** (1), 57–63.
29. Jovanov, E., Milenkovic, A., Otto, C., and de Groen, P.C. (2005) A wireless body area network of intelligent motion sensors for computer assisted physical rehabilitation. *J. NeuroEng. Rehabil.*, **2**, 6.
30. Gelfand, S.A. (1990) *Hearing: An Introduction to Psychological and Physiological Acoustics*, 2nd edn, Marcel Dekker, New York and Basel.
31. Fornage, B.D. and Deshayes, J. (1986) Ultrasound of normal skin. *J. Clin. Ultrasound*, **14**, 619–622.
32. Halperin, D., Heydt-Benjamin, T.S., Ransford, B., Clark, S.S., Defend, B., Morgan, W., Fu, K., Kohno, T., and Maisel, W. (2008) *Proceedings of the 2008 IEEE Symposium on Security and Privacy (SP '08)*, IEEE Computer Society, Washington, DC, pp. 129–142.
33. Denning, T., Borning, A., Friedman, B., Gill, B.T., Kohno, T., and Maisel, W.H. (2010) *Proceedings of the SIGCHI Conference on Human Factors in Computing Systems (CHI '10)*, ACM Press, New York, pp. 917–926.
34. Powermat (2013) Inductive Charging, http://www.powermat.com (accessed 7 September 2013).
35. Ni, T. and Baudisch, P. (2009) *Proceedings of the 22nd Annual ACM Symposium on User Interface Software and Technology (UIST '09)*, ACM Press, New York, pp. 101–110.
36. Starner, T. (2001) The challenges of wearable computing. *IEEE Micro*, **21** (4), 44–67.

9
Ultralow Power and Robust On-Chip Digital Signal Processing for Closed-Loop Neuro-Prosthesis

Swarup Bhunia, Abhishek Basak, Seetharam Narasimhan, and Maryam Sadat Hashemian

9.1
Introduction

Implantable microsystems have emerged as important and highly effective instruments for monitoring and manipulating internal activities within the body. With great advances in electronics and electrode technology over the last decades, it has become possible to implement complex implantable systems that interface with the biological organisms to monitor internal body signals and manipulate the activity of body parts using electrical/chemical/optical stimulation. One of the prominent success stories in the field of implantable biomedical devices is the cardiac pacemaker, which has been successfully implanted in a large population of patients across the globe. With emerging requirements of biomedical devices to interface with different body parts for therapeutic usage, pervasive implantable devices are rapidly becoming a reality.

Figure 9.1a shows some example applications of bioimplantable devices. These devices are used to recognize and treat symptoms of various diseases such as epilepsy, heart disease, Parkinson's disease, blindness, urinary incontinence, and so on. Several applications of biomedical devices are mentioned in [1]. Examples of implantable medical devices produced commercially by the leading medical technology company Medtronic are presented in [2].

A typical implantable system contains a sensing circuitry for recording biological signals, circuits for signal conditioning and analysis, and transceiver circuits for external communication, as shown in Figure 9.1b. The front-end electronics are usually analog/mixed-signal circuits for signal conditioning (preamplifier, filter, data converter, etc.) of the recorded biological signal. For untethered power and data transfer, the implantable systems tend to have radio frequency (RF) transceivers and portable energy sources either within the implant unit or as wearable modules connected to the implant. The back-end transceiver circuits [3] usually deal with wireless data communication, typically at low RF ranges, which are more effective in penetrating skin and other tissue, without causing tissue damage. The energy source can be a rechargeable battery powered through inductive or RF links or using alternative

Implantable Bioelectronics, First Edition. Edited by Evgeny Katz.
© 2014 Wiley-VCH Verlag GmbH & Co. KGaA. Published 2014 by Wiley-VCH Verlag GmbH & Co. KGaA.

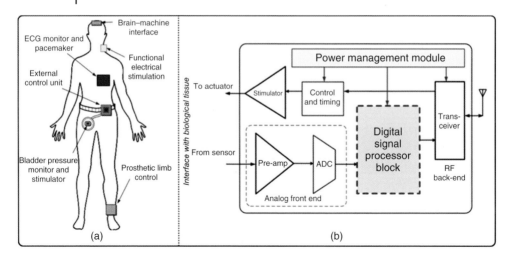

Figure 9.1 (a) Pervasive nature of bioimplantable systems, which are used for monitoring and/or controlling activities of different body parts. (b) Overall block diagram of a typical bioimplantable system, highlighting the digital signal processing (DSP) block.

sources of energy, commonly termed as *energy harvesting* or *energy scavenging*. The digital unit is used to provide control of different tasks and for digital signal processing (DSP) for on-board real-time data analysis. This block is often implemented using commercial microcontroller or DSP units, which are software-programmable, during the initial testing phase when algorithms are being developed.

The major design challenges for digital hardware in implantable systems are highlighted in Figure 9.2. The two parameters of *area* and *power* are the most

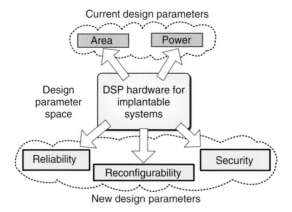

Figure 9.2 Design parameters for implantable systems. Area and power are traditional design considerations that need to be minimized. Performance is not a significant constraint for bioimplantable systems because of the low frequency content of the signals. Reliability, reconfigurability, and security are emerging as important design parameters.

significant. The implantable system needs to fit within a small area and hence, the chips inside the system should have a small form factor as well. The second design parameter is power. There are two reasons behind the quest for ultralow power. First, the implantable systems are equipped with a small battery (usually rechargeable at present) with limited capacity for supplying energy to all the active circuits. If the circuits consume too much power, the lifetime of the system becomes reduced. The second reason for using low-power circuits is to limit power dissipation and avoid associated problems such as tissue damage due to overheating or temperature-induced circuit reliability degradation. As biological signals typically have very low frequency content (in the range of fractions of hertz to a few kilohertz), the computational speed requirement of the processing electronics is very low compared to state-of-the-art microprocessors, which have maximum operating frequencies in the order of few gigahertz. It has been shown in [3] that the system power is dominated by the wireless transmission and that can be reduced drastically by using complex signal processing on-chip to reduce the amount of data to be transmitted. However, the power budget for the electronics is also extremely low. Moreover, the transmission frequencies of the implantable transceivers are limited to few megahertz, according to the regulations by the Federal Communication Commission (FCC) [1]. As the number of recording sites increases and more complex data analysis needs to be performed online, the DSP unit in the implantable system needs to have increased computational power. The DSP block serves two main purposes:

1) It recognizes meaningful patterns to trigger appropriate measures, for example, neuromodulation or drug delivery.
2) It performs data compression to reduce the amount of raw data to be transmitted wirelessly to an external control unit.

For example, neural recording from an array of 100 electrodes, sampled at 25 kHz per channel with 8-bit precision, yields an aggregate data rate of 20 Mbps, which is well beyond the reach of state-of-the-art wireless telemetry. Also, if the system keeps transmitting raw sampled data, it will drain the power source and seriously hamper the freely operating lifetime of the device. Other growing digital system requirements include *reconfigurability* to adapt to individual patient's condition and/or temporal changes, information, and device level *security* to protect the recorded data and *robustness* to ensure the long-term reliable operation of the system. It should be noted that design solutions for addressing these latter issues should not come at the expense of increase in area or power consumption. Hence, overhead is an important design parameter when considering different solutions for these challenges.

Microsystems which interface with the central or peripheral nervous system are a growing class of implantable systems. A computational task for real-time online multichannel neural signal analysis would greatly benefit from special-purpose, low-power, robust, and area-efficient hardware, since conventional microprocessor or DSP chips would dissipate too much power and are too large in size for an implantable device. For hardware implementation of this signal-processing block, nanoscale technologies offer great potential because of their tera-scale integration density, low switching power, and high performance. However, they also bring in number of design challenges, such as exponential increase in leakage power [4], reduced yield, and lack of robustness due to reduced noise margin [5]. It is important to develop circuit/architecture-level design solutions tailored to the computational algorithms for neural signal processing that can leverage the benefits of nanoelectronics while addressing its limitations. In summary, it can be seen that conventional design solutions for implementation of energy-efficient hardware can be combined with novel solutions that exploit the nature of the application for the implantable systems and its signal-processing algorithm in order to achieve efficient solutions in the target design parameter space. The next subsection gives a background about implantable systems, specifically dealing with neural interfacing.

9.1.1
Neural Interfaces

A major limitation of our current understanding of the nervous system comes from the limitations of existing techniques for monitoring its electrical and chemical activity. Currently, it is possible to precisely measure the activity of small numbers of neurons (e.g., using intracellular electrodes), or to obtain an overall measure of the activity of large neural assemblies (using techniques such as electroencephalography, EEG, or functional magnetic resonance imaging, fMRI). However, precisely recording and analyzing the activities of large number of neurons (on the scale of hundreds to thousands) in a normally behaving animal is not currently feasible. This is a significant limitation, because many important aspects of neural control occur because of the activities of populations of nerve cells at this scale.

The neurons are the communication pathways for the neural signals to travel to and from the central nervous system. However, these neurons are more than simple conducting wires for neural signals. The electrical signal within a neuron occurs as a spike (where a spike means a concentrated short-time voltage change with a high frequency band) in the intracellular membrane potential due to precisely defined electrochemical processes. These spikes are known as *action potentials*, which travel along the axon and provide electrochemical stimuli to other neurons in their vicinity. The neural interface systems typically require monitoring the activities of the neural network at the cellular level using microelectrodes capable of sensing neural action potentials. Considering the fact that a human nervous system typically consists of 10^{11} neurons each with diameters 4–100 µm, a realistic implantable microsystem should be capable of monitoring a large number of sites (in the range of hundreds or thousands) simultaneously with closely spaced electrodes and transmit the multichannel recorded data wirelessly for untethered measurements [6]. Similarly, microstimulation by insertion of electronic signals requires interfacing with the neural networks at the cellular level and controlling a large number of sites simultaneously. Researchers are increasingly using these implantable devices as interfaces to the central and peripheral nervous system to achieve better understanding of the mechanisms of neural communication and control. By studying simple organisms with tractable nervous systems, one can gain insight into the correlation between patterns of neural activity at the level of individual neurons and the resultant behavior of the organism [7]. Such behaviorally meaningful patterns can range from single spikes in a single neuron to timed bursts of neural spikes from a population of neurons, depending on the granularity of the behavior being studied.

Implantable neural interfaces have been explored in diverse contexts including neural stimulation, as in cardiac pacing and functional electrical stimulation (FES). FES of nerves or muscles is used to assist patients in grasping, standing, or urination, while deep brain stimulation (DBS) has been shown to be an effective treatment for Parkinson's disease. Cochlear implants are commercially available for treating deafness, while visual prostheses have had preliminary success in creating sensations of vision in human beings [6, 8]. Extensive research has been done on developing brain–computer interfaces (BCIs) [9] in which a tetraplegic person (whose limbs are paralyzed typically because of spinal cord injury) can control movement of a computer cursor or a robotic arm. Current implementations of these systems, however, do not perform *in situ* signal processing, although some of them use simple control algorithms based on external sensor data.

Neuro-electric electrodes can be of two types – intracellular, which are invasive but can pick up the voltage profile inside individual neurons, and extracellular, which are noninvasive (causing minimal neural tissue damage) but pick up neural signals of different strengths from different neurons in the vicinity along with lots of background noise. A major component of this noise is the biological noise (e.g., action potentials from neural cells far from the measuring electrodes) and other electrical noise (e.g., body movements). Figure 9.3 shows the activity recorded from the same neuron using intracellular and extracellular electrodes. The latter has lower amplitude and is corrupted by background noise. In order

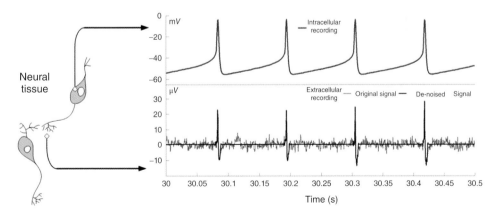

Figure 9.3 Intracellular and extracellular recordings from the same neuron in• *Aplysia californica* (sea-slug). De-noising and spike detection in the extracellular signal are performed using a wavelet-based adaptive thresholding [10].

to use the recorded information containing neuronal action potentials mixed with background noise, one needs to perform spike detection and alignment followed by classification of the action potential signals. From the neuroscientist's point of view, it is important to retain the spike topology and shape in the transmitted data, since they convey important information regarding the activity inside a neuron, and can be critical for identifying neuronal subtypes. In the context of neural signal processing, major tasks to be performed by the DSP unit are de-noising of the recorded signal, detection of spikes, spike sorting, detection of bursts, efficient representation of spike data (for data compression), and recognition of meaningful patterns (that relate to the overall behavior of an animal) from the multichannel recorded signal. Previous efforts for spike detection have been aimed at using simple thresholding schemes, with or without an adaptive threshold for detecting spikes in the recorded neural signal using analog comparators [7], with an on-chip data compression circuit [11], wavelet-based spike detection hardware implementation [12], and so on.

In vivo implantable device recordings from freely behaving animals in their natural environment are crucial to final validation of any neural-signal-processing algorithm under real environmental noise conditions. In order to allow chronic implantation and operation, the system needs to be more independent in terms of energy resource, data telemetry, and control for decision-making.

9.1.2
Closing the Neural Loop: Significance of On-Chip DSP

The need for a closed-loop neural system, which records from multiple neurons, analyzes the neural activity, and stimulates some neurons on the basis of the analysis, has been emphasized before [13]. Several researchers are working toward understanding how artificial stimulation can be made to mimic realistic neural signals, so that the feedback loop can be closed. However, most of the current neural interface systems employ sophisticated data analysis performed on an external computer, which receives recorded data and sends control signals using wireless communication with the implant unit, as shown in the left part of Figure 9.4. The need for real-time analysis, signal integrity during transmission, along with the limited bandwidth of wireless channels and constrained movements of the patient often hinder accurate closed-loop operation. Real-time closed-loop neural control can greatly benefit from in situ signal processing using low-power miniaturized implantable hardware. Such *in situ* processing is more important for chronic implantations as well as to facilitate untethered ambulatory movements of a patient in daily life scenarios. Besides, with on-chip signal analysis, more number of channels can be integrated into the interface, enhancing the accuracy of neural analysis and control. An important task of the embedded signal-processing module is to extract neural information at different levels, ranging from spike shapes in individual neurons to burst patterns in a group of behaviorally related neurons. This information is critical in performing closed-loop stimulation for neural prosthesis. It has been shown that a set of neurons that control specific muscles

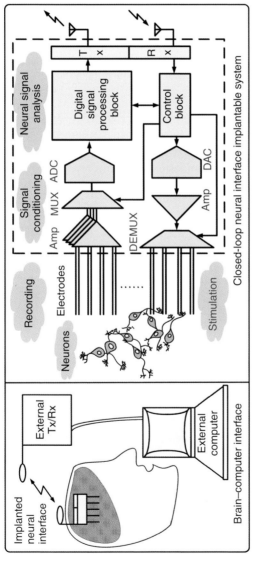

Figure 9.4 Existing brain computer interfaces use external computers to perform neural signal analysis. The proposed implantable system uses on-chip digital signal processing for neural pattern recognition to facilitate closed-loop neural recording and stimulation. Here, the wireless transceiver and external computer are only required during training and calibration.

can generate different neural patterns corresponding to different movements of body parts [14]. Moreover, the context-dependent shape of the action potential from the same neurons provides different information [15]. Hence, simple encoding of the presence or absence of a neural spike or the rate of firing of an individual neuron may miss behaviorally meaningful patterns from large sets of neurons. It requires both an efficient algorithm to reliably identify meaningful neural patterns and custom hardware realization of the algorithm to meet the tight area and power demands of an implantable system.

One of the major limitations of current BCI systems, as suggested before, comes from interfacing with the nervous system at the cellular level. As the number of recording channels increases, the transmission ability of the implantable telemetry device (in terms of bandwidth and power resource) becomes the bottleneck. For example, neural recording from an array of 100 electrodes sampled at 25 kHz per channel with 10-bit precision yields an aggregate data rate of 25 Mbps, out of reach of current telemetry systems. Hence, the implantable signal-processing unit as shown in the right part of Figure 9.4 can be used for real-time neural signal analysis to achieve data compression [16]. The compressed data packets from the neural signal-processing module are transmitted to the external receiver (hence to the computer, during only initial training and calibration) and/or sent to the control unit for appropriate stimulation.

The chapter describes a structured approach for integrating wavelet-based spike detection and sorting with higher level signal analysis so that we get closer to the realization of a closed-loop neural prosthesis system that can identify behaviorally significant neural patterns. Such a system can be initially used for on-chip multichannel spike analysis at various levels of compression and resolution, while the animals' behavior and feedback protocols are studied. The ultimate goal of the proposed system is to complete the loop between neural recording and stimulation inside the implant unit by eliminating interaction with an outside computer.

9.2
Algorithm: a Vocabulary-Based Neural Signal

9.2.1
Analysis

Closed-loop neural prosthesis systems typically involve analysis of multichannel recordings from behaviorally relevant neurons for predicting intended behavior before undertaking any preventive/corrective/assistive actions by appropriate neural stimulation/drug delivery. There are several design challenges to be addressed before such a system becomes a reality. When hundreds of electrodes are employed for parallel recording, the transmission ability of the telemetry device (e.g., its bandwidth) becomes insufficient and power-hungry [6, 8]. Therefore, it is extremely important to use on-chip electronics for preprocessing and compressing the recorded data that can be transmitted wirelessly with very low power dissipation. For closed-loop neural control, the implantable system must also perform real-time analysis of spike patterns from multiple channels and recognize behaviorally relevant patterns in the neural signal.

In order to use the recorded neural information containing action potentials (spikes in intracellular voltage) mixed with background noise (biological and electrical), we need to de-noise the signal and detect the spikes. Such a computational task for online, real-time data analysis and compression requires both efficient signal-processing algorithms and special-purpose customized hardware to implement them on-chip, as conventional microprocessor or DSPs would dissipate too much power and are too large in size for an implantable device.

In this section, we present a hardware-aware algorithm design approach for neural pattern recognition and data compression in implantable neural interface systems. By representing neural signals in terms of bursts, we can achieve large data compression, thus saving precious bandwidth and reducing power dissipation. We use multi-resolution wavelet analysis [17] of recorded signals to de-noise the data, detect and sort spikes, and then determine the on/off timing of a burst. Next, a hierarchical vocabulary is built to maintain spikes from different neurons, develop higher order symbols in terms of bursts, and recognize behaviorally meaningful burst patterns across multiple channels from in vivo neural and muscular recordings.

The proposed signal analysis approach is illustrated with the flow diagram depicted in Figure 9.5. Multi-resolution wavelet analysis [18] is used to decompose the recorded signal in time-frequency domain. This step helps to localize the action potentials using a thresholding scheme. It is worth noting that thresholding of the wavelet coefficients helps to de-noise the signal and detect spikes more accurately than purely time-domain thresholding approaches [10]. Next, a dynamic vocabulary is built by encoding distinct spike shapes as unique alphabet symbols. Other major steps of the algorithm involve spike disambiguation to identify different spike types on the same channel, identification, and representation of a "burst" of spikes and finally multichannel burst analysis for behaviorally meaningful pattern recognition.

The recorded signal is encoded in terms of individual spike shapes (alphabet symbols) or burst of spikes (words) or collection of bursts across channels (sentences), representing different levels of data compression and information content. Major signal-processing steps of the algorithm and the associated parameters are described below.

9.2.2
Spike-Level Vocabulary

We first detect spikes from the noisy extracellular recording and perform shape matching in the wavelet domain to create a spike-level vocabulary. By encoding each unique spike shape by its corresponding "alphabet symbol," we reduce the amount of data that needs to be transmitted outside. As new spikes are detected, they are compared with existing alphabet symbols in the vocabulary. If a match is found,

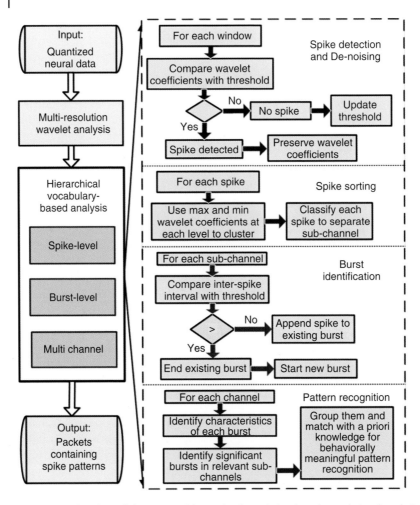

Figure 9.5 Flowchart of the proposed hierarchical vocabulary-based neural signal analysis algorithm [19].

we represent the new spike with the existing alphabet symbol; otherwise, a new alphabet symbol is created. The vocabulary is built dynamically for any recorded channel, which allows real-time compression and analysis of neural spikes. Next, we look at the different sub-steps involved in building the spike-level vocabulary and discuss the design challenges that shape our choice of parameters at each step.

9.2.3
Spike Detection

The neural signal is segmented into blocks of equal size, and wavelet transformation is applied to each block. The wavelet transform is similar to Fourier analysis, with the exponential (or sinusoidal) matching function replaced by a finite-time waveform, with a particular shape, called the *mother wavelet*. This transformation is performed by multi-resolution discrete wavelet transform (DWT), using Daubechies-3 (db3) basis function [20]. Wavelet transform decomposes the recorded signal in both time and frequency domains [21, 22], which turns out to be very useful in feature extraction [19, 23]. It generates a set of coefficients corresponding to the low-frequency components, called the *approximations*, and another set of coefficients corresponding to the high-frequency components, called the *details*, while preserving the timing information. If we consider only the approximations for reconstructing the neural signal by using an inverse wavelet transform, then we achieve our first level of data compression, along with de-noising. The parameters of importance are the mother wavelet used for the wavelet decomposition and the number of levels of decomposition performed on the signal. The former determines the efficiency of matching with the action potentials, which increases if the mother wavelet resembles a typical spike in shape. The latter signifies the amount of data compression achieved in this first step, without having to lose important information about the recorded signal.

Once the wavelet coefficients corresponding to the "windowed" signal for a particular mother wavelet and a certain level of decomposition are generated, we can perform a thresholding step. This helps in removing most of the noise from the wavelet domain, by preserving only the most significant coefficients (Figure 9.6). Even though we perform thresholding of the approximation coefficients, we

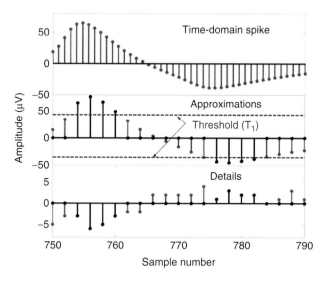

Figure 9.6 Wavelet coefficients of a neural spike: approximation coefficients larger than a threshold signify presence of a spike. The corresponding detail coefficients as well as the coefficients below threshold that are part of the spike are preserved in order to retain the shape information.

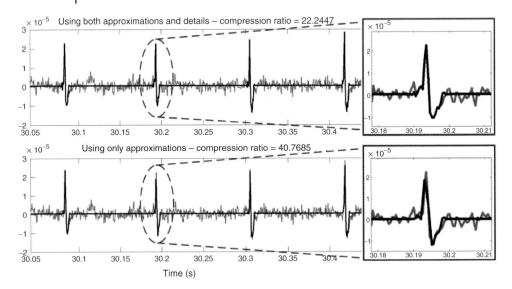

Figure 9.7 Using both approximation and detail coefficients to represent the spike gives better reconstruction quality.

preserve the corresponding detail coefficients, because they contain the high-frequency information corresponding to the sharp edges in the spike shapes, leading to better reconstruction quality, as shown in Figure 9.7. Further, as we perform the thresholding in the wavelet domain, the timing information corresponding to the location of spikes is maintained with negligible error. The parameter of importance here is the thresholding level. It can be either fixed or variable. If variable, then it can be recomputed for each window separately or once in a while. The thresholding step is a compromise between information retention and data compression. If we keep a high threshold, the latter will be very high, but we might miss out on some spikes. On the other hand, if the threshold is too low, we end up with lot of false positives and a lower compression ratio. Since it is important to avoid false negatives, the threshold values must be chosen optimally. Corresponding to each window of the signal, we have a set of wavelet coefficients with some nonzero values, which correspond to the spikes of the neural signal. We intend to create data packets from these coefficients such that we need to send only the information regarding location, duration, and amplitudes of a set of consecutive nonzero approximation coefficients. This "packetizing" algorithm is aimed at further data compression.

9.2.4
Spike Characterization and Sorting

Spike sorting is required to classify the spikes in an extracellular recording based on the source neuron. On the basis of the window discriminator algorithm [14] for spike sorting, we have developed a clustering approach (Figure 9.8) for spike disambiguation from the wavelet domain representation of the signal. By using four significant wavelet coefficients of the recorded signal as our dimensions of interest, we can sort spikes based on both amplitude and frequency components. This allows classifying spikes of similar shape and amplitude as a single alphabet symbol of the vocabulary. The thresholds for clustering depend on the noise level and are chosen using a supervised training process. As a result of spike sorting, each channel of extracellular data is decomposed into multiple sub-channels, each subchannel consisting of spikes of similar shape (i.e., belonging to a single cluster), as shown in Figure 9.8.

9.2.5
Burst-Level Vocabulary

The characteristics of bursts of neural activity, such as the frequencies of occurrence of particular spike shapes in a particular channel, contain significant information regarding the behavior of the specific neuron corresponding to that channel. This leads to a hierarchical formation of higher level symbols (referred as *words*) in the vocabulary. In the proposed approach, burst identification is based on monitoring the inter-spike interval or firing frequency for each spike shape that constitutes a particular sub-channel. If the inter-spike interval is less than a predefined threshold of 0.25 s (determined through an initial supervised training phase), the spikes are grouped together as a burst. Once the onset and offset of a burst are identified, we investigate the use of two types of burst representation. In the first one, we use the representative spike shape (alphabet symbol) for the burst, average firing frequency, and burst duration. In the second representation, we replace the average firing frequency by a more accurate trajectory of interspike intervals. Clearly, the first representation provides better compression, while the second one trades off compression for better reconstruction quality (Figure 9.9).

9.2.6
Multichannel Vocabulary for Behavior-Specific Patterns

Activity recorded from individual neurons may not convey significant information regarding the behavior of the animal. For instance, the difference between two types of swallowing activities in Aplysia lies in the timing and intensity of firing in different neurons controlling the excitation or inhibition of relevant muscles, which are responsible for jaw movement [14] (Figure 9.10). Hence, the words encoding activity in individual neurons (recorded across multiple channels) must be combined to extract behaviorally significant information. This leads to the next level of hierarchy in vocabulary development, which involves construction of syntactically and semantically correct "phrases" and "sentences" (again, encoded with a number of bits). Simultaneous encoding of bursts of spikes in multiple channels can lead to significantly higher data compression than existing approaches along with behavior-specific pattern recognition.

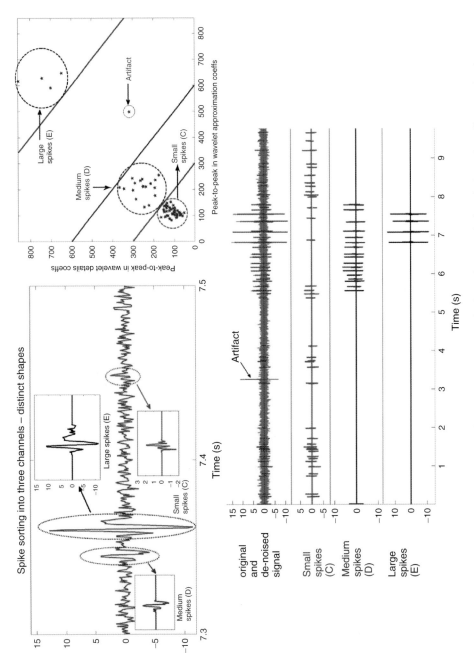

Figure 9.8 Clustering technique for sorting the detected spikes in a single channel into multiple sub-channels. Different spike shapes from the same channel denote different sources, which are translated to different alphabet symbols in the vocabulary. The thresholds for clustering are determined during a calibration phase for each channel. The alphabet symbols corresponding to each cluster are stored in the vocabulary and used for matching new spike shapes [10].

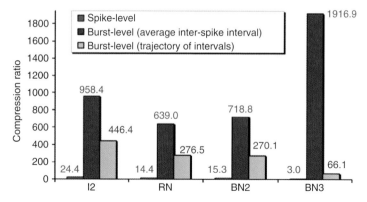

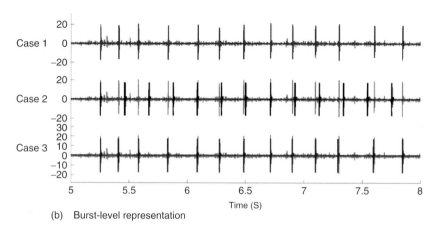

Figure 9.9 (a) Compression ratio values for multichannel recordings at different levels. (b) The reconstructed signal (superimposed over original noisy signal) for the largest spikes in channel BN3 is shown for the three cases to demonstrate the deterioration in reconstruction quality. Case 3 provides a good balance between compression and signal quality [10].

9.2.7 Output Packet Generation

A typical packet consists of the location, duration, signature, and amplitudes corresponding to the nonzero approximation coefficients. We also keep 1 bit for identifying whether a match for the identified spike has been found in the existing vocabulary or not. If yes, then we do not need to send the amplitudes again. If it is a newly generated signature, then the corresponding information has to be sent. Hence, the data packets have variable length, based on the duration and also on whether the signature is repeated or not. A frame identification bit is

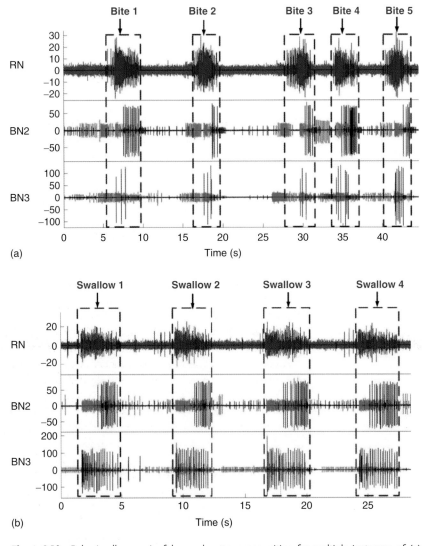

Figure 9.10 Behaviorally meaningful neural pattern recognition for multiple instances of (a) biting and (b) swallowing [10].

needed to signify the beginning of each window of the signal. This bit is also useful for synchronization purpose. In the initial stage, when the vocabulary is being constructed and updated, there will not be considerable data compression. Once the warm-up period is over, that is, when the vocabulary is built-up, the data compression is considerable. It can be increased by pre-computation of all possible signatures and embedding the vocabulary inside the chip and only updating the vocabulary whenever necessary.

9.3
Hardware Implementation

Miniaturized implantable systems provide an important interface with the central nervous system for interpreting and engineering its activity [6]. Neural prosthesis systems typically involve analysis of multichannel recordings from behaviorally relevant neurons for predicting intended behavior before undertaking any preventive/corrective/assistive actions. While previously reported implantable neural interface systems have included active components [6, 12], there have been relatively few efforts toward embedding significant computational components for intelligent processing of the acquired neural data in a closed-loop framework.

Most of the current neural interface systems employ sophisticated data analysis performed on an external computer. As the number of recording electrodes increases, the system power is dominated by the wireless transmission [3] and this can be reduced drastically by using on-chip signal processing to recognize patterns of activity in the neural signal for reducing the volume of transmitted data. However, in order to extract significant information from neural recordings, we need to perform complex signal processing based on fast Fourier transform (FFT) [18] or DWT [10], which requires large power and area to be implemented.

Commercial DSP and reconfigurable chips are too power hungry and have significant area overhead to be used for implantable systems. Hence arises the need for ultralow power, area efficient, and robust on-chip implementation of the neural-signal-processing algorithm, leveraging the benefits of nanoscale electronics.

Figure 9.11 shows the interface among different constituent blocks of a typical implantable system. The data acquired by the recording electrodes is conditioned using analog circuits and converted to digital signals, which are the inputs of the neural-signal-processing block. This block is responsible for analyzing the digitized data, compressing it, and extracting meaningful neural patterns. The relevant information from this block is then transmitted to an outside receiver through the transmitter or used for stimulation in a closed-loop framework. In this section, we evaluate the hardware realizability of the vocabulary-based analysis framework presented in Section 9.2. First, we describe the hardware components required

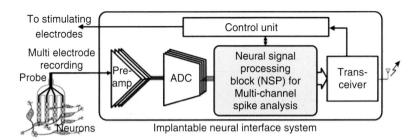

Figure 9.11 The high-level functional block diagram for a typical neural interface system. The digital signal processing block, which analyzes spike patterns on multichannel recorded data, constitutes a key part of the system [24].

to implement the neural signal analysis and choose appropriate architecture for each block. The signal-processing algorithm is composed of modular steps, which are amenable to efficient hardware implementation. At each step, our choice of architecture is guided by the primary design requirements of low area and low power. We observe the effect of scaling the operating frequency, supply voltage, and process technology on major design parameters. The focus of this section is on the system-architecture-circuit codesign choices, which exploit the nature of the neural-signal-processing principles for area-efficient and ultralow-power on-chip implementation of the algorithm.

The main functional blocks of the proposed neural interface system, along with the flow of data, are shown in Figure 9.12. The sampling and digitization hardware generate 10-bit input data points for the windowing module, which buffers up to six samples at a time, using a shift register. The window size is determined on the basis of the requirement of the DWT hardware using the "sym4" mother wavelet. The wavelet coefficients are computed sequentially, with the intermediate steps being executed in multiple cycles using the same computational block. Thresholding of the coefficients is performed before being transferred to the spike detection module, which has a buffer register to store the last window of coefficients. The vocabulary module matches with and updates the vocabulary tables (fixed-length register arrays) to generate a signature for the detected spike. This information is used to create packets of output data containing spike level information, which

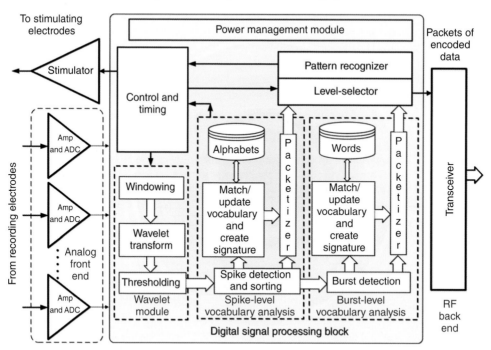

Figure 9.12 System architecture highlighting the digital signal-processing block.

are sent to the transceiver module. The windowing, wavelet transform, and thresholding operations are grouped into the *wavelet module*, while the spike detection, vocabulary, and packetizing operations are done by the *vocabulary module*. The vocabulary module has multiple levels for matching and identifying higher level patterns in the recorded neural signal. The wireless transceiver is capable of transmitting information at any level, trading off number of channels, to remain within the data-rate limit. There is also provision for feedback of the information into the control block, which can take decision on performing stimulation. The power management module controls the supply voltage and operating frequency for different modules, as well as provides power-gating controls for idle blocks.

9.3.1
Wavelet Module

The appropriate architecture for low-power DWT implementation has been investigated. One of the main contributions of wavelet theory has been to relate discrete-time filter banks to continuous-time function space [20], which has facilitated its efficient hardware implementation [21]. Recently, a lifting-based forward and reverse wavelet transform [21, 22] has been proposed, which requires far fewer computations and significantly less memory than previous approaches.

Equation 3.1 contains the basic five steps for computing an approximation "a" and a detail "d" coefficient from the even and odd input data samples, represented as f0 and g0, using one-level of DWT decomposition and the "sym4" wavelet basis function. g1, f1, and g2 are the intermediate results, stored in temporary registers, and C0–C7 are the eight constant filter coefficients [23].

$$\begin{aligned}
&\textbf{step}\,1 : g1(i) = g0(i) + C0^*f0(i) \\
&\textbf{step}\,2 : f1(i) = f0(i) + C1^*g1(i+1) + C2^*g1(i) \\
&\textbf{step}\,3 : g2(i) = g1(i) + C3^*f1(i) + C4^*f1(i-1) \\
&\textbf{step}\,4 : a(i) = f1(i) + C5^*g2(i) + C6^*g2(i-1) \\
&\textbf{step}\,5 : d(i) = g2(i) + C7^*a(i+1)
\end{aligned} \quad (9.1)$$

Since each step requires similar computational hardware, we can identify a "processing element" (PE) as the basic computational block for the DWT module. It consists of two signed multipliers and two signed adders, as shown in Figure 9.13. Since we are dealing with quantized integer representation of numbers, we need to make adjustments for the quantization of the filter coefficients at each step ("TRUNC"). These adjustments ensure that the inputs and outputs of a PE have the same number of bits and quantization level. On the basis of the analysis given in [23], 6 bits are chosen for the quantization of the multiplier coefficients, while we keep 10 bits for the signal input and wavelet output coefficients.

For the overall DWT architecture, a parallel implementation is looked at, as shown in Figure 9.14, where the windowing module buffers enough samples before the lifting operations take place. An overlapping window scheme (8 samples overlap in a 72 sample window) has been implemented, in order to avoid missing

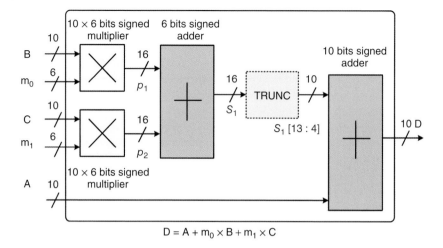

Figure 9.13 The main processing element for the lifting wavelet transform [24].

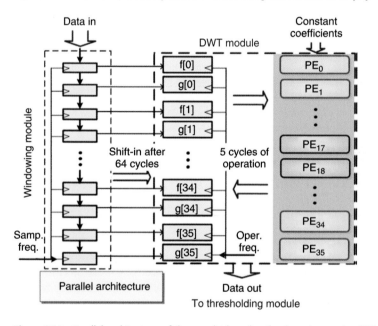

Figure 9.14 Parallel architecture of the wavelet-based spike detection engine [24].

spikes on the window edges. Since all necessary samples for computation of the five steps are ready, we need not consider the availability of the "future" ($i + 1$) samples in steps 2 and 5 of Equation 3.1. In-place computation is performed, where the number of registers is minimal – two for each sample, since the output of the PE is written back into one of the register inputs, for example, d, g2, and g1 overwrite g0, while a and f1 overwrite f0. It takes five clock cycles for the five steps to be computed. The remaining 64 − 5 = 59 cycles required for collecting all samples of the next window are used for other processing steps such as thresholding (32 cycles for comparing 32 approximation coefficients). Also, the cores can be reused to process the samples of other channels in a multichannel recording system. If different mother wavelets need to be used for different channels, the multiplier coefficients can be changed for each channel without incurring extra hardware overhead.

The second scheme [25] involves a sequential computation of the five lifting steps by folding them back onto the same PE. This requires less registers and the latency is also reduced. However, the five computational steps have to be skewed in order to wait for the "future" samples to be ready and this requires extra registers for storing the intermediate values. The actual computations in the ith clock cycle are given in Equation 3.2.

step 1 : $g1(i) = g0(i) + C0^*f0(i)$
step 2 : $f1(i − 1) = f0(i − 1) + C1^*g1(i) + C2^*g1(i − 1)$
step 3 : $g2(i − 1) = g1(i − 1) + C3^*f1(i − 1) + C4^*f1(i − 1)$
step 4 : $a(i − 1) = f1(i − 1) + C5^*g2(i − 1) + C6^*g2(i − 2)$
step 5 : $d(i − 2) = g2(i − 2) + C7^*a(i − 1)$ (9.2)

From Equation 3.2, we can observe that four values from previous computations are necessary for the current set of computations, namely, we need to store values of f0, g1, f1, g2. All the other values required for computation are available in the same cycle if the steps are followed in order. Register minimization techniques were applied. The structure of the control path is shown in Figure 9.15. It has been shown [25] that this scheme is suitable for multichannel and multilevel wavelet decomposition. For multichannel processing, the intermediate values for each channel can be stored in a memory array and proper care needs to be taken to load them in the registers prior to the same channel's next step of computations. It is worth noting that the PE needs to be operated at five times the clock speed in order to get one set of approximation and detail coefficients at the end of each cycle, resulting in higher clock speed and hence, higher power. However, the requirement of a large input buffer for windowing the signal and multiple PEs for processing all samples simultaneously is eliminated, resulting in huge area savings.

For most biological signal processing, the frequency range of interest and hence the sampling frequency are in the order of a few (∼10) kilo hertz. If we need to compute wavelet coefficients for 100 channels, the maximum clock frequency of operation can be estimated to be $(100 \times 5 \times 10\,kHz) = 5\,MHz$. The maximum delay path length has to be less than $1/(5\,MHz) = 0.2\,\mu s$, which is easily achievable in nanoscale technologies. The two architectures described above represent a trade-off between area and power dissipation. The second architecture, although area optimal, consumes higher power because of increased operating frequency, while the parallel architecture allows scaling of voltage (frequency) to achieve quadratic (linear) reduction in power. As a next step, we need to perform *thresholding* of

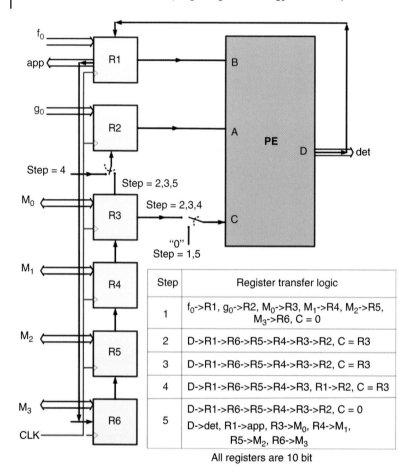

Figure 9.15 Register transfer logic for the sequential design [26].

the wavelet coefficients. The scheme for thresholding as well as the value of the threshold has to be chosen so as to optimize data compression, signal quality upon reconstruction as well as effectiveness of hardware realizability.

1) *Quantize*: Since we are dealing with quantized values, we can opt for a modified version of soft thresholding, in which we consider only the five most significant bits. However, the thresholds are constrained to powers of 2.
2) *Compare and threshold*: By using a comparator to identify coefficients that are larger than a fixed number, one can opt for any value of the threshold. The threshold value can be adaptively computed for each channel and assigned to a fixed value for future cycle.
3) *Compare and preserve*: In the previous scheme, the samples that are below a threshold and form part of a spike get eliminated. Hence, to get good reconstruction quality, we opt for a scheme that groups the "detected" coefficients

above threshold along with some buffered coefficients as a spike. For ease of hardware implementation, we consider a group of 32 coefficients to represent a single spike.

The three techniques were compared for signal reconstruction quality as shown in Figure 9.16. The signal reconstructed after thresholding is superimposed over the original signal. The samples were processed through the Verilog description of the hardware modules using a functional simulator (Synopsys VCS) and plotted using MATLAB. As observed from the figure, the last scheme gives the best reconstruction quality. The window number and location information are sent to the packetizer module and the wavelet coefficients are sent to the vocabulary module for further processing.

9.3.1.1 Vocabulary Module

The hardware implementation of the vocabulary scheme is challenging. It requires realization of two major tasks: (i) finding a match of a spike with existing alphabet symbols in the vocabulary and (ii) updating the vocabulary with new alphabet symbols. We implement the vocabulary scheme using an architecture that minimizes the area and power requirement. The vocabulary is realized using a bank of registers, which we refer to as *tables*. The incoming spike (in terms of the grouped approximation coefficients, or some extracted features) is compared with the existing symbols, in the table over multiple search and compare cycles. If a match is obtained, the corresponding address location is used to represent the alphabet symbol of the detected spike. If no match is found, the information is stored in a new location of the table. In case the vocabulary is full, a suitable update strategy is adopted, which will lead to less number of misses in the future. For this, one can store the frequencies of occurrence of each symbol (by using a simple counter) and employ a least frequently used (LFU) replacement scheme. However, an array of registers entails the requirement of significant area overhead. Hence, an alternate embedded super-threshold random access memory was looked at and compared with other architectures for the on-chip vocabulary module. With higher frequency of operation (much higher than the sampling frequency) along with extensive intelligent power gating (sleep transistors sized such that the memory cells do not lose the data) in the super-threshold domain to reduce standby leakage, the embedded static random access memory (SRAM) achieved a much lower area

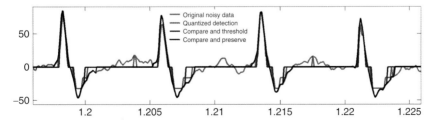

Figure 9.16 Comparison of various thresholding schemes implemented in hardware.

overhead and higher robustness of operation at almost iso-energy as compared to conventional subthreshold embedded memory [27].

As an example of hardware optimization to take advantage of the algorithm, the vocabulary search and update module can be implemented using a modified version of the hierarchical search-line-based content addressable memory (CAM) [28] architecture. Here, we implement a tunable "Distance Metric" based on equality of variable number of parameters with different precision, and use the "First Match" strategy by implementing a Priority Encoder and "binary encoding" strategy. By arranging the vocabulary tables as a memory array, we save large amount of area and power. Though the incorporation of matching logic increases the area by almost 2× per cell (compared to a 6-T SRAM cell), the overall savings in terms of power and area compared to an array of registers is nearly 10×. The data in the memory needs to be updated as new alphabet symbols are generated. Upon detection of a match between the incoming symbol and the stored data, it returns the memory address, which is essentially the alphabet symbol. Since the match does not need to be perfect, we use hierarchical search, where local search lines are used to search only a part of the array, which is selected on the basis of the duration of the spike and channel number. Also, since our alphabet level vocabulary needs to match with only some of the important coefficients, the matching is limited to certain rows/columns. In order to incorporate tunability of the distance metric, we propose to use the ternary CAM principle where certain bit line pairs are both assigned to logic "0" in order to signify do not care. By setting the less significant bits to do not care, the distance metric threshold can be varied easily. This optimization is particularly suited for the higher level vocabulary where the "words" and "sentences" can be symbolized by multiple lower level features, but even a partial matching of certain features symbolizes a similar burst or behavior, which can be encoded using the same symbol.

On the basis of the architecture of the wavelet engine and the vocabulary module, register-transfer level descriptions of the modules [29] were written using Verilog hardware description language. Each module was verified for correctness using functional simulation using prerecorded neural data from Aplysia. These modules were then synthesized using RTL Compiler from Cadence [30] and mapped to standard cell libraries at the TSMC 0.18 μm process technology node [31]. Next, the area, power, and critical path delay (that determines the maximum frequency of operation) were obtained using estimation tools integrated with RTL compiler platform. Table 9.1 enumerates the results for the synthesized modules. The power is dominated by the dynamic power, which is computed as $P = \alpha \times C \times V_{DD}^2 \times f$ (α is the average activity in the hardware, C is the switching capacitance, V_{DD} is the supply voltage, and f is the operating frequency). The power values reported in Table 9.1 are based on nominal V_{DD} of 1.8 V for the technology, worst-case switching activity, and a value of f determined by the critical path delay (17.3 ns for the wavelet engine and 10.1 ns for the vocabulary module).

Table 9.1 Area, power, and delay using the 0.18 μm technology ($V_{DD} = 1.8$ V).

Hardware block	Area (μm²)	Power (mW)			Delay (ns)
		Dynamic	Leakage	Combined	
Wavelet	55 089	11.165	0.000114	11.165	17.31
Vocabulary	808 028	44.321	0.001320	44.323	10.13
Combined	863 117	55.486	0.001434	55.488	27.44

9.3.2
Area, Power Reduction Methodologies

Miniature size and low energy consumption are two primary design requirements for implantable systems. Though the size of an implantable system is still limited by the size of battery, antenna, and off-chip components, the area of the signal processor can be significant with increasing requirement for on-chip processing of signals from multiple channels. While nanoscale (sub 100 nm) technologies provide high integration density, faster switching speed, and lower switching power per transition, they also bring new challenges. These include exponential increase in leakage current [4] and reliability issues due to process variations (PVs) as well as temporal parameter variations due to temperature and voltage fluctuations (collectively termed as *P-V-T variations*) [5]. Hence, in order to use nanoscale technologies for neural signal-processing electronics, we need to address the power and parameter variation induced reliability issues.

To decrease power consumption, one can use popular techniques such as *clock gating* and *supply voltage gating*. Clock gating saves dynamic power in the clock line that drives the sequential elements such as flip-flops and latches. Supply gating saves leakage power due to the stacking effect [4]. Both approaches require identification of idle cycles for a logic block during which it can be "gated." While computation of wavelet coefficients for one channel is taking place, the registers that are holding intermediate values for other channels are clock-gated. Similarly, blocks that are idle for sufficiently long and predefined periods of time are supply-gated to reduce leakage power as well as redundant switching power. For instance, the PEs in the parallel architecture, which are idle for many clock periods when they are waiting for the samples to be shifted-in, can be supply-gated during this pre-defined idle time. It is interesting to note that control circuitry overhead for the various gating schemes can be minimized by proper choice of algorithm and architecture. In our design, since the idle times of different blocks are determined during design time, the counter used for collecting data samples can be used to trigger SLEEP/ACTIVE modes in different modules. To take care of the time required for the virtual ground node to discharge, the SLEEP-to-ACTIVE transition should be triggered at least one clock cycle before the module needs to be active.

Another scheme to reduce dynamic power is to run the design at a slower clock frequency so that the processing is complete just in time to receive new data from the data collection registers. This allows us to perform *extensive supply voltage scaling* for further power reduction. Owing to the quadratic dependence of power on supply voltage, voltage scaling [32] has emerged as a popular low-power design approach. To achieve ultralow power operation, supply voltage can be scaled below the transistor threshold voltage, when the circuit starts operating in the subthreshold region. This leads to huge reduction in power dissipation with a corresponding increase in path delays, leading to low frequency operation. Subthreshold design [33, 34] is, therefore, more suited for low-sampling rate applications such as biomedical signal processing. However, it comes at the cost of increased vulnerability of a design to variation-induced failures leading to reduced reliability and high yield loss. At scaled voltage, digital circuits can suffer from functional failure due to variations in circuit delays. For example, in low-power DWT hardware, the time required for computation of the most significant approximation coefficients can be affected because of voltage scaling, causing the clocked storage elements (flip-flops) to latch wrong logic values, leading to loss in reliability.

Since the sampling frequency is only 25 kHz for our application, we can scale the frequency significantly, while still maintaining the real-time processing capability of the system. The interval between sampling two data points is 80 μs, which determines the time bound for processing time of the hardware. Hence, the required frequency of operation for wavelet and packetizer modules is only 12.5 kHz. Along with frequency scaling, it is also possible to incorporate commensurate scaling of the supply voltage (V_{DD}). Due to quadratic dependence of dynamic power on V_{DD}, supply voltage scaling results in large savings in dynamic power. Table 9.2 presents the revised power results obtained by scaling the operating frequency to 12.5 kHz and supply voltage to 1.2 V based on the critical path delay requirement in our application. Note that the power required for a single channel is reduced significantly, allowing the scheme to be extended to multichannel applications. We can use additional power gating to prevent leakage current from draining the battery during long periods of idle operation.

Table 9.2 Total power dissipation at 0.18 μm technology after voltage and frequency scaling ($V_{DD} = 1.2$ V, $f = 12.5$ kHz).

Block	1 channel (mW)	64 channels (mW)
Wavelet	0.00107	0.0687
Vocabulary	0.00248	0.1596
Combined	0.00355	0.2283

9.3.2.1 Subthreshold versus Super-Threshold Operation

Subthreshold design appears as an obvious choice for bioimplantable circuits because of its ultralow power operation at extremely low frequencies [1]. However, there are trade-offs between subthreshold and conventional super-threshold designs, depending on the particular circuit application. The latter can exploit the nature of the neural signal-processing algorithm to employ extensive power gating (clock and supply gating) to drastically reduce power. There is an optimal super-threshold voltage and frequency of operation where power gating can be used to match the power dissipation of its subthreshold counterpart. The super-threshold implementation also provides better robustness of operation and parametric yield under large PVs.

To evaluate the two approaches, the parallel implementation of the DWT architecture is chosen, where the windowing module buffers enough samples before the lifting operations take place. It takes five clock cycles for the five PE steps to be computed. The remaining (64 − 5 = 59) cycles, required for collecting all samples of the next window, are used for other processing steps such as thresholding (32 cycles for comparing 32 approximation coefficients). Considering a sampling frequency of 10 kHz, a new data sample for each channel arrives every 100 μs. If we consider 20 channels with time-multiplexed operation, the five-cycle DWT operations in each PE need to be complete in 5 μs, giving a minimum operating frequency target of 1 MHz. The remaining operations overlap with the computation of wavelet coefficients for other channels.

On one end of the spectrum, in the subthreshold domain (supply voltage V_{DD} less than transistor threshold voltage in the particular technology node), one can operate the PEs at 1 MHz operation and scale the supply voltage to 100 mV. Here, we reduce the dynamic power required for computation by reducing the voltage swing, but it leaves no idle time for supply gating. On the other hand, we can perform the computations at a faster rate of, say, 1 GHz with a supply voltage of 1 V and then "gate" the idle blocks for the remainder of the 5 μs period, thus reducing the leakage power through the blocks, at the cost of slight increase in dynamic power. The timing diagram for the five steps of the wavelet transform, shown in Figure 9.17, demonstrates the increased opportunity for power gating when operating at super-threshold voltage at high frequency, as opposed to the subthreshold operation at ultralow frequency. A simulation study was performed regarding the energy efficiency of the main PE of the DWT algorithm, over the five cycles (5 μs) with the scaling of supply voltage. The delay variation was also taken into consideration in the energy delay product (EDP). Simulations were performed in HSPICE at the 70 nm technology node [35]. Previous works on subthreshold digital circuits [1] have recommended avoiding gates with large stacks, pass transistors, and transmission gates, in order to avoid decrease in the effective drive strength of each device and to minimize sneak leakage paths. The gates should also be properly sized to ensure correct operation under such low

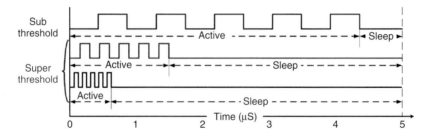

Figure 9.17 Timing diagram for executing the five steps of lifting wavelet transform. The steps can be performed at very low frequency and voltage (subthreshold). Alternatively, it can be completed in a shorter time by operating at higher frequency in super-threshold mode, which provides the opportunity for more power gating in the longer idle time [24].

supply voltage condition. The total energy of the circuit can be represented as

$$E_{tot} = P_{tot} \times T_{tot}$$
$$= P_{dyn} \times T_{active} + P_{leak} \times T_{tot} \qquad (9.3)$$

where the dynamic power P_{dyn}, total time T_{tot}, and active time T_{active} are given by

$$P_{dyn} = C_{tot} \times V_{DD}^2 \times f \qquad (9.4)$$

$$T_{tot} = T_{active} + T_{idle} \qquad (9.5)$$

$$T_{active} = \frac{1}{f} \times N_{op} \qquad (9.6)$$

In the above equations, C_{tot} is the total capacitance, V_{DD} is the supply voltage, f is the frequency, and N_{op} is the number of active cycles within the time period T_{tot}, after which the idle time T_{idle} begins.

The effect of voltage scaling on total energy consumption, critical path delay, and EDP is shown in Figure 9.18. It should be noted that the minimum energy point (see Figure 9.18a) has a much lower supply voltage than the point with the minimum EDP (see Figure 9.18c). Taking 0.6 V as the nominal super-threshold voltage, we increased the idle time interval for power gating by increasing the operating frequency. As seen from Figure 9.18e,f, the leakage power P_{leak} is considerably lower for a gated design than for a non-gated design and is independent of the frequency of operation, whereas the dynamic power P_{dyn} increases with the frequency. However, as f increases, the active time T_{active} decreases while the idle time T_{idle} increases. Since the leakage power is reduced only during the idle period, the total energy decreases considerably for a gated design at high frequencies. The above equations also explain why the total energy is independent of frequency for a non-gated design. For a power-gated design, at very low frequencies, there is almost no idle time and the power overhead due to the switching of the supply gating (or sleep) transistor can be considerable. As the frequency is increased, the energy consumption decreases considerably till it saturates. Hence, we can obtain comparable energy

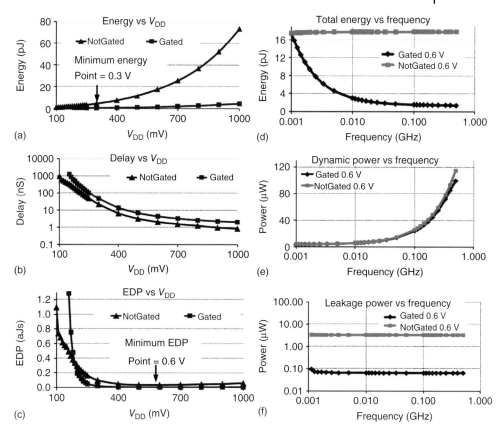

Figure 9.18 (a) Total energy, (b) delay, and (c) EDP of a PE with voltage scaling. Effect of increasing frequency on (d) energy consumption, (e) dynamic power, and (f) leakage power of a PE. The extra savings in energy at higher frequency can be attributed to the increase in idle period during which supply gating can be applied to decrease leakage power, even though switching power increases with frequency [24].

savings for subthreshold operation at ultralow frequency and super-threshold operation at higher frequencies with extensive power gating applied. We also performed simulations with power gating applied to the subthreshold design, but the functionality of the circuit was hampered at extreme low voltages (~0.10 V) and the delay target was violated at slightly higher voltages (~0.15 V).

9.3.3
Impact of Process Variations on Yield

Since both subthreshold design and super-threshold design with power gating have similar energy consumption, one can employ either technique to implement the neural signal-processing algorithm. However, at nanoscale technologies, PVs can cause wide variations in circuit parameters such as critical path delay, leading to delay failures. The effects of PVs are exacerbated in designs with low-power techniques applied, especially in the subthreshold region because of the exponential change in subthreshold drive current with change in V_{th} [36]. An analysis was performed of how PVs (modeled as V_{th} variations) impact critical path delay of a PE at different supply voltages, with and without supply gating transistors (which are kept on during Active mode, but still affect performance). The sleep transistors were sized to have maximum power savings, allowing up to 30% performance degradation under nominal conditions. The increase in critical path delay, T_{crit} (normalized with respect to delay of non-gated design at same supply voltage) is plotted for increasing V_{th} variations in Figure 9.19. The delay values of the three cases under nominal process conditions are as follows:

1) Super-threshold NotGated Design at 0.6 V: = 3.34 ns
2) Super-threshold Gated Design at 0.6 V: = 4.39 ns
3) Subthreshold NotGated Design at 0.15 V: = 732.40 ns.

It is observed that under large PVs, the delay of a subthreshold design can increase by up to 3.5× of that under nominal conditions, causing significant loss in parametric yield, while a super-threshold design with supply gating has minimal impact under variation. To compute the yield under PVs, Monte Carlo simulations were performed for both gated and non-gated versions of the PE in HSPICE using 30% inter-die and 25% random intra-die variations. The resultant delay distribution for 10 000 dies for a super-threshold (0.6 V) gated design and a subthreshold (0.15 V) non-gated design are shown in Figure 9.20. Given a delay target of 1 μs, the yield in the case of the subthreshold design is 79.94%, while all super-threshold designs pass the delay target of 30 ns (corresponding to a target operating frequency of 33.33 MHz). However, some of the low V_{th} dies might have high leakage current, which can cause them to exceed the energy budget under extreme variations because of more active leakage, even though they are gated for the same idle period. It is worth noting that control circuitry overhead for the various gating schemes can be minimized by proper choice of algorithm and architecture. In our design, since the idle times of different blocks are determined at design time, the counter used for triggering collection of different data samples

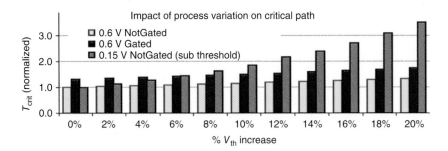

Figure 9.19 Effect of process variation on normalized critical path delay of the subthreshold design and super-threshold design with and without supply gating [24].

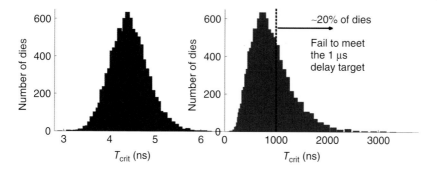

Figure 9.20 Yield results for (a) super-threshold (0.6 V gated) and (b) subthreshold (0.15 V non-gated) designs [24].

in a window is used to trigger Sleep/Active modes in different modules. The NMOS sleep transistor sizing was performed so that the virtual ground node floats to a high enough value during Sleep mode, without causing huge performance overhead during Active mode. To take care of the time required for the virtual ground node to discharge, the Sleep-to-Active transition is triggered one clock cycle before the module needs to be active. The energy consumption overhead when the sleep transistor switches on/off was also taken into account when computing the total energy.

9.3.3.1 Preferential Design

Choice of super-threshold design provides a higher delay margin and reduced vulnerability to PVs while use of power gating gives similar energy reduction as subthreshold operation. Moreover, it opens avenues for architecture-level design optimization techniques. In this subsection, we propose a novel preferential design approach, which exploits the nature of the algorithm to achieve further reduction in area and power while maintaining robustness.

In a preferential design scheme, we first identify the critical PEs that contribute more to the output signal quality than others. These critical PEs need to be designed conservatively to provide them larger delay margin, thus ensuring their robust operation. On the other hand, the noncritical PEs can be designed to reduce area and power. These PEs can experience delay failure when the supply voltage is scaled aggressively to reduce power. However, by ensuring that the failure is always confined to noncritical PEs, we achieve graceful degradation in output quality. We note that conservative design of the critical PEs and aggressive design of the noncritical ones allows significant area reductions over the worst-case design approach, while achieving better output signal quality than area-optimal or nominal design [24]. The skewing of delay margins across critical and noncritical components can be realized by using a constraint-driven logic synthesis or gate-sizing approach. The proposed design paradigm can be applied hierarchically at different levels. Along with the application of different timing margins to different PEs, one can assign different timing margins to different output bits inside a

PE. Since the most significant bits (MSBs) of a PE contribute more toward the output quality compared to the least significant bits (LSBs), we can assign higher margin to the MSBs by upsizing gates or restructuring logic. In this way, we can confine the delay failures to the LSBs of the least significant components, thus ensuring minimal impact on output with voltage scaling or parameter variations. The preferential design paradigm, applied to the parallel DWT architecture, is shown in Figure 9.21. It is to be noted that the proposed preferential design approach can be used for other signal-processing blocks such as spike sorting and burst recognition, where we can isolate the critical computing blocks from noncritical ones.

The main PE of the DWT was synthesized using Synopsys Design Compiler and LEDA 250 nm standard-cell library, under different delay constraints to generate eight different gate-level netlists with increasing area and power values and decreasing maximum path delay values. The variation in energy with different delay constraints for two values of supply voltages is shown in Figure 9.22. To implement the preferential design, the 36 instances of the PE were chosen with

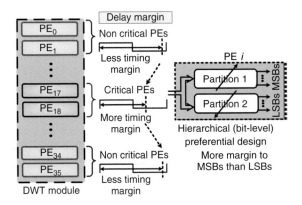

Figure 9.21 Proposed hierarchical preferential design approach for low power and robust neural signal-processing system. With voltage scaling and parameter variations, the system undergoes graceful degradation in performance [24].

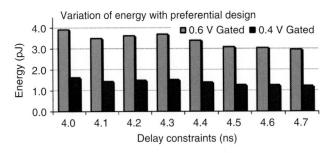

Figure 9.22 Preferential design for low power: variation in energy at different supply voltages for different versions of the same design [24].

different timing margins, with the more critical blocks (the ones in the middle are more critical compared to the ones at the edges, because of overlapping windows) assigned with maximum timing margin (see Figure 9.21). It is observed that the preferential design has similar area and energy as the nominal design case [24].

To investigate the effect on the signal quality, we used a 5 s train of noise-free extracellular neural spikes (Hodgkin Huxley model) sampled at 10 kHz with an amplitude range of +75 to −42 µV (using 8-bit signed quantization) and simulated the hardware-emulating algorithm in MATLAB. The impact of voltage scaling and PVs is simulated by introducing uniform random noise ((−0.5 to +0.5)* V_{noise}) at each step of computation through the PE. The value of V_{noise} depends on the level of voltage scaling and degree of PV. The signal reconstructed after spike detection and vocabulary creation in the nominal case is considered as the reference signal s and the reconstructed signal in all other cases is considered as signal + noise ($sn = s + n$). The quality degradation due to the different cases is computed as follows:

$$\text{Qual} = 10^* \log 10 \left(\frac{\left(\sum s(i)^{\wedge} 2 \right)}{\left(\sum (sn(i) - n(i))^{\wedge} 2 \right)} \right)$$

where i is from 1 to T, T being the number of samples in the signal.

For four different cases, namely, Case A (area-optimal), Case C (nominal design), Case E (signal quality optimal), and Case F (preferential design), the degradation

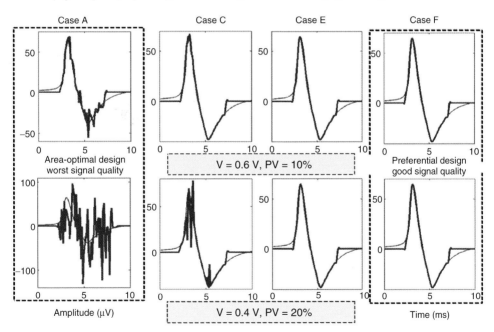

Figure 9.23 Spike reconstruction quality for different design methodologies. Case F (preferential design) gives much better quality under voltage scaling and process variations compared to Case C (nominal design) or Case A (area-optimal design) [24].

in signal quality can be observed in Figure 9.23, where one randomly chosen reconstructed neural spike (superimposed over the original signal in red) is shown for $V = 0.6\,\text{V}$ and PV of 10%, as well as for the case where $V = 0.4\,\text{V}$, PV = 20%. From the results in [24], it is inferred that Case A is poor in terms of signal quality, Case E is poor in terms of area, while preferential design (Case F) has better signal quality compared to the nominal (Case C) at almost iso-area and iso-power.

Hence, we can use preferential design technique to achieve better output quality under iso-area and iso-power as the nominal design, under process variations. On the other hand, it can be used to achieve lower power by further voltage scaling under iso-quality as the nominal design. Along these lines, it is observed from [24] that the value of signal quality is similar for preferential design at $V_{DD} = 0.4\,\text{V}$ and PV of 20% and nominal design at $V_{DD} = 0.6\,\text{V}$ and PV of 10%. Besides, a bit level preferential design can also be applied hierarchically to the parallel architecture by designing the MSBs of the less critical PEs to be more robust than the LSBs, using delay constrained logic synthesis.

9.3.4
Overall Design Flow

For neural signal-processing hardware implementation, the co-design of algorithm, architecture, and circuit is important to meet the design requirements of ultralow power and miniature area. As seen earlier, a super-threshold design at optimal voltage/frequency can be equally energy efficient as a subthreshold implementation, when extensive power gating is employed by using the idle time information from

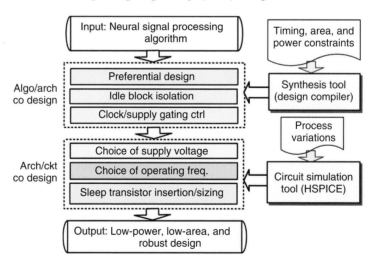

Figure 9.24 Overall design flow for ultralow-power and robust signal processing in implantable biomedical systems. Algorithm/architecture/circuit co-design helps in achieving ultralow-power operation while maintaining robustness and yield under increasing process variations.

the algorithm. A *preferential design* approach can be used to achieve graceful degradation of robustness, allowing the application of power reduction techniques such as voltage scaling in the super-threshold domain. These design steps can be integrated into an automatic synthesis flow applicable to most biomedical signal-processing algorithms, as shown in Figure 9.24. The design flow also considers the impact of process variations on the yield and robustness of a low-power design.

Such an integrated design flow can be very effective in satisfying the design constraints of implantable biomedical systems. Here, we evaluate the overall impact of the proposed design flow on the neural signal-processing algorithm. The variation in total energy with different power reduction schemes is shown in

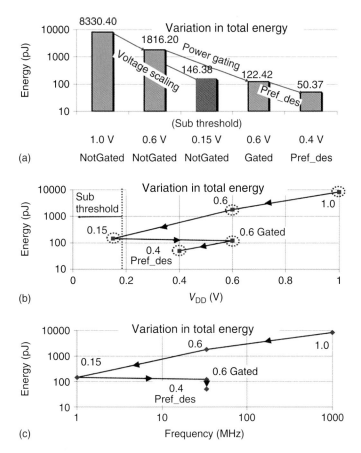

Figure 9.25 (a) Variation in total energy with different power reduction schemes, along with corresponding changes in (b) V_{DD}, and (c) frequency. Using extensive power gating in super-threshold design, we can achieve similar power reduction as a subthreshold design while operating at a higher supply voltage and frequency. Additional energy savings at iso-quality (under parameter variations) can be achieved by employing a preferential design approach, which allows further voltage scaling at the same frequency.

Figure 9.25a. It can be readily observed that the total energy reduces with voltage scaling, and subthreshold (0.15 V) operation achieves orders of magnitude lower energy consumption compared to super-threshold (1.0 V) operation. However, super-threshold operation at an intermediate optimal voltage (0.6 V) can achieve similar reduction in energy by taking advantage of the power-gating opportunities in the design. The variation in total energy with supply voltage scaling and commensurate frequency scaling are shown in Figure 9.25b,c, respectively. Finally, we can see that application of preferential design allows further voltage scaling (to 0.4 V) with decrease in total energy at the same frequency while maintaining iso-quality under extreme process variations.

Finally, we note that although, we considered 20 channels of data with time-multiplexed sharing of the PEs, we can easily extend the analysis to larger number of channels. Increasing the number of channels requires increasing the operating frequency, which, in turn, would require designing the circuits with tighter delay constraints leading to slight increase in area and power. The comparatively high operating frequency can be obtained on-chip by using ultralow power relaxation oscillators or by extracting the clock from the carrier frequency used for wireless transmission.

Recently, researchers have articulated security concerns about implantable medical devices that use wireless communication protocols [37]. The lack of

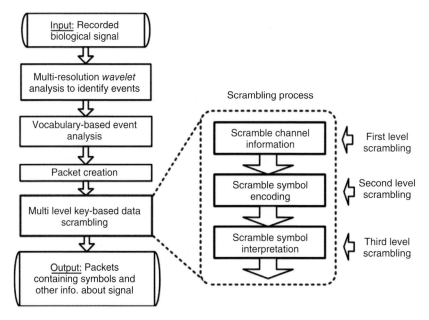

Figure 9.26 Main steps for a biological signal-processing algorithm. The inclusion of security as a part of the main flow enables us to use low-overhead design approaches to achieve security against possible attacks. The key-based scrambling for data obfuscation can be performed at three levels to achieve higher protection against snooping attacks [38].

authentication and integrity mechanisms put patients at risk from attack by anyone with a transceiver. However, data protection in implantable medical systems needs to come at extremely low power and area overhead, which is difficult to achieve by using conventional data encryption solutions.

We propose a novel low-cost multilevel key-based obfuscation technique [38] that exploits the nature of the recorded biological signal to protect it from possible attacks. The main steps of the algorithm are shown in Figure 9.26.

9.4
Summary

In this chapter, we have presented an integrated design methodology for low-power, robust, and area-efficient implantable neural signal-processing hardware. The hardware implementation of neural pattern recognition can exploit the benefits of nanoscale technologies to minimize implant size and power. Such an integrative algorithm–hardware design approach can be effective for closed-loop neural control for neural prosthesis, involving analysis of increasingly large number of channels. In order to achieve the low-power and low-area requirements of such applications, we can exploit the nature of the algorithm as well as signal acquisition characteristics. Although subthreshold design appears to be the natural choice for hardware implementation of such algorithms, a well-optimized super-threshold design can achieve comparable energy efficiency while maintaining significantly higher robustness and yield. This is possible by employing extensive clock and supply gating in the design that leverages the speed-gap between slow signal acquisition frequency and fast super-threshold operation. Using an example design for the wavelet transform engine for the neural signal-processing algorithm, we have shown that an optimal operating voltage and frequency can be derived that minimizes the power dissipation in super-threshold region.

The proposed methodology also exploits the nature of the signal-processing algorithm to isolate performance-critical computing blocks and preferentially assign more delay margin to the critical components. This allows further voltage scaling under iso-signal quality as the nominal design scenario, while improving robustness of operation under process variations. Such design approaches can be seamlessly integrated into an automatic design synthesis framework for biomedical signal-processing applications.

References

1. Chandrakasan, P., Verma, N., and Daly, D.C. (2008) *Annu. Rev. Biomed. Engg.*, **10**, 247–274.
2. Allan, R. (2003) *Electron. Des.*, **1**, 41–46.
3. Chandler, R.J., Gibson, S., Karkare, V., Farshchi, S., Markovic, D., and Judy, J.W. (2009) Annual International Conference of the IEEE Engineering in Medicine and Biology Society, pp. 5525–5530.
4. Roy, K., Mukhopadhyay, S., and Mahmoodi-Meimand, H. (2003) *Proc. IEEE*, **91**, 305–327.

5. Borkar, S., Karnik, T., Narendra, S., Tschanz, J., Keshavarzi, A., and De, V. (2003) Proceedings of the 40th Annual Design Automation Conference, pp. 338–342.
6. Wise, K.D., Anderson, D.J., Hetke, J.F., Kipke, D.R., and Najafi, K. (2004) *Proc. IEEE*, **92**, 76–97.
7. Harrison, R.R. (2003) Proceedings of the 25th Annual International Conference of the IEEE Engineering in Medicine and Biology Society, pp. 3325–3328.
8. Sawan, M., Yamu, H., and Coulombe, J. (2005) *IEEE Circuits Syst. Mag.*, **5** (1), 21–39.
9. Black, M.J., Bienenstock, E., Donoghue, J.P., Serruya, M., Wu, W., and Gao, Y. (2003) First International IEEE EMBS Conference on Neural Engineering, pp. 580–583.
10. Narasimhan, S., Cullins, M., Chiel, H.J., and Bhunia, S. (2008) 30th Annual International Conference of the IEEE Engineering in Medicine and Biology Society, pp. 38–41.
11. Olsson, R.H. III, and Wise, K.D. (2005) *IEEE J. Solid-State Circuits*, **40**, 2796–2804.
12. Chae, M.S., Yang, Z., Yuce, M.R., Hoang, L., and Liu, W. (2009) *IEEE Trans. Neural Syst. Rehabil. Eng.*, **17**, 312–321.
13. Nicolelis, A. (2001) *Nature*, **409**, 403–407.
14. Ye, H., Morton, D.W., and Chiel, H.J. (2006) *J. Neurosci.*, **26**, 1470–1485.
15. Bean, B.P. (2007) *Nat. Rev.: Neurosci.*, **8**, 451–465.
16. Gosselin, B., Ayoub, A.E., Roy, J.-F., Sawan, M., Lepore, F., Chaudhuri, A., and Guitton, D. (2009) *IEEE Trans. Biomed. Circuits Syst.*, **3**, 129–141.
17. Strang, G. and Nguyen, T. (1996) *Wavelets and Filter Banks*, Wellesley Cambridge Press.
18. Mallat, S. (1999) *A Wavelet Tour of Signal Processing*, Academic Press.
19. Rao, R. and Bopardikar, A. (1998) *Wavelet Transforms: Introduction to Theory and Applications*, Addison-Wesley, Reading, MA.
20. MathWork Matlab Wavelet Toolbox, Version 2.1, Mathworks, Natick, MA, http://www.mathworks.com/products/wavelet (accessed 6 September 2013).
21. Kim, K.H. and Kim, S.J. (2003) *IEEE Trans. Biomed. Eng.*, **50**, 999–1011.
22. Hu, X., Wang, X.X., and Ren, X.M. (2005) *Comput. Meth. Programs Biomed.*, **79** (3), 189–195.
23. Daubechies, I. (1992) *Ten Lectures on Wavelets*, SIAM, Philadelphia, PA.
24. Narasimhan, S., Chiel, H.J., and Bhunia, S. (2011) *IEEE Trans. Biomed. Circuits Syst.*, **5**, 169–178.
25. Oweiss, K.G., Mason, A., Suhail, Y., Kamboh, A.M., and Thomson, K.E. (2007) *IEEE Trans. Circuits Syst.*, **54**, 1266–1278.
26. Narasimhan, S., Chiel, H.J., and Bhunia, S. (2009) Annual International Conference of the IEEE Engineering in Medicine and Biology Society, pp. 6383–6386.
27. Hashemian, M.S. and Bhunia, S. (2013) Proceedings of 2013 26th International Conference on VLSI Design, pp. 66–71.
28. Pagiamtzis, K. and Sheikholeslami, A. (2006) *IEEE J. Solid-State Circuits*, **41**, 712–727.
29. Narasimhan, S., Zhou, Y., Chiel, H.J., and Bhunia, S. (2007) Biomedical Circuits and Systems Conference, pp. 134–137.
30. Cadence RTL Compiler, Cadence, http://www.cadence.com/products/digital_ic/rtl_compiler (accessed 6 September 2013).
31. The Mosis Service TSMC 0.25 um and 0.18 um Technology, http://www.mosis.org/products/fab/vendors/tsmc (accessed 6 September 2013).
32. Gonzalez, R., Gordon, B.M., and Horowitz, M.A. (1997) *IEEE J. Solid-State Circuits*, **32**, 1210–1216.
33. Calhoun, B.H. and Chandrakasan, A. (2004) Proceedings of the International Symposium on Low Power Electronics and Design, pp. 90–95.
34. Raychowdhury, A., Paul, B.C., Bhunia, S., and Roy, K. (2005) *IEEE Trans. Very Large Scale Integr. VLSI Syst.*, **13**, 1213–1224.
35. Predictive Technology Model http://www.eas.asu.edu/~ptm/ (accessed 6 September 2013).

36. Zhai, B., Hanson, S., Blaauw, D., and Sylvester, D. (2005) Proceedings of the 2005 International Symposium on Low Power Electronics and Design, pp. 20–25.
37. Stanford, V. (2002) *IEEE Pervasive Comput.*, **1**, 8–12.
38. Narasimhan, S., Wang, X., and Bhunia, S. (2010) Annual International Conference of the IEEE Engineering in Medicine and Biology Society.

10
Implantable CMOS Imaging Devices
Jun Ohta

10.1
Introduction

Charge-coupled device (CCD) image sensors are conventionally used in consumer, industrial, and scientific cameras. Recently, a complementary metal-oxide semiconductor (CMOS) image sensor has emerged and its rapid growth has opened a wide variety of applications such as in entertainment, automobiles, security/surveillance, biometrics, and biomedical technology in addition to the applications in which a CCD imager has been used [1]. The performance of CMOS image sensors has been continuously improving and it is now comparable to that of CCD image sensors. The fabrication process for CMOS image sensors is based on conventional CMOS technology, which makes it relatively easy to implement state-of-the-art ultrafine CMOS fabrication techniques. In contrast, CCD image sensors are fabricated using specially customized methods, making it difficult to introduce such state-of-the-art techniques. In addition, CMOS image sensors have advantages over CCD image sensors in terms of their on-chip integration of functional circuits, low power consumption, and multimodal functions, all of which are derived from CMOS technology. Such features are especially suitable for biomedical applications, as shown in Figure 10.1.

Biomedical devices that use image sensors can be classified into three configurations, as shown in Figure 10.2. The first type is a conventional configuration, such as in an optical microscope, where an image sensor is attached to the optics of the microscope (Figure 10.2a). Conventional sensors can be used in this configuration. An example of this type is a fluorescent microscope equipped with a high-speed CMOS image sensor for measuring neural activities using a voltage-sensitive dye (VSD) [2]. Another example is an ultra-high-speed CMOS image sensor with a nanosecond response for fluorescent lifetime imaging microscopy (FLIM) [3]. The second type of device is a hermetic device, in which an image sensor is installed in an enclosure. A capsule endoscope [4] is a typical example of this type of device, as illustrated in Figure 10.2b. In this configuration, the image sensor is required to be small enough to fit into small enclosures, such as a capsule that can be swallowed. Recently, a third type of device has emerged, in which biomaterials or

Implantable Bioelectronics, First Edition. Edited by Evgeny Katz.
© 2014 Wiley-VCH Verlag GmbH & Co. KGaA. Published 2014 by Wiley-VCH Verlag GmbH & Co. KGaA.

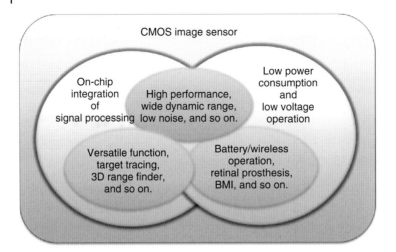

Figure 10.1 Features of CMOS image sensors.

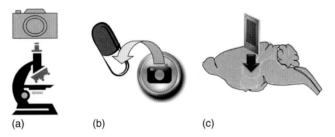

Figure 10.2 Three types of biomedical devices using image sensors: (a) conventional type, (b) enclosure type, and (c) direct implantation type.

living tissues containing fluorophores are in direct contact with the surface of an image sensor, which allows more compact measurement systems to be realized [5]. Such a compact system can also be implanted into a living body, as shown in Figure 10.2c. The implantation of an imaging system also allows for new clinical device applications. A typical example of such a clinical application of this type is a retinal prosthesis [6], which is an imaging device implanted into the eye.

In developing such biomedical devices, several challenges are faced, as shown in Figure 10.3 [5]. The surface of the device can affect living tissue or cells, and these effects can include cytotoxicity. For example, in order to implant a chip, its surface must be treated (for example, by coating it with Parylene) in order to prevent packaging materials and/or electrode materials dissolving into the surrounding tissue and cells. In addition, during stimulation, electrochemical reactions can occur when the stimulation voltage exceeds the voltage window, and pH changes and/or bubbling can have harmful effects on neural cells [7]. Moreover,

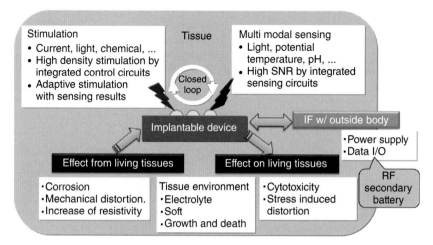

Figure 10.3 Advantages and challenges of implantable biomedical devices. (Reproduced from Ref. [5] with permission.)

devices that come into contact with living materials are also affected by the living environment, which is composed primarily of saline solution, and so a CMOS chip with no coating may get damaged. Therefore, highly watertight packaging, such as Parylene, is necessary. Of course, this material must be biocompatible. An implanted device is stressed by living tissue, and therefore, the device may get deformed or broken. For example, although a thin CMOS chip can easily be implanted, such a chip may break owing to stress from the tissue because thin Si is fragile. Living tissue may grow and die, and the configuration between the device and the tissue may change gradually. This may cause impedance changes between the electrode and living cells. The implanted devices must be designed recognizing that the configuration of the device may change.

Power supply is also an important issue. Some devices use wireless transmission of power electromagnetically and/or optically, while some others have a battery installed inside the body [8]. In all of these cases, the total power supply is very limited. Biofuel cell technology promises to alleviate this issue [9], although it is not yet ready for implantation.

This chapter describes the fundamentals and applications of such implantable CMOS imaging devices. In Section 10.2, the fundamentals of CMOS image sensors are presented, and the sensor structure and characteristics are explained, including its photosensing and noise reduction properties. Two typical examples of implantable CMOS imaging devices are shown in Sections 10.3 and 10.4. One is an artificial retinal device, and the other is a brain-implantable imaging device. Finally, in Section 10.5, this work is summarized, and the future direction of implantable CMOS imaging devices is discussed.

10.2
Fundamentals of CMOS Imaging Devices

10.2.1
Photosensors

A photodiode (PD) is a basic photosensor consisting of a pn-junction diode. The operating principle of a PD is quite simple. In a pn-junction diode, the forward current I_F is expressed as

$$I_F = I_s \left[\exp\left(\frac{eV}{nk_B T} \right) - 1 \right] \quad (10.1)$$

where e is the electron charge, n is an ideal factor, I_s is the saturation current or diffusion current, and k_B and T are the Boltzmann constant and absolute temperature, respectively. Light incident on the PD produces a photocurrent I_{ph}, which is expressed as

$$I_{ph} = R_{ph} P \quad (10.2)$$

where R_{ph} is the photosensitivity and P is the incident light power. The maximum photosensitivity $R_{ph,max}$, where the quantum efficiency is 100%, is determined as

$$R_{ph,max} = \frac{e\lambda}{hc} \quad (10.3)$$

where e, λ, h, and c are the electron charge, the wavelength of the incident light, Planck's constant, and the speed of light in vacuum, respectively. The total current from the pn-junction PD I_L under bright conditions is expressed as follows:

$$I_L = I_{ph} - I_F = I_{ph} - I_s \left[\exp\left(\frac{eV}{nk_B T} \right) - 1 \right] \quad (10.4)$$

Figure 10.4 illustrates I–V curves of a pn-PD under dark and illuminated conditions. There are three modes for the bias conditions: solar cell mode, PD mode, and avalanche mode, as shown in Figure 10.4.

In solar cell mode, no bias voltage is applied to the PD. Under light illumination, the PD acts as a battery; that is, it produces a voltage across the pn-junction. In Figure 10.4, the open-circuit voltage V_{oc} is shown. In the open-circuit condition, the voltage V_{oc} can be obtained from $I_L = 0$ in Equation (10.4), and thus,

$$V_{oc} = \frac{k_B T}{e} \ln\left(\frac{I_{ph}}{I_s} + 1 \right) \quad (10.5)$$

This shows that the open circuit voltage does not linearly increase with input light intensity.

The second mode is the PD mode. When a PD is reverse biased (that is, $V < 0$), the exponential term in Equation (10.4) can be neglected, and thus, I_L becomes

$$I_L \approx I_{ph} + I_s \quad (10.6)$$

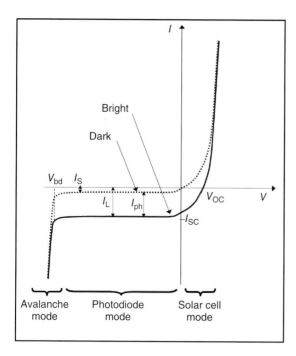

Figure 10.4 Current–voltage curve for a photodiode.

This shows that the output current of the PD is equal to the sum of the photocurrent and diffusion current. Thus, the photocurrent linearly increases with input light intensity.

The third mode is the avalanche mode, where a reverse bias voltage greater than the breakdown voltage V_{bd} is applied. In avalanche mode, the amplification of photo-generated carriers occurs with a fast response. Thus, the mode is used in such applications as low-light detection and high-speed optical communications. Since the gain is not uniform over a large area, avalanche mode has not yet been used in image sensors. It is, however, used in CMOS image sensors in the form of a single-photon avalanche diode (SPAD), which is discussed later.

10.2.2
Active Pixel Sensor

A PD in a CMOS image sensor is usually operated in accumulation mode or integration mode [1]. In this mode, weak light signals are detectable. It should be noted that the pixel size in a CMOS image sensor is generally small, for example, a few square micrometers. Thus, the photocurrent from a single pixel is so small that it is difficult to detect within a large number of pixels. The accumulation time or integration time in a conventional image sensor is set to about 30 ms, which the human visual system cannot detect.

In accumulation mode, a PD is electrically floated, and when light illuminates it, photocarriers are generated and are swept to the surface because of the potential well in the depletion region. The potential voltage decreases when photocarriers (normally electrons) accumulate. By measuring the voltage drop, the total light intensity can be determined. It should be noted that the accumulation of electrons is interpreted as the process of discharge from a charged capacitor by the generated photocurrent, as shown in Figure 10.5. In this figure, a PD is expressed as an equivalent circuit that consists of a light-controlled current source and a capacitor C_{PD}. The capacitor originates from the depletion region of the pn-junction in the PD. First, the PD is reset to V_{dd} and is then floated. The capacitor C_{PD} is charged by this reset process. When light with a power of P impinges on the PD, a photocurrent I_{ph} is produced. This photocurrent I_{ph} discharges the capacitor C_{PD}. Thus, the node voltage of the anode in the PD V_{PD} decreases in proportion to the input light intensity P according to Equation 10.2, and is expressed as

$$V_{PD}(t) = V_{dd} - \frac{I_{ph}}{C_{PD}} t = V_{dd} - \frac{R_{ph} \cdot P}{C_{PD}} t \tag{10.7}$$

where t is the time from the reset action. Actually, C_{PD} depends on the PD voltage, and so Equation 10.7 changes slightly. For most cases, however, Equation 10.7 can be used. A detailed explanation appears in Ref. [1]. It should be noted that the capacitor C_{PD} is so small in a conventional CMOS image sensor that the value of $\frac{I_{ph} t}{C_{PD}}$ is detectable even if I_{ph} is small.

In an active pixel sensor (APS), a source follower circuit is used as a readout circuit. The circuits are shown in Figure 10.6, where the gate of a source follower transistor is connected to the anode of a PD followed by a select transistor for controlling the pixel output. The current load is located in each column, that is, outside a pixel. There are a total of three transistors, and thus, the configuration of this APS is called *3T-APS*. To decrease noise, a 4T-APS is used [1]. The pixel array is usually scanned using vertical and horizontal scanners, as shown in Figure 10.7.

The advantages of a CMOS image sensor over a CCD image sensor include lower power consumption, integration of on-chip signal processing, and easy introduction of the latest CMOS technology. These features are suitable for implantable biomedical devices that require batteries or wireless transmission power delivery with limited installation volume inside the body as previously shown in Figure 10.1.

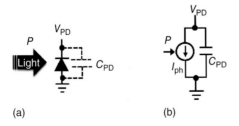

Figure 10.5 (a,b) Photodiode and equivalent circuits.

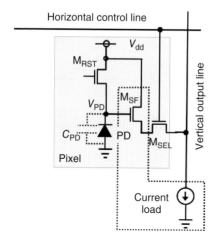

Figure 10.6 3T-APS circuits.

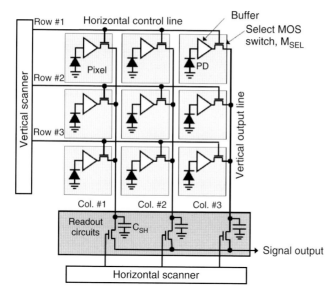

Figure 10.7 Block diagram of a CMOS image sensor.

10.2.3
Log Sensor

A conventional image sensor responds linearly to the input light intensity, as shown in Equation 10.7. A log sensor is based on the subthreshold operation mode of MOSFET, and responds logarithmically to the input light intensity [1]. The main application of a log sensor is a wide dynamic range image sensor. Figure 10.8 shows the basic pixel circuit of a logarithmic CMOS image sensor. In the subthreshold

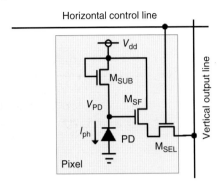

Figure 10.8 Pixel circuits of a log sensor.

region, the MOSFET drain current I_d is very small and increases exponentially with the gate voltage V_g.

$$I_d = I_0 \exp\left[\frac{e}{mk_B T}\left(V_{gs} - V_{th}\right)\right] \quad (10.8)$$

where I_0 and m are constants. In Figure 10.8, M_{SUB} operates in subthreshold mode in Equation 10.8, and thus the variation of the anode voltage in PD, ΔV_{PD}, responds logarithmically to the photocurrent I_{ph} as

$$\Delta V_{PD} = \Delta V_{gs} = \frac{mk_B T}{e} \ln\left(\frac{I_{ph}}{I_0}\right) \quad (10.9)$$

It is noted that a log sensor does not work in accumulation mode as does an APS; the output signal can be detected in any timing or asynchronously.

Although a log sensor has a wide dynamic range of over 100 dB, it has some disadvantages, such as low photosensitivity (especially in the low illumination region), and both a slow response and a relatively large variation in device characteristics due to subthreshold operation.

10.2.4
Pulse Width Modulation Sensor

In APS, photo-generated carriers are integrated and converted into voltage signal with a source follower. This means that the signal voltage is read when the time reaches a prefixed time or the inverse value of the frame rate. It is possible that the time is read when the signal voltage reaches a prefixed voltage. This type of a sensor is called a *pulse width modulation* (PWM) sensor [1]. The pulse width corresponds to the time when the signal voltage reaches a threshold voltage. It is noted that this scheme is asynchronous and can operate without a clock signal, while a conventional APS requires a clock signal.

When the signal voltage reaches a threshold voltage and then resets the PD, this scheme is called a *pulse frequency modulation sensor* [1]. The value of the pulse

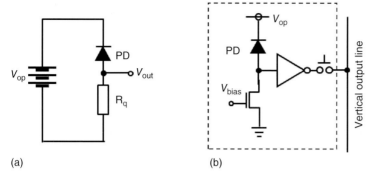

Figure 10.9 SPAD: (a) fundamental circuits of SPAD and (b) pixel circuits of SPAD.

frequency corresponds to the input light intensity; stronger light intensity becomes higher output pulse frequency.

10.2.5
SPAD Sensor

As mentioned previously, avalanche mode is difficult to use in an image sensor because of the nonuniformity of the gain over a large area. Recently, a SPAD has been developed [10]. A SPAD consists of a PD and a quenching resistance, as shown in Figure 10.9a. The PD is subjected to a bias voltage greater than the breakdown voltage. When a PD is illuminated by light, it quickly switches to avalanche mode and generates a large photocurrent. It then returns to the initial state because of the voltage drop due to the quenching resistance. This mode is called *Geiger mode* since it acts similarly to a Geiger counter. In a SPAD sensor, each pixel consists of a SPAD or PD with a quenching resistance and pulse-shaping circuits. The basic pixel circuit is shown in Figure 10.9b. Thus, the output of the SPAD array is a pulse train.

Since a SPAD can be made using standard CMOS technology, a SPAD-based image sensor has been developed. Low-intensity fluorescence detection in the field of biotechnology is one of the major applications of SPAD [10].

10.3
Artificial Retina

10.3.1
Principle of Artificial Retina

This section describes a retinal prosthetic device for use as an implantable image sensor. With some eye diseases, such as retinitis pigmentosa (RP) and aged macular degeneration (AMD), the photoreceptor cells are dysfunctional, whereas most of the other retinal cells, such as ganglion cells, remain healthy (unless the disease is in the terminal stage) [11]. Consequently, by stimulating the remaining retinal cells, visual sensation or phosphenes can be evoked. This is the principle of the retinal prosthesis or artificial vision. On the basis of this principle, a retinal prosthetic device stimulates retinal cells with a patterned electrical signal so that a blind patient can sense a phosphene pattern, or something similar to an image. Stimulating the optic nerve and the visual cortex can also restore visual sensation, but they require more complicated surgical operations.

Retinal prostheses are classified into three types based on where the stimulator is implanted: epi-retinal, sub-retinal, and suprachoroidal places. Figure 10.10 illustrates the three types of retinal prosthesis. The total system of the retinal prosthesis is shown in Figure 10.11, where power/data is sent by wireless transmission through a coil system [12]. A typical system is shown in Figure 10.12.

A retinal prosthesis requires an imaging device. Two types of imaging systems have been developed: extraocular and intraocular. In the extraocular imaging system, a CMOS camera is installed outside the body, for example, in a pair of eyeglasses. Epi-retinal and suprachoroidal systems use the extraocular imaging system. In the intraocular imaging system, a photosensor is integrated with a stimulus electrode, which acts as a photoreceptor cell. A sub-retinal system belongs to the intraocular imaging system. In this chapter, we refer to this device as an artificial retina.

10.3.2
Artificial Retina Based on CMOS Imaging Device

In order to realize better vision through a retinal prosthesis, at least 1000 electrodes are needed. When increasing the number of electrodes, there are problems

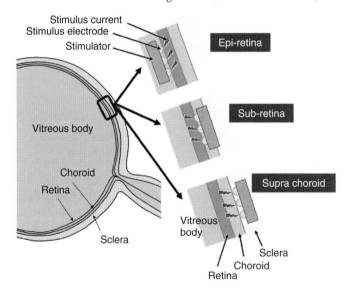

Figure 10.10 Three types of retinal prosthesis. (Reproduced from Ref. [12] with permission.)

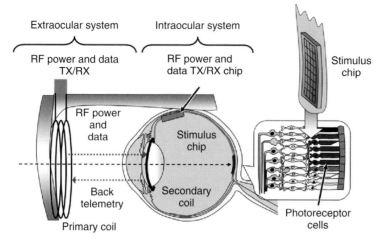

Figure 10.11 Total system of retinal prosthesis. (Reproduced from Ref. [20] with permission.)

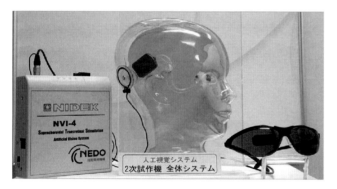

Figure 10.12 Example of a retinal prosthesis system. (Courtesy from Nidek Co. Ltd.)

associated with the interconnection of electrodes and external lead wires with good mechanical flexibility. Preferably, the stimulator should be bent to match the curvature of the eyeball. To solve this problem, artificial retinas based on CMOS imaging devices are introduced. In the following, four types of artificial retinal devices are described.

The first artificial retinal device is a micro photodiode array (MPDA) that operates in the solar cell mode described in Section 10.2.1 [13, 14]. In this device, the topmost electrode of the PD is utilized as a stimulus electrode, as shown in Figure 10.13. Because solar cell mode requires no power supply, only an MPDA is implanted into the sub-retinal space. Figure 10.13 shows one pixel of the artificial retinal device based on the MPDA. The voltage of the PD in the solar cell mode is about 0.6 V, and in order to increase the output voltage, three serially connected PDs are used in the device. The return electrode is placed near the PDs so that the stimulation current is localized near the electrode.

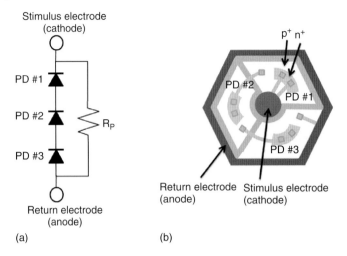

Figure 10.13 Artificial retina device based on MPDA. (a) Equivalent circuits and (b) layout.

Since the photocurrent from a solar cell under normal illumination conditions is not enough to stimulate retinal cells, an infrared (IR) light pattern illuminates the MPDA so as to produce a sufficient photocurrent. The IR light is produced from an image taken by a camera and is converted into a pulse train for a stimulus current pulse. The biphasic current pulse, which is required to neutralize the injected charges, is realized by a pulsed IR light as follows. In the first step, the light turns on, followed by photocurrent generation (that is, the stimulus current flows into the retina). In this step, the photocurrent charges an electrical double layer as a capacitor. In the second step, the light turns off, and the electrical double layer is discharged by a forward current through the PDs. Actually, as shown in Figure 10.13, a shunt resistance is placed in parallel with the PDs to discharge quickly. The device is simple, but it is difficult to realize charge balance.

The second artificial retinal device is based on a log sensor, as described in Section 10.2.3. Since the response of the human photoreceptor is logarithmic, it is natural to use a log sensor for an artificial retina. The disadvantages of a log sensor, which are mentioned in Section 10.2.3, are not so harmful when it is used in an artificial retina because at present, the resolution is not so high, and a high-speed response is not required. Figure 10.14 shows the pixel circuits used with a log sensor. In this architecture, one log sensor is used to obtain an ambient illumination level and the other measures the local illumination level. The pixel output is the difference between the local illumination level and the ambient or global illumination level [15, 16].

This process mimics the response of a human retina, where the response is adaptive to the ambient illumination level. Therefore, the artificial retina achieves an ultrawide dynamic range, covering light intensity levels from a star at night to midsummer daylight. Figure 10.15 shows a microphotograph of the fabricated chip.

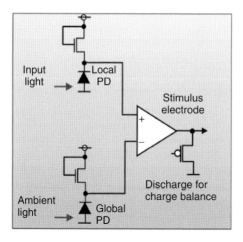

Figure 10.14 Pixel circuits of the log sensor for artificial retina device.

Figure 10.15 Microphotograph of the log sensor for artificial retina device. (Reproduced from Ref. [16] with permission.)

The third artificial retinal device is based on a modification of an APS, which is mentioned in Section 10.2.2. Multiple microchip architecture is applied to the device as shown in Figure 10.16. The basic circuits of the photosensor are shown in Figure 10.17a. The pulse width from CONT corresponds to the accumulation time of the PD, that is, the duration of ON in CONT controls the light sensitivity. When the anode voltage of the PD reaches the threshold of the inverter, the input to the D-FF turns ON or HIGH. Meanwhile, when CONT turns OFF or LOW, the D-FF latches the input value. If the light signal is weak, the OUT in the photosensor is LOW. In that case, by increasing the pulse width of CONT, then the light-sensitivity increases, the OUT can turn to HIGH.

This photosensor is employed in a new type of smart stimulator that consists of a number of CMOS-based microchips distributed on a flexible substrate, as shown

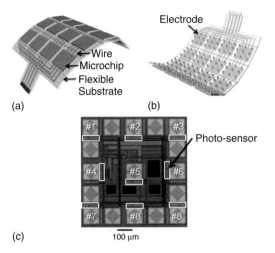

Figure 10.16 Artificial retina device based on multiple microchip architecture. The illustrations of the stimulator in the backside (a) and the front-side (b). (c): a photo of a microchip. (Reproduced from Ref. [16] with permission.)

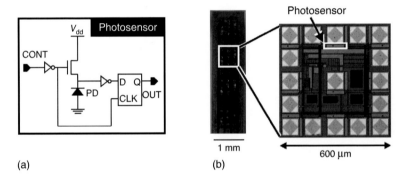

Figure 10.17 Light-sensing function of the microchip (a) the photosensor circuits and (b) a photograph of the fabricated artificial retina device based on multiple microchip architecture and the microchip. (Reproduced from Ref. [19] with permission.)

in Figure 10.17b [17–20]. Each microchip incorporates several stimulus electrodes, which can be externally controlled such that the electrodes can be turned on and off through an external control circuit. In addition to solving the interconnection issue, CMOS-based stimulators offer several advantages, such as signal processing. In order to allow flexibility, several microchips are placed on a substrate in a distributed manner.

Each microchip has nine photosensors, as shown in Figure 10.17b [19]. The stimulus current is controlled by the light impinging on the microchip. When the light intensity reaches a threshold value in the chip, the stimulus current control switch is turned on, and the stimulus current flows into the retinal cells. After chip fabrication, nine Pt bumps were formed on the Al pads of the chip as stimulus

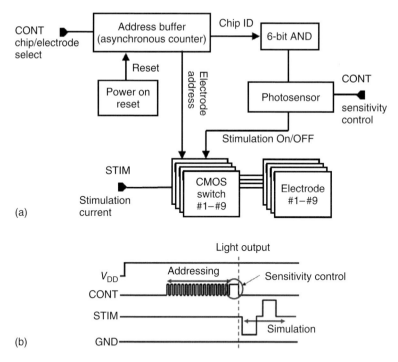

Figure 10.18 Block diagram of the artificial retina device based on multiple microchip architecture (a) and its timing diagram (b). (Reproduced from Ref. [19] with permission.)

electrodes. The four microchips were assembled on a flexible polyimide substrate using a flip-chip bonding technique specially developed for this distributed architecture. The block diagram and the timing diagram of the microchip are shown in Figure 10.18. The Nth photosensor controls the Nth CMOS switch, which in turn makes a stimulus current injected from the Nth electrode, where N is the number of electrodes in one chip.

The fabricated stimulator was implanted into a pocket formed in the sclera of a rabbit eye. A near-infrared (NIR) light-emitting diode (LED) array illuminated the eye at the location where the stimulator was implanted. Note that NIR light cannot evoke photoreceptors and can penetrate the epithelium and the sclera of a rabbit eye to some distance. The electrically evoked potential (EEP) signal was measured through screw electrodes set in the visual cortex. After implanting the stimulator, we confirmed that no visually evoked potential (VEP) signal was produced by the NIR light before measurement of the EEP signal. When NIR light was incident on the eye, a clear EEP signal was obtained, as shown in Figure 10.19, and thus, the stimulation of retinal cells was successfully demonstrated [19].

The last type of artificial retinal device is different from the others; the device is based on a micro camera system that is implanted in a crystalline lens [20]. The output of the camera is connected to the stimulator, which stimulates the optic nerve. In this case, the entire camera system is implanted so that a conventional

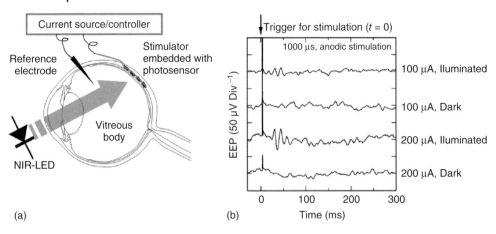

Figure 10.19 EEP signals obtained with the artificial retina device. (Reproduced from Ref. [19] with permission.)

camera system can be utilized. Although this type may be applied to all types of retinal prostheses, the connection between the image sensor and the electrodes must be considered if the number of pixels is large.

10.4
Brain-Implantable CMOS Imaging Device

10.4.1
Measurement Methods for Brain Activities

The measurement of neural activity in the brain is important not only for studying learning and memory in basic neuroscience but also for clinical evaluation of brain diseases, such as Parkinson's disease, epilepsy, depression, and schizophrenia. Although electroencephalography (EEG), functional magnetic resonance imaging (fMRI), positron emission tomography (PET), and functional near-infrared spectroscopy (fNIRS) or optical tomography are very powerful noninvasive tools for investigating brain activity in humans, it is difficult to use these tools on small animals such as mice without tethering the animal. For small experimental animals, invasive measurements are acceptable, and electrical measurements by inserting electrodes into the brains of untethered animals are conventionally used. Besides simple metal electrodes, several types of sophisticated Si probes have been developed (such as the Utah probe [21], the Michigan probe [22], and the Toyohashi probe [23]) to measure brain activity at multiple points. It is, however, noted that electrical measurements are nonspecific, and thus, in many experiments, optical measurement using fluorescence is required, because such measurements are specific. In addition, an optical method using fluorescence can measure a large amount of data over a wide area. A fluorescence microscope is conventionally used

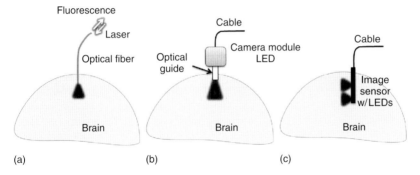

Figure 10.20 Imaging devices for measuring neural activities in the brain. (a) Fiber endoscope, (b) head-mountable device, and (c) the brain-implantable device.

for the measurements, and to some extent, it can be used to measure brain neural activity of an awake and head-restraint mouse [24]. The mouse can walk on an air-supported styrofoam ball with its head fixed. It is, however, difficult to apply this to freely moving rodents.

To overcome this difficulty, three types of imaging devices have been developed: a fiber endoscope [25–29], a head-mountable device [28–30], and a brain-implantable device [31–36], as shown in Figure 10.20. In the following sections, these devices are introduced, and the brain-implantable device is described in detail.

10.4.2
Fiber Endoscope and Head-Mountable Device

In these devices, an image sensor or a photodetector with a scanning mirror, optics, and excitation light sources are placed outside the body of an animal, as shown in Figure 10.22. In a fiber endoscope, guiding optics such as a single-mode fiber [25] and graded index (GRIN) lenses [26, 27] are used to introduce excitation light and gather fluorescence. The light is guided to an imaging device, such as a CCD [27] or a PMT (photo multiplier tube) with a scanner [25, 26]. Since this method is frequently used in combination with two-photon microscopy, the spatial resolution is high enough to measure individual neural activities in regions a few hundred micrometers deep. The drawback of this method is that although the optical fiber is flexible, it is still limited in the extent to which it can bend. In addition, the field of view is not so large.

Recently, a miniaturized microscope system used with a CMOS image sensor, which is also mounted on a mouse head [28–31], has been developed and was demonstrated to be effective for a freely moving mouse. The head-mountable device is composed of lenses, a CMOS image sensor, and an LED for the excitation light source, as shown in Figure 10.22b. All components are assembled in a single housing case. The guiding optics is inserted into the brain. The total weight of the system is less than 2 g, so that it is lighter than an endoscope system [28]. In addition, electrical wires are more flexible than optical fibers, so that an animal with the head-mountable device can move more smoothly than one with the endoscope system.

However, it is still difficult to observe neural activity in deep regions of the brain, such as the hippocampus. Both the endoscope and the head-mountable device can take a surface image of the brain, but not an image along the depth direction. In order to address this issue, an implantable CMOS imaging device has been proposed and demonstrated. This device has a submillimeter/sub-second spatiotemporal resolution and has successfully been used to monitor the time course of serine protease activity inside a mouse hippocampus [31–36]. In the next section, the brain-implantable CMOS imaging device is discussed.

10.4.3
Brain-Implantable CMOS Imaging Device

The brain-implantable CMOS imaging device is based on a dedicated CMOS image sensor fabricated using standard CMOS technology. The pixel structure is 3T-APS with a parasitic PD composed of n-wells and p-substrate junctions. The number of inputs/outputs is limited to four in order to minimize the constraint on the implanted animals due to the external wires [35]. The pixel size was designed to be $7.5 \times 7.5\,\mu m^2$, which is sufficient for image neural cells because the spatial resolution is directly related to the pixel size. In this setup, only objects in the vicinity of the sensor can be clearly observed because the device has no imaging optics. The frame rate of the device can be varied by external control. The frame rate depends on the input light power; if the input light power is large, the frame rate can be higher. The device is thin and long enough to observe the corpus striatum in the basal ganglion with minimal invasion. In addition, the dimensions of the device are expected to allow implantation into deeper regions of the brain and to enable stable imaging in deep brain regions of a freely moving mouse.

The device is compactly packaged in a polyimide substrate with LEDs and an excitation light filter for fluorescent imaging and improved biocompatibility. The polyimide substrate is fully flexible and has been shown to be biocompatible for the surgical insertion of implanted devices into the living body. The color filter blocks the excitation light, allowing only fluorescent emission to reach the image sensor. LEDs are also implemented on the polyimide substrate. In order to protect the chip during the experiment, the chip is coated with the optically transparent and waterproof Parylene-C. The thickness of the assembled CMOS imaging device is approximately 350 μm. The fabricated device is shown in Figure 10.21, where LEDs for exciting fluorophore and a CMOS image sensor are assembled on a flexible substrate. Figure 10.22 shows two examples of these devices. One device is used for implantation in the surface of the brain or planar type, whereas the other device is used for implantation into the deep brain or needle type. These devices have different size.

In this experiment, a device was implanted into the basal ganglion of the mouse brain. The device was fixed to the skull with dental cement and was connected to a wiring harness through a slip-ring connector, as shown in Figure 10.23.

10.4 Brain-Implantable CMOS Imaging Device

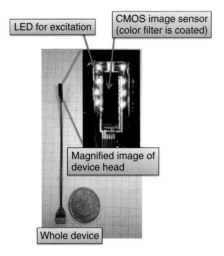

Figure 10.21 Brain-implantable device. (Reproduced from Ref. [19] with permission.)

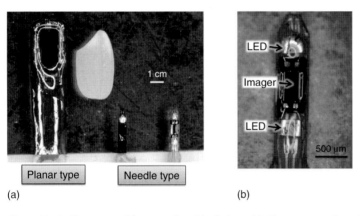

Figure 10.22 Two types of brain-implantable devices. (a) Planar type and needle type and (b) close-up view of a needle type. (Reproduced from Ref. [19] with permission.)

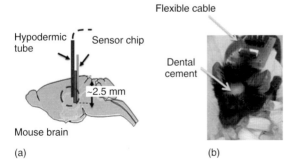

Figure 10.23 (a,b) The measurement system of the brain activities of freely moving mouse implanted with an imaging device.

As a fluorophore, a chemical substance was introduced into the mouse brain through a hypodermic tube. The substance changes from a non-fluorophore into a fluorophore when neural activity occurs in the hippocampus, which means that the sensor can detect a specified molecular activity.

It has been confirmed that the device implanted deep into the mouse brain could be successfully operated, and the mouse implanted with the device was alive and able to move freely 1 month after the implantation. On the basis of a previous experiment, it was confirmed that the mouse brain was intact (except for the region in which the device was inserted) and that the brain tissue was not severely damaged. Little foreign-body reaction was observed. Figure 10.24 shows typical experimental measurement results for deep-brain neural activity obtained using a device implanted in a mouse hippocampus. Spatiotemporal neural activity was successfully measured.

Since the device has no optics, the spatial resolution is only a few tens of micrometers but is still enough to observe neural activity over a wide field. The biggest advantage over the previous two methods is the ability to measure deep

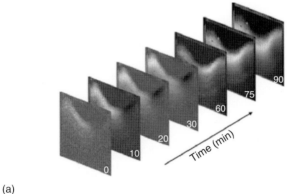

(a)

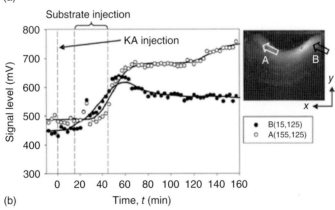

(b)

Figure 10.24 (a,b) Experimental results obtained with the brain-implantable imaging device. (Reproduced from Ref. [31] with permission.)

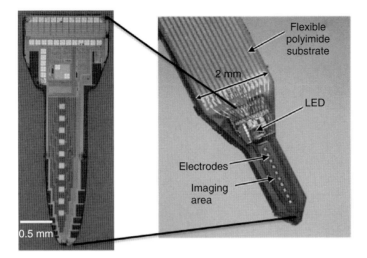

Figure 10.25 The brain-implantable device with electrodes. (Reproduced from Ref. [35] with permission.)

brain regions in the vertical direction. Another advantage is that it can integrate with electrical measurement and/or stimulation functions of the device. Figure 10.25 shows an implantable CMOS imaging device integrated with electrodes [35, 36].

To reduce the invasiveness when it is implanted into the brain, a CMOS imaging device with LEDs embedded on the chip has been developed to decrease the total thickness of the device. The device requires no substrate, so it can be thinner than the previous device. Figure 10.26 shows three types of the device configuration. Type (a) is a device where LEDs are placed on a flexible polyimide substrate as well as a CMOS image sensor. The thickness of the device is about 350 μm. It is difficult to reduce the thickness because a flexible substrate is required to hold the sensor and LEDs, which in turn increases the thickness of the device. Type (b) is a device where an LED is placed on the surface of a CMOS image sensor. The substrate is only used to support the sensor. Even in this type, it is still difficult to reduce the thickness of the device. Finally, type (c) is a device where an LED is embedded into the sensor. A reservoir is fabricated on an image sensor chip to accommodate an LED on the chip. The surface of the fabricated device is almost flat because of the process. The total thickness is determined by the thickness of the sensor chip. In the present device, the thickness is about 150 μm, but it can be reduced. Figure 10.27 shows a scanning electron microscopic (SEM) image of the fabricated device [37].

10.5
Summary and Future Directions

Implantable CMOS imaging devices were introduced for artificial retinal devices and brain-implantable CMOS imaging devices. Before presenting the two devices,

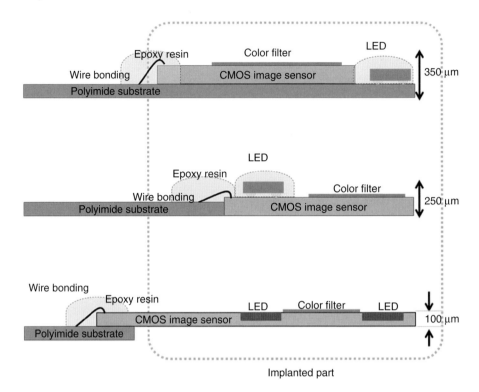

Figure 10.26 Three types of the brain-implantable devices; (a) an LED and a sensor are both placed on a substrate, (b) an LED is placed on a sensor, and (c) a reservoir is fabricated to embed an LED. All sensors are covered with Parylene.

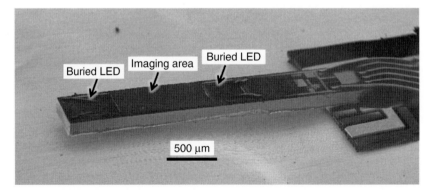

Figure 10.27 SEM photograph of the fabricated brain-implantable device with the thickness of about 150 μm. (Reproduced from Ref. [37] with permission.)

the fundamentals of CMOS image sensors were described. Several features of CMOS image sensors, such as low power consumption and on-chip signal processing functions, are suitable for implantable devices. For artificial retinal devices, a solar cell mode device, a log sensor device, and a multiple microchip device were introduced. For brain-implantable CMOS imaging devices, a fiber-based endoscope and a head-mountable imaging device were introduced to measure the neural activity in an untethered rodent brain. In addition to the two devices, a CMOS imaging device that is completely implanted into the mouse brain was introduced. The device can measure neural activity deep in the brain, although the spatial resolution is lower than that for the other devices.

These devices can be improved by introducing the state-of-the-art CMOS image sensor technologies. Several such technologies are being developed, such as backside illumination, shared pixels, and column-parallel signal processing. One of the most urgent issues to be considered is biocompatible coating materials for direct implantation of a CMOS imaging device. In particular, medical devices are required to remain implanted for a long period of time. Although Parylene is one of the most suitable materials in terms of biocompatibility and water resistance, some studies found that it is questionable for long-term implantation. In the future, more biocompatible and durable coating materials will need to be developed, in addition to a less invasive structure for the next generation of implantable CMOS imaging devices.

Acknowledgments

The author would like to thank Profs. T. Tokuda, K. Sasagawa, and T. Noda in his laboratory, the Photonic Device Science Laboratory, the Graduate School of Materials Science, and the Nara Institute of Science and Technology for their collaboration on implantable devices; Prof. T. Fujikado and Dr. Y. Terasawa for their collaboration on artificial retinal devices; and Prof. S. Shiosaka for his collaboration on animal experiments. Part of this work was supported by the Core Research for Evolutional Science and Technology (CREST) from the Japan Science and Technology Agency (JST) and the Strategic Research Program for Brain Sciences from the Ministry of Education, Culture, Sports, Science and Technology of Japan.

References

1. Ohta, J. (2007) *Smart CMOS Image Sensors and Applications*, CRC Press, Boca Raton, FL.
2. Brain Vision Inc. http://www.brainvision.co.jp/manual/ultima_overview.pdf (accessed 10 September 2013).
3. Yoon, H.J., Ito, S., and Kawahito, S. (2009) *IEEE Trans. Electron Devices*, **56**, 214–221.
4. Iddan, G., Meron, G., Glukhovsky, A., and Swain, P. (2000) *Nature*, **405**, 417.
5. Ohta, J., Tokuda, T., Sasagawa, K., and Noda, T. (2009) *Sensors*, **9**, 9073–9093.

6. Humayun, M.S., Weiland, J.D., Chader, G., and Greenbaum, E. (eds) (2007) *Artificial Sight*, Springer, New York, NY.
7. Agnew, W.F. and McCreery, D.B. (eds) (1990) *Neural Prostheses*, Prentice Hall, Englewood Cliffs, NJ.
8. Puers, R. and Thoné, J. (2011) Short distance wireless communications, in *Bio-Medical CMOS ICs* (eds H.J. Yoo and C. van Hoof), Springer, New York, NY.
9. Miyake, T., Oike, M., Yoshino, S., Yatagawa, Y., Haneda, K., Kaji, H., and Nishizawa, M. (2009) *Chem. Phys. Lett.*, **480**, 123–126.
10. Charbon, E. and Fishburn, M.W. (2011) Monolithic single-photon avalanche diodes: SPADs, in *Single-Photon Imaging* (eds P. Seitz and A.J.P. Theuwissen), Springer, Berlin.
11. Humayun, M.S., Prince, M., de Juan, E., Barron, Y., Moskowitz, M., Klock, I.B., and Milam, A.H. (1999) *Invest. Ophthalmol. Visual Sci.*, **40**, 143–148.
12. Ohta, J. (2011) Artificial retina IC, in *Bio-Medical CMOS ICs* (eds H.J. Yoo and C. van Hoof), Springer, New York, NY, p. 481.
13. Mathieson, K., Loudin, J., Goetz, G., Huie, P., Wang, L., Kamins, T., Galambos, L., Smith, R., Harris, J.S., Sher, A., and Palanker, D. (2012) *Nat. Photonics*, **6**, 391–397.
14. Wang, L., Mathieson, K., Kamins, T.I., Loudin, J.D., Galambos, L., Goetz, G., Sher, A., Mandel, Y., Huie, P., Lavinsky, D., Harris, J.S., and Palanker, D.V. (2012) *J. Neural Eng.*, **9**, 046014.
15. Dollberg, A., Graf, H.G., Hoefflinger, B., Nisch, W., Schulze Spuentrup, J.D., Schumacher, K., and Zrenner, E. A Fully Testable Retina Implant, (2003) Proceedings of the International Conference on Biomedical Engineering, Salzburg, Austria, pp. 255–260.
16. Rothermel, A., Wieczorek, V., Liu, L., Stett, A., Gerhardt, M., Harscher, A. and Kibbel, S. (2008) A 1600-pixel Subretinal Chip with DC-free Terminals and 2V Supply Optimized for Long Lifetime and High Stimulation Efficiency, *Dig. Tech. Papers Int'l Solid-State Circuits Conf. (ISSCC)*, pp. 144–145, San Francisco, CA, February.
17. Tokuda, T., Pan, Y.L., Uehara, A., Kagawa, K., Nunoshita, M., and Ohta, J. (2005) *Sens. Actuators, A*, **122** (1), 88–98.
18. Ohta, J., Tokuda, T., Kagawa, K., Sugitani, S., Taniyama, M., Uehara, A., Terasawa, Y., Nakauchi, K., Fujikado, T., and Tano, Y. (2007) *J. Neural Eng.*, **4** (1), S85–S91.
19. Tokuda, T., Hiyama, K., Sawamura, S., Sasagawa, K., Terasawa, Y., Nishida, K., Kitaguchi, Y., Fujikado, T., Tano, Y., and Ohta, J. (2009) *IEEE Trans. Electron Devices*, **56** (11), 2577–2585.
20. Ng, D.C., Furumiya, T., Yasuoka, K., Uehara, A., Kagawa, K., Tokuda, T., Nunoshita, M., and Ohta, J. (2006) *IEEE Trans. Circuits Syst. II*, **53** (6), 487–491.
21. Campbell, P.K., Jones, K.E., Huber, R.J., Horch, K.W., and Normann, R.A. (1991) *IEEE Trans. Biomed. Eng.*, **38** (8), 758–768.
22. Wise, K.D. and Najafi, K. (1991) *Science*, **254**, 1335–1342.
23. Kawano, T., Kato, Y., Tani, R., Takao, H., Sawada, K., and Ishida, M. (2004) *IEEE Trans. Electron Devices*, **51** (3), 415–420.
24. Dombeck, D.A., Khabbaz, A.N., Collman, F., and Tank, D.W. (2007) *Neuron*, **56**, 43–57.
25. Sawinski, J., Wallace, D.J., Greenberg, D.S., Grossmann, S., Denk, W., and Kerr, J.N.D. (2009) *Proc. Natl. Acad. Sci. U.S.A.*, **106** (46), 19557–19562.
26. Flusberg, B., Nimmerjahn, A., Cocker, E., Mukamel, E., Barretto, R., Ko, T., Burns, L., Jung, J., and Schnitzer, M. (2008) *Nat. Method*, **5** (11), 935–938.
27. Saunter, C.D., Semprini, S., Buckley, C., Mullins, J., and Girkin, J.M. (2012) *Biomed. Opt. Express*, **3** (6), 1274–1278.
28. Ghosh, K.K., Burns, L.D., Cocker, E.D., Nimmerjahn, A., Ziv, Y., Gamal, A.E., and Schnitzer, M.J. (2011) *Nat. Methods*, **8** (10), 871–878.
29. Park, J.H., Platisa, J., Verhagen, J.V., Gautam, S.H., Osman, A., Kim, D., Pieribone, V.A., and Culurciello, E. (2011) *J. Neurosci. Methods*, **201** (2), 290–295.
30. Murari, K., Etienne-Cummings, R., Cauwenberghs, G., and Thakor, N. (2010) An integrated imaging microscope for untethered cortical imaging in

freely-moving animals. Engineering in Medicine and Biology Society (EMBC), 2010 Annual International Conference of the IEEE, 2010, pp. 5795–5798.
31. Ng, D.C., Nakagawa, T., Mizuno, T., Tokuda, T., Nunoshita, M., Tamura, H., Ishikawa, Y., Shiosaka, S. and Ohta, J. (2008) *IEEE Sensors J.*, **8** (1), 121–130.
32. Tamura, H., Ng, D.C., Tokuda, T., Naoki, H., Nakagawa, T., Mizuno, T., Hatanaka, Y., Ishikawa, Y., Ohta, J., and Shiosaka, S. (2008) *J. Neurosci. Methods*, **173** (1), 114–120.
33. Kobayashi, T., Tamura, H., Hatanaka, Y., Motoyama, M., Noda, T., Sasagawa, K., Tokuda, T., Ishikawa, Y., Shiosaka, S., and Ohta, J. (2011) Functional neuroimaging by using an implantable CMOS multimodal device in a freely-moving mouse. IEEE BioCAS, San Diego, CA, November 10–12, 2011.
34. Kobayashi, T., Motoyama, M., Masuda, H., Ohta, Y., Haruta, M., Noda, T., Sasagawa, K., Tokuda, T., Tamura, H., Ishikawa, Y., Shiosaka, S., and Ohta, J. (2012) *Biosens. Bioelectron.*, **38** (1), 321–330.
35. Tagawa, A., Minami, H., Mitani, M., Noda, T., Sasagawa, K., Tokuda, T., Tamura, H., Hatanaka, Y., Ishikawa, Y., Shiosaka, S., and Ohta, J. (2010) *Jpn. J. Appl. Phys.*, **49**, 01AG02.
36. Tagawa, A., Higuchi, A., Sugiyama, T., Sasagawa, K., Tokuda, T., Tamura, H., Hatanaka, Y., Shiosaka, S., and Ohta, J. (2009) *Jpn. J. Appl. Phys.*, **48** (4), 04C195-1–04C195-5.
37. Ohta, J., Kitsumoto, C., Noda, T., Sasagawa, K., Tokuda, T., Motoyama, M., Ohta, Y., Kobayashi, T., Ishikawa, Y., and Shiosaka, S. (2012) A micro imaging device for measuring neural activities in the mouse deep brain with minimal invasiveness. IEEE BioCAS, Hsinchu, Taiwan.

11
Implanted Wireless Biotelemetry
Mehmet Rasit Yuce and Jean-Michel Redoute

11.1
Introduction

Biomedical implants are used to interface with biological systems inside the human body to perform tasks such as vision and hearing, neural recording and stimulation, and physiological signal sensing and monitoring. As integrated circuit technology advances, implantable biotelemetry devices will handle more complex functions. Recent implantable systems have focused on more complex microelectronics systems with the use of the advanced integrated circuit technology. Some milestones of implantable devices are given below:

- *1957*, first wearable pacemaker (in 1958 fully implant).
- *1984*, the Australian cochlear implant approved by the U.S. FDA (Food and Drug Administration).
- *2000*, first clinical trial wireless endoscope (camera based). Since 1957, researchers have been working on wireless endoscopy devices.
- *2013*, the visual prosthetic device-Argus II designed by Second Sight has received the FDA approval. The device was previously approved for use in Europe in 2011.

In addition to these projects, there is a vast area of different implantable systems being designed by designers such as cardioverter defibrillators, microelectrode arrays (MEAs) for detection and stimulation of brain functions, and electronic pills for drug delivery systems. Moreover, there are implantable biosensor developments to conduct animal studies. More implantable biomedical devices will be developed in the future to undertake various body tasks and to diagnose the operation of body organs. A wireless telemetry system is necessary for information transfer from these devices inserted inside the body for the purpose of controlling and monitoring. Wireless telemetry for implantable devices should form an intelligent communication network such as a body area network in order to work efficiently and safely in the same environment [1].

It is desirable to construct highly integrated and miniaturized communication systems because of the physical limitation of implantable devices. The ability to develop a low-power device to achieve a reliable communication is a critical and

Implantable Bioelectronics, First Edition. Edited by Evgeny Katz.
© 2014 Wiley-VCH Verlag GmbH & Co. KGaA. Published 2014 by Wiley-VCH Verlag GmbH & Co. KGaA.

important aspect of such small biomedical devices [2]. Low power consumption is paramount to enable a long operating life.

Electronics for the wireless telemetry portion of implantable devices is developed on the basis of very strict design requirements. Wireless technologies such as ZigBee, wireless local area networks (WLANs), and Bluetooth (IEEE 802.15.1) developed in the commercial domain cannot be used directly in medical implants because of the following reasons: (i) they have been optimized for general use, (ii) the device size exceeds the required size limitation of the current implant technology, (iii) their frequency bands are very crowded, and finally (iv) they do not meet the safety criteria related to radiation. The existing advanced wireless systems operate at 2.4 GHz ISM (the industrial, scientific, and medical) band and suffer from strong interference from each other when they are located in the same environment [3]. Thus, an implant system should have a different transmission band for an interference-free wireless link as the transmitted information could be related to patients' critical conditions.

Advances in the development of implanted devices have been significant and the use of implanted devices is also increasing. Recent successful examples are the wireless capsule endoscope, cochlear implant, retinal implants (i.e., bionic eye), and implantable defibrillators. Among implantable biomedical devices listed in Table 11.1, except for retina implants, neural recording, and wireless endoscopy devices, the remaining devices require low data transmission rates. The wireless endoscopy device requires the highest data transmission rate, around 10 Mbps or more to transmit good quality video from inside the body to a monitoring device [4].

Table 11.1 List of bioimplantable devices.

Devices	Signal types
Retina implants	Camera images (data rates: 2–5 Mbps)
Wireless capsule endoscope (electronic pill)	Camera images (data rates: ~10 Mbps)
Neural recording	Multichannel neural signal (~1–10 Mbps)
Implantable defibrillators	Stimulation pulses
Cochlear implants	Audio signal
Deep brain stimulator	Stimulation
Pacemaker	ECG (physiological signal)/stimulation pulses
Temperature	Physiological signal
Pulse oximeter	Physiological signal
Blood pressure	Physiological signal
Oxygen, pH value	Physiological signal
Glucose sensor	Glucose level
Implantable drug delivery	Control signal

11.2
Biotelemetry

In implantable telemetry systems, the dual band (i.e., two-inductive links) has been used to optimize the power and data telemetry links independently [5]. The power and data signals are carried on two wireless inductive links. For implantable systems such as the cochlear implant and bionic eye, one wireless data link (i.e., forward data link) is used to transmit data from the external unit to inside the skin to stimulate the nerves in order to help patients restore hearing and sight (they are known as *prosthetic devices*). These devices also require the transmission of wireless power to supply energy to the implant device. In addition to the forward data link and wireless power link, these devices may require the transmitting of status information regarding the operation of the implant in the body, which is done on the basis of a third link-back telemetry. This third wireless link is very important for implanted devices for safety reasons that can be used to control, monitor, and adjust the implant device inside the body.

In-body medical systems such as pacemakers, electronic pill, and implantable biosensors send physiological data such as electrocardiogram (ECG), blood pressure, or video images from inside the body to the external unit and then to the monitoring station in order to track a patient's condition. These implantable or in-body medical systems may or may not require the forward data link, depending on whether a control signal is required from the external unit to control the implantable system.

Figure 11.1 shows a general block diagram representing building blocks used in most implantable telemetry systems. An implantable device consists of sensors/electrodes that detect biological signals from the body, a forward data telemetry link receiving external data message, and a back telemetry link to transmit information from the device in the body to the external device. The power signal is received and then regulated to provide power supply for the wireless data links, the sensors/electrodes, and the signal acquisition unit that processes and amplifies the sensors' data. The external device operates with a battery that transmits and receives signals from inside to outside or outside to inside. The external unit may contain another wireless link that will control the communication between a monitoring station and the implant device for remote monitoring and diagnosis.

As described earlier, implantable systems such as the cochlear implant and the bionic eye use the forward data link to transmit data from the external unit to the inside of the skin to stimulate the nerves in order to restore hearing and sight. The transmitting of status information regarding the operation of the device is done by the back data telemetry wireless link. In order to monitor the inner organs as well as the status of medical implants such as pacemakers and defibrillators, a wireless link (similar to back data telemetry) from implant to external device is also necessary. A frequency around 400 MHz has been used as a popular transmission band for this link in recent systems [6].

Treating a large number of patients wearing implanted systems in the same environment (e.g., hospital) requires a reliable wireless networking in order to

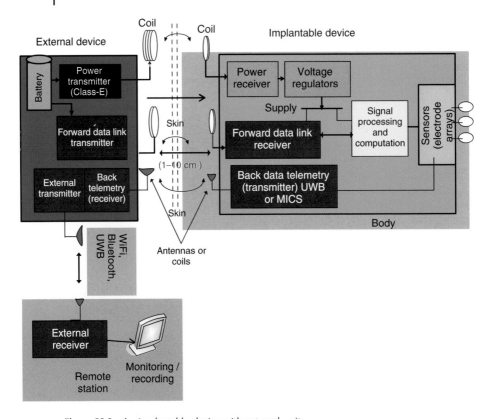

Figure 11.1 An implantable device with external units.

monitor and differentiate each individual implanted device and patient. Thus, implanted wireless nodes in a patient's body should form a wireless link so that one or more implanted devices inserted in the bodies of a number of patients in a hospital environment can be controlled with minimum complexity. A back data telemetry link is most likely to operate using the MICS (medical implantable communication service) band or using UWB (ultrawide band) or other specific bands [7].

Wireless technologies such as ZigBee (IEEE 802.15.4), Wi-Fi (IEEE 802.11), and Bluetooth (IEEE 802.15.1), developed in the commercial domain, cannot be used directly in the devices operating in the body because the sizes of these devices exceed the required size limitation of the current in-body sensor devices. A wireless telemetry system for in-body sensor devices should have a dedicated transmission band for an interference-free wireless link as the transmitted information could be related to patients' critical conditions.

Most of implantable systems such as retina prostheses and cochlear implants have a very short transmission distance for wireless telemetry. Thus, the forward data link for these biomedical devices has been based on an inductive link. Consequently, a wireless system with a very simple communication scheme such as binary ASK (amplitude shift keying), FSK (frequency shift keying) [8], and PSK (phase shift keying) modulators and demodulators have been employed [5, 9]. The frequency 13.56 MHz in ISM band is usually the most common for such inductive link telemetry systems and is also used for RFID (radio frequency identification) applications. A low transmission frequency, usually less than 20 MHz, is utilized mainly because of the simplicity in the design and to avoid the use of power-hungry blocks such as mixer and oscillators.

In addition to the forward data link, in implantable systems such as the cochlear implant and the bionic eye, the inductive link has also been utilized to transfer power from the external device to inside the skin to supply power for electronics. In order to transfer enough energy, inductive links require the external unit to be very close to the implanted unit. Ideally, the external device should be placed on the body for an efficient wireless power.

The back-telemetry link can have a longer transmission range if it is designed on the basis of antennas rather than inductive link using coils. Antennas are used because of the higher transmission frequencies (i.e., MICS: 402–406 MHz). High-frequency transmission link is the only option to dedicate enough spectrums for a reliable communication. This will be useful for some of advanced medical implants such as pacemaker, implantable cardioverter defibrillators, and wireless endoscope as a much longer range is required for their wireless telemetry link.

11.2.1
Inductive Link for Forward Data

In earlier inductive link designs, the same inductive link would be used to transfer both data and power to an implant. The wireless power is separated from data transmission in the recent systems to optimize power and forward data links separately [6]. The choice of a dedicated frequency for power transmission simplifies the design of the power transmitter, rectification, and voltage regulation circuits. Inductive links are constructed to operate at a frequency lower than 20 MHz for a forward data transmission. A wireless data link design for such a low frequency does not involve the power-hungry RF (radio frequency) blocks such as oscillators and mixers. A block diagram of a receiver for a forward data link can be illustrated as in Figure 11.2. A simple bandpass filter (BPF) together with a limiting amplifier is sufficient at the front end. The received signal is passed through a BPF first to reject interference and the out-of-band signals. Then, the limiting amplifier provides enough gain to digitize the input analog signal. A reference signal (sampling clock) is required in the digital part of the design (i.e., DSP unit) to sample, process, and demodulate the transmitted signal. This reference clock should be a part of the implant device either using a clock recovery circuit to recover a reference signal from the received signal or can be generated in the implanted device using one of the oscillator circuit topologies.

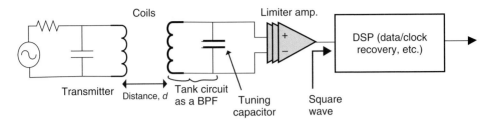

Figure 11.2 An inductive link transmission/reception for forward data telemetry link.

11.2.2
Wireless Power Link

A critical challenge for implantable devices is the limited energy source. It is not always possible to use batteries. If implantable batteries are used, it will require periodic surgery to replace batteries, which is not desired. The solution is to successfully develop a wireless power link to transfer energy to the human body via an inductive link. This is known as *magnetic induction* and involves having two coils. One is the transmitting coil and the other is the receiving coil. The coils must both be exactly aligned for a transfer of power to occur. The most effective transfer of wireless power occurs at lower frequencies [8–10] in order to avoid the penetration loss occurring in the human body tissue.

A block diagram of an inductive-link-based wireless power system is depicted in Figure 11.3. The power signal is sent from the primary site to the secondary site through the inductive coils. A class E amplifier has been used to achieve a highly efficient power transfer [8]. The wireless power link shown in Figure 11.3 has been implemented to transmit power through a biological muscle environment using 2 cm thick beef. The received power successfully powered the telemetry device implanted in the biological tissue [11]. A crystal oscillator operating at 27 MHz is used in this circuit to provide an accurate transmission signal frequency that

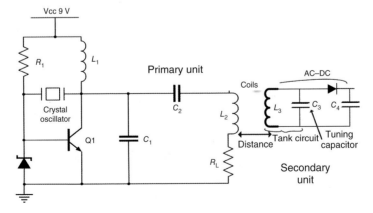

Figure 11.3 An inductive link using class-E amplifier for wireless power transfer.

enables an easier demodulation process at the implant site. This design combines an oscillator and a class E amplifier into a self-oscillating power transmitter circuit for greater power efficiency.

The transistor device (i.e., Q1 transistor) acting as a switch is driven by a digital signal at the carrier frequency f_c. The network resonant frequency should be the transmission frequency and is given by Equation 11.1. The capacitor C_2 connected in series with the primary coil (L_2) is acting as a DC blocking capacitor and its value should be small enough to be nearly resonant with L_2 at the carrier frequency f_c.

$$f_c = f_{osc} = \frac{1}{(2\pi\sqrt{L_2 C_1})} = \frac{1}{2\pi\sqrt{L_3 C_3}} \qquad (11.1)$$

As described in Figure 11.3, the supply of power via a wireless inductive link involves several considerations, from power transmitter to inductive transmission with coil optimization, and rectification and regulation within the implant. Wireless power approaches for several implanted devices are listed in Table 11.2. In addition to directly powered implants, it is also possible to charge the implanted batteries externally. As seen in the table, wireless power has been implemented in the kilohertz to megahertz range. Inductive power transfer is more efficient at lower frequencies. However, a low transmissions frequency requires larger circuit components and a larger coil. Space is important for implantable devices and this requires the design of highly efficient power transfer at higher frequencies. It is important to point out that the distance between two coils plays an important role for power efficiency.

Table 11.2 Wireless power design examples.

Powering method	Implant type	Frequency	Distance (mm)	Power transmitter	Energy/efficiency
Inductive battery charging [12]	Cochlear	152 kHz	10	A power transistor used to switch on-off the DC supply	(75 mAh)
Inductive battery charging [10]	General	4 MHz	—	—	6.15 mW
Direct inductive transfer [13]	Retinal	1 MHz	7–15	Class-E amplifier	100 mW/30%
Direct inductive [11]	General	27 MHz	20	Class-E self-oscillation	5.82 mW (transmit) 0.36 mW (receive)/0.06%
Direct inductive [14]	Animal	4 MHz	10	Class-E amplifier	4.1 mW/0.07%

11.2.3
Implantable Telemetry Links

Medical telemetry can be categorized into two groups: high data rate and low data rate systems. For example, the wireless neural recording, wireless endoscope, and the retina prosthesis are the biomedical systems requiring a large amount of data that should be delivered outside of or into the body. Multi-implantable neural recording systems and multichannel monitoring of continuous signals such as ECG and electroencephalography (EEG) also necessitate a high data rate communication. As an example, for brain–computer interfaces it is required to achieve the recording of more than 100 channels in order to simultaneously record brain functions. This necessitates a data rate more than 10 Mbps [15, 16]. A similar figure is also useful for a wireless endoscopy device for higher resolution pictures and images. The medical implant band (MICS) has channels with 300 kHz width and thus cannot provide such data rates. For high data rate and short-range applications, wideband communication is an ideal physical layer solution by achieving a data rate equal to or higher than 100 Mbps.

11.2.3.1 Wideband Telemetry Link
Wideband telemetry links have become popular because of high data rate requirements from some of biomedical implant systems including wireless endoscope, retinal implants, and multichannel continuous biological signal monitoring. Wideband technology can also find applications in biomedical monitoring, especially for multichannel continuous signal monitoring such as EEG, ECG, and electromyography (EMG). In addition to the high-capacity wireless link, the wideband telemetry link can have some additional benefits because it operates on the basis of pulses rather than sinusoidal carrier signal, enabling the low-power transmitter to increase the battery life and with less interference effect on the other wireless systems in medical centers.

There have been different methods of generating transmission pulses to obtain wideband [7, 17]. Figure 11.4 shows a very low power circuit technique to generate narrow pulses and hence a wideband spectrum. As illustrated in Figure 11.4, the oscillator signal is digitized and its delayed replica is passed through the XNOR gate to obtain a narrow square pulse (i.e., UWB pulse). After the signal is mixed with the medical data, the modulated signal is then passed through a BPF. The signal at the output of the BPF is a wideband signal that can carry a very large amount of data (Figure 11.4b). Narrow pulses can also be configured to be transmitted over an inductive link via coils rather than using antennas [17].

11.2.3.2 Multichannel Neural Recording Systems
Brain is the most complex organ of the body that has become an attractive source of research. Many efforts are being undertaken by scientists to understand the functions of the brain, which may be useful in diagnosing and treating some nervous system disorders. EEG recording has been used more than 50 years in neurological diagnosis to measure brain activity. The current technology of EEG

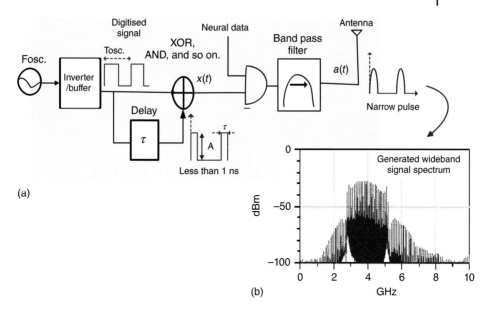

Figure 11.4 (a,b) A wideband signal generation technique.

recording has used wires to communicate between the multiple electrodes and the recording and monitoring devices so far in medical and research centers. As the electrodes placed over the subject's head, this technology restricts the subject to a particular location. In order to record and treat subjects while they are doing their daily activities, wireless technology should be involved [18]. However, due to the limitation in size, weight, and power, the commercially available wireless technology has only been successful for limited number of simultaneous recordings [19, 20]. There is still research needed on wireless recording devices of EEG to increase the number of channels to an adequate level, such as 100 channels and more.

It would be desirable to detect brain signal directly from the surface of the brain to improve the quality of the recording signal. Since their development around 1990s, using MEAs to measure the electrical activity from the brain tissue has thus been very attractive research area recently. This implantable technology allows us to detect brain signals more precisely. For such a deep implantation, using wires is impractical. The current goal in biotelemetry systems for neural recording and monitoring are made wireless so as to reduce the risk of infection and to increase the subject's comfort [21]. Simultaneous neural signal recording from many neurons has contributed to many medical applications including biological neural networks, brain-controlled neural prostheses, spinal cord injuries, stroke, sensory deficits, and neurological disorders [22, 23].

Wireless transmission of the recorded signals combined with signal processing techniques such as on-chip spike detection and sorting (sensing the activity of an ensemble of neurons) has been the main research focus in neural recording systems recently. All these critical functions should be low power and wirelessly powered or operated on a rechargeable battery. To understand the activities of real neurons, scientist need new tools and systems that will enable wideband, multichannel, and long-duration neural recording.

Wideband link (i.e., UWB) is suitable for this biomedical telemetry application as the number of neuron channels monitored increases; the data rate will reach a speed of 100 Mbps and higher [15]. Figure 11.5 shows the block diagram of a multichannel neural recording system. A general multichannel wireless neural recording system consists of preamplifiers, filters (low pass filters, LPFs), an analog multiplexer, a second amplifier, an analog-to-digital converter (ADC), and a wireless telemetry. The amplifiers are used to amplify the weak signals measured by an MEA. Neural signals obtained from the MEA have amplitude levels in the range of 40–500 μV with a bandwidth of 0.1 Hz to 10 kHz. The front-end amplifiers and LPFs are designed according to these parameters to successfully capture neural signals.

There have been significant advances in MEAs and multichannel neural recording systems. Some state-of-art neural recording designs are listed in Table 11.3. The recent reported devices for neural recording have features of simultaneous multiple channel recording, wireless data telemetry, and wireless power. It is clearly seen that the wireless neural recording devices target at least 100 channels, and thus a wideband telemetry link is needed in order to simultaneously monitor these neural signals. More research effort is required to realize a fully implantable recording system with simultaneous channels including a wireless power capability. A single device with the capability to record, process, and wirelessly transmit multiple neural signals real time will contribute significantly in the area of brain–computer interfaces.

11.2.3.3 Wireless Endoscope

A small miniaturized wireless endoscopy device can reach areas such as the small intestine and deliver real-time video images wirelessly to an external console. The

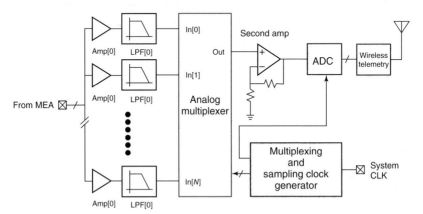

Figure 11.5 A multichannel neural recording systems with UWB transmitter [7] (Copyright @IEEE 2009).

11.2 Biotelemetry

Table 11.3 State-of-art wireless neural recording systems.

Neural recording device	# of channels	Data telemetry	Power source	Recording method/data rate
[15]	128	4 GHz, wideband	Battery, 6 mW	Spike-sorting/90 Mbps
[24]	100	915 MHz, FSK	Inductively power (2.765 MHz), 8 mW	Spike data/157 kbps
[25]	64	70–200 MHz, on–off keying (OOK)	Inductively power (4/8 MHz), 14.4 mW	Spike data/2 Mbps
[26]	96	UWB	30 mW	Raw/30 Mbps
[27]	100	3.2–3.8 GHz, FSK	90.6 mW	Raw/48 Mbps
[28]	32	433/915 MHz PWM/ASK/OOK	Inductively powered (13.56 MHz), 5.85 mW	Raw/58–709 kbps

wireless endoscopy device travels through the digestive system to collect image data and transfers them to a nearby computer for display with a distance 1 m or more. A high-resolution video-based capsule endoscope produces a large amount of data that should be delivered over a high-capacity wireless link (Figure 11.6) [4].

One of current state-of-the-art technologies for electronic pills is the commercially available PillCam by the company "Given Imaging" [29]. The pill uses the Zarlink's RF chip [6] for wireless transmission based on the MICS band.

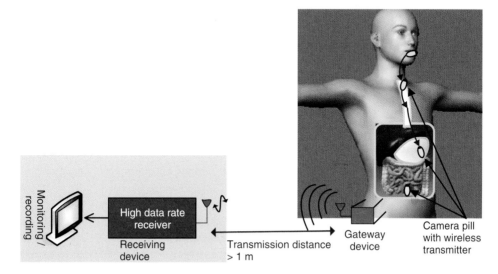

Figure 11.6 A wireless endoscope monitoring system.

An endoscope capsule requires high data rate to increase the resolution and image quality. For high data rate and short-range applications of this kind, UWB communication is an ideal physical layer solution by achieving a data rate equal or higher than 100 Mbps. For such a high data rate capacity, a wideband electronic pill can transmit raw video data without any compression, resulting in low-power transmission, less delay in real time, and increased picture resolution. With a high-definition camera, such as 2 megapixels, UWB telemetry can send up to 10 fps. The existing wireless endoscope devices use small batteries to power the electronics. A wireless power mechanism can also be used to recharge the battery of a wireless endoscopy device [30].

11.3
Microelectrode Arrays and Interface Electronics

Both recording and stimulation in a biotelemetry system require electrodes: microelectrodes, in particular, penetrate their target and are manufactured in dense arrays. Because of their very small sizes and very sharp tips, which are in the order of magnitude of a few micrometers, very dense MEAs can be manufactured, which allow stimulation and recording of a comparatively small volume of tissue [31]. This improves the stimulation selectivity and spatial resolution, and allows the recording of more neural data.

MEAs are categorized according to their function (neural recording or stimulation). To understand the functional requirements of MEAs, five essential properties should be considered, namely, the electrode site where the electrode makes contact with the tissue that is in close proximity of the cells that should be stimulated or recorded; the leads that interconnect each electrode to the electronics interface; the dielectric that electrically insulates each individual lead from the surrounding tissue; the electrode substrate providing the structural integrity; and the optional coating layer [32]. An arbitrary electrode can, in its simplest form, be represented by the equivalent electrical circuit depicted in Figure 11.7, where, Vneuron is the ideally recorded neural activity voltage, Rs is the spreading resistance, namely, the tissue resistance between the neuron and the electrode tip, Re and Ce represent the double layer impedance (Ce is the capacitance between the metal tip and the tissue, while Re is the latter's leakage resistance), Rm is the ohmic resistance of the electrode itself, and Cs is the total shunt capacitance to ground from the tip of the electrode to the input of the amplifier [33].

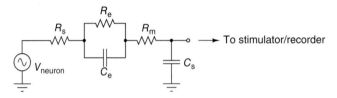

Figure 11.7 Electrode equivalent circuit.

Popular MEMS base electrode array types include microwire arrays, the Utah array type, and the Michigan array type. Microwire arrays use insulated wires with diameters ranging from 10 to 100 μm, and are typically fabricated of tungsten, platinum, platinum/iridium, epoxylite, or teflon [32]. The Utah array is a three-dimensional "bed of nails" type of structure consisting of 10 by 10 electrodes of 1.5 mm long with a 400 μm spacing: it is used to stimulate as well as to record [34]. A sputtered iridium oxide film process is used to metallize the electrode tips so as to facilitate the ionic current, and coated with Parylene-C in order to improve its biocompatibility. The Utah array has been developed in 1991, and it has been subsequently refined ever since [34]. The Michigan array includes a single silicon shank, with multiple electrode sites along the top surface, as well as the addition of on-chip circuitry for amplifying, multiplexing, and buffering neural signals recorded from 10 recording electrode sites [35]. The typical electrode length was 1.5–3 mm: the shank was tapered (from a width of 25 μm or less near the tip to 75–100 μm near the base) and had a thickness of 8–15 μm. In this way, multiple electrode sites are hereby distributed along a number of electrode shanks: this has the advantage of reducing the displacement of tissue: additionally, they can be manufactured in almost any reasonable one-, two-, or three-dimensional layout using the electrode sites across one or more planar shanks. In 1994, the first three- dimensional Michigan MEA was fabricated by combining multiple two-dimensional planar structures [36].

11.3.1
Stimulation Front Ends

Neural stimulator devices are used in applications such as the cochlear implants, deep-brain stimulation, muscle stimulation, and cortical as well as retinal prostheses. By generating an electric (ionic) current through body tissue between two electrodes, nearby cell membranes are depolarized, which generates action potentials. Because of the MEAs' high impedances, the voltage compliance at the output of the stimulating circuit needs to increase in order to source or sink a fixed current to the electrode. This effect is further exacerbated by the fact that the electrode impedances are heavily nonlinear and dependent on whether an electrode has just been stimulated or not, electrode geometries, materials used, the location of each electrode relative to the cells that need to be stimulated, the electrode array density, and the degree of foreign body response that is experienced [31].

Stimulation can be delivered using current or voltage pulses. Constant current stimulation can be integrated using a current-source output driver and has the advantage of more easily controlling charge injection as electrode impedances change over time. It is, therefore, the most preferred stimulation form at present. Depending on the stimulation target as well as on the MEA that is being used, different stimulation waveforms are required in order to generate an optimal cell excitation. The most commonly used strategy for electrical neural stimulation is to use charge balanced, constant current biphasic pulses, as these tend to cause less neuronal loss than monophasic pulses [31]. The electrode can be current stimulated using two schemes, namely using one interconnect lead per site (Figure 11.8a), or using two (Figure 11.8b) [37]. In the former case, a dual supply voltage rail is required in order to generate the biphasic current pulses. In addition, a return electrode is also required in order to complete the current loops. When two interconnect leads per site are used, only one supply voltage is required. However, two bondpads per stimulated electrode are necessary, which drastically reduces the number of stimulator drivers that can be placed on a chip. For these reasons, many circuits reported in the literature are based on a current source-type output stage using a dual supply voltage rail, which has been schematically depicted in Figure 11.8a.

Classical current mirror-based designs replicating and scaling the output currents of current-steering digital to analog converters (DACs), as depicted in Figure 11.9, work properly as long as the output transistors are biased in saturation. A current-steering DAC provides the (scaled) output current: this current is then copied using low- and high-side current mirrors. Two transistor switches are connected to the HI/LO signals in order to source/sink current to the electrode. Depending on the voltage levels as well as on the integration process, different implementations are possible. For instance, these transistor switches can be replaced with one transistor switch in series with the electrode provided that the appropriate current source transistor is turned off.

Since high output impedance is required, cascode devices are often used, meaning that the voltage headroom available to stimulate the electrode is reduced dramatically. Because of the required high voltage compliance, topologies based on wide-swing cascode transistors [38] as well as linearized voltage-controlled resistors have been introduced [39]. Both circuits are schematically depicted in Figure 11.10a and Figure 11.10b, respectively. These circuits allow an output voltage compliance, which lies very close to the supply rails, while maintaining the required high

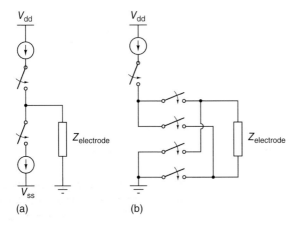

Figure 11.8 Current stimulation schemes using one (a) or two (b) interconnect leads per electrode.

11.3 Microelectrode Arrays and Interface Electronics

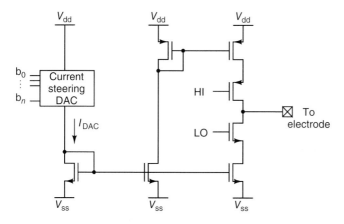

Figure 11.9 Current output driver.

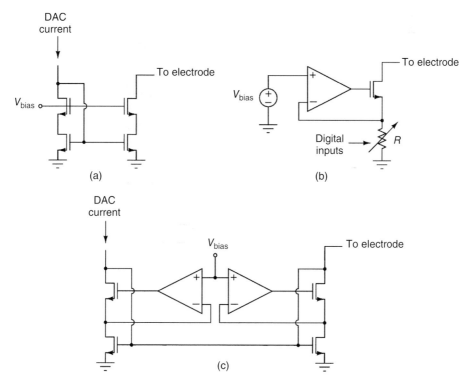

Figure 11.10 Design topologies increasing the output impedance with a minimal voltage headroom – (a) a wide swing cascode current mirror, (b) the linearized voltage-controlled resistor [39], and (c) the stimulator using a linearized current controlled resistor structure presented in [40].

output impedance. The principle behind the linearized voltage-controlled resistors is briefly explained here.

A small constant value for V_{bias} is chosen, and the resistor R is realized as a transistor that is biased in its triode region. By keeping V_{bias} constant but by controlling the resistive value of the triode-biased resistor, an increased as well as constant voltage compliance is obtained for a wide range of output currents. The opamp provides gain boosting by ensuring that the drain-source voltage of the triode-biased transistor stays at V_{bias} independently of the output current, which boosts the output impedance. However, since the voltage-controlled resistor is realized as a transistor biased in its triode region, complicated schemes must be used in order to linearize the voltage–current relationship and the threshold voltage variation of that transistor. An alternative approach has been taken in [40], where a design using a demultiplexed (driving four electrodes) and biphasic current driver with a two-reference 5-bit current-steering DAC has been described. This circuit has an inbuilt compliance monitor performing a level detection of the control output voltage, and its output driver design is based on linearized voltage-controlled resistors [35]. However, the voltage-mode DAC has been replaced by a current-steering DAC, which allows for a more straightforward implementation (Figure 11.10c). The bottom transistors are biased in their triode region, and the high current matching is achieved by adopting gain boosting to both cascode transistors, thus regulating their drain source voltage to the same value.

Since current stimulation is used, resorting to a current-steering DAC is the obvious choice. The major disadvantage of DACs for implantable stimulators is an implementation area that grows exponentially with the resolution: 2^N-1 W/L-sized transistors are necessary in order to design a current-steering DAC with a resolution of N bits. The required area can be relaxed using the multi-bias current steering DAC design proposed in [41], which reduces the W/L-sized transistor count to N. The resolution of the DAC determines the minimum current level at which each electrode can be stimulated. In addition, in the event that the circuit can stimulate with asymmetrical cathodic and anodic pulses, the DAC's resolution allows a closer matching between the sourced and sinked current, which improves charge balancing and takes the pressure of the charge-balancing circuit. Charge balance should be observed so as to reduce a detrimental net DC current which induces irreversible reduction and oxidation reactions leading to electrolysis in the cell tissues, thereby degrading the electrode, damaging tissue, or otherwise limiting the charge that can be delivered in a stimulation pulse [31]. The safe voltage range is known as the *water window*, and ranges from −0.6 to 0.8 V [42].

Clearly, charge balance is necessary to maintain the electrode potential within a range that does not cause adverse effects. One possibility is to use blocking capacitors in series with the electrodes [43]. Reverting to blocking capacitors results in large values that cannot be integrated on-chip, since they need to be much larger than the electrode's double layer capacitance in order to integrate the stimulation current. An alternative approach is to use passive charge balancing, by relying on electrode shortening [42] or on the use of discharge resistors [44]. However, in presence of a high voltage at the electrode, the current might potentially have a high value, thereby causing tissue damage. In [45], a resistive discharge path is created using a cascode connected NMOS and PMOS transistor: the advantage of this circuit is that one of these transistors is always biased as a current source irrespective of the charge polarity present at the electrode.

Charge-balancing strategies involving active devices have been investigated and designed. A possible approach consists of measuring the voltage difference between a common electrode and a reference electrode (counter electrode) when the circuit is not stimulating, and then applying a burst of charge packets of suitable polarity to the electrode until the voltage difference at this circuit's inputs falls below a specified level [46]. Because of the pulsed nature of this charge compensation circuit, it is referred to as a *pulse insertion charge balance circuit*: the latter acts instantaneously (Figure 11.11b). Another active charge balance topology consists in providing an offset current that nullifies the effect of charge buildup: identified as offset regulation charge balancing, this approach provides a continuous current, which removes the charge imbalance in the long term (Figure 11.11c) [46]. The effects of both charge balancing schemes in the presence of an asymmetrical stimulating current pulse are schematically depicted in Figure 11.11.

In [40], both previously described active charge balance circuits were designed and measured: it has been reported that both methods manage to reduce the voltage imbalance level at an arbitrary electrode below |100 mV|. The pulse injection circuit used an injected charge of 12 nC per spike and about 12 spikes per biphasic stimulation cycle were needed: on the other hand, the offset regulation circuit's control loop requires a settling time before cancelling the excess charge below the 100 mV safe window. Finally, the test chip measurements conducted in [40] reported that a passive charge balance circuit cannot cancel the charge mismatch sufficiently, resulting in a resting potential of approximately 250 mV. As mentioned by the authors, these results are heavily dependent on the individual as well as process-related mismatch conditions, and so these conclusions cannot be generalized. However, they do illustrate that active charge-balancing circuits have a superior behavior compared to passive solutions such as electrode shortening as well as to the other schemes.

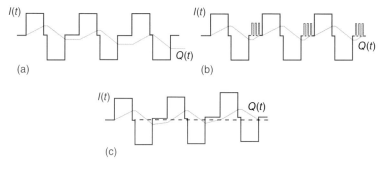

Figure 11.11 The effect of asymmetric current pulses on the total injected charge – (a) no charge balancing circuit, (b) using a pulse injection balancing circuit, and (c) using an offset regulation charge balancing circuit.

Finally, a topology resorting to dynamic current matching has been proposed in [47]. This circuit uses a dynamic current-matching calibration phase in order to match the source current to the sinked one (active charge balancing): additionally, the electrode is shorted out when it is not stimulating (passive charge balancing). A charge-imbalance current error of 6 nA and a respective residual voltage error of 12 mV were reported by the authors.

It should be reiterated that the proposed measurements and circuits depend very heavily on the application, which dictates the waveform shape, the current magnitude, and the amplitude as well as the required time resolution.

11.3.2
Recording Front-Ends

ECG, EEG, and electromyograms (EMG), are all biopotential signals that are often measured noninvasively. Neural field potentials, in contrast, are measured using medical implants [48]. It has been reported that neural spike trains are mostly located in the 300 Hz to 5 kHz energy spectrum, meaning that amplifiers should be designed so as to suppress signals outside of this band. Low-frequency signals in the 1–100 Hz are identified as local field potentials. These represent the synchronous firing of many neurons throughout a region several hundred microns from the electrode [48].

The signal quality provided by the recording circuit as well as the long-term recording stability are related to multiple factors including distance from the electrode to the neuron, the electrode surface area, the electrode to tissue interface impedance, the insertion trauma, and the foreign body response [49]. Since the peak-to-peak amplitude of a transmembrane action potential is around 70 mV, the corresponding neuronal signal levels that are recorded by the implanted MEAs have typical amplitudes in the range of 50 µV to 1 mV, with a spectral bandwidth spanning from sub-hertz values (ECG, EMG, local field potentials) to 10 kHz (for recording action potentials directly) [25].

Because of these small recording signal levels, the analog front-end circuitry should have a very low noise figure while amplifying the recorded signal with a high gain. Since the signal-to-noise ratio of the recorded signal should be maximized, it is preferable for the electrodes to have low impedances. This is understood by considering that the thermal noise power spectral density generated by an electrode is given by Johnson–Nyquist's formula:

$$\overline{v}_n^2 = 4k_B T R$$

where k_B is Boltzmann's constant, T is the temperature, R is the resistance of the electrode, and B is the recording bandwidth. As shown in the equivalent circuit depicted in Figure 11.7, the resistance of the electrode includes the resistance of the metallic portion of the microelectrode, the resistive component of the double-layer capacitance, as well as the seal resistance. In addition, high electrode impedance reduces its high frequency behavior and increases the required instrumentation amplifier's input impedance. It has been mentioned in [50] that the input-referred noise of the instrumentation amplifier should be kept below the background noise of the recording site, which can be around 5–10 µV.

Because of the high source (electrode) impedance levels, a high level of noise as well as common mode interference will be superimposed on the recorded biopotential signal. While the former reduces the signal-to-noise ratio, thereby impacting the effective resolution of the recorded signal, a fraction of the latter will be converted to a differential signal owing to the finite common-mode rejection ratio (CMRR). A high disturbing signal means that the dynamic range of the instrumentation amplifier should increase accordingly, which, in turn, leads to increased power consumption. As an example, the power line interference that is received differentially on the human body can be as large as 5 mV_{ptop} differentially [51]. If the system is devised to measure a biopotential signal of 5 μV_{ptop}, the increase in dynamic range will be equal to 30 dB.

The CMRR of an amplifier is defined as the ratio of the differential mode gain over the common mode gain. This spurious signal will distort the recorded signal. Finally, due to electrochemical effects at the electrode–tissue interface, DC offset voltages in the range of 1–2 V are present at the electrode–electrolyte interface, meaning that the amplifier needs to be capacitively coupled [52]. All these factors complicate the design of the analog front-end.

A biopotential recording analog front-end consists, in its simplest form, of an instrumentation amplifier, followed by an LPF that cancels out the high frequency noise, followed by an ADC.

11.3.2.1 Instrumentation Amplifier

Because of the DC offset voltage present at the electrode–tissue interface, a widely used topology is the capacitively coupled negative-feedback amplifier [52]. This circuit is depicted in Figure 11.12. The mid-band voltage gain of this amplifier is set by the ratio C_1/C_2. The input is capacitively coupled through C_1, so that any DC offset from the electrode–tissue interface is removed. In a practical setup, C_2 is much smaller than C_1. In addition, for the mid-band frequencies, the impedance presented by RF is larger than the one presented by C_1. This means that for the

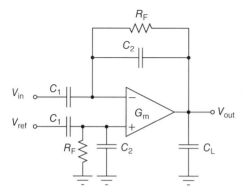

Figure 11.12 Neural integrated recording amplifier proposed in [52].

frequencies of interest, the amplifier is connected in an inverting voltage amplifier configuration, which, in turn, sets the input impedance of this circuit equal to C_1. For this reason, C_1 should be made much smaller than the electrode impedance so as to minimize signal attenuation. Feedback resistance R_F biases the circuit at DC, while setting the lower cutoff frequency in combination with C_2 ($f_L = 1/R_F \cdot C_2$). The higher cutoff frequency depends on the transconductance Gm of the amplifier and on the load capacitance ($f_H = (C_1/C_2) \cdot (Gm/C_L)$).

Clearly, R_F should have a very high value in order to decrease the low frequency pole. As an example, assuming that the impedance of the microelectrode is equal to 1 MΩ at 1 kHz, C1 should be 16 pF in order to make the input impedance 10 MΩ [53]. This means that in order to achieve a lower cut-off frequency equal to 1 Hz, the feedback resistance RF needs to be equal to 10 GΩ. Realizing a resistance of this magnitude on chip across a sufficient input voltage range is a considerable challenge. In [52], a PMOS transistor connected in the following configuration was used. This MOS-parasitic BJT (bipolar junction transistor) configuration is known as *pseudoresistance*, and is depicted in Figure 11.13a. In case the voltage V_X is negative, both transistors behave as forward-biased PMOS diode connected transistors (Figure 11.13b). In case V_X is positive, the transistors are off as there is no inversion channel: however, the parasitic drain to bulk diodes will be forward biased, since the bulks are connected to the sources (Figure 11.13c). Observe that as long the voltage swings do not forward bias the respective diode (MOS or drain-bulk), a very high impedance is obtained with this configuration.

Other recording amplifiers have used various schemes utilizing the properties of subthreshold biased transistors [15, 54]. The resistance obtained using pseudo-resistors has been reported to be superior to the one obtained with subthreshold transistors [50]. Finally, some works use the large off-resistance of reset switches [26].

In [52] a symmetrical operational transconductance amplifier (OTA) has been used. The transistors are scaled accordingly so as to make their thermal noise component small, while avoiding a significant power sacrifice. The input transistors are PMOSs in order to minimize their $1/f$ noise contribution, as the latter can be up to 30–60 times higher for NMOS transistors. The study in [52] reported an input-referred noise of 2.2 μV_{rms} over a bandwidth of 7.2 kHz, while requiring a power consumption of 80 μW per channel.

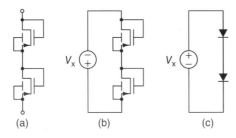

Figure 11.13 (a–c) The pseudoresistor used in [52] in order to generate a very high impedance.

In [15] a fully differential and self-biased OTA has been presented. The overall instrumentation amplifier topology using the fully differential and self-biased OTA are depicted in Figure 11.14a, and Figure 11.14b, respectively. This front-end circuit has been implemented in a 0.35 μm CMOS test chip, which allows 128-channel neural recording with on-the-fly spike feature extraction and wireless telemetry using UWB. Sixteen instrumentation amplifiers are shared through time-multiplexing.

A possible approach in order to suppress the detrimental DC offset voltage, as well as to mitigate the effect of $1/f$ noise, is to use a chopper-stabilized instrumentation amplifier [55]. This principle was subsequently refined in [56, 57]. The principle of chopping-stabilized amplifiers is depicted in Figure 11.15. The input signal is mixed with the chopper frequency. After the modulation, the upconverted signal is amplified, and the $1/f$ noise contribution of the amplifier is summed on top of

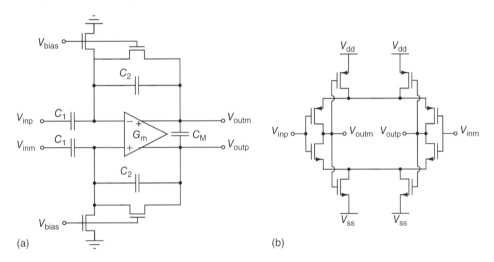

Figure 11.14 Neural integrated recording amplifier proposed in [15] – (a) the recording amplifier and (b) the fully differential and self-biased OTA.

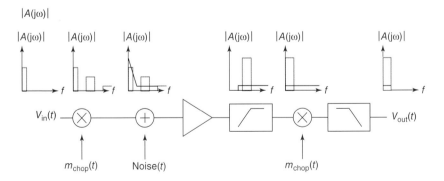

Figure 11.15 Principle schematic of chopping-stabilized amplifiers.

this signal. However, since the wanted signal has been upconverted, a high pass filter will remove the unwanted low frequency components ($1/f$ noise as well as DC offset). After the signal has been downconverted to the original signal band, an LPF ensures that the out-of-band high frequency noise is filtered. Therefore, the $1/f$ noise contribution of the amplifier, as well as the DC offset level at the input, is effectively removed.

A common problem with chopper-stabilized amplifiers is the chopping spikes generated at the output because of the switching of the chopper: the high frequency components can mix into the base band and cause spurious signals and spikes [56]. This is overcome by adding a chopping spike filter in order to suppress these spurious signals. Other undesirable effects include the finite bandwidth in the signal path of the chopper-stabilized instrumentation amplifier, which ultimately results in distortion and sensitivity issues, as described in [57]. This effect is exacerbated by the low bandwidth of the instrumentation amplifier, which is kept in that range since the biopotential signals to be measured lie at relatively limited frequencies, resulting in reduced power consumption. In addition, the DC offsets that are present at the electrode–tissue interface need first to be amplified by the chopping amplifier before being removed by the LPF: a significant DC offset, combined with a high voltage gain, may, therefore, saturate the amplifier. In [57], these issues have been solved using a multipath feedback loop, which is added in order to suppress distortion.

11.4
Conclusion

There is considerable demand in the development of miniaturized, low-power wireless communication systems for implantable biomedical devices. Wireless telemetry will continue to be an essential part of biomedical devices operating inside the human body. In this chapter, we have discussed implementation issues and presented details of techniques for design of wireless implantable telemetry. Current research projects in the area of wireless implantable telemetry are targeting to extend the performance, reduce the power consumption, and increase data rate. In addition, the design of wideband telemetry link has been discussed. Wideband telemetry link provides some advantages for multichannel neural recording and wireless endoscopy systems. A high-efficiency wireless power transmission is a primary requirement for implantable systems to eliminate the use of batteries.

References

1. Yuce, M.R. and Khan, J. (2011) *Wireless Body Area Networks: Technology, Implementation and Applications*, Pan Stanford Publishing ISBN: 978-981-431-6712.
2. Webster, J.G. (1998) *Medical Instrumentation Application and Design*, John Wiley & Sons, Inc., Hoboken, NJ.
3. Yuce, M.R. (2010) Implementation of wireless body area networks for

healthcare systems. *Sens. Actuators, A: Phys.*, **162**, 116–129.

4. Yuce, M.R. and Dissanayake, T. (2012) Easy-to-swallow wireless telemetry. *IEEE Microwave Mag.*, **13**, 90–101.

5. Liu, W. et al. (2005) Implantable biomimetic microelectronic systems design. *IEEE Eng. Med. Biol. Mag.*, **24**, 66.

6. Bradley, P. (2006) An ultra low power, high performance medical implant communication system (MICS) transceiver for implantable devices. Proceeding of the IEEE Biomedical Circuits and Systems Conference, pp. 158–116.

7. Yuce, M.R., Ho, C.K., and Chae, M. (2009) Wideband communication for implant-able and wearable systems. *IEEE Trans. Microwave Theory Tech.*, **57**, 2597–2604.

8. Troyk, P.R. and DeMichele, G.A. (2003) Inductively-coupled power and data link for neural prostheses using a class-E oscillator and FSK modulation Proceeding of the IEEE International Conference, Engineering in Medicine and Biology Society, September, 2003, pp. 3376–3379.

9. Zhou, M., Yuce, M.R., and Liu, W. (2008) A non-coherent DPSK data receiver with interference cancellation for dual-band transcutaneous telemetries. *IEEE J. Solid State Circuits*, **43**, 2203–2012.

10. Li, P. and Bashirullah, R. (2007) A wireless power interface for rechargeable battery operated medical implants. *IEEE Trans. Circuits Syst. II*, **54**, 912–916.

11. Laskovski, A. and Yuce, M.R. (2011) A MICS telemetry implant powered by a 27MHz ISM inductive link. The International Conference of the IEEE Engineering in Medicine and Biology Society, August-September 2011, pp. 290–2912.

12. Lim, H., Yoon, Y., Lee, C., Park, I., Song, B., and Cho, J.H. (2005) Implementation of a transcutaneous charger for fully implantable middle ear hearing device, IEEE-EMBS 2005. 27th Annual International Conference of the, Engineering in Medicine and Biology Society, pp. 6813–6816.

13. Wang, G., Liu, W., Sivaprakasam, M., Zhou, M., Weiland, J. D., and Humayun, M.S.A (2006) Dual band wireless power and data telemetry for retinal prosthesis. EMBS '06. 28th Annual International Conference of the IEEE, pp. 4392–4395.

14. Zimmerman, M.D., Chaimanonart, N., and Young, D.J. (2006) In Vivo RF powering for advanced biological research. 28th Annual International Conference of the IEEE, 2006 Engineering in Medicine and Biology Society, pp. 2506–2509.

15. Chae, M., Yang, Z., Yuce, M.R., Hoang, L., and Liu, W. (2009) A 128-channel 6 mW wireless neural recording ic with spike feature extraction and UWB transmitter. *IEEE Trans. Neural Syst. Rehabil. Eng.*, **17**, 312–321.

16. Harrison, R.R. (2008) The design of integrated circuits to observe brain activity. *Proc. IEEE*, **96**, 1203–1216.

17. Inanlou, F. and Ghovanloo, M. (2011) Wideband near-field data transmission using pulse harmonic modulation. *IEEE Trans. Circuits Syst. I*, **58**, 186–195.

18. Babiloni, F. et al. (2007) Simultaneous tracking of multiple brains activity with high resolution EEG hyperscannings. Proceeding of the IEEE International Conference on Functional Biomedical Imaging, NFSI-ICFBI, October 12–14, pp. 196–199.

19. Farshchi, S., Pesterev, A., Nuyujukian, P., Modi, I., and Judy, J. (2006) A TinyOS-enabled MICA2-based wireless neural interface. *IEEE Trans. Biomed. Eng.*, **53**, 1416.

20. Modarreszadeh, M., and Schmidt, R.N. (1997) Wireless, 32-channel, EEG and epilepsy monitoring system. Proceeding of the 19th International Conference IEEE/EMBS, Chicago, IL, October 30–November 2, p. 1157.

21. Sodagar, A.M., Wise, K.D., and Najafi, K. (2007) A fully integrated mixed-signal neural processor for implantable multi-channel cortical recording. *IEEE Trans. Biomed. Eng.*, **54** (6), 1075–1088.

22. Schwartz, A.B. (2004) Cortical neural prosthetics. *Annu. Rev. Neurosci.*, **27**, 487–501.

23. Loeb, G.E., Peck, R.A., Moore, W.H., and Hood, K. (2001) BION system for

23. distributed neural prosthetic interfaces. *Med. Eng. Phys.*, **23** (1), 9–18.
24. Harrison, R.R., Kier, R.J., Chestek, C.A., Gilja, V., Nuyujukian, P., Ryu, S., Greger, B., Solzbacher, F., and Shenoy, K.V. (2009) Wireless neural recording with single low-power integrated circuit. *IEEE Trans. Neural Syst. Rehab. Eng.*, **17** (4), 322–329.
25. Sodagar, A.M., Perlin, G.E., Yao, Y., Najafi, K., and Wise, K.D. (2009) An implantable 64-channel wireless microsystem for single-unit neural recording. *IEEE J. Solid-State Circuits*, **44**, 2591–2604.
26. Gao, H., Walker, R.M., Nuyujukian, P., Makinwa, K.A.A., Shenoy, K.V., Murmann, B., and Meng, T.H. (2012) HermesE: A 96-channel full data rate direct neural interface in 0.13 um CMOS. *IEEE J. Solid State Circuits.*, **47** (4), 1043–1054.
27. Yin, M., Borton, D.A., Aceros, J., Patterson, W.R., and Nurmikko, A.V. (2012) A 100-channel hermetically sealed implantable device for wireless neurosensing applications. Proceeding of the IEEE International Symposium on Circuits and Systems, May 2012, pp. 2629–2632.
28. Lee, S.B., Lee, H.-M., Kiani, M., Jow, U.-M., and Ghovanloo, M. (2010) An inductively powered scalable 32-channel wireless neural recording system-on-a-chip for neuroscience applications. *IEEE Trans. Biomed. Circuits Syst.*, **4**, 360–371.
29. Givenimaging (2013) http://www.givenimaging.com/ (accessed 11 September 2013).
30. Puers, R., Carta, R., and Thone, J. (2011) Wireless power and data transmission strategies for next-generation capsule endoscopes. *J. Micromech. Microeng.*, **21**, 054008.
31. Cogan, S.F. (2008) Neural stimulation and recording electrodes. *Annu. Rev. Biomed. Eng.*, **10** (1), 275–309.
32. Otto, K.J., Ludwig, K.A., and Kipke, D.R. (2012) Acquiring brain signals from within the brain, in *Brain-Computer Interfaces: Principles and Practice*, Oxford University Press.
33. Robinson, D.A. (1968) The electrical properties of metal microelectrodes. *Proc. IEEE*, **56** (6), 1065–1071.
34. Bhandari, R., Negi, S., and Solzbacher, F. (2010) Wafer-scale fabrication of penetrating neural microelectrode arrays. *Biomed. Microdevices (Springer)*, **12** (5), 797–807.
35. Najafi, K., Wise, K., and Mochizuki, T. (1985) A high-yield IC-compatible multichannel recording array. *IEEE Trans. Electron Devices*, **22**, 230–241.
36. Hoogerwerf, A.C. and Wise, K.D. (1994) A three-dimensional microelectrode array for chronic neural recording. *IEEE Trans. Biomed. Eng.*, **41** (12), 1136–1146.
37. Sivaprakasam, M., Liu, W., Wang, J.S., Weiland, D., and Humayun, M.S. (2005) Architecture Tradeoffs in High-Density Microstimulators for Retinal Prosthesis. *IEEE Transactions on Circuits and Systems I*, **52** (12).
38. Jones, K.E. and Normann, R.A. (1997) An advanced demultiplexing system for physiological stimulation. *IEEE Trans. Biomed. Eng.*, **44** (12), 1210–1220.
39. Ghovanloo, M. and Najafi, K. (2005) A compact large voltage-compliance high output-impedance programmable current source for implantable microstimulators. *IEEE Trans. Biomed. Eng.*, **52** (1), 97–105.
40. Noorsal, E., Sooksood, K., Xu, H., Hornig, R., Becker, J., and Ortmanns, M. (2012) A neural stimulator frontend with high-voltage compliance and programmable pulse shape for epiretinal implants. *IEEE J. Solid-State Circuits*, **47** (1), 244–256.
41. DeMarco, S.C., Liu, W., Singh, P.R., Lazzi, G., Humayun, M.S., and Weiland, J.D. (2003) An arbitrary waveform stimulus circuit for visual prosthesis using a low-area multibias DAC. *IEEE J. Solid-State Circuits*, **38** (10), 1679–1690.
42. Ghovanloo, M. and Najafi, K. (2007) A wireless implantable multichannel microstimulating system-on-a-chip with modular architecture. *IEEE Trans. Neural Syst. Rehabil. Eng.*, **15** (3), 449–457.
43. Suaning, G. and Lovell, N.H. (2001) CMOS neurostimulation ASIC with 100

channels, scalable output, and bidirectional radio-frequency telemetry. *IEEE Trans. Biomed. Eng.*, **48** (2), 248–260.

44. Liu, W., Vichienchom, K., Clements, M., DeMarco, S.C., Hughes, C., McGucken, E., Humayun, M.S., de Juan, E., Weiland, J.D., and Greenberg, R. (2000) A neuro-stimulus chip with telemetry unit for retinal prosthetic device. *IEEE J. Solid-State Circuits*, **35** (10), 1487–1497.

45. Sivaprakasam, M., Liu, W., Humayun, M. S., and Weiland, J. D. (2005) A variable range bi-phasic current stimulus driver circuitry for an implantable retinal prosthetic device, *IEEE Journal of Solid-State Circuits*, **40**, (3), pp. 763–771.

46. Ortmanns, M., Rocke, A., Gehrke, M., and Tiedtke, H.-J. (2007) A 232-channel epiretinal stimulator ASIC. *IEEE J. Solid-State Circuits*, **42** (12), 2946–2959.

47. Sit, J.-J. and Sarpeshkar, R. (2007) A low-power blocking-capacitor-free charge-Balanced electrode-stimulator chip with less than 6 nA DC Error for 1-mA full-scale stimulation. *IEEE Trans. Biomed. Circuits Syst.*, **3**, 184–192.

48. Harrison, R.R., Watkins, P.T., Kier, R.J., Lovejoy, R.O., Black, D.J., Greger, B., and Solzbacher, F. (2007) A low-power integrated circuit for a wireless 100-electrode neural recording system. *IEEE J. Solid-State Circuits*, **42** (1), 123.

49. López, C.M., Welkenhuysen, M., Musa, S., Eberle, W., Bartic, C., Puers, R., and Gielen, G. (2012) Towards a noise prediction model for in vivo neural recording, International Conference of the IEEE Engineering in Medicine and Biology Society, 2012.

50. Wattanapanitch, W., Fee, M., and Sarpeshkar, R. (2007) An energy-efficient micropower neural recording amplifier. *IEEE Trans. Biomed. Circuits Syst.*, **1** (2), 136–147.

51. Bohorquez, J.L., Yip, M., Chandrakasan, A.P., and Dawson, J.L. (2011) A biomedical sensor interface with a sinc filter and interference cancellation. *IEEE J. Solid-State Circuits*, **46** (4), 746–756.

52. Harrison, R.R. and Charles, C. (2003) A low-power low-noise cmos amplifier for neural recording applications. *IEEE J. Solid-State Circuits*, **38** (6), 958–965.

53. Chae, M.S., Yang, Z., and Liu, W. (2010) Microelectronics of recording, stimulation, and wireless telemetry for neuroprosthetics: design and optimization, in *Implantable Neural Prostheses Part II*, Springer, New York, pp. 253–330.

54. Mohseni, P. and Najafi, K. (2004) A fully integrated neural recording amplifier with DC input stabilization. *IEEE Trans. Biomed. Eng.*, **51** (5), 832–837.

55. Dagtekin, M., Liu, W., and Bashirullah, R. (2001) A multi channel chopper modulated neural recording system. IEEE International Conference of the Engineering in Medicine and Biology Society, 2001.

56. Yazicioglu, R.F., Merken, P., Puers, R., and Van Hoof, C. (2007) A 60 uW 60 nV/ sqrt(Hz) readout front-end for portable biopotential acquisition systems. *IEEE J. Solid-State Circuits*, **42** (5), 1100–1110.

57. Denison, T., Consoer, K., Santa, W., Avestruz, A.-T., Cooley, J., and Kelly, A. (2007) A 2 uW 100 nV/sqrt(Hz) chopper-stabilized instrumentation amplifier for chronic measurement of neural field potentials. *IEEE J. Solid-State Circuits*, **42** (12), 2934–2945.

12
Nano-Enabled Implantable Device for *In Vivo* Glucose Monitoring

Esteve Juanola-Feliu, Jordi Colomer-Farrarons, Pere Miribel-Català, Manel González-Piñero, and Josep Samitier

12.1
Introduction

12.1.1
Nanotechnology

Nanotechnology is seen to represent a new industrial revolution, boosting the biomedical devices market. Nanosensors may well provide the tools required for investigating biological processes at the cellular level *in vivo* when embedded into medical devices of small dimensions, using biocompatible materials, and requiring reliable and targeted biosensors, high-speed data transfer, safely stored data, and even energy autonomy.

A *nanobiosensor or nanosensor* is generally defined as a nanometer-scale measurement system comprising a probe with a sensitive biological recognition element, or bioreceptor, a physicochemical detector component, and a transducer in between. Two types of nanosensors with potential medical applications are cantilever array sensors and nanotube/nanowire sensors and nanobiosensors, which can be used to test nanoliters or less of blood for a wide range of biomarkers. In our work, a nanobiosensor with three electrodes has been selected to explain and develop the system.

To achieve the ambitious goal of bridging the gap from the public-funded research to the market, we need to improve the performance of each dimension in the "knowledge triangle": education, research, and innovation. Indeed, recent findings point to the importance of strategies of adding value and marketing during R&D processes so as to leverage the successful commercialization of new technology-based products [1]. Moreover, in a global economy in which conventional manufacturing is dominated by developing economies, the future of industry in the most advanced economies must rely on its ability to innovate in those high-tech activities that can offer a differential added value, rather than on improving existing technologies and products. It seems quite clear, therefore, that the combination of

Implantable Bioelectronics, First Edition. Edited by Evgeny Katz.
© 2014 Wiley-VCH Verlag GmbH & Co. KGaA. Published 2014 by Wiley-VCH Verlag GmbH & Co. KGaA.

health (medicine) and nanotechnology in a new biomedical device is very capable of meeting these requisites.

Nanotechnology provides breakthroughs that support endless sources of innovation and creativity at the intersection between medicine, biotechnology, engineering, the physical sciences, and information technology, and the discipline is opening up new directions in R&D, knowledge management, and technology transfer. A number of nanotech products are already in use and analysts expect markets to grow by hundreds of billions of euros during the present decade. After a long R&D incubation period, several industrial segments are already emerging as early adopters of nanotech-enabled products [2]; in this context, surprisingly rapid market growth is expected and high mass market opportunities are envisaged for targeted research subsegments. Findings suggest that the bio and health market will provide some of the greatest advances over the next few years and that, as a result, the applications of nanoscience and technology to medicine will benefit patients by providing new prevention assays, early diagnosis, nanoscale monitoring, and effective treatment via mimetic structures. It is doubtless that there are considerable challenges in the design of nanostructures that can operate reliably over extended timescales in the body.

The scale-length reduction that has been achieved through nanosynthesis (bottom-up technology) and nanomachining (top-down technology) has the potential to interact with the biological world as never before. Bio-nanotechnologies operate at the interface between organized nanostructures and biomolecules, which are key control routes for achieving new breakthroughs in medicine; dentistry and therapeutics; in food of animal and vegetable origin; and in daily care products such as cosmetics. According to the GENNESYS White Paper (2009), this new field of research will provide significant breakthroughs in the near future in the realms of bioreactors, biocompatible materials, and lab-on-chip technologies.

12.1.2
Nanomedicine

For the purpose of this document, *nanomedicine* is defined as the application of nanotechnology to health. It exploits the improved and often novel physical, chemical, and biological properties of materials at the nanometric scale. Nanomedicine has a potential impact on the prevention, early and reliable diagnosis, and treatment of diseases. In the nanomedicine case, there is a wide range of technologies that can be applied to medical devices, materials, procedures, and treatment modalities. A closer look at nanomedicine introduces emerging nanomedical techniques such as nanosurgery, tissue engineering, nanoparticle-enabled diagnostics, and targeted drug delivery. Still in its infancy, much of the work in the discipline involves R&D and it is, therefore, crucial that health institutions, research institutes, and manufacturers work together efficiently. In particular, multidisciplinary research groups and technology transfer offices are playing a key role in the development of new nano-enabled implantable biomedical devices through advanced understanding of the microstructure/property relationship for biocompatible materials and of their effect on the structure/performance of these devices. To proceed further, a general framework is required that can facilitate an understanding of the technical and medical requirements so that new tools and methods might be developed. Moreover, in medicine there is a pressing need to ensure close cooperation between university-hospital-industry-administration while specific tools and procedures are developed for use by clinicians. Drawing on the experience of the authors, in this case study we seek to demonstrate the importance of cooperation and collaboration between these four stakeholders and the citizens involved in the innovation process leading to the development of new biomedical products ready for the market.

The interaction between medicine and technology allows the development of diagnostic devices to detect or monitor pathogens, ions, diseases, and so on. Today, the integration of rapid advances in areas such as microelectronics, microfluidics, microsensors, and biocompatible materials allows the development of implantable biodevices such as the lab-on-chip and the point-of-care devices [3]. As a result, continuous monitoring systems are available to develop faster and cheaper clinical tasks – especially when compared with standard methods. It is in this context that we present an integrated front-end architecture for *in vivo* detection.

12.2
Biomedical Devices for *In Vivo* Analysis

12.2.1
State of the Art

Many different problems need to be overcome in obtaining the ideal implantable device [4]. First and foremost, the device must be biocompatible to avoid unfavorable reactions within the body. Secondly, the medical device must provide long-term stability, selectivity, calibration, miniaturization, and repetition, as well as power in a downscaled and portable device. In terms of the sensors, label-free electrical biosensors are ideal candidates because of their low cost, low power, and ease of miniaturization. Recent developments in nanobiosensors provide suitable technological solutions in the field of glucose monitoring [5], and pregnancy and DNA testing [6]. Electrical measurement, when the target analyte is captured by the probe, can exploit voltmetric, amperometric, or impedance techniques. Ideally then, the device should be able to detect not just one target agent or pathology but rather different agents, and it should be capable of working in a closed-loop feedback, as described by Wang [7] in the case of glucose monitoring.

Several biomedical devices for *in vivo* monitoring are currently being developed. Thus, highly stable, accurate intramuscular implantable biosensors for the simultaneous continuous monitoring of tissue lactate and glucose have recently been produced, including a complete electrochemical cell-on-a-chip. Moreover, with the parallel development of the on-chip potentiostat and signal processing, substantial progress has been made toward a wireless implantable glucose/lactate sensing biochip [8]. Elsewhere, implantable bio-micro-electromechanical systems (bio-MEMSs) for the *in situ* monitoring of blood flow have been designed. Here, the aim was to develop a smart wireless sensing unit for noninvasive early stenosis detection in heart bypass grafts [9]. Interestingly, this study examines the use of surface coatings in relation to biocompatibility and the non-adhesion of blood platelets and constituents. In this case, nanotechnology presents itself as a useful tool for improving the biocompatibility of silicon bio-MEMS structures when nanoscale metallic titanium layers are used, since it enhances biocompatibility.

The next step involves developing a configurable application-specific integrated circuit (ASIC) working with a multiplexed array of nanosensors designed to be reactive for a set of target agents (enzymes, viruses, molecules, chemical elements, molecules, etc.). Multiple sensors of the array can then be used for one specific target, while other arrays can be prepared for the other targets, while also seeking a redundant response. Thus, a panel of biomarkers needs to be developed. In this way, the reproducibility and accuracy can be improved for each target, and different targets can be assayed simultaneously.

The configuration capacity of the ASIC should also be defined in terms of the type of measurement that is to be conducted [10]: it could be amperometric, measuring current variations and detecting threshold values [11], or it could be electrochemical impedance spectroscopic, for a fixed frequency, detecting both impedance variations crossing threshold values and anomalies. The combination of both techniques of measurement could be used to obtain a more reliable method of detection. Power and communications are also key features in the design of implantable devices. The former is concerned with methods of transferring sufficient energy to the devices, whereas the latter involves the integration of instrumentation and communication electronics to control the sensors and to send the information provided by the sensors through human skin. However, if the detection of vital signs or the threshold detections are sufficient for monitoring purpose, it is not necessary to measure and send raw data with a high degree of accuracy from the user to an external data processing unit. Indeed, local processing within the same sensor would reduce power and communication requirements.

Radio frequency (RF) power harvesting through inductive coupling is an increasingly used alternative for transmitting energy to the implanted device, as opposed to using batteries or wires [12]. Furthermore, this alternative permits a bidirectional communication to be established between the implanted device and an external base or reader. A number of implantable telemetry circuits based on inductive coupling can be found in the literature [13]. By contrast, several studies have developed integrable electronics for *in vivo* monitoring. Examples of this are provided in some studies where femtoampere-sensitivity applications for conductometric biosensors are used [14], and where a signal processing unit based on a current-to-frequency converter and a communication protocol is presented [15].

12.2.2
The Innovative Biomedical Device

The innovation introduced is a system envisaged to be implanted under the human skin. Powering of and communication between this device and an external primary transmitter are based on an inductive link. The architecture presented is designed for two different approaches: defining a true/false alarm system based on either amperometric or impedance nano-biosensors. It is the aim of this work to focus on diabetes, among the diseases that might be monitored by *in vivo* analysis, given that its incidence and prevalence are increasing worldwide, reflecting lifestyle changes and aging populations. Specifically, this growing prevalence is closely linked to that of obesity, creating significant market opportunities as reported in the World Diabetes Market Analysis 2010–2025 [16], and, especially, because the World Health Organization estimates that the number of diabetics will exceed 350 million by 2030.

For this *in vivo* implantable biomedical device, we also examine an ambitious approach that covers the entire value chain (from basic research through engineering and technology to industry), the infrastructure required, and the implications for society of these and similar current market challenges. In this instance, the entire value chain is hosted by the university system, which highlights the social turnover of public research investment. We also consider the extent to which recent technological innovations in the biomedical industry have been based on academic research, and the time lags between investment in such academic research projects and the industrial application of their findings – that is, so as to estimate the social rate of return from academic research. Because the results of academic research are so widely disseminated and their effects so fundamental, subtle, and widespread, it is often difficult to identify and measure the link between academic research and industrial innovation. Nevertheless, there is convincing evidence, particularly from industries such as drugs, instruments, and information processing, that the contribution of academic research to industrial innovation has been considerable [17].

12.2.3
Architecture of the Implantable Device

At this juncture, the architecture presented represents an initial approach for the development of applications based on biosensors aimed at detecting the presence or absence of certain levels of proteins, antibodies, ions, oxygen, glucose, and so on. These *in vivo* detection circuits, or true/false applications [14], work as an alarm. When the concentration level under analysis falls outside a range of accepted values, a threshold value activates the alarm. For instance, in the case of glucose monitoring, the detection of a threshold decrease in glucose levels would be mandatory for avoiding critical situations such as hypoglycemia [15, 18]. Such detection would be achieved when the amplitude of the measured signal falls below a specified threshold value.

Various approaches have been developed for the continuous monitoring of glucose [19]. These range from commercial solutions such as the blood glucose tester marketed by Cygnus Inc. to subcutaneous Minimed Medtronic and Abbott Inc. solutions. These devices, placed just under the skin, have a closed-loop control to deliver insulin and enjoy autonomy up to 5 days.

The generic implantable, front-end architecture [20] is based on inductive coupling for the *in vivo* monitoring of the presence or absence of pathogens, ions, oxygen concentration levels, and so on.

The system in Figure 12.1 shows a platform with a true/false alarm for the monitoring of different targets [21]. The data are transferred to a central database where all the inputs can be personalized for each patient. The data collected can be measured in different scenarios: when the patient is at rest, is undertaking a certain type of physical activity, and so on, depending on the parameter of interest; hence, an accurate diagnosis can be obtained and prognosis predicted [22]. The system has a research application in the constant monitoring of patients as they carry out their daily activities under normal conditions.

Figure 12.2 shows the designed implantable prototype with two full custom integrated circuits (ICs), one dedicated to power and communications and the other one with the sensor's interface electronics. A PCB (printed circuit board)-transponder antenna (30 mm × 15 mm), tuned to 13.56 MHz, in the basic industrial, scientific, and medical (ISM) bandwidth is placed near the power and communications IC to provide the power and communication link. The three connections for the amperometric biosensor, which has been the approach followed for the sensor, are also shown. The proposed architecture is analyzed at this stage as a threshold detector for just one sensor, working amperometrically, and includes all on-chip electronics, which are presented in Figure 12.3, implemented in this case using two different ICs. Future approaches will define just one IC implementation.

The three amperometric electrodes making up the sensor are (i) the working electrode (W), which serves as a surface on which the electrochemical reaction takes place; (ii) the reference electrode (R), which measures the potential at the

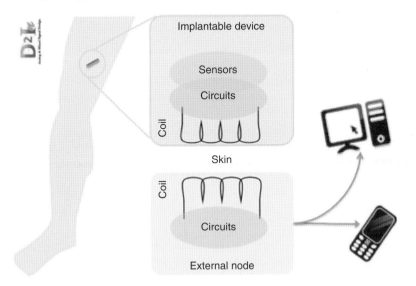

Figure 12.1 Conception of the implantable device. (Adapted with permission from [21]; Copyright (2011) IEEE, USA.)

Figure 12.2 Implantable prototype (30 mm × 6 mm × 1.5 mm). (Reproduced from [20] with kind permission from Springer Science+Business Media B.V.)

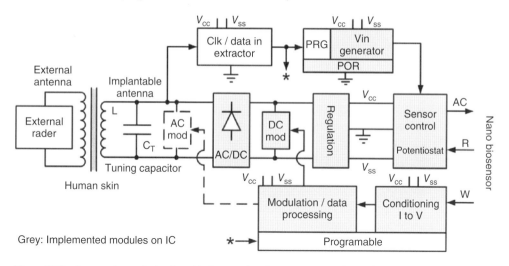

Figure 12.3 Proposed generic implantable front-end architecture. (Reproduced from [20] with kind permission from Springer Science+Business Media B.V.)

W electrode; and (iii) the auxiliary or counter electrode (A/C), which supplies the current required for the electrochemical reaction at the W electrode.

The system is designed as a wirelessly powered active radio frequency identification (RFID) tag [23] where the inductively coupled link, generated by the implantable and the external antenna, is able to supply enough energy to power the entire system and to provide wireless bidirectional communication through the human skin.

12.2.4
Implantable Front-End Architecture for *In Vivo* Detection Biosensor Applications

Nowadays, evolution in the microelectronics facilities and the medical needs are opening up new trends in medical technologies. Future patients will have different types of biosensors, placed externally and also internally in the body, for monitoring purposes. This is the concept of pervasive monitoring [24]. The conception of a front-end subcutaneous electronics is presented for the event-detector implantable device envisaged, inspired by the solution in [25], where a artificial pancreas is presented.

The approaches for the external sensors or the *in vivo* implantable sensors are quite different in terms of two key aspects in their design: powering and communications. In the case of external sensors that define the body–wireless sensor network, no major powering concerns are involved in their design; but for implantable modules this is quite different. The first problem that must be addressed is the way to transfer enough energy to power such devices. Then, the particular placement of the sensor will define which way could be followed to power the implanted device. The other main constraint is the size of the device. The integration of the necessary instrumentation and communication electronics to control the sensors and to send the information provided by the sensors through human skin defines the powering budget and complexity in the design, that is, its level of intelligence. In this approach, the implantable device is placed beneath the skin. Then, an inductive coupling RF power harvesting module is used for transmitting energy to the implanted devices. Furthermore, this mechanism permits the establishment of bidirectional communication between the implanted device and an external interface (base or reader). The basics of this mechanism are the same as the RFID tags, where the implanted device acts as a tag. Some implantable telemetry circuits based on inductive coupling can be found in the literature such as Ref. [13]. As has been stated before, this envisaged device is conceived as an event-detector module. This *in vivo* true/false detector [14] works as an alarm: when the analyzed concentration level exceeds or is under a threshold value or the system detects the alarm condition. The system sends to an external reader a signal indicating the fault or showing the value as in some pregnancy tests. The system detects, but is not conceived to have an exact measurement with great accuracy. For instance, in the case of glucose monitoring, the detection of a threshold decrease in the glucose level is mandatory to avoid critical situations such as hypoglycemia [18]. The proposed architecture, depicted in Figure 12.3, includes the on-chip biasing, the potentiostat to drive and read the sensor, and the modulation and communication modules, introduced in the following sections.

12.2.4.1 Architecture of the Envisaged Subcutaneous Device
The envisaged implantable architecture is presented in Figure 12.3 and the main blocks are introduced. From the external reader the electromagnetic energy that is recovered by the implantable device through the implanted antenna and the AC/DC module is generated. The system has a power-on-reset module that activates

the electronics when enough energy has been recovered through the inductive coupling, and the voltage regulator, the bandgap reference circuit, generates the suitable voltage. The *sensor control* module drives the sensor, by an integrated potentiostat amplifier, with the voltage provided by the V_{in} *generator*. The three electrodes configuration is the typical one defined by a potentiostat amplifier, as a driving circuit to control the voltage applied at the biosensor. These electrodes are the working electrode (W), the reference electrode (R), and the auxiliary or counter electrode (A/C). The W electrode is the surface where the electrochemical reaction takes place; R, is used to measure the potential at the W electrode, and the A/C electrode supplies the current required for the electrochemical reaction at the W electrode. From the reaction that takes place in the W electrode, a current proportional to the reaction is generated and it is measured and conditioned by the *sensor conditioning* module. The measured signal is forwarded to the *modulation/data processing* module. In this block, the absence/presence detection takes place as also the modulation process to send the information to the external reader using a backscattering method through the inductive link, which can be AC or DC modulation. When one threshold value is detected, the module activates the communication. In this way, the external reader can be quickly advised every time the desired substance exceeds the programmed threshold level or levels.

Antenna and Rectification Power Module The antenna is used for two different purposes: as a wireless communication link and as the main element for the power harvesting energy source, based on inductive powering. The commercial Texas Instrument® TRF7960 is used as an external reader, which is used to transmit the power to the integrated antenna, and it receives the data from it. From the AC signal that is transmitted by the external reader, an AC/DC rectifier generates a nonregulated DC signal, which is treated afterwards by an integrated low dropout (LDO) regulator. The antenna must assure enough communication range to send and receive data through the human skin. In our case, the study has been carried out through the air medium. In order to design an implantable antenna several possibilities could be explored. It is possible to use a small coiled antenna with extremely small size such as those used in some RFID human or animal identifiers by Verichip® Corp. Moreover, it is possible to integrate an antenna using a MEMS CMOS compatible process or using a modern organic technology as in Ref. [13c].

In our case, a planar rectangular coil of 5.5 mm × 14.5 mm with a thickness of 0.5 mm has been designed, as a proof of concept, which is depicted in Figure 12.2. It has seven turns with a conductor width of 0.2 mm. The design presents an inductance of 400 nH and a series resistance of 340 mΩ. The antenna is tuned to work at the ISM low-frequency of 13.56 MHz, a range suitable for subcutaneous placement.

An AC voltage is produced by the coupled antenna and an AC/DC rectifier is placed immediately after that. From this, an unregulated DC voltage level is generated, which is then introduced in a regulation block that produces a regulated DC voltage to drive all the on-chip electronics. A bipolar power scheme able to supply a regulated differential voltage of ±1.2 V and a maximum current of ±1.5 mA has been implemented.

The AC/DC block is based on a half-bridge NMOS rectified with a bulk control voltage. This circuit maintains the bulk voltage at the highest potential to avoid parasitic conductance. The regulation block is based on the use of two LDO regulators [20]. A low-power bandgap circuit is implemented to generate the reference voltage needed for the regulation. It is also used to define the threshold values for the detector stage, placed in the modulation/data processing module (Figure 12.3). It is quite important to have an approach to the generated voltage that can be obtained in the device in terms of the distance between the emitter antenna and itself. The on-chip voltage generated at various air coupling distances has been measured. The desired on-chip regulated voltages of ±1.2 V are obtained for a distance up to 20 mm on air between coupling antennas. A more accurate study with human tissue is beyond the scope of this study.

Driving and Control Modules In this section, the electronic modules used to drive the sensor and to generate and program the driving signals are introduced. The first module is defined by the clock and the data extraction circuitry. This block has the role of detecting the incoming data provided by the external reader and, if it is the case, program some modules of the scheme; besides, it recovers and produces the necessary clocks used by the electronic modules. In any case, there are several possibilities to design this module [13b,c].

The second module is the generation circuitry. This block is used to generate the voltages to be applied to the sensor. Depending on the type of the integrated biosensor and its operation, this block can generate a DC or AC signal to properly drive the biosensor. This module could be programmed by the external reader in order to add more functionality to the system. In this implementation, three internal voltages of 0.6, −0.6, and 0.5 V can be selected. These signals are generated from the regulation module, based on the implemented bandgap reference circuit.

Finally, there is the potentiostat amplifier. The voltage that is applied to the amperometric sensor must be controlled by a specific circuitry, which is the potentiostat amplifier. This circuitry fixes the voltage provided by the V_{in} *Generation Module* – V_{in} between the sensor cell electrodes, that is, between the *working* (W) and *reference* (R) electrodes. In this approach, the adopted architecture is based on three operational amplifiers [20]. This architecture controls the voltage at the W and R electrodes, such that $V_{in} = -V_R$ where, V_R is the voltage at the *Reference* electrode. In order to keep this condition, the current at the R electrode should be ideally zero and no current should flow through it. The potentiostat is designed to work with signals from DC up to 2 kHz and with V_{in} voltages up to ±1.0 V with driving cell signals up to 30 mV; in this way, tissue necrosis around the implanted device or any subcutaneous damage resulting from the detection would be avoided for *in vivo* implants. The size of each operational amplifier (OA) is designed such that each one can supply the right current for the worst conditions of load, defined by the electrochemical model [26]. The design looks for a very small size implementation and very low power consumption. Individual amplifiers are 50 μm in height and 160 μm in width, with a power consumption of 12.8 μW under nominal conditions. Its open-loop gain is 60 dB at low frequencies and 50 dB at 1 kHz. The potentiostat amplifier occupies an area of 327 μm · 260 μm, and has an average power consumption of 51.2 μW, which is smaller than the solution introduced in [27], which has an area of 0.16 mm², and a power dissipation of 600 μW, or in [28] which has a power dissipation greater than 150 μW.

Signal Conditioning and Communication Modules The conditioning module measures the current that is generated in the amperometric sensor, which is proportional to the electrochemical reaction generated at the working electrode, and adapts it to be treated by the modulation and data processing module. The adopted solution is a current-to-voltage converter based on a transimpedance amplifier (TIA).

Its input resistance at design is 1 GΩ at DC, allowing a current detection up to 1 nA. The current-to-voltage conversion is defined by $V_{TIA} = -I_W R_{TRANS}$, where I_W is the current through the working electrode and R_{TRANS} is the externally selected gain resistance. A second gain stage based on an inverter configuration follows the TIA and adapts the voltage values for the next stage.

From the conditioning module, which converts the current generated to a voltage signal, the signal is placed in the modulation and data processing module. This module is used to process and prepare the transmission for the measurements. Depending on the application, a complex analysis is needed using some digital processing, which increases the consumption of the module. In this case, where the objective it to detect threshold values as the basis for an event-detector device, a simple approach has been followed. It is based on the use of comparators, which is a suitable solution in terms of silicon area and power consumption, to detect one or several threshold values of medical interest. In the implemented approach some comparators able to detect three different threshold voltages (V_{th1}, V_{th2}, and V_{th3}) have been implemented. All three voltages are on-chip generated and programmed externally using S_3 and S_4 in this prototype. These values are used to define a simple AM modulation protocol. To transmit the information, various codifications and modulations, such as frequency shift keying (FSK), amplitude shift keying (ASK), and NZR, used in RFID applications, could be implemented. In our case, to validate the communication, looking for a low power consumption and reduced size cost, an AM modulation protocol has been integrated as follows: The signal is always a high level "1" but when a threshold value is achieved then a "0" level is generated. This functionality is based on the use of the comparators, monostable flip-flops, and a very simple digital circuitry. As soon as there is enough voltage, the power-on-reset module generates a signal that activates the circuitry and the antenna starts to transmit continuously a series of "1." When the first threshold level is achieved the system transmits one zero (T_{th1}), if the second is reached two zeros are transmitted (T_{th2}), and when the third is achieved a series of three zeros are sent (T_{th3}). A zero time slot interval is defined as 250 ms ($T_{th1} = 250$ ms). This time is controlled by an external capacitor, C_{PRG}, connected at the monostable based circuit. The communication between the coupled antennas is via backscattering, which is a method where the impedance of the integrated antenna is changed in such a way that the external antenna can detect this variation thanks to the activation of a switch. Depending on the position of the load modulation, it could be AC (in front of the rectifier) or DC (behind the rectifier) [13c] (Figure 12.3). DC modulation using a PMOS ($W/L = 3000\,\mu m/2\,\mu m$) transistor connected directly behind the rectifier is used. This modulation presents a higher Q-factor modulation than the AC [13c].

12.2.4.2 Implementation and Results

The implemented circuits have been designed using a commercial 130 nm technology. As has been introduced, two different full-custom ICs have been implemented: one for the managing the powering of the device [29], and the other one for the integrated instrumentation and processing (Figure 12.2), with dimensions of 5.5 mm × 29.5 mm. On this PCB is placed one antenna, a sensor connection, and finally the external capacitors and resistors. The Texas Instrument® TRF7960 is used as external reader with a maximum emission power of 200 mW at 13.56 MHz. A commercial sensor AC1.W1.R1 form BVT Technologies® is used to validate the detection architecture as described later. Full characterization of the AC/DC modules and the potentiostat amplifier is available in [20].

A key aspect to analyze is the voltage, and then the powering, that the implantable is able to recover. This analysis has been carried out in terms of the distance (Z-axis) between the external antenna and the coil designed in the PCB, which defines the full implantable, that is, between the reader and the implantable, recovering enough energy up to 20 mm of distance between coils in order to assure ±1.2 V to bias on-chip electronics. The recovered power is in the range of 1.1 mW to 800 µW from 1 to 20 mm, respectively.

However, it is also necessary to have an approach to the misalignments between both antennas in the XY plane.

Figure 12.4 depicts the rectified voltage (V_{rec}) distribution in function of the XY misalignment for Z distances of 10, 15, and 20 mm. It can be noticed that the farther the antenna is placed from the center the lower the rectified voltage is. From this analysis, the concept of a safety operating area (SOA) is derived – that is, the area that indicates where the implantable can be placed and misalignments can take place, with a suitable rectified and regulated voltage for the implantable device. As can be expected, the size of the SOA is reduced if the Z axis distance increases. For a distance of 20 mm the area is reduced at a minimum, while for a distance of 10 mm, the area is at a maximum.

After the power conditioning module, defined by the AC/DC converter and the regulation, we focus our interest on the instrumentation.

The instrumentation and the communication protocol are validated using several concentrations of $K_4[Fe(CN)_6]$ in phosphate-buffered saline (PBS). In this case, a commercial sensor is used [20]. Several cyclic voltammetries (CVs) [30] were carried out in order to verify the performance of the control and conditioning modules. These measurements were compared with those obtained with a

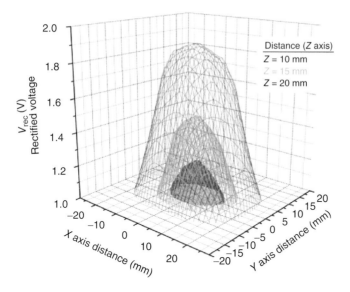

Figure 12.4 Distribution of the V_{rec} voltage in the XY plane for three different distances: 10, 15, and 20 mm. (Reproduced from [20] with kind permission from Springer Science+Business Media B.V.)

commercial potentiostat amplifier, the CH 1232A from CHInstruments®. These measurements were also used in order to calibrate the setup and check the obtained values of the measured current peaks for the oxidation peak (around 240 mV) and the reduction peak (around 170 mV) for each concentration of $K_4[Fe(CN)]_6$ in PBS tested, from 1 to 5 mM. In addition, this setup was used to validate the measured CV shapes obtained by the commercial equipment and the full-custom implementation. After the CV characterization, some amperometric tests [30] were done for different concentrations of $K_4[Fe(CN)]_6$ in PBS: 1, 2, 3, 4, and finally 5 mM. Then, an experiment was carried out where the concentration was changed from 1 to 5 mM in time, with a voltage of 500 mV fixed in the sensor. This voltage is not defined at the oxidation peak. For this value of voltage applied in the sensor, the current varies from an average current of 3 µA (1 mM) to 16 µA (3 mM), up to 28 µA (5 mM). This experiment was carried out to validate the detection protocol, for a particular case implemented on the basis of three threshold values. These values were programmed to detect the variations in the concentrations, defined by V_{th1}, V_{th2}, V_{th3}, as is depicted in Figure 12.5, taking into account the current expected for each concentration case, defining different windows of comparison. When the first threshold is detected, a first zero is transmitted, with a programmed width of 250 ms. When the second threshold level is reached, two zeros are transmitted, in this case with an amplitude of 500 ms. Finally, in the particular case that a threshold V_{th3} is defined to detect the highest concentration level, the modulation and data processing module will generate the longest transmission of zeros, in this case, three zeros with a total width of 750 ms.

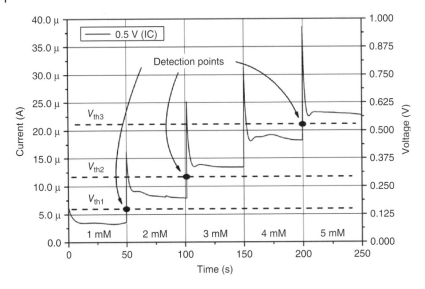

Figure 12.5 Amperometries measured with the prototype with three current levels programmed. (Reproduced from [20] with kind permission from Springer Science+Business Media B.V.)

12.2.5
The Diabetes Care Devices Market

As diabetes is the fastest growing disease in the world, the nano-enabled implantable device for *in vivo* biomedical analysis needs to be introduced into the global diabetes care devices market. Once thought of as primarily a childhood disease – sometimes referred to as *juvenile diabetes*, now mostly Type 1 diabetes – the obesity crisis linked to the adoption of a high-fat, high-carbohydrate, high-calorie American diet has resulted in skyrocketing rates of diabetes, particularly Type 2 diabetes, among adults across the world. As the World Health Organization has estimated the number of diabetics to exceed 350 million by 2030, governments and other healthcare providers around the world are investing in health education, diagnosis, and treatments for this chronic, debilitating – but controllable – disorder.

As such, the global market for blood glucose testing products is undergoing a significant transition, driven by the advent of new analytical technologies and developments in the treatment of diabetes. Although the blood glucose testing segment of the *in vitro* diagnostics (IVDs) industry is mature, certain segments of the market, such as home blood glucose monitoring devices for diabetes management, are set to exhibit strong growth. What is more, noninvasive testing now represents a major new area for the application of IVD testing. In addition, direct access testing – or over-the-counter testing, which allows consumers to order tests themselves without visiting a doctor – has emerged as a strong force in the blood glucose testing segment.

According to market research information provided in the report World Diabetes Market Analysis, 2009–2023 (Bharat Book Bureau, USA, February 2009), in 2007, the diabetes treatment market worldwide was worth over $25 billion, and had a double-digit growth from the year before. Consequently, it is one of the largest sectors in the global healthcare industry. All segments of the diabetes therapy market have been expanding, but the area that is expected to grow most rapidly in forthcoming years is the needle-free devices for the diagnosis of diabetes, especially for periodic, regular glucose level monitoring and insulin dispensing.

Thus, the Department of Electronics at the University of Barcelona, in collaboration with the Institute for Bioengineering of Catalonia (IBEC) and the Biomedical Research Networking Centre in Bioengineering, Biomaterials and Nanomedicine (CIBER-BBN), have promoted an alliance with the aim of developing a cutting-edge multidisciplinary research and commercial team covering the entire range from applied research to clinical diagnosis. In this way, research centers, hospitals, firms, health policies, and citizens share the same goal: launching onto the market safe, reliable, and affordable biomedical devices for the diagnosis and therapy of diabetes, improving the quality of life of those who have to control their blood sugar level by online monitoring and keeping the requested glucose level constant (by using an RF-activated clock-alarm or with exact doses administered directly from an insulin reservoir). The aim of this chapter was to describe the huge market for a wide product segment of implantable biomedical devices; the medical conditions and indications themselves have not been considered in this chapter as they lie beyond its scope.

12.3
Conclusions and Final Recommendations

In this chapter, the design of a generic *in vivo* implantable biomedical device capable of detecting threshold values for targeted concentrations (i.e., detection of glucose levels) has been presented. In this case, the solution combines communication, powering, and sensor instrumentation modules in the same design and all modules have been fabricated using CMOS technology.

Given the speed with which diabetes can spread and the improvements that are possible in its diagnosis and control if needle-free systems are available, the medical device introduced in this chapter is designed to reach a huge market over the next few years. Moreover, when the entire value chain is publicly funded, this means that the goals of technology transfer from university to industry and the social returns on the public investment have been fully realized. Thus, a successful model for research, innovation, and technology transfer can be introduced to a particular scenario typified by the convergence of technologies and disciplines, as well as by the convergence of various stakeholders combining representatives from research centers, hospitals, market, policy centers, and citizens as well.

The design of a generic front-end architecture detection system for *in vivo* implanted devices is presented. The implemented module is envisaged in such way that because of its simplicity and size, it could be placed under the human skin. It is oriented to work with three electrode amperometric sensors, thanks to the low-power biopotentiostat amplifier, which has a suitable input voltage range between +1 and −1 V. The electronics has power consumption around 340 µW for a power voltage of ±1.2 V. At this stage the sensors are based on commercial solutions. The system has a size of 5.5 mm × 29.5 mm, following approaches such as Jadelle® or Implanon®, but in our case the envisaged device is conceived as a subcutaneous event detector. The full system can be reduced in size with a better package and by just placing all the electronics in one ASIC. In addition, the system will be based on bio-nanosensors. The resulting biomedical device is nano-enabled in a dual sense: when miniaturizing the system (fluidics, electronic, energy autonomy) and when new functional structures are included (nanobiosensors developed by the IBEC). Still an emerging technology, the future ASIC will work with an array of nanobiosensors with different targets, and it will define the configuration of the measurement method. Each array will be used to detect a specific type of target, and the multiplexed system will be used to analyze each array focusing on a particular target. Then, top-down approaches using nanoengineering and nanofabrication and bottom-up approaches using supramolecular chemistry can produce novel diagnostics, which will increasingly focus on delivering a personalized solution based on the analysis of array data in real time and where appropriate, applying this decision to deliver an automated therapy (theranostics).

In conclusion, closer and more regular collaboration between industry, hospitals, and scientific facilities is required to guarantee the commercial success of biomedical research. Clearly, the speed of change being recorded in markets, products, technologies, competitors, regulations, and even in society means there are significant structural variations and R&D challenges in commercializing nano-enabled implantable devices for *in vivo* biomedical analysis. Biomedical devices represent a strategic gamble for the future scientific and technological policy areas as they seek accelerated economic growth within the knowledge-based society. In this way, it is necessary to strengthen the network links between R&D agents – science and technology parks, institutes and research centers, hospitals, technology platforms, and business incubators – as they explore and confront the new scientific and market challenges presented by the nanobiotech sciences.

References

1. Juanola-Feliu, E., Colomer-Farrarons, J., Miribel-Català, P., Samitier, J., and Valls-Pasola, J. (2012) *Technovation*, **32** (3–4), 157–256.
2. Fuji-Keizai USA (2007) Worldwide Market Research: Nanotechnology-based Product Market and Business Opportunities-Current and Future Outlook.
3. (a) Ghafar-Zadeh, E. and Sawan, M. (2008) Toward fully integrated CMOS based capacitive sensor lab-on-chip. International Workshop on Medical Measurements and Applications, Vol. 9 (10), pp. 77–80; (b) Barretino, D. (2006) Design considerations and recent advances in CMOS-based microsystems for point-of-care clinical diagnosis. Proceedings of the IEEE International Symposium on Circuits and Systems, pp. 4362–4365.
4. Sadik, O.A., Aluoch, A.O., and Zhou, A. (2009) *Biosens. Bioelectron.*, **24**, 2749–2765.

5. Nim Choi, H., Hoon Han, J., Park, A., Mi Lee, J., and Won-Yong, L. (2007) *Electroanalysis*, **19** (17), 1757–1763.
6. Erdem, A., Karadeniz, H., and Caliskan, A. (2009) *Electroanalysis*, **21** (3–5), 461–471.
7. Wang, J. (2008) *Talanta*, **75** (3), 636–641.
8. Rub, A., Rahman, A., Justin, G., and Guiseppi-Elie, A. (2009) *Biomed. Microdevices*, **11**, 75–85.
9. Steeves, C.A., Young, Y.L., Liu, Z., Bapat, A., Bhalerao, K., Soboyejo, A.B.O., and Soboyejo, W.O. (2007) *J. Mater. Sci.: Mater. Med.*, **18**, 25–37.
10. (a) Hassibi, A. and Lee, T.H.A. (2006) *IEEE Sens. J.*, **6** (6), 1380–1388. (b) Beach, R.D., Conlan, R.W., Godwin, M.C., and Moussy, F. (2005) *IEEE Trans. Instrum. Meas.*, **54** (1), 61–72.
11. Gosselin, B. and Sawan, M. (2008) Proceedings of the IEEE International Symposium on Circuits and Systems, pp. 2733–2736.
12. (a) Zierhofer, C.M. and Hachmair, E.S. (1996) *IEEE Trans. Biomed. Eng.*, **43**, 708–714.(b) Sawan, M., Yamu, H., and Coulombe, J. (2005) *IEEE Circuits Syst. Mag.*, **5**, 21–39.
13. (a) Sauer, C., Stanacevic, M., Cauwenberhs, G., and Thakor, N. (2005) *IEEE Trans. Circuits Syst.*, **52** (12), 2605–2613. (b) Li, Y. and Liu, J.A (2005) IEEE International Symposium on Circuits and Systems, pp. 5095–5098; (c) Myny, K., Van Winckel, S., Steudel, S., Vicca, P., De Jonge, S., Beenhakkers, M. J., Sele, C.W., van Aerle, N.A.J.M., Gelink, G.H., Genoe, J., and Heremans, P. (2008) IEEE International Solid-State Circuits Conference, pp. 290–614.
14. Gore, A., Chakrabartty, S., Pal, S., and Alocilja, E. (2006) 28th IEEE Engineering in Medicine and Biology Science Conference, pp. 6489–6492.
15. Haider, M.R., Islam, S.K., and Zhang, M. (2007) *Sens. Transducers J.*, **84** (10), 1625–1632.
16. Visiongain Ltd http://www.visiongain.com/Report/453/World-Diabetes-Market-Analysis-2010–2025 (accessed 21 March 2013)
17. Mansfield, E. (1991) *Res. Policy*, **20**, 1–12.
18. Wolpert, H.A. (2007) *J. Diabetes Sci. Technol.*, **11** (1), 146–150.
19. Newman, J.D. and Turner, A.P.F. (2005) *Biosens. Bioelectron.*, **20**, 2435–2453.
20. Colomer-Farrarons, J. and Miribel-Catala, P. (2011) *A CMOS Self-Powered Front-End Architectures for Subcutaneous Event-Detector Devices: Three-Electrodes Amperometric Biosensor Approach*, Springer.
21. Colomer-Farrarons, J., Miribel-Català, P., Rodrı́guez, I., and Samitier, J. (2009) Proceedings of the 35th International Conference of Industrial Electronics IECON, pp. 4401-4408.
22. Lin Tan, E., Pereles, B.D., Horton, B., Shao, R., Zourob, M., and Ghee Ong, K. (2008) *Sensors*, **8**, 6396–6406.
23. Tesoriero, R., Gallus, J.A., Lozano, M., and Penichet, V.M.R. (2008) *IEEE Trans. Consum. Electron.*, **54** (2), 578–583.
24. Garcia-Morchon, O.,Falck, T., Heer, T., and Wehrle, K. (2009) International Workshop Mobile and Ubiquitous Systems: Networking and Services, pp. 1–10.
25. Ricotti, L., Assaf, T., Menciasii, A., and Dario, P. (2011) Proceeding of the International IEEE Conference Engineering in Medicine and Biology Society, pp. 2849–2853.
26. Lasia, A. (1999) *Modern Aspects of Electrochemistry*, Vol. 32, Kluwer Academic/Plenum Publisher, New York, p. 143.
27. Chung, W.Y., Paglinawan, A.C., Wang, Y.H., and Kuo, T.T. (2007) IEEE Conference on Electron Devices and Solid-State Circuits, pp. 1087–1090.
28. Ahmadi, M.M. and Jullien, G.A. (2005) Proceedings of the Fifth International Workshop on System-On-Chip for Real-Time Applications, pp. 184–189.
29. Colomer-Farrarons, J., Miribel-Catala, P., Saiz-Vela, A., Puig-Vidal, M., and Samitier, J. (2008) *IEEE Trans. Ind. Electron.*, **55** (9), 3249–3257.
30. Zoski, C.G. (2007) *Handbook of Electrochemistry*, Elsevier, Amsterdam.

13
Improving the Biocompatibility of Implantable Bioelectronics Devices
Gymama Slaughter

13.1
Introduction

From the vantage point of the human body, implantable bioelectronics is truly magnificent – a synergistic convergence between biology and electronics. In 1959, the Nobel-Prize-winning physicist Richard Feyman [1] presented a vision of the future in a lecture to the American Physical Society, entitled, "There's Plenty of Room at the Bottom". He described the development of microelectronic devices and micromachines that can maneuver at the micro- and even nanolevel. It is possible that we have the capability to make these extremely tiny devices or machines do what we want by mimicking the extremely small biological system, which consists of tiny cells. These active cells can maneuver through their environment effortlessly, produce various chemical substances, store information, and perform all kinds of awe-inspiring things at a very small scale. Feyman's vision of the future has inspired many scientists to pursue miniaturization of semiconductor devices in order to transform our world by creating new opportunities in biomedical research and medicine. We can develop these micro or nanodevices with the capability to assemble anything similar to a cell, atom by atom, and even repair our bodies from the inside, cell by cell. In this case, healthcare cost could be drastically reduced because nanoscale bioelectronics will be an enabler for the development of molecular-based personalized medicine and the threat of disease, aging, and even death will be eliminated. It will be something of the past.

Realizing the promise of implantable bioelectronics requires the understanding of molecular/cell–electronics interfaces, cellular responses, and the variability in cellular response to various stimulations, such as electrical, mechanical, chemical, and thermal. Scientists must also have the capability to develop implantable devices that can actively sense their environment, detect different biomarkers, acquire and analyze essential data on the state of the biological system as well as deliver appropriate therapeutic agents and stimuli in real time. The safe use of implantable bioelectronics in the human body demands that the bioelectronic device and the material from which it is constructed must not be degraded by

Implantable Bioelectronics, First Edition. Edited by Evgeny Katz.
© 2014 Wiley-VCH Verlag GmbH & Co. KGaA. Published 2014 by Wiley-VCH Verlag GmbH & Co. KGaA.

the environment in which it is implanted and its presence must not inhibit or cause deleterious effects on the biological system [2]. In a large part, the material used to construct the bioelectronic device plays a significant role in determining the biocompatibility of an implantable bioelectronic device, although other factors can play a role. One of the most critical issues is to understand how bioelectronic devices can disturb the cellular responses in the local environment in which they are implanted. Often, the discovery of better biocompatible materials relies on a variety of screening methodologies instead of design. A number of screening tests must be conducted to determine the performance of the bioelectronic device surface materials in contact with body fluids, cells, tissues, and organs under appropriate conditions [3–5] and for a significant length of time (e.g., for duration of more than a few hours). Evaluation of the biocompatibility of implantable bioelectronic device materials is performed by assessing nonspecific protein adsorption (biofouling), cytotoxicity, mutagenesis, carcinogenesis, and cell function [3, 5]. *In vitro* biocompatibility assessment is often used to screen tissue compatibility of implantable bioelectronic devices, whereas *in vivo* assessment of tissue compatibility of a device is to understand the chemical composition of the material and the conditions of tissue exposure. Table 13.1 describes the general guidelines that may be applied to the evaluation of bioelectronic devices [3] for *in vivo* assessment of tissue compatibility. Table 13.2 provides the ISO 10993-1 and US Food and Drug Administration (FDA) categories for selection of biological response test that are categorized by body contact and duration [3, 6, 7]. Biocompatibility of future implantable bioelectronics with cells,

Table 13.1 General guidelines for the evaluation of bioelectronic devices for *in vivo* assessment of tissue compatibility.

Material(s)
Additives, process contaminants, and residues
Leachable substances
Degradation products
Properties and characteristics of the final product

Table 13.2 ISO 10993-1 and FDA categories for selection of biological response test.

Tissue contact	Tissue contact	Contact duration
Surface devices	Implant devices	Limited, $\leq 24\,h$ prolonged, $> 24\,h$ and < 30 days permanent, > 30 days
Skin	Blood path, indirect	
Mucosal membrane	Circulating blood	
Compromised surfaces	Blood	
External communicating devices	Tissue/bone	

tissue, and organs can be drastically improved via surface chemistry (surface conditioning, surface treatment, and device geometry) to enable devices such as insulin delivery systems to be implanted in the body [8]. The surface modification strategies used to enhance device biocompatibility must allow for adequate environmental sampling for accurate real-time measurements and must not hinder the operational functionality of the bioelectronic device. With the advent of microfabrication and semiconductor technology, intelligent implantable bioelectronics can be created. However, the main obstacle of biofouling, which affects the biofunctionality of implantable bioelectronic device by altering the electrical properties of the device, still remains, thereby making it difficult to sense and actuate.

13.2
Implantable Bioelectronic Device Materials

The requirements for the materials used in implantable bioelectronic devices are particularly high since these devices are incorporated into a very challenging physiological environment. There are important factors that must be considered when selecting materials for the implantable bioelectronic device, substrate, or encapsulation layer to ensure optimal performance and durability, and minimal tissue or blood response is evoked by implantable bioelectronics. The key challenges with implantable bioelectronics include (i) implementation of materials that do not harm cells, tissues, or organs, while not compromising device biofunctionality; (ii) limitation of device rejection by the body and maintenance of the intrinsic structure of biological systems; and (iii) minimization of tissue heating occurrences that result from electromagnetic radiation used during signal transmission [9]. Therefore, the major prerequisite for implantable bioelectronics is that the device must be biocompatible. There are many definitions of biocompatibility that are very descriptive [4, 10]; here, the focus is on using a broad definition of biocompatibility – "the ability of a material to perform with an appropriate host response in a specific application" [11]. This more inclusive definition comprises aspects of biological, chemical, and physical properties of the implantable bioelectronics.

Although the concept of biocompatibility is strongly dependent on the application field, in many cases surface properties of implantable bioelectronics, such as surface charge, surface free energy, chemical group distribution, heterogeneity, surface texture, porosity, and smoothness, are key factors that affect device biocompatibility. Under the influence of maneuverability, physical properties such as size, geometric shape, and stiffness become key determinants of biocompatibility [12]. Biocompatibility is not limited only to nontoxicity but includes both chemical and physical properties of the device materials as well as its complete behavior in the physiological environment.

Device implantation is a traumatic ordeal that requires that the implantable bioelectronic device satisfy two criteria: (i) biosafety – where the devices does not cause any deleterious effects on the host or be susceptible to attack by biological fluids, metabolic substances, proteases, or macrophages and (ii) biofunctionality – where the device functions as it is intended in the host system. Long-term blood contacting implantable bioelectronic devices have been shown to be continuously platelet reactive and ultimately fail because of thrombus accumulation [13–16]. In addition, degradation of device materials may inadvertently harm the host system in terms of biostability and biocompatibility. It is important that the implantable bioelectronic devices with sharp edges be flexible enough to enable multiple bending, especially when the device is implanted subcutaneously to minimize damage to surrounding tissues and cells. Although biocompatibility and tissue function preservation are highly important, the biofunctionality of many implantable bioelectronic devices may be limited by biofouling. Bioelectronic devices with rough surface materials promote the adsorption of proteins, cell adhesion, and proliferation, which frequently results in device fouling and ultimately failure. Surface chemical modification strategies of device substrate materials have been developed to minimize the device–physiological environment issue to produce devices that do not elicit biofouling while retaining their biocompatibility. Polymeric coatings have been developed to prevent device corrosion and enhance the long-term performance of the implantable bioelectronic device in terms of biofunctionality [17] and biocompatibility [18]. In this instance, the insulating or encapsulating polymeric materials used must maintain their insulating properties throughout the lifetime of the implantable bioelectronics and must not leach any substances. However, there are occasions in which antibiotics, hormones, and growth factors may be incorporated in the encapsulating or insulating material that could be released in order to promote tissue repair and wound healing [19]. In all these cases, particular emphasis must be placed on preventing protein adsorption and platelet adhesion, which could hinder the release of substances.

There are numerous materials that meet the requirement for biocompatibility and are used in the fabrication of implantable bioelectronics. A silver–silver chloride electrode (Ag/AgCl) is widely utilized for surface measurements, particularly in the development of various biosensor and neurological applications; however, it cannot be used for implantable applications because of the rapid dissolution of the AgCl layer into tissues resulting in severe inflammation in the surrounding tissue [20]. Silver has been used extensively as antibacterial material in various applications [21]; however, silver and AgCl have been demonstrated to be extremely toxic to tissues when implanted [22].

The most commonly utilized implantable electrode materials are the noble metals gold (Au) and platinum (Pt) because of their excellent corrosion resistance and biocompatibility [9, 23, 24]. Pt and Au are utilized as electrode materials in numerous biosensors, biological and chemical sensors, and neurological applications such as glucose biosensors, pH sensors, gas sensors, cardiac pacemaker, and cochlear implants. The amount of Pt or Au ions released into surrounding tissue is negligible even after long-term implantation. In addition, optically transparent indium tin oxide (ITO) is extensively employed in electrode materials used in electrochemical sensors along with Pt and Au. The biocompatibility of ITO for long-term implantation is not well understood. Recent studies focusing on cell adhesion and cell proliferation [25] as well as protein adsorption [26] have shown ITO to be biocompatible. Titanium (Ti) has excellent tissue compatibility and it has the ability to form an oxide layer on the surface, which makes it even more susceptible to corrosion [23, 27, 28], and has been shown to exhibit properties that are favorable for blood contact [29]. Carbon fibers are electrically conductive and are also used as electrode materials because of their biocompatibility and biostability; however, they exhibit higher surface roughness than metal electrodes and are rarely used as an implantable electrodes.

Naturally occurring materials such as dextran [30, 31], collagen [32, 33], and chitosan [34–36] are commonly used in the fabrication of implantable bioelectronics. Polymeric materials are used as carrier and encapsulation materials. The most commonly used polymeric materials are polydimethylsiloxane (PDMS), polytetrafluoroethylene (PTFE), epoxy resins, cellulose acetate, polyimide, polymethyl methacrylate (PMMA), polyacylonitride, nylon, polycarbonate, polyurethane, low-temperature isotropic pyrolytic carbons [37], polyethylene glycol (PEG) [38–40], poly(2-hydroxyethyl methacrylate) (pHEMA) [41], polylactic acid (PLA) and poly(lactic-co-glycolic acid) (PLGA) [41–47], and poly(vinyl alcohol) (PVA). Although these materials are considered to be biocompatible, electrically insulating, and highly stable, studies have shown that bioelectronic devices made from these materials exhibit biocompatibility issues [4, 48]. In addition, the bulk properties of the polymeric materials can be modified to a certain degree, and also surface modification procedures can be performed to enhance biocompatibility.

13.3
Surface Composition

Implantable bioelectronics material research can broadly be organized into studying two surface properties: composition and texture. Surface composition of materials intended for implantable bioelectronics are widely studied in order to enhance biocompatibility, reduce the adhesion and proliferation of macrophages, and prevent inflammatory reactions. Implantable bioelectronic devices are typically coated with biocompatible encapsulation layers to improve implantable device/host tissue interactions and improve device functionality and durability [7–51]. Biocompatible polymeric materials are used as coating or encapsulation materials for implantable bioelectronic devices because of their inherent ability to mask the underlying device surface. The chemical composition of the encapsulation and insulation material of devices plays a key role in the long-term biostability and biofunctionality of the implantable bioelectronics. It determines how well the device will support the biological system by inhibiting the attraction and adhesion of macrophages and white blood cells to the implantable bioelectronic device in order to minimize inflammatory reactions. These biocompatible materials are used to promote tissue cells such as fibroblasts to adhere to the device surfaces in order to incorporate them into the host system, thereby minimizing tissue reaction induced by the device implantation.

Major research efforts are dedicated toward the development of materials with improved biocompatibility by producing materials that prevent thrombosis via surface modification strategies. The objective is to enhance blood compatibility and administer therapeutic substances to prevent deposition of fibrin in order to elicit the appropriate host response [52]. Surfaces of incompatible bioelectronics can be modified using biological or nonbiological strategies [52] to elicit a biocompatible response [8]. Biological surface modification strategies involve coating of the device surfaces with heparin, enzymes, growth factors, hyaluronic acid, polysaccharides, lipids, or peptides. These methods are very similar to the macromolecular substances that the biological environment recognizes and operates with metabolically. However, there are several disadvantages to biological surface modification. These surfaces are often immunogenic and decompose or undergo pyrolytic modification at temperatures [7], thereby preventing their use in high temperature thermoplastic processing methods.

Nonbiological surface modification strategies employs functional groups such as carboxyl, amine, and hydroxyl groups, sulfonates, *n*-alkyl chains, and polymers (polyethylene oxide, pHEMA, polyacrylamide, and poly(*n*-vinyl pyrrolidone)). Polymeric materials such as polyimide, benzocyclobutene (BCB), PDMS, parylene, PLGA, and epoxy resins have been recently studied as encapsulating and insulating materials in implantable bioelectronics [53–55]. Polyimide and BCB have been utilized in microelectronics because of their excellent resistance to solvents, strong adhesion to metals and metal oxides, and good dielectric properties [56]. One major disadvantage of polyimide is its permeability to moisture and ions, which can lead to short circuiting and reduction in the electrode's durability. BCB, on the other hand, has low moisture absorption and good chemical stability, and has been recently demonstrated to be a potential material for use in implantable bioelectronics [57]. Parylene and PDMS are well-known hydrophobic and optically transparent materials, which are biocompatible and have been approved by FDA for use as implanted materials in medical devices. Parylene is a thermoplastic polymer that can be vapor deposited at room temperature to form coatings that are stress free, chemically and biologically inert, and stable as well as being minimally permeable to moisture.

The most biocompatible substance available is water and water is a hydrophilic substance. From this viewpoint, a device surface that is wettable would be advantageous in terms of eliciting biocompatibility. SiO_2 is a good example of a hydrophilic substance. In many cases, hydrophilic surfaces can be obtained via surface modification strategies such as functionalization with polar or ionic groups such as polysaccharides, proteins, or polymers [58, 59]. Coating of bioelectronic device surfaces with dextrans has been recognized to be advantageous because each monomer within the chain carries up to three hydroxyl groups, thereby minimizing nonspecific protein adsorption [60, 61]. Hydrogel-based modifications have been applied to a broad range of bioelectronic devices [62–66] and include HEMA [64–67], PEG [64–66, 68], and PVA [69–74]. These modifications utilize the attachment of PEG or its analogs to the substrate material in order to inhibit protein adsorption, thereby minimizing biofouling [75]. Hydrogels are three-dimensional polymeric networks with the capability to adsorb and retain large amounts of water while permitting small molecules such as glucose to diffuse through the water-swollen layer. The degree of analyte diffusion can be controlled by the cross-linking density of the gel [7, 49] and the mechanical properties of the hydrogels are very similar to those of biological tissues in that they are permeable to small molecules and possess low interfacial tension [5, 7]. However, hydrogels have poor device substrate adhesion. The properties and applications of the most commonly used polymers in bioelectronic devices are summarized in Table 13.3.

Bioelectronic device surfaces are chemically modified in order to not activate the intrinsic blood coagulation system, or attract platelets or leukocytes when the device comes into contact with the host system (blood component or tissue). Blood coagulation is dictated by the delicate balance between the three factors illustrated in Virchow's triangle (Figure 13.1): blood flow conditions, the nature of the bioelectronic device material, and the blood itself. The interaction involves intermolecular forces (e.g., London dispersion forces, hydrogen bonds, dipole–dipole interactions, and acid–base interactions) that developed at the material – blood interface. In order to prevent undesired protein adsorption, hydrophilic surfaces have been obtained by coating the surface of the bioelectronic device with polyethylene oxide, heparin, and albumin. In this case, a reduced thrombogenicity accompanied by a reduced plasma protein adsorption and platelet adhesion was observed [76] because of a steric repulsion mechanism that results when the surface molecules act as entropic "springs" [77]. Albumin-coated surfaces have been shown not to elicit platelet attraction and adhesion [78, 79], whereas γ-globulin and fibrinogen coatings have been shown to elicit platelet adhesion and aggregation as well as the release of platelet constituents [80].

It has been hypothesized that material surfaces with an interfacial energy of zero tend to be highly thromboresistant. Thus, polymeric materials such as hydrogels,

Table 13.3 Commonly used polymers properties and applications.

Component	Properties	Applications
Phospholipid-based biomimicry	Mimics cell membrane Fragile and difficult to deposit	Coating
Albumin	Immobilization of glucose oxidase	Glucose biosensor
Cellulose	Decrease complement activation	Coating
Nafion	Decrease biofouling	Coating
PEO–PPO–PEO	Decrease biofouling sensor passivation	Coating
Hydrogels-pHEMA	Negligible protein adsorption	Coating
Hydrogels-PEO, PEG	Negligible protein adsorption	Coating
PVA	Surfactant and gel-forming	Emulsifier in drug encapsulation process
PLA and PLGA	Negligible protein adsorption	Coating

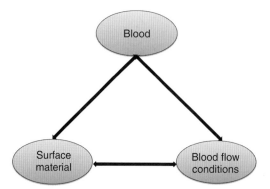

Figure 13.1 Virchow's triangle applied to implantable bioelectronic devices in which blood coagulation is dictated by blood flow conditions, the nature of the bioelectronic device material, and the blood itself.

which are known to have minimal interfacial energy, possess a relatively low thrombus adherence, thus enhancing the validity of this hypothesis. Although high water content hydrogels are non-thromboadhesive, they are platelet consumptive and embolic [81, 82]. Contact angle measurements have been used to characterize and calculate the surface free energies of many polymers and it has been postulated that a polymer with a critical surface tension of approximately 20–30 dyn cm^{-1} is highly blood compatible [83]. Several methods have been developed for the immobilization of polymers onto substrate surfaces. The two most common methods utilized are grafting by way of photoinduced polymerization and chemical surface attachment [84–86]. Cross-linking has been extensively used with glutaraldehyde or carbodiimide for arterial prostheses coated with cross-linked gelatin [87, 88], albumin [89, 90], or collagen [91–93]. Cross-linking can also be achieved via the derivatization of molecules to be deposited with photodimerizable groups, such as thymine, cinamate, or coumarin. Upon UV irradiation, these groups bind and the derivatized molecules are deposited onto the surface, as demonstrated for the deposition of chondroitin sulfate, hyaluronic acid, and gelatin onto Dacron or polyurethane [94]. To further understand blood–material interaction, ^{60}Co γ-irradiation has been investigated as a means to obtain similar surface energy by grafting various hydrogels onto Angioflex, and a series of well-characterized radiation grafted hydrogels were developed to evaluate the water interface for enhanced blood compatibility [95–97]. Since water is "blood compatible" the higher the water content within a hydrogel grafted to a surface, the more "water-like" the surface contacting the blood would be. It is apparent from the results that the grafting of PEG onto the substrate demonstrated not only the adsorption of albumin but also less adhesion of platelets. This reflects the relevance of the chemical nature of the substrate besides surface energy parameters.

In addition, PTFE surfaces have been modified using different functional groups [98] where it was determined that cells adhesion and cytokine release were inhibited best when the modification was performed with amide groups to obtain neutral hydrophilic surfaces. Hydrophilic surfaces can also be obtained by treating the surface of the material with oxygen, ammonia, or water plasma [99, 100]. Organic molecules such as methacrylates or siloxanes and inert materials such as PTFE and polystyrene can undergo plasma treatment, since plasma treatment can be used to form reactive groups for further binding to other molecules. PEGs can also be deposited to inhibit protein adsorption and cellular adhesion, as shown with tetraethylene glycol dimethyl ether [101]. Plasma-polymerized siloxanes have been shown to provide an anti-thrombogenic surface, which is important for blood-contacting bioelectronic devices [102]. Regardless of the nature of the underlying substrate, modified surfaces are usually hydrophilic.

In addition, the texture of a device surface has been shown to dictate cell behavior. Evidence suggests that restricting biocompatible regions into particular geometric features or submicron roughness can also control cellular behavior [103] such as cell viability, differentiation, and axonal outgrowth. Smooth surfaces are advantageous when minimal cell adhesion is desired, whereas rough surfaces are advantageous when the intended use is to attract cells and promote cell adhesion in order to achieve good incorporation of bioelectronic device into the host system. Since roughness is an intrinsic property of a material, there is a significant challenge in controlling surface roughness in the fabrication of nonplanar three-dimensional surface topographies with submicron dimensions. Roughening can be achieved by utilizing microfabrication and micromachining techniques, where particular patterns are applied to the smooth material using photolithography or micromechanical procedures. In many cases, various shaped grooves are etched or milled into the material in order to align the growth of fibroblasts and to fix the implantable bioelectronic device within the tissue.

Pyrolytic carbon and glassy carbon have a turbostratic structure, which leads to the microscopically heterogeneous surface distribution of electron density and microscopic roughness [104–106]. These surfaces enable increased surface functionality and provide more elaborate control over cellular behavior because of the combinatorial nature of the hydrophobic surface. This results in the promotion of protein adsorption, and the surface roughness in turn elicits cell adhesion. The fabrication of these complex micro- and nanostructures is currently limited by the conventional planar microfabrication and nanofabrication processes.

13.4
Response to Implantation

When bioelectronic devices are implanted in the human body, proteins instantaneously adsorb onto the material surface with a certain rate, concentration, and degree of selectivity along with activatable clotting factors. Figure 13.2 shows the schematic representation of the coagulation cascade, which describes the

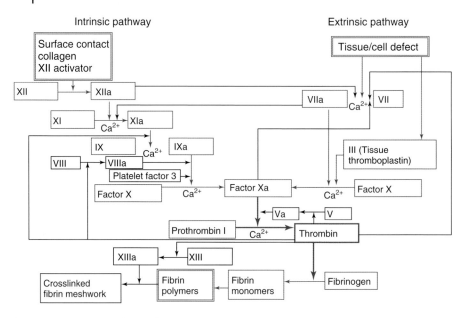

Figure 13.2 Blood coagulation cascade *in vivo*, which describes the central role played by thrombin.

complexity of blood material interactions at the material–blood interface. Table 13.4 provides a list of the blood coagulation factors. The first event is that high molecular weight kininogen (HMWK) displaces the fibrinogen layer within a few minutes of device implantation. In the case where there is insufficient HMWK to displace the fibrinogen layer deposited, platelet adhesion is promoted to compete with HMWK in interacting with fibrinogen [107, 108]. The protein adsorption process depends

Table 13.4 Blood coagulation factors.

Factor	Name
I	Fibrinogen
II	Prothrombin
III	Thromboplastin
IV	Calcium ion
V	Proaccelerin
VI	Does not exist
VII	Proconvertin
VIII	Antihemophilic Factor A
IX	Antihemophilic Factor B
X	Stuart Prower factor
XI	Antihemophilic Factor C
XII	Hageman factor
XIII	Fibrin stabilizing factor

13.4 Response to Implantation

on the surface properties of the device and the nature of the adsorbed protein as well as the composition of the biological environment. The proteins may bind to a surface via electrostatic bonding, hydrogen bonding, or hydrophobic interactions. Since proteins, platelets, and cells have a net negative zeta potential of -8 to -13 mV [109], a not highly charged implantable bioelectronic device surface may attract proteins such as albumin, collagen, or fibronectin. The adsorbed protein forms a passivating layer, which can promote the adhesion of tissue cells [110, 111], thereby rendering the device material relatively more hydrophilic and biocompatible as a result of the protein adsorption. The "masking" layer produced by the adsorbed protein on the device surface ultimately makes the device material less susceptible to adverse reactions of the immune system, thereby resulting in the successful incorporation of the implantable bioelectronic device into the human body.

For highly charged surfaces, even if protein adsorption does not take place, blood and tissue components may experience some amount of damage. For example, activate platelets may aggregate as microemboli and deposit in various organs, causing the implantable bioelectronic device to be unsuccessful [112]. It has also been demonstrated that high water content hydrogels generated showers of emboli [113, 114], which roughly correlated with the water content of the gel. Although blood elements do not adhere to the hydrogels, these hydrated surfaces cause platelet activation leading to aggregation downstream. The biofunctionality and longevity of biosensors can be compromised by biofouling of the surface, and the fibrotic encapsulation process around the device greatly decreases both the transport of analyte from the tissue to the sensor and the diffusion of reaction products from the sensor to surroundings environment [115, 116]. This results in the loss of device sensitivity and/or instability of measurement [13–15, 116–120]; therefore, it is critically important to control the fibrotic encapsulation at the implant site in order to enhance the biofunctionality and longevity of the biosensor *in vivo*. Growth factors can be used to promote angiogenesis by inducing new blood vessel formation [121–123] in the surrounding tissue.

Biocompatibility of implantable bioelectronic devices remains a critical issue in limiting device functionality and longevity. A foreign body response to implantable bioelectronic devices also presents a tremendous risk to patients, resulting from either normal homeostatic response to the implantation injury, tissue or blood/device interface interactions, or the lack of biocompatibility [11, 124–128]. When blood comes into contact with a bioelectronic device, the intrinsic pathway in the coagulation system gets activated with the adsorption of HMWK, prekallikrein, and Factor XII proteins, where the Hageman factor is activated to XIIa. Factor XIIa converts prekalikrien to kalikrien and, with HMWK as a cofactor, activates Factor XI to Factor XIa. Factor IXa with calcium ions activates Factor VIII to VIIIa. Through similar reactions, Factor X is converted to Xa and V is converted to Va. Following this cascade of reactions, prothrombin is converted to thrombin with phospholipid and calcium ion. Then thrombin converts fibrinogen to fibrin and Factor XIII to XIIIa. In the presence of calcium ions, fibrin is stabilized into cross-linked fibrins.

In the case of extrinsic pathway, Factor XIIa along with thrombin activates Factor VII and Factor VIIa with tissue factor and calcium ions. Factor VIIa is involved in Factor X activation to Factor Xa. Thereby, the extrinsic and intrinsic pathways are dependent on each other as a number of the coagulation factors play a key role in the entire coagulation system. In addition, the nature of protein adsorbed may also affect the activation of the blood coagulation cascade, complement, and fibrinolytic system.

Leukocytes have a strong propensity to adhere to implantable bioelectronic device surfaces where they interact with coagulation, complement, and fibrinolytic systems. They are major mediators of inflammatory response. Leukocyte adhesion appears to be mediated by complement components. Complement activation via the classical or alternative pathways (Figure 13.3) occurs when implantable bioelectronic device comes into contact with tissue or blood, where an enzyme catalyzes the formation of the C3 convertase. This in turn generates the C5 convertase, which permits the assembly of the terminal complement complex. Complement activation also releases C3a, C4a, and C5a, which are humoral messengers that bind to specific receptors on mast cells, monocytes, neutrophils, macrophages, and smooth muscle cells. They induce mast cell histamine release, cytokine (interleukin 1 and/or 4) release from monocyte, smooth muscle contraction, and platelet aggregation, and act as mediators of the surrounding inflammatory process [129]. The coagulation and complement cascades systems interact significantly to modulate each other's activity. For example, the classical complement can be activated via the cleavage of C1s by Factor XIIa and kallikrein. C3, C5, C6, and Factor B are activated

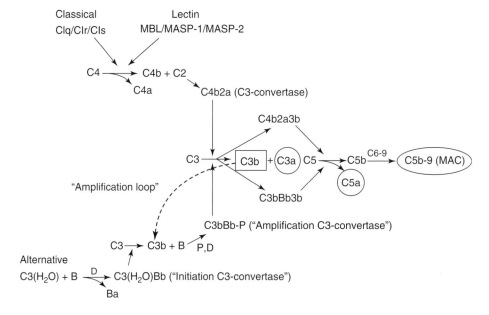

Figure 13.3 Complement activation pathways – the complement system can be activated by the classical, lectin, or alternative pathways, resulting in the formation of C3-convertases.

by thrombin, whereas C5 and Factor B are cleaved by kallikrein. In addition, Factor XIIa also cleaves C3 [130].

In recent years, a number of groups have made important advances in biocompatibility assessment. Yet, we still have no widely recognized test systems, nor do we have an agreed upon list of blood compatible materials. Figure 13.4 describes the alternative scenarios that can be applied for interpreting results of blood–material interaction assays [131]. Failures in implantable bioelectronic devices are due to surface blood-clot formation, which can be overcome by blocking the coagulation cascade in the vicinity of the device surface. Given that thrombin enzyme plays a very important role in the coagulation cascade system, its inhibition has become the focal point for blood contacting material development [132, 133]. Through the incorporation of thrombin inhibitors on device surface materials, blood compatibility can be significantly enhanced. Figure 13.5 provides a schematic representation of the development of a non-thrombogenic polymer-modified device by surface grafting with thrombin inhibitors, which is characterized by high specificity and affinity toward thrombin, the key factor in blood coagulation cascade. The thrombin inhibitors graft on the surface of polymer to bind thrombin, and thus avoid its reaction with fibrinogen that leads to the fibrin clot formation. This requires a number of advances in the synthesis and characterization of novel graftable inhibitors, the understanding of polymer surface reactivity, and the effect of bioactive molecules fixation. The final success of implantable bioelectronic device materials will arise from positive results in each of these research areas.

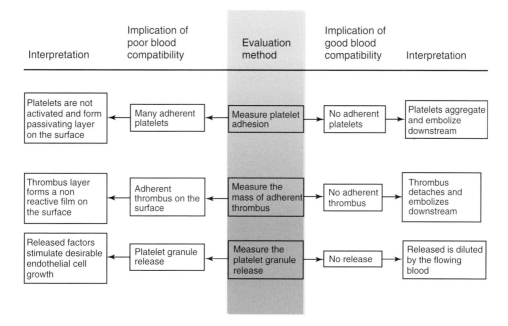

Figure 13.4 Interpretation of blood–material interaction.

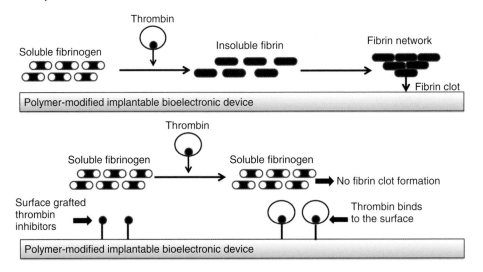

Figure 13.5 Polymer-modified implantable bioelectronic device – improving blood compatibility. The presence of thrombin on the polymer surface inhibits the conversion of fibrinogen to fibrin, thereby preventing the formation of fibrin clot.

13.5
Conclusion

Bioelectronics implantation is accompanied by tissue injury, and the surface of the device interacts with the body for a long period of time. This results in the initiation and maintenance of the foreign body reaction. Thereby, considerable attention must be given to the selection of the appropriate biocompatible coating materials for use with implantable bioelectronic devices in order to minimize the foreign body response while maintaining implantable device biofunctionality and longevity. The device material–blood interface created causes an inflammatory response that involves complex interactions between specific cells and molecular mediators. The intensity of the inflammation depends on the encapsulation material, which induces nonspecific adsorption of blood and tissue fluid proteins. The extent of the nonspecific protein adsorption can be used to evaluate the degree of the implantable bioelectronic device biocompatibility. Following nonspecific protein adsorption, monocytes, leukocytes, and platelets intervene in order to protect the body from the foreign object. At the end of the foreign body reaction, the device is walled off by a vascular fibrous capsule, which confines the implantable bioelectronic device and prevents it from interacting with the surrounding tissues by restricting the transport of low molecular weight analytes such as glucose to the device surface [134–141]. Surface modification strategies using antithrombin, antibiotics, anesthetics, and antihypertensive drugs have been shown to encourage albumin absorption and reduce platelet adhesion. The membrane poly(2-methacryloyloxyethyl phosphorylcholine-co-n-butyl methacrylate) (MPC) specifically developed for medical application is similar in behavior and structure to a cell membrane [142]. *In vivo* tests of MPC-coated devices for implantation in rats and humans demonstrated improved biocompatibility of the implantable device. Moreover, maintaining implantable bioelectronic device biofunctionality in an *in vivo* environment remains a key challenge. Despite outstanding advances in the *in vitro* functionality of bioelectronic devices, a reliable long-term *in vivo* bioelectronic device has not yet been achieved because of the gradual loss of bioelectronic device biofunctionality following implantation. However, using MPC as a grafting membrane and surface modifier may potentially improve the long-term stability and biofunctionality of implantable bioelectronic devices.

The combination of medicinal chemistry products with implantable bioelectronic devices provides an exciting path forward for controlling and localizing delivery of tissue-modified drugs, such as anti-inflammatory agents released at implantation site to prevent inflammation and fibrosis. Growth factors can also be released at the implantation site to induce the growth of blood vessels in order to ensure adequate healing process and analyte transport. These efforts to develop biocompatible materials for implantable bioelectronic devices promise to improve the quality of life for many patients suffering from diseases and pain.

References

1. Feyman, R.P. (1992) *J. Microelectromech. Syst.*, **1**, 60–66.
2. Park, J.B. and Lakes, R.S. (2007) *Biomaterials: An Introduction*, Springer, New York.
3. Ratner, B.D., Northup, S.J., and Anderson, J.M. (2004) in *Biomaterials Science: An Introduction to Materials in Medicine* (eds B.D. Ratner, F.J. Schoen, and J.E. Lemons), Elsevier, San Diego, CA, pp. 355–360.
4. Anderson, J.M. and Langone, J.J. (1999) *J. Controlled Release*, **57**, 107–113.
5. Saad, B., Abu-Hijleh, G., and Suter, U.W. (2003) in *Introduction to Polymeric Biomaterials*, The Polymeric Biomaterials Series (ed. R. Arshady), Citus Books, London, pp. 263–299.
6. ISO (1992) ISO 10933. *Biological Evaluation of Medical Devices*, International Standards Organizations, Geneva.
7. Cooper, S.L., Visser, S.A., Hergenrother, R.W., and Lamba, N.M.K. (2004) in *Biomaterials Science: An Introduction to Materials in Medicine* (eds B.D. Ratner, F.J. Schoen, and J.E. Lemons), Elsevier, San Diego, CA, pp. 67–127.
8. Kasemo, B. (2002) *Surf. Sci.*, **500**, 656–677.
9. Geddes, L.A. and Roeder, R. (2003) *Ann. Biomed. Eng.*, **31**, 879–890.
10. Schoen, F.J. and Anderson, J.M. (2004) in *Biomaterials Science: An Introduction to Materials in Medicine* (eds B.D. Ratner, F.J. Schoen, and J.E. Lemons), Elsevier, San Diego, CA, pp. 293–296.
11. Williams, D.F. (1987) *Progress in Biomedical Engineering*, Elsevier Science, Amsterdam, pp. 54.
12. Boss, J.H., Shajrawi, I., Aunullah, J., and Mendes, D. (1995) *Isr. J. Med. Sci.*, **31**, 203–209.
13. Frost, M. and Meyerhoff, M.E. (2006) *Anal. Chem.*, **78**, 7370–7377.
14. Vaddiraju, S., Tomazos, I., Burgess, D.J., Jain, F.C., and Papadimitrakopoulos, F. (2010) *Biosens. Bioelectron.*, **25**, 1553–1565.
15. Frost, M.C. and Meyerhoff, M.E. (2002) *Curr. Opin. Chem. Biol.*, **6**, 633–641.
16. Salthouse, T.N. (1976) *J. Biomed. Mater. Res.*, **10**, 197–229.

17. Justin, G., Finley, S., Abdur Rhman, A.R., and Guiseppi-Elie, A. (2009) *Biomed. Microdev.*, **11**, 103–115.
18. Ahmad, F., Christenson, A., Bainbridge, M., Yusof, A.P.M., and Ghani, A. (2007) *Biosens. Bioelectron.*, **22**, 1625–1632.
19. Park, H. and Park, K. (1996) *Pharmaceut. Res.*, **13**, 1770–1776.
20. Moussy, F. and Harrison, D.J. (1994) *Anal. Chem.*, **66**, 674–679.
21. Kawashita, M., Tsuneyama, S., Miyaji, F., Kozuka, T., and Yamamoto, K. (2000) *Biomaterials*, **21**, 393–398.
22. Yuen, T.G., Agnew, W.F., and Bullara, L.A. (1987) *Biomaterials*, **8**, 138–141.
23. Campbell, P.K. and Jones, K.E. (1992) *Materials Science and Technology, A Comprehensive Treatment: Medical and Dental Materials*, Wiley-VCH Verlag GmbH, Weinheim, pp. 345–372.
24. Stensaas, S.S. and Stensaad, L.J. (1978) *Acta Neuropathol.*, **41**, 145–155.
25. Bogner, E., Dominizi, K., Hagl, P., Bertagnolli, E., Wirth, M., Gabor, F., Brezna, W., and Wanzenboeck, H.D. (2006) *Acta Biomater.*, **2**, 229–237.
26. Selvakumaran, J., Keddie, J.L., Ewins, D.J., and Hughes, M.P. (2008) *J. Mater. Sci.: Mater. Med.*, **19**, 143–151.
27. Dymond, A.M., Kaechele, L.E., Jurist, J.M., and Crandalll, P.H. (1970) *J. Neurosurg.*, **33**, 574–580.
28. Sul, Y.-T., Johanasson, C.B., Jeong, Y., and Albrektsson, T. (2001) *Med. Eng. Phys.*, **23**, 329–346.
29. Zhang, F., Zheng, Z., Chen, Y., Liu, X., Chen, A., and Jiang, Z. (1998) *J. Biomed. Mater. Res.*, **42**, 128–133.
30. Draye, J.P., Delaey, B., Van de Voorde, A., Van Den Bulcke, A., De Reu, B., and Schacht, E. (1998) *Biomaterials*, **19**, 1677–1687.
31. Cadée, J.A., Brouwer, L.A., den Otter, W., Hennink, W.E., and van Luyn, M.J. (2001) *J. Biomed. Mater. Res.*, **56**, 600–609.
32. Geiger, M., Li, R.H., and Friess, W. (2003) *Adv. Drug Delivery Rev.*, **55**, 1613–1629.
33. Sano, A., Hojo, T., Maeda, M., and Fujioka, K. (1998) *Adv. Drug Delivery Rev.*, **31**, 247–266.
34. Borchard, G. and Junginger, H.E. (2001) *Adv. Drug Delivery Rev.*, **52**, 103.
35. Khor, E. and Lim, L.Y. (2003) *Biomaterials*, **24**, 2339–2349.
36. Uchegbu, I.F., Schätzlein, A.G., Tetley, L., Gray, A.I., Sludden, J., Siddique, S., and Mosha, E. (1998) *J. Pharm. Pharmacol.*, **50**, 453–458.
37. Rihova, B. (1996) *Adv. Drug Delivery Rev.*, **21**, 157–176.
38. Shen, M. and Horbett, T.A. (2001) *J. Biomed. Mater. Res.*, **57**, 336–345.
39. Dalsin, J.L., Hu, B.H., Lee, B.P., and Messersmith, P.B. (2003) *J. Am. Chem. Soc.*, **125**, 4253–4258.
40. Tsai, W.B., Grunkemeier, J.M., McFarland, C.D., and Horbett, T.A. (2002) *J. Biomed. Mater. Res.*, **60**, 348–359.
41. Royals, M.A., Fujita, S.M., Yewey, G.L., Rodriguez, J., Schultheiss, P.C., and Dunn, R.L. (1999) *J. Biomed. Mater. Res.*, **45**, 231–239.
42. Daugherty, A.L., Cleland, J.L., Duenas, E.M., and Mrsny, R.J. (1997) *Eur. J. Pharm. Biopharm.*, **44**, 89–102.
43. Shive, M.S. and Anderson, J.M. (1997) *Adv. Drug Delivery Rev.*, **28**, 5–24.
44. Athanasiou, K.A., Niederauer, G.G., and Agrawal, C.M. (1996) *Biomaterials*, **17**, 93–102.
45. Ronneberger, B., Kao, W.J., Anderson, J.M., and Kissel, T. (1996) *J. Biomed. Mater. Res.*, **30**, 31–40.
46. Lunsford, L., McKeever, U., Eckstein, V., and Hedley, M.L. (2000) *J. Drug Target*, **8**, 39–50.
47. Ronneberger, B., Kissel, T., and Anderson, J.M. (1997) *Eur. J. Pharm. Biopharm.*, **43**, 19–28.
48. Mendes, S.C., Reis, R.L., Bovell, Y.P., Cunha, A.M., van Blitterswijk, C.A., and de Bruijn, J.D. (2001) *Biomaterials*, **22**, 2057–2064.
49. Shastri, V.P. (2003) *Curr. Pharm. Biotechnol.*, **4**, 331–337.
50. Göpferich, A. (1996) *Eur. J. Pharm. Biopharm.*, **42**, 1–11.
51. Wisniewski, N. and Reichert, M. (2000) *Colloids Surf., B Biointerfaces*, **18**, 197–219.

52. Ratner, B.D., Hoffman, A.S., Schoen, F.J., and Lemons, J.E. (2004) *Biomaterials Science: An Introduction to Materials in Medicine*, Elsevier, San Diego, CA.
53. Cheung, K.C. (2007) *Biomed. Microdev.*, **9**, 923–938.
54. Lutolf, M.P. and Hubbell, J.A. (2005) *Nat. Biotechnol.*, **23**, 47–55.
55. Vozzi, G., Flaim, C., Ahluwali, A., and Bhatia, S. (2003) *Biomaterials*, **24**, 2533–2540.
56. Stieglitz, T., Beutel, H., Schuettler, M., and Meyer, J.-U. (2000) *Biomed. Microdev.*, **2**, 283–294.
57. Lee, K., He, J., Clement, R., Massia, S., and Kim, B. (2004) *Biosens. Bioelectron.*, **20**, 404–407.
58. Kishida, A., Iwata, H., Tamada, Y., and Ikada, Y. (1991) *Biomaterials*, **12**, 786–792.
59. Arshady, R. (1993) *Biomaterials*, **14**, 5–15.
60. Osterberg, E., Bergstrom, K., Holmberg, K., Riggs, J.A., Vam, A.J.M., Schumann, T.P., Burns, N.L., and Harris, J.M. (1993) *Colloids Surf., A: Physicochem. Eng. Aspects*, **77**, 159.
61. Marchant, R.E., Yuan, S., and Szakalas-Gratzl, G. (1994) *J. Biomater. Sci., Polym. Ed.*, **6**, 549–564.
62. Ravin, A.G., Olbrich, K.C., Levin, L.S., Usala, A.L., and Klizman, B. (2001) *J. Biomed. Mater. Res.*, **58**, 313–318.
63. Kost, J. and Langer, R. (1997) in *Hydrogels in Medicine and Pharmacy* (ed. N.A. Peppas), CRC, Boca Raton, FL, pp. 95–108.
64. Guiseppi-Elie, A., Brahim, S., Slaughter, G., and Ward, K.R. (2005) *IEEE Sens. J.*, **5**, 345–355.
65. Brahim, S.I., Slaughter, G.E., and Guiseppi-Elie, A. (2003) in *Smart Structures and Materials*, International Society for Optics and Photonics, pp. 1–12.
66. Slaughter, G. (2010) *J. Diabetes Sci. Technol.*, **4**, 320.
67. Quinn, C.P., Pathak, C.P., Heller, A., and Hubbell, J.A. (1995) *Biomaterials*, **16**, 389–396.
68. Espadas-Torre, C. and Meyerhoff, M.E. (1995) *Anal. Chem.*, **67**, 3108–3114.
69. Hickey, T., Kreutzer, D., Burgess, D.J., and Moussy, F. (2002) *J. Biomed. Mater. Res.*, **61**, 180–187.
70. Patil, S.D., Papadimitrakopoulos, F., and Burgess, D.J. (2004) *Diabetes Technol. Ther.*, **6**, 887–897.
71. Galeska, I., Kim, T.K., Patil, S.D., Bhardwaj, U., Chatttopadhyay, D., Papadimitrakopoulos, F., and Burgess, D.J. (2005) *AAPS J.*, **7**, E231–E240.
72. Patil, S.D., Papadimitrapouuos, F., and Burgess, D.J. (2007) *J. Controlled Release*, **117**, 68–79.
73. Bhardwaj, U., Sura, R., Papadimitrakopoulos, F., and Burgess, D.J. (2007) *Diabetes Sci. Technol.*, **1**, 8–17.
74. Bhardwaj, U., Radhacirsshana, S., Papadimitrakopoulos, F., and Burgess, D.J. (2010) *Int. J. Pharm.*, **484**, 78–86.
75. Snellings, G.M.B., Vansteenkiste, S.O., Corneillie, S.I., Davies, M.C., and Schacht, E.H. (2000) *Adv. Mater.*, **12**, 1959–1962.
76. Amiji, M. and Park, K. (1993) *J. Biomater. Sci., Polym. Ed.*, **4**, 217–234.
77. Milner, S.T. (1991) *Science*, **251**, 905–914.
78. Kaeble, D.H. and Moacanin, J. (1977) *Polymer*, **18**, 475.
79. Lyman, D.J., Brash, J.L., Chaikin, S.W., Klien, K.G., and Carini, M. (1968) *Trans. Am. Soc. Artif. Org.*, **11**, 301.
80. Vroman, L. and Adams, A.L. (1971) *J. Polym. Sci.*, **34**, 159.
81. Cholakis, C.H. and Sefton, M.V. (1989) *J. Biomed. Mater. Res.*, **23**, 399.
82. Ip, W.F. and Sefton, M.V. (1991) *J. Biomed. Mater. Res.*, **25**, 875.
83. Baier, R.E., Gott, V.L., and Feruse, A. (1970) *Trans. Am. Soc. Artif. Org.*, **16**, 50.
84. Papra, A., Bernard, A., Juncker, D., Larsen, N.B., Michel, B., and Delamarche, E. (2001) *Langmuir*, **17**, 4090–4095.
85. Sharma, S., Popat, K.C., and Desai, T.A. (2002) *Langmuir*, **18**, 8728–8731.
86. Bearinger, J.P., Castner, D.G., Golledge, S.L., Rezania, A., Hubchak, S., and Healy, K.E. (1997) *Langmuir*, **13**, 5175–5183.

87. Drury, J.K., Ashton, T.R., Cunningham, J.D., Maini, R., and Pollock, J.G. (1987) *Ann. Vasc. Surg.*, **1**, 542–547.
88. Bordenave, L., Caix, J., Basse-Cathalinat, B., Baquey, C., Midy, D., Baste, J.C., and Constans, H. (1989) *Biomaterials*, **10**, 235–242.
89. Guidoin, R., Snyder, R., Martin, L., Botzko, K., Marois, M., Awad, J., King, M., Domurado, D., Bedros, M., and Gosselin, C. (1984) *Ann. Thorac. Surg.*, **37**, 457–465.
90. Cziperle, D.J., Joyce, K.A., Tattersall, C.W., Henderson, S.C., Cabusao, E.B., Garfield, J.D., Kim, D.U., Duhamel, R.C., and Greisler, H.P. (1992) *J. Cardiovasc. Surg.*, **33**, 407–414.
91. Noishiki, Y. and Chvapil, M. (1987) *Vasc. Surg.*, **21**, 401–411.
92. Guidoin, R., Marceau, D., Couture, J., Rao, T.J., Merhi, Y., Roy, P.E., and Dela Faye, D. (1989) *Biomaterials*, **10**, 156–165.
93. Slaughter, G.E., Bieberich, E., Wnek, G.E., Wynne, K.J., and Guiseppi-Elie, A. (2004) *Langmuir*, **20**, 7189–7200.
94. Kito, H. and Matsuda, T. (1996) *J Biomed. Mater. Res.*, **30**, 321–330.
95. Ratner, B.D., Hoffman, A.S., Hanson, S.R., Harker, L.A., and Whiffen, J.D. (1979) *J. Polym. Sci., Polym. Symp.*, **66**, 363.
96. Sasaki, T., Ratner, B.D., and Hoffman, A.S. (1976) *ACS Symp. Ser.*, **31**, 283.
97. Ratner, B.D. and Hoffman, A.S. (1974) *J. Appl. Polym. Sci.*, **18**, 3183.
98. Yun, J.K., DeFife, K., Colton, E., Stack, S., Azeez, A., Cahalan, L., Verhoeven, M., Cahalan, P., and Anderson, J.M. (1995) *J. Biomed. Mater. Res.*, **29**, 257–268.
99. Morosoff, N. (1990) in *Plasma Deposition, Treatment, and Etching of Polymers* (ed. R. d'Agostino), Academic Press, San Diego, CA, p. 1–84.
100. Piskin, E. (1992) *J. Biomater. Sci., Polym. Ed.*, **4**, 45–60.
101. Lopez, G.P., Ratner, B.D., Tidwell, C.D., Haycox, C.L., Rapoza, R.J., and Horbett, T.A. (1991) *J. Biomed. Mater. Res.*, **26**, 415–439.
102. Hu, C.-Z., Dolence, E.K., Person, J., and Osaki, S. (1997) in *Surface Modification of Polymeric Biomaterials* (eds B.D. Ratner and R.D.G. Castner), Plenum Press, New York, pp. 61–68.
103. Jung, D.R., Kapur, R., Adams, T., Giuliano, K.A., Mrksich, M., Craighead, G.H., and Taylor, D.L. (2001) *Crit. Rev. Biotechnol.*, **21**, 111–154.
104. Jenkins, G.M., Kawamura, K., and Ban, L.L. (1972) *Proc. R. Soc. London, Ser. A*, **327**, 501–517.
105. Wege, E. (1984) in *Process Mineralogy of Ceramic Materials* (eds W. Baumgart, A.C. Dunham, and G.C. Amstutz), Enke, Stuttgart, pp. 125–149.
106. Oberlin, A. (1989) in *Chemistry and Physics of Carbon* (ed. B.A. Thrower), Marcel Dekker, New York, pp. 1–143.
107. Vroman, L. (1984) *Biomater. Med. Dev. Artif. Organs*, **121**, 307.
108. Andrade, J.D. and Hlady, V. (1987) *Ann. N. Y. Acad. Sci.*, **516**, 158–172.
109. Sawyer, P.N. and Pate, J.W. (1954) *Surgery*, **34**, 491.
110. Seeger, J.M. and Klingman, N. (1988) *J. Vasc. Surg.*, **8**, 476–482.
111. Drumheller, P.D., Elbert, D.L., and Hubbel, J.A. (1994) *Biotechnol. Bioeng.*, **43**, 772–780.
112. Sharma, C.P. and Szycher, M. (1991) *Blood Compatibility Materials and Devices: Perspective towards 21st Century*, Technomic Publishing Co., Inc., Lancaster, PA.
113. Reynolds, L.O., Horbett, T.A., Ratner, B.D., and Hoffman, A.S. (1982) *Trans. Soc. Biomater.*, **5**, 37.
114. Reynolds, L.O. and Simon, T.L. (1980) *Transfusion*, **20**, 669.
115. Kyrolainen, M., Rigsby, P., Eddy, S., and Vadgama, P. (1995) *Acta Anaesthesiol. Scand. Suppl.*, **104**, 55–60.
116. Gerritsen, M., Jansen, J.A., Kros, A., Vriezema, D.M., Sommerdijk, N.A., Nolte, R.J., Lutterman, J.A., Van Hovell, S.W., and Van der Gaag, A. (2001) *J. Biomed. Mater. Res.*, **54**, 69–75.
117. Gerritsen, M., Jansen, J.A., and Lutterman, J.A. (1999) *Neth. J. Med.*, **54**, 167–179.
118. Abel, P.U. and von Woedtke, T. (2002) *Biosens. Bioelectron.*, **17**, 1059–1070.
119. Service, R.F. (2002) *Science*, **297**, 962–963.

120. Ward, W.K., Jansen, L.B., Anderson, E., Reach, G., Klein, J.C., and Wilson, G.S. (2002) *Biosens. Bioelectron.*, **17**, 181–189.
121. Tanihara, M., Suzuki, Y., Yamamoto, E., Noguchi, A., and Mizushima, Y. (2001) *J. Biomed. Mater. Res.*, **56**, 216–221.
122. Ward, W.K., Quinn, M.J., Wood, M.D., Tiekotter, K.L., Pidikiti, S., and Gallagher, J.A. (2003) *Biosens. Bioelectron.*, **19**, 155–163.
123. Kedem, A., Perets, A., Gamlieli-Bonshtein, I., Dvir-Ginzberg, M., Mizrahi, S., and Cohen, S. (2005) *Tissue Eng.*, **11**, 715–722.
124. Laurencin, C.T. and Elgendy, H. (1994) in *Site Specific Drug Delivery* (eds A. Domb and M. Maniar), John Wiley & Sons, Inc., New York, pp. 27–46.
125. Anderson, J.M. (2001) *Annu. Rev. Mater. Res.*, **31**, 81–110.
126. Fournier, E., Passirani, C., Montero-Menei, C.N., and Benoit, J.P. (2003) *Biomaterials*, **24**, 3311–3331.
127. Arshady, R. (2003) in *Introduction to Polymeric Biomaterials*, The Polymeric Biomaterials Series (ed. R. Arshady), Citus Books, London, pp. 1–62.
128. Ratner, B.D. and Horbett, T.A. (2004) in *Biomaterials Science: An Introduction to Materials in Medicine* (eds B.D. Ratner, F.J. Schoen, and J.E. Lemons), Elsevier, San Diego, CA, p. 237.
129. Hakim, R.M. (1992) *Cardiovasc. Pathol.*, **2**, 187S–197S.
130. Blajchman, M.A. and Ozge-Anwar, A.H. (1986) *Prog. Hematol.*, **14**, 149–182.
131. Hanson, S.R. (2004) in *Biomaterials Science, An Introduction to Materials in Medicine* (eds B.D. Ratner, F.J. Schoen, and J.E. Lemons), Elsevier, San Diego, CA, p. 332.
132. Lewis, S.D., Assunta, S., Lyle, E.A., Mellott, M.J., Appleby, S.D., Brady, S.F., Stauffer, K.J., Sisko, J.T., and Mao, S. (1995) *Thromb. Haemost.*, **74**, 1107–1112.
133. Sanderson, P.E.J.S. and Naylor-Olsen, A.M. (1998) *Curr. Med. Chem.*, **5**, 289–304.
134. Sharkawy, A.A., Klitzman, B., Truskey, G.A., and Reichert, W.M. (1997) *J. Biomed. Mater. Res.*, **37**, 401–412.
135. Wisniewski, N., Klitzman, B., Miller, B., and Reichert, W.M. (2001) *J. Biomed. Mater. Res.*, **57**, 513–521.
136. Freeman, C.L., Mayhan, K.G., Picha, G.J., and Colton, C.K. (1987) *A Study of Mass Transport Resistance of Glucose Across Rat Capsular Membranes*, Materials Research Society Symposium, Materials Research Society, Boston, MA, pp. 773–778.
137. Woodward, S.C. (1982) *Diabetes Care*, **5**, 278–281.
138. Reach, G. and Wilson, G.S. (1992) *Anal. Chem.*, **64**, 381A–386A.
139. Bobbioni-Harsch, E., Rohner-Jeanrenaud, F., Koudelka, M., de Rooij, N., and Jeanrenaud, B. (1993) *J. Biomed. Eng.*, **15**, 457–463.
140. Rebrin, K., Fischer, U., Hahn von Dorsche, H., von Woetke, T., Abel, P., and Brunstein, E. (1992) *J. Biomed. Eng.*, **14**, 33–40.
141. Gilligan, B.J., Shults, M.C., Rhodes, R.K., and Updike, S.J. (1994) *Diabetes Care*, **17**, 882–887.
142. Ishihara, K. (1997) *Trends Polym. Sci.*, **5**, 401–407.

14
Abiotic (Nonenzymatic) Implantable Biofuel Cells
Sven Kerzenmacher

14.1
Introduction

14.1.1
The History of Implantable Abiotic Fuel Cells

Ever since the introduction of the cardiac pacemaker as the first fully implantable medical device in 1958 [1, 2], efficient and reliable supply of medical implants with electric power has been a challenge for researchers and engineers [3, 4]. Today, batteries are still the only practically available power supply solution for medical implants. Inherently limited, battery capacity often defines the overall lifetime of the implant. Usually, spent batteries necessitate replacement of the complete implanted device in a surgical procedure, which is associated with extra patient burden, risk, and cost.

A fascinating alternative to batteries is the use of the body's own energy to theoretically indefinitely supply the medical implant. Among the envisioned concepts are implantable biofuel cells, which, in contrast to mechanical or thermoelectric generators, promise a constant energy supply independent of body movements or temperature gradients. Biofuel cells generate electrical energy from the oxidation of a fuel at the anode, coupled to the reduction of a terminal electron acceptor (usually oxygen) at the cathode. The electron flow between the two electrodes is directly available as electrical energy to drive implantable electronic circuits. In this context, glucose is almost exclusively considered as fuel because of its abundance in body fluids (see Table 14.1). Fuel cells typically require electrode catalysts to enable the thermodynamically feasible, but often kinetically hindered electrode reactions. With enzymes, nature has found efficient and highly specific protein structures with excellent biocatalytic activity under physiological conditions. Employed at the electrodes of an implantable fuel cell operating, for example, on glucose and oxygen, they enable fast reactions with high turnover rates. This directly translates into high current densities at low polarization losses, and thus high electrical power densities. However, the drawback of enzymes is their limited long-term stability [5]. Their catalytic activity strongly depends on correct folding and formation of

Implantable Bioelectronics, First Edition. Edited by Evgeny Katz.
© 2014 Wiley-VCH Verlag GmbH & Co. KGaA. Published 2014 by Wiley-VCH Verlag GmbH & Co. KGaA.

Table 14.1 The standard free energy of formation ΔG_f^0 of selected reactants [7].

Substance	ΔG_f^0 at 25 °C (kJ mol^{-1})
O_2	0
H_2O	−237.178
H^+	0
Glucose	−917.22
Gluconate	−1128.3

their three-dimensional protein structure, and through oxidation and the breaking of bonds enzymes tend to lose a significant share of their catalytic activity over time. While current research approaches thus try to stabilize the correct folding, for example, by the introduction of additional intramolecular bonds or the immobilization of enzymes in stabilizing polymer structures, most of these approaches only delay the natural degradation of enzymes [5]. So far, a maximum lifetime of up to 40 days has been realized in the context of implantable biofuel cells [6].

A viable alternative to enzymes is the use of *abiotic catalysts* such as platinum or other noble metals to enable the anodic oxidation of glucose in a fuel cell [8]. Compared to enzymes, abiotic catalysts are less specific and less active under physiological conditions, and the power densities of such systems are up to two orders of magnitude lower. However, their robust nature renders them particularly attractive for the construction of implantable glucose fuel cells with extended lifetimes. Researchers in the late 1960s had already developed this kind of fuel cell to power implantable devices and demonstrated that the energy-autonomous operation of, for instance, a cardiac pacemaker (demanding below 100 µW of electricity) is well within the range of feasibility. In the first successful *in-vivo* trials the continuous operation of such fuel cells implanted in a dog for periods of up to 5 months has been demonstrated [9].

This chapter is intended to summarize operation principles, design considerations, and the current state-of-the-art and future research trends in the field on implantable biofuel cells based on abiotic catalysts. Compared to a previous publication by the author [10] this chapter is an updated and expanded treatise, focused specifically on the recent developments and challenges. A comprehensive review on abiotic implantable glucose fuel cells with a detailed description of the historical development is available elsewhere [8].

14.2
Basic Principles

In the following sections, the operation principle of abiotic glucose fuel cells and the corresponding thermodynamics and electrochemical principles are introduced. For

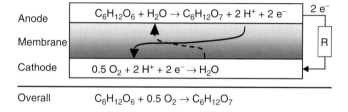

Figure 14.1 General electrode reactions of an implantable glucose–oxygen fuel cell, assuming a proton-conducting membrane and gluconic acid as the reaction product, according to [26]. (Reprinted from [8] with permission from Elsevier.)

further reading the textbooks *Electrochemistry* by Hamann et al. [11], *Electrochemical methods* by Bard and Faulkner [12], and *Fuel cell systems explained* by Larminie and Dicks [13] are suggested.

14.2.1
Electrode Reactions and Theoretical Potentials

In Figure 14.1, the basic operation principle of an abiotic glucose fuel cell based on platinum electrodes is illustrated. At the cathode, oxygen is reduced to water. At the anode, glucose is predominantly oxidized to glucono-lactone, releasing two electrons and two protons. Gluconolactone undergoes hydrolysis to form gluconic acid [14]. A number of studies showed that gluconic acid can be further oxidized, but at comparably slower reaction rates [15–19]. A detailed treatise on glucose electro-oxidation under different conditions is available elsewhere [20–25].

The maximum theoretical voltage that such a fuel cell can deliver under standard conditions (pH 0; $T = 298$ K; concentration of all species at 1 mol l^{-1}) corresponds to the *standard redox potential* (or *standard electromotive force*) ΔE_r^0 of the fuel cell reaction. It is calculated from the *standard free energy* of the overall reaction ΔG_r^0 according to

$$\Delta E_r^0 = -\frac{\Delta G_r^0}{nF} \qquad (14.1)$$

Herein, $n = 2$ is the number of electrons transferred in the reaction and $F = 96\,500$ C mol^{-1} is the Faraday constant. The *standard free energy* of the overall reaction ΔG_r^0 is available from tabulated values (see for instance Table 14.1) of the *standard free energy of formation* ΔG_f^0 using the relation

$$\Delta G_r^0 = \sum \Delta G_f^0(\text{Products}) - \sum \Delta G_f^0(\text{Educts}) \qquad (14.2)$$

For the overall reaction of the glucose fuel cell

$$C_6H_{12}O_6 + 0.5 O_2 \rightarrow C_6H_{12}O_7 \qquad (14.3)$$

the standard *free energy* ΔG_r^0 and the standard potential ΔE_r^0 thus calculated are

$$\Delta G_r^0 = (-1128.3)\,\text{kJ mol}^{-1} - (-917.22 + 0.5 \times 0)\,\text{kJ mol}^{-1}$$
$$= -211.08\,\text{kJ mol}^{-1}$$

$$\Delta E_r^0 = -\frac{\Delta G_r^0}{nF} = \frac{211.08\,\text{kJ}\,\text{mol}^{-1}}{2\cdot 96500\,\text{C}\,\text{mol}^{-1}} = 1.094\,\text{V}$$

The potentials of the half-cell reactions can be calculated accordingly. In this case, the direction of the anode reaction must be reversed, so that the products are the reduced species [27]. This leads to the following potentials, referenced with respect to the standard hydrogen electrode (SHE):

$$\text{Anode}: C_6H_{12}O_7 + 2H^+ + 2e^- \rightarrow C_6H_{12}O_6 + H_2O$$
$$\Delta G_{\text{anode}}^0 = ((-917.22) + (-237.178)) - (-1128.3) + 2\times 0)\,\text{kJ}\,\text{mol}^{-1}$$
$$= -26.098\,\text{kJ}\,\text{mol}^{-1}$$

$$\Delta E_{\text{anode}}^0 = -\frac{\Delta G_{\text{anode}}^0}{nF} = \frac{26.098\,\text{kJ}\,\text{mol}^{-1}}{2\cdot 96500\,\text{C}\,\text{mol}^{-1}} = 0.135\,\text{V versus SHE} \quad (14.4)$$

$$\text{Cathode}: 0.5O_2 + 2H^+ + 2e^- \rightarrow H_2O$$
$$\Delta G_{\text{cathode}}^0 = (-237.178\,\text{kJ}\,\text{mol}^{-1}) - (0 + 2\times 0)$$
$$= -237.178\,\text{kJ}\,\text{mol}^{-1}$$

$$\Delta E_{\text{cathode}}^0 = -\frac{\Delta G_{\text{cathode}}^0}{nF} = \frac{237.178\,\text{kJ}\,\text{mol}^{-1}}{2\cdot 96500\,\text{C}\,\text{mol}^{-1}} = 1.229\,\text{V versus SHE} \quad (14.5)$$

In the context of implantable fuel cells, it has to be considered that in biological systems the reaction conditions differ substantially from standard conditions. Consequently, the potential of the overall reaction under nonstandard conditions $\Delta E^{0\prime}$ has to be calculated using the Nernst equation, which takes into account the concentration of the products and educts to the power of their respective stoichiometric coefficients p and e according to

$$\Delta E_r^{0\prime} = \Delta E_r^0 - \frac{RT}{nF}\cdot \ln\frac{[\text{products}]^p}{[\text{educts}]^e} \quad (14.6)$$

For a glucose fuel cell operating under the physiological condition of body tissue the concentrations of glucose and oxygen can be assumed to be 3×10^{-3} and $0.06\times 10^{-3}\,\text{mol}\,\text{l}^{-1}$, respectively [28], whereas at neutral pH the H^+ concentration is $10^{-7}\,\text{mol}\,\text{l}^{-1}$. Since the concentration of gluconic acid in body tissue is not readily available, a value of $0.5\times 10^{-3}\,\text{mol}\,\text{l}^{-1}$ may be assumed, which is in the range of concentration values assessed for the lens of the human eye [29]. With the above concentrations, the potential of the overall reaction under nonstandard conditions is calculated as

$$\Delta E_r^{0\prime} = \Delta E_r^0 - \frac{RT}{nF} \cdot \ln \frac{[C_6H_{12}O_7]}{[C_6H_{12}O_6] \cdot [O_2]^{0.5}}$$

$$= 1.094\,\text{V} - \frac{8.31 \cdot 298}{2 \cdot 96500}\,\text{JC}^{-1} \cdot \ln \frac{[0.0005]}{[0.003] \cdot [6 \cdot 10^{-5}]^{0.5}}$$

$$= 1.055\,\text{V}$$

The Nernst equation can also be applied to calculate the individual electrode potentials under nonstandard conditions. According to the above concentrations this yields

$$\text{Anode}: C_6H_{12}O_7 + 2H^+ + 2e^- \rightarrow C_6H_{12}O_6 + H_2O$$

$$\Delta E_{\text{anode}}^{0\prime} = \Delta E_{\text{anode}}^0 - \frac{RT}{nF} \cdot \ln \frac{[C_6H_{12}O_6]^1}{[C_6H_{12}O_7] \cdot [H^+]^2}$$

$$= 0.135\,\text{V} - \frac{8.31 \cdot 298}{2 \cdot 96500}\,\text{JC}^{-1} \cdot \ln \frac{[0.003]^1}{[0.0005]^1 \cdot [10^{-7}]^2}$$

$$= -0.301\,\text{V versus SHE}$$

$$\text{Cathode}: 0.5O_2 + 2H^+ + 2e^- \rightarrow H_2O$$

$$\Delta E_{\text{cathode}}^{0\prime} = \Delta E_{\text{cathode}}^0 - \frac{RT}{nF} \cdot \ln \frac{[H_2O]^1}{[H^+]^2 \cdot [O_2]^{0.5}}$$

$$= 1.229\,\text{V} - \frac{8.31 \cdot 298}{2 \cdot 96500}\,\text{JC}^{-1} \cdot \ln \frac{[1]^4}{[10^{-7}]^2 \cdot [6 \cdot 10^{-5}]^{0.5}}$$

$$= 0.753\,\text{V versus SHE}$$

14.2.2
Practical Fuel Cell Voltage, Power Density, and Efficiency

In practice, the fuel cell voltages are considerably lower than the theoretical maximum derived from thermodynamics. The corresponding current density–voltage behavior plot (also called *polarization plot* or *load curve*) of a fuel cell shows a distinct nonlinear shape and is depicted exemplarily in Figure 14.2. The polarization behavior of the individual fuel cell electrodes is analogous to the cell voltage. However, upon polarization the cathode potential shifts toward more negative values, whereas the anode potential is increased toward positive potentials. The current density–potential curve of a fuel cell electrode can be mathematically described using the so-called Butler–Volmer equation as described elsewhere [11, 12, 30]. As indicated in Figure 14.2 and explained in the following, the curve can be divided into the sections A, B, and C.

- *Section A*: Already under open circuit conditions (no current flow through the external load circuit) and in the low current density range a potential loss (polarization) occurs. This potential loss stems from the activation energies necessary to drive the electrochemical reactions at the respective electrodes and can be related to electrocatalytic activity.

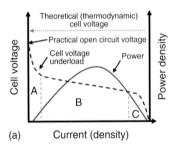 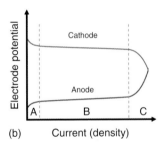

(a) Current (density) (b) Current (density)

Figure 14.2 Schematic representation of the polarization behavior of a complete fuel cell (a) and the individual electrode potentials (b). The thermodynamic reversible potential (indicated by a gray line) differs from the practical open circuit potential. With increasing current (density) losses occur due to (A) activation over potential, (B) ohmic resistances, and (C) mass transport limitations. See text for further explanations.

- *Section B*: In its linear range the polarization curve is dominated by ohmic conduction losses in the electrochemical system, caused by the electrical resistance of the electrolyte and electrodes, as well as the electrical connections in the circuit.
- *Section C*: With further increasing current densities, voltage losses related to the limited transport of reactants and their depletion at the electrode occur. Consequently, a rapid drop in cell voltage is observed. Upon further increase in current density a different electrode reaction (e.g., water splitting) has to take place to sustain the higher currents.

For practical application, the electrical power output of an implantable fuel cell is of importance. It is calculated as the product of cell voltage ΔE_{FC} and current I:

$$P = \Delta E_{FC} I \tag{14.7}$$

As can be seen from Figure 14.2, the power output of the fuel cell goes through a maximum, referred to as the *maximum power point* (*MPP*). To gain the most power of the device, it is thus essential to operate the fuel cell at the associated optimum current/voltage pair. To compare the performance of differently sized fuel cells it is useful to normalize power and current to the projected area or the volume of the electrode or total fuel cell. Given that fuel cells are essentially two-dimensional devices, power densities and current densities are commonly reported in microwatts per square centimeter or microamperes per square centimeter.

The conversion efficiency η of a fuel cell, for example, the degree to which the energy content of the fuel is actually converted into electrical energy, is calculated from the molar enthalpy ΔH_{fuel} of the fuel and its mass flow according to

$$\eta = \frac{(\Delta E_{FC} \cdot I)}{(\Delta H_{fuel}) \cdot (\text{mass flow})} \tag{14.8}$$

In the case of implantable fuel cells, it will in most cases be difficult to draw up accurate mass and energy balances. An alternative way to estimate the fuel cell's electrical efficiency is to calculate its voltage efficiency η_V. This value relates the operation voltage of the fuel cell to the theoretical maximum voltage, which would be available if all of the fuel's energy content (heating value or molar enthalpy) was converted into electricity [13]. Considering the heating value of glucose at 2805 kJ mol^{-1} [31] and taking into account 0.4 V as the typical operation voltage this yields

$$\eta_V = \frac{\Delta E_{FC}}{\frac{\Delta H_{fuel}}{nF}} = \Delta E_{FC} \cdot \frac{2 \cdot 96500\,\text{C} \cdot \text{mol}^{-1}}{2805\,\text{kJ} \cdot \text{mol}^{-1}} = \Delta E_{FC} \cdot 0.0689\,\text{V}^{-1} \quad (14.9)$$

$$= 0.4\,\text{V} \cdot 0.0689\,\text{V}^{-1} = 0.28 = 28\% \quad (14.10)$$

When calculating the voltage efficiency, we assume that all of the fuel flowing into the fuel cell takes part in the electrochemical reaction. Because side reactions occur in practice also not all of the fuel is converted into electricity. The voltage efficiency is thus always an upper estimate of the fuel cell's overall conversion efficiency.

14.2.3
Reliable Characterization of Implantable Glucose Fuel Cells

From an application perspective, the most useful method to characterize an implantable glucose fuel cell is the recording of polarization curves (Figure 14.2), as the electrical power output of the fuel cell can be directly calculated from these. Furthermore, the shape of the polarization curve enables the identification of underlying loss mechanisms such as activation polarization, ohmic losses, and mass-transport polarization (see Section 14.2.2). In any case, it is useful to include a reference electrode in the measurement setup. This way, the individual polarization of anode and cathode can be monitored, and the polarization losses and mechanisms can be attributed to the individual electrodes – prerequisite for systematic development of an optimized device.

An easy-to-implement technique to record polarization curves is the use of passive load resistors. However, this approach mandates that always a complete fuel cell be characterized. This can become time-consuming and costly, in particular when a large number of experiments are performed to optimize a specific component of the fuel cell or to screen a catalyst material library. Since passive load resistors do not allow for the forced operation of the fuel cell beyond the point where the overall cell potential reaches zero, care must be taken to ensure that the fuel cell performance is not limited by a fuel cell component that is not in the focus of the experiment.

As an alternative that is less costly than dedicated potentiostats, researchers have also developed a number of specialized experimental setups dedicated to performing characterization experiments in a highly parallel manner. One example is the setup for high-throughput material screening developed by Bruce Logan and coworkers [32], in which individual electrochemical cells are connected in parallel to a single standard laboratory power supply as voltage source. A different setup with individually controllable electronic loads has been specifically developed for use with abiotic implantable glucose fuel cells [33].

For a reliable and meaningful characterization of fuel cell performance, it is indispensable to record polarization curves under quasi-steady-state conditions. As can be seen from the experimental results shown in Figure 14.3a, implantable glucose fuel cells exhibit a slow response to changes in load current. It can take hours before steady-state conditions are reached after a current step. Under such circumstances, a too rapid increase in load current (or alternatively cell potential) during the recording of polarization curves leads to an overestimation of performance [33]. A useful way to prevent performance estimation by very fast scan rates is the stepwise recording of polarization curves as illustrated in Figure 14.3b. Here, only the stabilized potential values upon a change in load current are considered. The dependence of the apparent fuel cell power density on scan rate is depicted in Figure 14.3c. The steady-state power density of the fuel cell in this example amounts to $5\,\mu W\,cm^{-2}$, but with very fast scan rates wrongful power densities that are orders of magnitude too high can be obtained.

14.3
Abiotic Catalyst Materials and Separator Membranes

14.3.1
Electrocatalysts for Glucose Oxidation

In the early works, platinum has been mainly employed as catalyst for the anodic glucose oxidation [35]. Later, other noble metals such as gold, rhodium, iridium, as well as platinum–ruthenium [36, 37], platinum–bismuth [38], and platinum–tungsten alloys [18] have been successfully used. Remarkably, there is a meeting abstract that reports on electrodeposited platinum, gold–platinum, and gold–palladium (Pd content of less than 50 atom%) alloys that exhibit electrocatalytic selectivity for glucose oxidation in the presence of oxygen [39]. Reportedly, the addition of lead acetate to the plating solution was responsible for the observed oxygen tolerance, but more details have not been disclosed. In a more recent study on platinum–gold alloys, it was shown that Pt_{20}–Au_{80} has a higher electrocatalytic activity for glucose oxidation performance than $Pt_{50}Au_{50}$ or Pt and Au alone [40, 41].

Besides noble metals, metal chelates such as $Mo\text{-}O_2\text{-}4,4',4'',4'''$-tetrasulfophthalocyanine-Ca^{2+} have also been used as anode catalyst and have shown remarkably high current densities of up to $1\,mA\,m^{-2}$ with practically no electrode polarization [42]. Recently, it was also demonstrated that silicon nanoparticles of 1–3 nm diameter exhibit catalytic activity toward glucose oxidation and can be used as electrocatalyst in neutral phosphate buffer [43]. However, these catalysts have not yet been used in complete abiotic implantable glucose fuel cells.

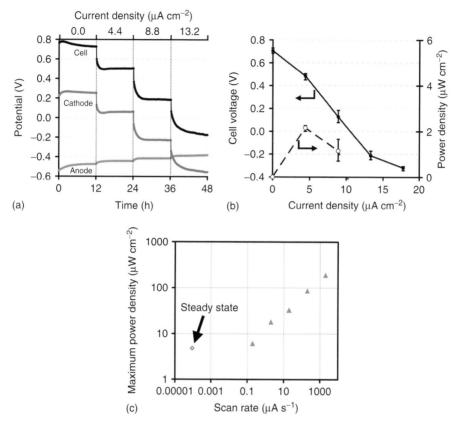

Figure 14.3 (a) Exemplary evolution of electrode potentials during the recording of a polarization curves with stepwise increase in load current. (b) Corresponding polarization curve constructed from stabilized potentials at the end of a current step. (Reproduced from [33] with kind permission from Springer Science and Business Media.) (c) Dependence of maximum power density on scan speed. The power density under steady-state conditions is marked with an arrow. (Adapted from [34] with permission; Copyright (2012) American Chemical Society.)

14.3.2
Electrocatalysts for Oxygen Reduction

At the cathode, noble metals such as platinum, palladium, gold, and silver have also been investigated. Among these, platinum exhibited the highest oxygen reduction activity in neutral buffer containing no glucose [44]. In contrast, silver has the advantage that it shows no catalytic activity for glucose oxidation and is thus not prone to mixed potential formation in the presence of glucose. However, silver shows considerable activation losses, with an approximately 400 mV more negative onset potential compared to platinum [44]. In this context, activated carbon is more favorable. Similarly to silver, it is insensitive toward glucose, but shows a reduced open circuit potential compared to platinum by only 100 mV. A similar behavior is also reported for ferric phthalocyanine deposited on pyrolytic graphite [45, 46]. However, no corresponding fuel cell performance has been reported. Recently, reduced graphene oxide was compared to activated carbon as cathode. The use of graphene effectively doubled the power density from 1 to 2 μW cm^{-2} [47].

Largely unexplored for use in implantable fuel cells are the abiotic cathode materials that are currently being developed for microbial fuel cells [48] that operate under similar conditions. Examples are *manganese oxide* [49] and *metal macrocycles* adsorbed to carbon and pyrolyzed [50]. Similarly, numerous non-noble oxygen reduction catalysts are currently being developed for hydrogen and methanol fuel cells, for example, based on carbon supported *transition metal macrocycles* [51] or *nitrogen-containing carbon nanotubes* [52].

14.3.3
Separator Membranes

The main function of the separator membrane in an implantable fuel cell is to provide electrical insulation between the electrodes, while at the same time enabling ion transport to close the electrochemical circuit of the fuel cell. Given that the typical operating current densities of implantable glucose fuel cells are comparably low, it is usually sufficient to apply a simple spacer structure such as a filter membrane as separator, with its pores filled with body fluid as electrolyte. However, for fuel cells constructed with a hydrophobic membrane to shield the cathode from glucose interference (see Section 14.4.2), the use of anion exchange membranes was reported to be preferable. In anion exchange membranes, charge transport is achieved via OH$^-$ ions and the reaction water is generated at the anode; thus, flooding of the cathode and formation of detrimental water pockets inside the fuel cell was prevented [16, 53].

For implantable glucose fuel cells assembled in *depletion design* (cathode mounted in front of the anode; see Section 14.4.2) it is essential that the separator membrane enables the diffusion of glucose and its oxidation products to and from the interior anode. As suitable materials cuprophane, sulfonated (hydrophilic) PTFE, and dialysis and cellulose membranes are reported [54–56]. Also, porous filter membranes made from polyethersulfone [28, 34] or anodized alumina in combination with photoresist spacers [57] as well as a cation-selective Nafion-membrane [58] have successfully been employed. Hydrogels based on poly(vinyl alcohol)–poly(acrylic acid) and glycol methacrylate showed signs of degradation through hydrolysis or the oxidative effect of electrocatalysts after prolonged fuel cell operation [56]. To form ultrathin separator membranes with improved mass-transport characteristics, mesoporous silica has been suggested as an inorganic separator material [47].

14.4
Design Considerations

14.4.1
Site of Implantation

An important boundary condition regarding implantable biofuel cells is the intended site of implantation. An obvious choice is the direct connection of the fuel cell to the blood stream, which promises a high rate of reactant supply. However, blood stream implantation comes along with a number of challenges. In particular, major surgery will be required to connect a flow-through type fuel cell device to a blood vessel. Alternatively, the implementation of the fuel cell similarly as a stent can be envisaged [47]. In any case, the fuel cell parts that are in direct contact with the blood stream have to be designed in a way that the risk of thrombi formation is alleviated. This concerns both a proper fluidic design to omit areas with reduced flow velocity [59] and the choice of materials with good hemocompatibility, in particular with respect to coagulation [60].

A different possibility is the implantation of the fuel cell in a body tissue environment, for example, the subcutaneous tissue, into which pacemakers are also implanted. Prime advantage of this approach is that the fuel cell can be integrated together with the medical implant. This way no additional surgical procedure or potentially fault-prone intrabody wire connections are required. However, tissue implantation comes at the price of reduced (solely diffusive) reactant supply to and from the fuel cell. This puts an upper limit on the fuel cell's maximum current density and power. Reactant supply is further limited by the formation of tissue capsules resulting from foreign body reaction. In a theoretical and experimental study, it was shown that under conditions of the subcutaneous tissue and considering a typical tissue capsule as diffusion resistance, the availability of dissolved oxygen in the tissue actually limits the maximum sustainable current density to values in the range of $16\,\mu A\,cm^{-2}$ [28].

To increase oxygen transfer, researchers considered the design of an air-breathing fuel cell design in which only the anode of the fuel cell is in contact with the tissue, whereas the cathode is exposed to air through a percutaneous opening [61]. Obviously, the safe implementation and successful operation of such a fuel cell design will be challenging under *in vivo* conditions. Related experimental studies are not available.

Recently, Rapoport *et al.* [58] envisioned the application of glucose fuel cells in the brain cavity to power implantable brain–machine interfaces. They concluded that in theory the reactant supply around the human brain would permit operation of a glucose fuel cell with an electrical output power in the range of 1 mW. However, so far no experimental data concerning the actual operation of a glucose fuel cell in real or realistically composed artificial cerebrospinal fluid (CSF) is available.

The choice of implantation site also affects the available concentrations of reactants (glucose and oxygen) as well as the presence of potential endogenous

Table 14.2 Typical composition of physiological fluids: blood (plasma or serum), tissue (interstitial fluid), and cerebrospinal fluid (here, the maximum mean concentration among the given references is listed).

Component	Plasma/serum (μM)	Adipose/muscle tissue (μM)	Cerebrospinal fluid (μM)
Glucose	~5000 [63]	~3000–4000 [63]	~3400 [64]
Alanine	244 [63]	333$^{(a)}$ [63]	57 [65]
Arginine	131 [66]	—	53 [65]
Asparagine	45 [63]	56$^{(m)}$ [63]	35 [65]
Aspartic acid	9 [63]	62$^{(a)}$ [63]	6 [65]
Citrulline	28 [63]	51$^{(a)}$ [63]	3 [67]
Cystine	49 [63]	34$^{(a)}$ [63]	3 [68]
Glutamine acid	42 [63]	66$^{(m)}$ [63]	27 [65]
Glutamine	506 [63]	672$^{(m)}$ [63]	714 [65]
Glycine	194 [63]	565$^{(a)}$ [63]	15 [65]
Histidine	112 [66]	—	22 [69]
Isoleucine	53 [63]	70$^{(m)}$ [63]	8 [68]
Leucine	128 [63]	165$^{(m)}$ [63]	15 [67]
Lysine	232 [66]	—	42 [65]
Methionine	20 [63]	18$^{(a)}$ [63]	14 [65]
Ornithine	80 [66]	—	11 [65]
Phenylalanine	50 [63]	72$^{(m)}$ [63]	10 [68]
Proline	185 [67]	—	4 [68]
Serine	98 [63]	195$^{(a)}$ [63]	41 [65]
Taurine	29 [63]	165$^{(m)}$ [63]	14 [65]
Threonine	124 [63]	185$^{(m)}$ [63]	39 [65]
Tryptophane	65 [66]	—	2 [68]
Tyrosine	53 [63]	62$^{(m)}$ [63]	14 [68]
Valine	216 [63]	257$^{(m)}$ [63]	21 [67]

a, Adipose tissue; m, muscle tissue.

interferents (such as amino acids, which tend to deactivate platinum catalysts; see Section 14.5.2). In Table 14.2, the typical concentrations of glucose, and the physiological amino acids present in blood (plasma/serum), interstitial fluid, and CSF are compared. While the concentration of glucose is comparable in all three physiological environments, the concentration of amino acids differs drastically. For instance, the concentration of aspartic acid in adipose tissue is more than six times higher than in serum or CSF. Similarly, glycine, serine, and taurine show significantly higher concentrations in the tissue than in serum. On the other hand, the amino acid concentration in CSF is generally lower than in tissue or serum. Most distinctively, the levels of citrulline, histidine, cystine, glycine, leucine, proline, tryptophane, and valine are lower in CSF. Since in particular histidine leads to pronounced poisoning of platinum electrodes [62] operation of an implantable glucose fuel cell in the brain cavity may thus be a promising approach to lower the poisoning of the platinum electrodes (see Section 14.5.2).

14.4.2
Strategies to Cope with the Presence of Mixed Reactants

A particular challenge in the realization of abiotic implantable fuel cells is the simultaneous presence of glucose and oxygen in physiological environments. In enzymatic fuel cells, the high reactant specificity of the biocatalysts prevents cross-sensitivity between the electrodes (e.g., an enzymatic glucose fuel cell can be easily constructed from two electrodes, coated with different enzymes and inserted into the same electrolyte solution [70, 71]). In contrast, abiotic catalysts such as platinum or other noble metals are nonspecific, and exhibit good catalytic activity toward glucose oxidation and oxygen reduction. In the presence of both reactants, glucose oxidation and oxygen reduction occur simultaneously at the same electrode. The resulting intraelectrode electron flow leads to polarization and the formation of a mixed electrode potential, its value being between the individual redox potentials of the anode and the cathode (see for instance Chapter 8.2 in [30]). This can significantly reduce the cell voltage and overall electrical output of the fuel cell. To minimize mixed potential formation in abiotic glucose fuel cells, three different reactant separation concepts have been developed, as outlined in the following.

In the early design introduced by Drake, a *stacked assembly fuel cell with hydrophobic barrier* in front of the noble metal cathode prevents glucose access [72] (Figure 14.4a). At the anode, the outer layer of the porous electrode serves as a sacrificial layer where oxygen is consumed by direct reaction with glucose. Given its lower concentration compared to that of glucose, oxygen is quickly consumed and depleted from the reactant mixture. This results in a predominantly anoxic interior of the anode where glucose can be efficiently electro-oxidized. The advantage of this concept is that because of the hydrophobic protection of the cathode highly active, but at the same time glucose-sensitive, catalysts such as platinum can be used. However, this fuel cell construction mandates reactant access to the fuel cell from two sides, which means that it can only be realized as a standalone device. In the two following construction concepts reactant access to the fuel cell from only one side is sufficient, which enables the advantageous integration of these fuel cells directly on the impermeable surface of the implant casing.

The *depletion design fuel cell* originates from the early 1970s and features a glucose-insensitive but permeable cathode (e.g., made from hydrogel-dispersed activated carbon or silver [56]) that is mounted in front of the anode, as shown in Figure 14.4b. At the exterior cathode, oxygen is consumed by the electrochemical reaction and removed from the reactant mixture. This leads to predominantly anoxic conditions at the interior anode where glucose is oxidized [37, 73].

The more recent concept of the *single-layer fuel cell* (Figure 14.4c) exploits the fact that the sensitivity of a platinum electrode toward oxygen or glucose can be controlled by the porosity and specific surface area of the electrode, as shown schematically in Figure 14.5 [74, 75]. In the case of a relatively thick porous platinum electrode with a large interior surface area, oxygen is fully consumed

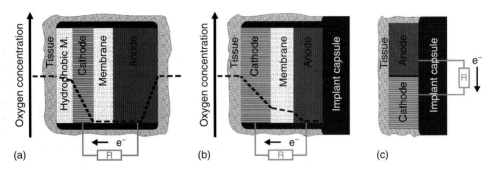

Figure 14.4 Three different designs to enable reactant separation in abiotically catalyzed glucose fuel cells. The dashed line illustrates the oxygen concentration inside the fuel cell layers. (a) *Stacked assembly with a hydrophobic cathode membrane* to prevent glucose interference [72]. (b) *Depletion design* with an oxygen-consuming cathode mounted in front of the oxygen-sensitive anode [37, 73]. (c) *Single-layer design* with an oxygen-tolerant anode placed next to the cathode [74]. (Reproduced from [74] with permission.)

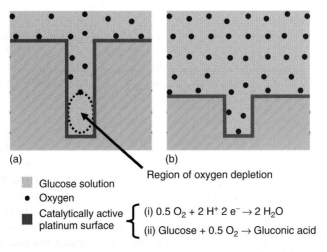

- Glucose solution
- Oxygen
- Catalytically active platinum surface
  - (i) $0.5\ O_2 + 2\ H^+ + 2\ e^- \rightarrow 2\ H_2O$
  - (ii) $Glucose + 0.5\ O_2 \rightarrow Gluconic\ acid$

Figure 14.5 Schematic illustration of oxygen availability in porous platinum cathodes of different thickness. (a) High porosity: oxygen is fully consumed in the outer region of the pore either by (i) electroreduction to water, or by (ii) direct chemical reaction with glucose. In the oxygen-depleted interior of the electrode oxygen influence is minimized and the potential is dominated by the redox potential of glucose oxidation. (b) Low porosity: within the complete pore, oxygen is available and the electrode potential is dominated by the redox potential of oxygen reduction. (Reprinted from [75] with permission from Elsevier.)

in the outer region of the pores by either (i) electroreduction to water or (ii) direct chemical reaction with glucose on the catalytically active surface. This leads to oxygen depletion in the interior of the electrode, where the local potential is dominated by the redox potential of glucose oxidation. In contrast, a comparably thin porous electrode allows for sufficient oxygen to diffuse into the pore and thus prevent the formation of anoxic regions. Consequently, the overall electrode potential is dominated by the redox potential of oxygen reduction. Using platinum electrodes with different porosity and specific surface area a fully platinum-based fuel cell with anode and cathode placed next to each other can be realized [74]. This design obviates the need for an elaborate stacking of the individual electrodes and offers a high degree of freedom in terms of integrating the fuel cell together with the implantable device. The drawback is that the placement of anode and cathode next to each other comes with an increased geometric footprint of the fuel cell, which reduces the overall power density. Since the size of anode and cathode can be freely adjusted to counterbalance electrode polarization, the geometric footprint of the device can be optimized. The power density of their optimized single-layer fuel cell is thus only 34% lower compared to a fuel cell built from the same electrodes but in stacked assembly [74].

14.5
State-of-the-Art and Practical Examples

In the following section, the current state-of-the-art in implantable abiotic glucose fuel cells is summarized. Special importance is given to experimental results obtained under physiological *in vitro* and *in vivo* conditions, as compared in Table 14.3. Investigations performed under nonphysiological conditions are not considered.

14.5.1
Comparison of Fuel Cell Designs and Their Power Densities

Among the early fuel cells investigated under near-physiological *in vitro* condition (modified Ringer-solution containing 5×10^{-3} mol l^{-1} glucose) is the blood-stream implantable fuel cell of Malachesky *et al.* presented in 1972. Their device was constructed from a platinum cathode covered with a hydrophobic layer to effect reactant separation, and a platinum black anode. Operated in flow-through mode mimicking blood stream implantation, it delivered considerable power densities of up to $50\,\mu\text{W cm}^{-2}$ [16]. A similarly constructed device but operated under conditions resembling tissue implantation (diffusive reactant supply) delivered almost an order of magnitude lower average power density of $4.4\,\mu\text{W cm}^{-2}$ over a period of up to 428 h [72]. Comparable power densities in the range up to $5\,\mu\text{W cm}^{-2}$ (based on cathode area) were achieved with fuel cells constructed in the advantageous *depletion design* [28, 34]. Here, an oxygen-selective cathode is mounted in front of the anode to effect reactant separation, which enables the implementation of the fuel cell on an impermeable surface such as the implant casing (see Section 14.4.2).

A drawback of early fuel cells in depletion design was the polymer binder required to fabricate glucose permeable and oxygen-selective cathode from activated carbon particles. This hydrogel binder was prone to oxidative and hydrolytic attack [37]. To alleviate this problem Kerzenmacher *et al.* developed a polymer-less glucose fuel cell design featuring micro-machined Raney-platinum film electrodes. Here, a thin-film Raney-platinum cathode was deposited on a silicon substrate with slits [28] that enabled reactant access to the anode (Figure 14.7). Compared to previous works [18], a Pt-Raney film anode with zinc as a non-noble metal partner in alloy formation was employed instead of the potentially problematic nickel. For performance characterization in phosphate-buffered saline containing glucose and oxygen at concentrations resembling body tissue, these fuel cells were equipped with a diffusion barrier to mimic the additional diffusion resistance expected from tissue capsule formation. Despite the limited reactant supply, the fuel cell exhibited maximum power densities of $4.4\,\mu W\,cm^{-2}$ at steady state.

Using the same electrode technology, Kloke *et al.* [74] presented a novel *single-layer fuel cell* design in which the oxygen-tolerant Raney-Pt anode is placed next to the cathode. As outlined in Section 14.4.2, this approach has the advantage that no elaborate stacking of the individual fuel cell components is required. This significantly simplifies the envisioned integration of the fuel cell as exterior coating of the implant capsule. Naturally, the power density of the single layer fuel cell is lower because when compared to a stacked assembly a larger base area is required for the electrode. However, the geometric area of the anode and the cathode can be freely adjusted toward an optimized power density, which is reduced compared to stacked assembly by only 34%.

While the *single-layer fuel cell* of Kloke *et al.* was assembled from two separately fabricated electrodes, Oncescu and Erickson [76] fabricated their single-layer fuel cell from micromachined Raney-Pt thin films deposited directly on a single chip (Figure 14.6) made of fused silica. To increase oxygen depletion over the anode and thus decrease mixed potential formation, they also implemented an interdigitated interface between the anode and the cathode. In terms of geometric power density the interdigitated device yields $2\,\mu W\,cm^{-2}$, compared to only $1.3\,\mu W\,cm^{-2}$ without interdigitation. An integrated fuel cell unit for high volumetric power density was assembled as a stack of 12 non-interdigitated fuel cells, yielding $16\,\mu W\,cm^{-3}$. However, no diffusion barrier to mimic tissue capsule formation was included in these experiments.

A particularly facile way to fabricate porous platinum electrodes by electrodeposition was introduced by Kloke *et al.* [34, 77, 78]. Their process roots on the alternating deposition of a Pt–Cu alloy and the subsequent selective dissolution of Cu during cyclic voltammetry scans. This leaves behind a porous platinum structure, its specific surface (or roughness factor, which describes the ratio of interior electrode surface to the geometric footprint) adjustable by choosing the appropriate number of deposition cycles. The process requires only an electrically conductive substrate for deposition and is applicable in the fabrication

14.5 State-of-the-Art and Practical Examples | 301

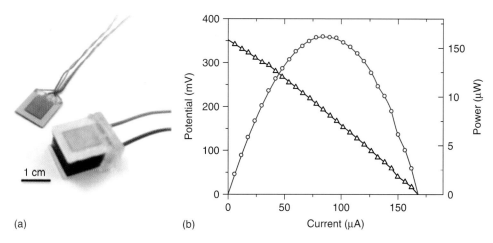

Figure 14.6 (a) Individual single-layer glucose fuel cell along with a stack of 12 cells connected in parallel. (b) Polarization curve of the corresponding stack. The maximum power density of the device amounts to 1.3 µW cm^{-2} (16 µW cm^{-3}). (Reprinted by permission from Macmillan Publishers Ltd: *Scientific Reports* [76], copyright (2013).)

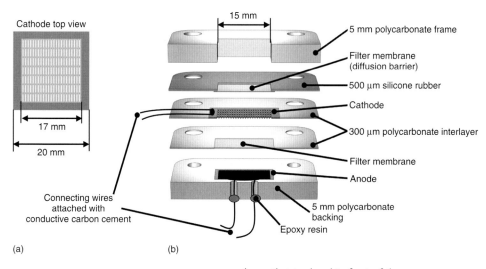

Figure 14.7 Abiotically catalyzed glucose fuel cell presented in [28]. (a) Shows the silicon substrate with 50 µm wide slits onto which a 500 nm thin Raney-platinum film cathode is deposited. (b) To scale assembly drawing of the fuel cell. Note the filter membrane that is placed in front of the cathode to account for the expected tissue capsule formation. A highly porous 30 µm thick Raney-platinum film is used as the anode. (Reprinted from [28] with permission from Elsevier.)

Table 14.3 Characteristics of implantable abiotic glucose fuel cells.

References	Electrodes	Test conditions	Performance ($\mu W\,cm^{-2}$)	Remarks
In vitro				
[72]	Cathode: platinum black laminated on one side of a hydrophobic carbon/PTFE matrix Anode: nonspecified noble metal alloy	Tyrode's solution, 5 mM glucose	(a) 4.4 (b) 11.6	Average performance over (a) 428 h and (b) 167 h
[16]	Cathode: Teflon-bonded Pt black Anode: Pt black	Modified Ringer solution, 5 mM glucose	~ 50	Flow through cell with forced reactant flow; performance monitored over a period of 32 h
[28]	Cathode: Raney-Pt thin film	Phosphate-buffered saline, 3 mM glucose	4.4	Stacked assembly: permeable cathode placed in front of the anode. Diffusion barrier in front of the fuel cell
[74]	Cathode: Raney-Pt thin film Anode: Raney-Pt thick film	Phosphate-buffered saline, 3 mM glucose	2.9	Single-layer assembly, large cathode placed next to smaller anode. Diffusion barrier in front of the fuel cell
[34]	Cathode and anode: Electrodeposited porous Pt (roughness factors: 92 and 3070, respectively)	Phosphate-buffered saline, 3 mM glucose, 7% O_2-sat.	(a) 5.1 (b) 1.5	Stacked assembly: permeable cathode placed in front of the anode. Diffusion barrier in front of the fuel cell. (a) Initial peak performance (b) After 90 days at 10 $\mu A\,cm^{-2}$
[58]	Cathode: carbon nanotube mesh Anode: thin-layer Raney-Pt	Phosphate-buffered saline, 10 mM glucose, 21% O_2-sat.	3.4	Stacked assembly: permeable cathode placed in front of the anode

Ref.	Electrodes	Electrolyte/Environment	Power	Comments
[47]	Cathode: reduced graphene oxide Anode: thin-layer Pt	Phosphate-buffered saline, 5 mM glucose, 21% O_2-sat.	5.3	Stacked assembly: permeable cathode placed in front of the anode
[57, 76]	Cathode: Pt sputtered on carbon paper Anode: thin-layer Raney-Pt	Phosphate-buffered saline, 5 mM glucose, 7% O_2-sat.	1.5	Stacked assembly: permeable cathode placed in front of the anode
[57, 76]	Cathode and anode: Raney-Pt thin film	Phosphate-buffered saline, 5 mM glucose, 7% O_2-sat.	2	Single-layer coating on a single chip, large cathode surrounding a smaller anode, interdigital interface
[37]	Cathode: activated carbon Anode: platinized carbon Both electrode catalysts embedded in PVA–PAA hydrogel	Tyrode's solution containing 5 mM glucose	4 (per cathode area)	Fuel cell with central anode, sandwiched between two cathodes
In vivo				
[72]	Cathode: platinum black laminated on one side of a hydrophobic carbon/PTFE matrix Anode: nonspecified noble metal alloy	Subcutaneously in dog	2.2	Average performance over a period of 30 days; no evidence of necrosis, hemorrhage, abscess formation, or overt degenerative change upon explanation after 78 days
[9]	Cathode: activated carbon on metal screen; Anode: Pt–Ni Raney catalyst	Subcutaneously in dog	1.6 (per cathode area)	Fuel cell with central anode, sandwiched between two cathodes; stable performance after 150 days of operation
[47]	Cathode: reduced graphene oxide Anode: thin-layer Pt	Right ventricle of a pig	2.1	Constant performance over a period of 50 min

See text for explanation.

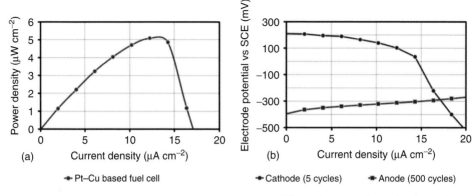

Figure 14.8 (a) Power density of a glucose fuel cell with electrodes cyclic electrodeposition of Pt–Cu alloy, recorded at 7% oxygen saturation and 3×10^{-3} mol l^{-1} glucose in phosphate-buffered saline and with a diffusion barrier in front. (b) Shows the corresponding electrode potentials of the cathode fabricated with 5 deposition cycles, and anode fabricated with 500 deposition cycles. (Adapted from [77] with permission.)

of both anode and cathode. Corresponding glucose fuel cells (see Figure 14.8) yielded maximum power densities of 5.1 µW cm^{-2} (operated in phosphate-buffered saline containing glucose and oxygen in concentrations resembling body tissue, including diffusion barrier to mimic tissue capsule formation in front of the device) [34].

Regarding the *in vivo* performance of abiotic glucose fuel cells over longer periods of time, only two studies from the early 1970s are presently available. Both Weidlich *et al.* and Drake *et al.* implanted their fuel cell devices subcutaneously in dogs. Although they used different reactant separation concepts (*hydrophobic barrier* vs *depletion design*; see Section 14.4.2), both studies yielded similar power densities in the range of 2 µW cm^{-2} [9, 72]. These fuel cells were successfully operated for periods of 30 days and more than 5 months, respectively. During these experiments, no negative tissue response caused by the implanted fuel cell was observed. In a more recent *in vivo* study in the blood-stream of a pig, a power density of 2.1 µW cm^{-2} was demonstrated over a period of 50 min. Despite this short operational period, the formation of blood clots and fibrinogenic coating occurred [47], presumably because of the lack of a hemocompatible coating (Table 14.3).

14.5.2
Factors Affecting Long-Term Operation

For practical application as reliable implant power supply, the long-term functionality of implantable glucose fuel cells is of primary importance. A well-known phenomenon is the gradual deactivation (poisoning) of platinum electrodes when operated for anodic glucose oxidation in neutral buffers. Commonly, this observation is attributed to the adsorption of reaction products and chemisorbed intermediates [18, 19]. Generally, there is no agreement in literature regarding the nature of the poisoning species [79]. Early studies indicated the reaction product gluconic acid [80] to be the poisoning species, whereas later works identified linear CO and bridged CO as adsorbates, depending on the pH [81, 82].

Anode poisoning causes an increasing polarization at the anode that leads to a significant performance loss over time already upon operation of the fuel cell in phosphate-buffered saline-containing glucose. As reported in literature, this performance loss can amount to 70% over operation periods ranging from 90 to 234 days [34, 38]. On the other hand, stable performances of approximately $0.3\,\mu\text{W}\,\text{cm}^{-2}$ after 650 days and $55\,\mu\text{W}$ (fuel cell size not specified) after 100 days of operation are reported [56]. In their early studies on Raney-platinum electrodes Gebhardt *et al.* [18] have earlier demonstrated that platinum alloys can exhibit significantly improved catalytic activity and poisoning resistance [18]. Following the search for catalysts for amperometric glucose sensors, a number of highly active and poisoning-resistant platinum alloys and other noble metals have been developed [25, 79, 83]. However, in the recent literature such improved catalysts have not yet been used at the anode of an implantable glucose fuel cell.

Of particular relevance for long-term operation in a physiological environment is the adsorption of endogenous amino acids [84], which are known to quickly and drastically decrease the glucose oxidation activity of platinum-based catalysts. In contrast to larger molecules such as proteins, the amino acids are of similar size as glucose and can thus not be effectively shielded from the electrode by the use of, for example, dialysis membranes. The effect of amino acids in an artificial tissue fluid on the performance of operational Raney-platinum film electrodes is shown in Figure 14.9 [75, 85]. In this example, degradation due to amino acid adsorption is less pronounced at the Raney-platinum cathode. Compared

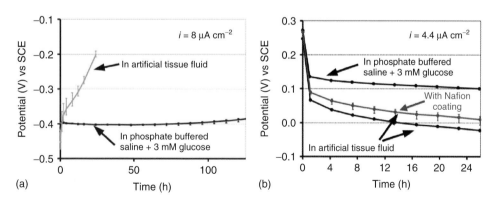

Figure 14.9 Comparison of electrode behavior in phosphate-buffered saline containing $3 \times 10^{-3}\,\text{mol}\,\text{l}^{-1}$ glucose, and artificial tissue fluid containing additional amino acids at physiological concentration. Shown is the evolution of electrode potential (vs the saturated calomel electrode, SCE) at a constant current density over time. (a) Raney-platinum film anode. (b) Raney-platinum thin film cathode with and without Nafion coating. (Data taken from [75, 85]. Reproduced with permission from www.woodheadpublishing.com [10].)

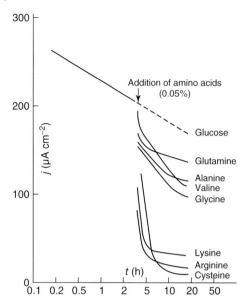

Figure 14.10 Effect of individual amino acids upon the current density versus time curve of glucose oxidation on a Raney-type Pt–Ni foil electrode polarized at a constant oxidation potential of 0.4 V versus RHE. Recorded in phosphate buffer solution at pH 7, containing 0.1% glucose and 0.05% amino acids. (Reproduced from [26] (Figure 22, page 314), with kind permission from Springer Science + Business Media B.V.)

to operation in neutral buffer with glucose and oxygen, the cathodes exhibit a potential loss of approximately 120 mV. The potential loss can be further reduced by coating the electrode with a protective layer of the cation exchange polymer Nafion. In contrast, the anodes exhibit a drastic and progressing poisoning effect in the presence of amino acids, which underlines the need for further research and optimization.

The poisoning effect of amino acids on glucose oxidation at platinum electrodes has been subject of several investigations. In Figure 14.10, the decrease of glucose oxidation current upon the addition of some physiological amino acids (at nonphysiological concentration) is shown. Already in buffered solution a degradation of electrode performance can be noticed, attributed to the self-poisoning of the platinum catalyst. Upon addition of amino acids the oxidation current is further reduced, which is attributed to the adsorption of amino acids on the platinum surface and the associated blockage of reaction sites for glucose oxidation [84]. As can be seen, the poisoning strength of the individual amino acids differs significantly. Furthermore, the current values tend to stabilize within 5–15 h after amino acid addition. Despite amino acid adsorption, the blockage of the electrode surface is only partial so that some reaction sites for glucose oxidation remain and a current can be sustained. It is thus expected that using electrodes with sufficiently larger specific surface can sustain current densities of practical use also in the presence of amino acids [84, 85]. A ranking of the amino acids with regard to their poisoning effect on a platinized platinum electrode was assembled by Giner *et al.* [62]. At physiological concentrations histidine, arginine, and glutamine proved to be the most strongly poisoning. A similar study on plain platinum electrodes has been undertaken by Hibbert *et al.* [86]. Despite the number of investigations regarding the poisoning of platinum electrodes by amino acids, a comprehensive study including all physiological amino acids under conditions of applicational relevance (porous electrodes with high specific surface area and potential range of relevance for fuel cell operation) is not available.

14.6
Conclusion and Outlook

In the following section, the state-of-the-art, technological challenges, and future research directions in the field of implantable abiotic glucose cells are summarized and discussed.

14.6.1
State-of the-Art

Over the last 10 years, there has been a noticeable increase in research on implantable glucose fuel cells based on abiotic catalysts, a concept that has gained only little attention since the late 1970s. This can be explained by the prevailing lack of a reliable implant power supply, coupled with an increasing number of medical implants currently being developed.

Under physiological concentrations of glucose and oxygen, implantable glucose fuel cells with abiotic catalyst can exhibit up to $50\,\mu W\,cm^{-2}$ when operated with a forced reactant flow corresponding to the location of blood stream implantation. Nevertheless, nowadays tissue implantation is mainly considered, presumably because of the complications that arise from blood stream implantation (*major surgery, intra-body wire connections*). Under these conditions, purely diffusive reactant supply prevails, leading to significantly lower power densities in the range of up to $5\,\mu W\,cm^{-2}$ *in vitro*. *In vivo* experiments with implantable abiotic glucose fuel cells are only available from the late 1960s and early 1970s. In two independent studies, fuel cells were implanted subcutaneously in the dog and yielded performances in the range of $2\,\mu W\,cm^{-2}$ for periods up to 150 days [9]. In these experiments, no inflammatory tissue responses were observed. Nevertheless, dedicated studies regarding the cytotoxicity and biocompatibility of abiotic glucose fuel cells have not yet been reported.

Regarding their initial power density, abiotic glucose fuel cells clearly fall behind the performance of their enzymatic counterparts. Under physiological *in vitro* conditions ($5 \times 10^{-3}\,mol\,l^{-1}$ glucose, Figure 14.3b in the reference), these yield up to $250\,\mu W\,cm^{-2}$ [70], which is almost two orders of magnitude higher. However, enzyme denaturation leads to a significant performance loss over time. The longest lifetime reported under *in vitro* conditions does not exceed 40 days, the associated power density of $1.2\,\mu W\,cm^{-2}$ being well comparable to abiotic glucose fuel cells [6]. *In vivo*, power densities in the range of $194\,\mu W\,cm^{-2}$ after 6–8 days of operation have been achieved [87], but lead failure did not allow for a clear statement on the lifetime of the device. In light of the presently available data, enzymatic glucose fuel cells can thus be regarded favorable for short-term implantable applications, whereas abiotic fuel cells are more promising to supply implantable devices over extended periods of time.

An interesting aspect is the comparison of glucose fuel cells to other energy-harvesting approaches for implantable devices, such as mechanical and thermoelectric harvesters. The prime advantages of these devices are good biocompatibility due to their hermetically sealed construction. Their power output is typically in the range of implantable fuel cells (see for instance [88] and references therein), but depending on the device and its placement, mechanical generators can extract significantly higher power densities in the range of $1\,mW\,cm^{-3}$ from body motion [89]. In any case, the power output of mechanical and thermoelectric generators is dependent on nonconstant external factors such as body movement or temperature gradient within the body. Depending on the intended application and in particular the type of implant, an implantable glucose fuel cell will be of advantage since it provides a reliable and continuous power supply. Furthermore, implantable glucose fuel cells offer a large degree of freedom in terms of their placement in the body and integration together with the medical implant. It is obvious that mechanical and thermoelectric generators have to be placed at a certain location in the body for optimum performance. In most cases, this location will not be the location of the medical implant, mandating the use of intrabody wire connections.

14.6.2
Applications

An important point is the type of medical implants that may benefit from an implantable abiotic glucose fuel cell as sustainable power supply. A reasonable estimate is that this will be low-power medical implants with a power demand of up to $25\,\mu W$. This calculation is based on a fuel cell power density of $2\,\mu W\,cm^{-2}$, a conversion efficiency of 50% to step up its voltage to the level required by implant electronics [90], and the typical surface area of a pacemaker casing ($25\,cm^2$) onto which the fuel cell may be applied. Today, cardiac pacemakers and recording channels of brain–machine interfaces [91] fall into that category. In addition, wireless sensors systems with power demands in this range are under development [92, 93], and could, for instance, be used to monitor vital parameters such as body temperature, blood pressure, or heart rate. With the current trend toward energy-efficient electronic circuits, further applications, such as the supply of electronic contact lenses [94], are likely to become feasible.

14.6.3
Challenges and Future Trends

The successful application of abiotic glucose fuel cells as autonomous implant power supply clearly depends on their biocompatibility and functionality in a physiological environment. Future studies should thus also consider operational parameters such as physiological concentrations of glucose and oxygen or the formation of tissue capsules that can drastically limit reactant supply and fuel cell performance. Eventually, it will not suffice to demonstrate and improve fuel cell performance in an idealized neutral buffer containing physiological amounts glucose and oxygen. To further advance the technology, experimental characterization of abiotic glucose fuel cells in real or realistically approached body fluids is required. In particular, there is a need for dedicated investigations to clarify the poisoning behavior and mechanism of amino acids and other endogenous substances. This will be essential to develop catalysts with improved resistance against poisoning through endogenous substances and functionality over extended periods of time. In this respect, biomimetic catalysts [95] may open up a possibility to combine the long-term stability of abiotic catalysts with the high reaction rate and specificity of enzymes. Furthermore *in vitro* and *in vivo* studies of biocompatibility will be gaining importance to ensure the nontoxicity and safety of abiotic glucose fuel cells.

At present, interfacing the fuel cell to implant electronics is associated with drastic performance losses of up to 60% [90]. DC–DC converters are required, since the body fluid as a common electrolyte prohibits the serial connection of individual cells to step up their output voltage. In the future, the advancement of low-voltage electronic circuits with operating voltages below 200 mV [96, 97] may completely circumvent the need for a boost converter and enable directly using the output voltage of the fuel cell as power supply.

References

1. Hayes, D.L. and Furman, S. (2004) Cardiac pacing: how it started, where we are, where we are going. *Pace-Pacing Clin. Electrophysiol.*, **27**, 693–704.
2. Senning, A. (1983) Cardiac pacing in retrospect. *Am. J. Surg.*, **145**, 733–739.
3. Gold, R.D. (1984) Cardiac pacing – from then to now. *Med. Instrum.*, **18**, 15–21.
4. Parsonnet, V. (1972) Power sources for implantable cardiac pacemakers. *Chest*, **61**, 165–173.
5. Rubenwolf, S., Kerzenmacher, S., Zengerle, R., and von Stetten, F. (2011) Strategies to extend the lifetime of bioelectrochemical enzyme electrodes for biosensing and biofuel cell applications. *Appl. Microbiol. Biotechnol.*, **89**, 1315–1322.
6. Cinquin, P., Gondran, C., Giroud, F., Mazabrard, S., Pellissier, A., Boucher, F., Alcaraz, J.P., Gorgy, K., Lenouvel, F., Mathé, S., Porcu, P., and Cosnier, S. (2010) A glucose BioFuel cell implanted in rats. *PLoS ONE*, **5**, e10476.
7. Thauer, R.K., Jungermann, K., and Decker, K. (1977) Energy-conservation in chemotropic anaerobic bacteria. *Bacteriol. Rev.*, **41**, 100–180.
8. Kerzenmacher, S., Ducrée, J., Zengerle, R., and von Stetten, F. (2008) Energy harvesting by implantable abiotically

catalyzed glucose fuel cells. *J. Power Sources*, **182**, 1–17.
9. Weidlich, E., Richter, G., von Sturm, F., and Rao, J.R. (1976) Animal experiments with biogalvanic and biofuel cells. *Biomater. Med. Dev. Artif. Org.*, **4**, 277–306.
10. Kerzenmacher, S. (2013) in *Implantable Sensor Systems for Medical Applications* (eds A. Inmann and D. Hodgins), Woodhead Publishing, Oxford, pp. 183–212.
11. Hamann, C.H., Hamnett, A., and Vielstich, W. (2007) *Electrochemistry*, Wiley-VCH Verlag GmbH, Weinheim.
12. Bard, A.J. and Faulkner, L.R. (2001) *Electrochemical Methods*, John Wiley & Sons, Inc., New York.
13. Larminie, J. and Dicks, A. (2000) *Fuel Cell Systems Explained*, John Wiley & Sons, Ltd, Chichester.
14. Ernst, S., Heitbaum, J., and Hamann, C.H. (1980) Electrooxidation of glucose in phosphate buffer solutions - kinetics and reaction-mechanism. *Ber. Bunsenge. Phys. Chem. Chem. Phys.*, **84**, 50–55.
15. Yao, S.J., Appleby, A.J., Geisel, A., Cash, H.R., and Wolfson, S.K. Jr., (1969) Anodic oxidation of carbohydrates and their derivatives in neutral saline solution. *Nature*, **224**, 921–922.
16. Malachesky, P., Holleck, G., McGovern, F., and Devarakonda, R. (1972) Parametric studies of the implantable fuel cell. Proceedings of the 7th Intersociety Energy Conversion Engineering Conference, pp. 727–732.
17. Kokoh, K.B., Leger, J.M., Beden, B., and Lamy, C. (1992) On line chromatographic analysis of the products resulting from the electrocatalytic oxidation of D-glucose on Pt, Au and adatoms modified Pt electrodes.1. Acid and neutral media. *Electrochim. Acta*, **37**, 1333–1342.
18. Gebhardt, U., Rao, J.R., and Richter, G.J. (1976) Special type of raney-alloy catalyst used in compact biofuel cells. *J. Appl. Electrochem.*, **6**, 127–134.
19. Lerner, H., Giner, J., Soeldner, J.S., and Colton, C.K. (1979) Electrochemical glucose oxidation on a platinized platinum-electrode in Krebs-Ringer solution. 2. Potentiostatic studies. *J. Electrochem. Soc.*, **126**, 237–242.
20. Demele, M.F.L., Videla, H.A., and Arvia, A.J. (1982) Comparative-study of the electrochemical-behavior of glucose and other compounds of biological interest. *Bioelectrochem. Bioenerg.*, **9**, 469–487.
21. Demele, M.F.L., Videla, H.A., and Arvia, A.J. (1982) Potentiodynamic study of glucose electrooxidation at bright platinum-electrodes. *J. Electrochem. Soc.*, **129**, 2207–2213.
22. Demele, M.F.L., Videla, H.A., and Arvia, A.J. (1983) The electrooxidation of glucose on platinum-electrodes in buffered media. *Bioelectrochem. Bioenerg.*, **10**, 239–249.
23. Luna, A.M.C., Bolzan, A.E., de Mele, M.F., and Arvia, A.J. (1991) The voltammetric electrooxidation of glucose and glucose residues formed on electrodispersed platinum-electrodes in acid electrolytes. *Pure Appl. Chem.*, **63**, 1599–1608.
24. Vassilyev, Y.B., Khazova, O.A., and Nikolaeva, N.N. (1985) Kinetics and mechanism of glucose electrooxidation on different electrode-catalysts. 1. Adsorption and oxidation on platinum. *J. Electroanal. Chem.*, **196**, 105–125.
25. Vassilyev, Y.B., Khazova, O.A., and Nikolaeva, N.N. (1985) Kinetics and mechanism of glucose electrooxidation on different electrode-catalysts. 2. Effect of the nature of the electrode and the electrooxidation mechanism. *J. Electroanal. Chem.*, **196**, 127–144.
26. Rao, J.R. (1983) in *Bioelectrochemistry I: Biological Redox Reactions* (eds G. Milazzo and M. Blank), Plenum Press, New York, pp. 283–335.
27. Logan, B.E. (2008) *Microbial Fuel Cells*, John Wiley & Sons, Inc., Hoboken, NJ.
28. Kerzenmacher, S., Kräling, U., Metz, T., Ducrée, J., Zengerle, R., and von Stetten, F. (2011) A potentially implantable glucose fuel cell with raney-platinum film electrodes for improved hydrolytic and oxidative stability. *J. Power Sources*, **196**, 1264–1272.
29. van Heyningen, R. (1964) Gluconic acid in the human lense. *Biochem. J.*, **92**, 14C–16C.

30. Paunovic, M. and Schlesinger, M. (2006) *Fundamentals of Electrochemical Deposition*, John Wiley & Sons, Inc., New York, Weinheim.
31. Ponomarev, V.V. and Migarskaya, L.B. (1960) The heats of combustion of some amino acids. *Russ. J. Phys. Chem. (Engl. Transl.)*, **34**, 1182–1183.
32. Call, D.F. and Logan, B.E. (2011) A method for high throughput bioelectrochemical research based on small scale microbial electrolysis cells. *Biosens. Bioelectron.*, **26**, 4526–4531.
33. Kerzenmacher, S., Mutschler, K., Kräling, U., Baumer, H., Ducrée, J., Zengerle, R., and von Stetten, F. (2009) A complete testing environment for the automated parallel performance characterization of biofuel cells: design, validation, and application. *J. Appl. Electrochem.*, **39**, 1477–1485.
34. Kloke, A., Köhler, C., Zengerle, R., and Kerzenmacher, S. (2012) Porous platinum electrodes fabricated by cyclic electrodeposition of PtCu alloy: application to implantable glucose fuel cells. *J. Phys. Chem. C*, **116**, 19689–19698.
35. Wolfson, S.K., Gofberg, S.L., Prusiner, P., and Nanis, L. (1968) Bioautofuel cell – a device for pacemaker power from direct energy conversion consuming autogenous fuel. *Trans. Am. Soc. Artif. Int. Org.*, **14**, 198–203.
36. Appleby, A.J. and Van Drunen, C. (1971) Anodic oxidation of carbohydrates and related compounds in neutral saline solution. *J. Electrochem. Soc.*, **118**, 95–97.
37. Rao, J.R., Richter, G., von Sturm, F., and Weidlich, E. (1973) Biological fuel-cells as power source for implantable electronic devices. *Ber. Bunsenges. Phys. Chem. Chem. Phys.*, **77**, 787–790.
38. Kerzenmacher, S., Ducrée, J., Zengerle, R., and von Stetten, F. (2008) An abiotically catalyzed glucose fuel cell for powering medical implants: reconstructed manufacturing protocol and analysis of performance. *J. Power Sources*, **182**, 66–75.
39. Fishman, J.H. and Henry, J.F. (1973) Electrodeposited selective catalysts for implantable biological fuel-cells. *J. Electrochem. Soc.*, **120**, C115.
40. Habrioux, A., Sibert, E., Servat, K., Vogel, W., Kokoh, K.B., and Alonso-Vante, N. (2007) Activity of platinum-gold alloys for glucose electrooxidation in biofuel cells. *J. Phys. Chem. B*, **111**, 10329–10333.
41. Hussein, L., Feng, Y.J., Alonso-Vante, N., Urban, G., and Krüger, M. (2011) Functionalized-carbon nanotube supported electrocatalysts and buckypaper-based biocathodes for glucose fuel cell applications. *Electrochim. Acta*, **56**, 7659–7665.
42. Arzoumanidis, G.G. and O'Connell, J.J. (1969) Electrocatalytic oxidation of D-glucose in neutral media with electrodes catalyzed by 4,4′,4,″4‴-tetrasulfophthalocyanine-molybdenum dioxide salts. *J. Phys. Chem.*, **73**, 3508–3510.
43. Choi, Y.K., Wang, G., Nayfeh, M.H., and Yau, S.T. (2009) A hybrid biofuel cell based on electrooxidation of glucose using ultra-small silicon nanoparticles. *Biosens. Bioelectron.*, **24**, 3103–3107.
44. Kozawa, A., Zilionis, V.E., and Brodd, R.J. (1970) Electrode materials and catalysts for oxygen reduction in isotonic saline solution. *J. Electrochem. Soc.*, **117**, 1474–1478.
45. Kozawa, A., Zilionis, V.E., and Brodd, R.J. (1970) Oxygen and hydrogen peroxide reduction at A ferric phthalocyanine-catalyzed graphite electrode. *J. Electrochem. Soc.*, **117**, 1470–1474.
46. Schaldach, M. and Kirsch, U. (1970) In-vivo electrochemical power generation. *Trans. Am. Soc. Artif. Int. Org.*, **16**, 184–192.
47. Sharma, T., Hu, Y., Stoller, M., Feldman, M., Ruoff, R.S., Ferrari, M., and Zhang, X. (2011) Mesoporous silica as a membrane for ultra-thin implantable direct glucose fuel cells. *Lab Chip*, **11**, 2460–2465.
48. Harnisch, F. and Schröder, U. (2010) From MFC to MXC: chemical and biological cathodes and their potential for microbial bioelectrochemical systems. *Chem. Soc. Rev.*, **39**, 4433–4448.
49. Roche, I. and Scott, K. (2009) Carbon-supported manganese oxide nanoparticles as electrocatalysts for Oxygen

Reduction Reaction (Orr) in neutral solution. *J. Appl. Electrochem.*, **39**, 197–204.

50. Yu, E.H., Cheng, S., Scott, K., and Logan, B. (2007) Microbial fuel cell performance with non-Pt cathode catalysts. *J. Power Sources*, **171**, 275–281.

51. Bezerra, C.W.B., Zhang, L., Lee, K.C., Liu, H.S., Marques, A.L.B., Marques, E.P., Wang, H.J., and Zhang, J.J. (2008) A review of Fe-N/C and Co-N/C catalysts for the oxygen reduction reaction. *Electrochim. Acta*, **53**, 4937–4951.

52. Kundu, S., Nagaiah, T.C., Xia, W., Wang, Y.M., Van Dommele, S., Bitter, J.H., Santa, M., Grundmeier, G., Bron, M., Schuhmann, W., and Muhler, M. (2009) Electrocatalytic activity and stability of nitrogen-containing carbon nanotubes in the oxygen reduction reaction. *J. Phys. Chem. C*, **113**, 14302–14310.

53. Wolfson, S.K., Yao, S.J., Geisel, A., and Cash, H.R. (1970) A single electrolyte fuel cell utilizing permselective membranes. *Trans. Am. Soc. Artif. Int. Org.*, **16**, 193–198.

54. Rao, J.R., Richter, G., von Sturm, F., Weidlich, E., and Wenzel, M. (1974) Metal-oxygen and glucose-oxygen cells for implantable devices. *Biomed. Eng.*, **9**, 98–103.

55. Rao, J.R. and Richter, G. (1974) Implantable bioelectrochemical power sources. *Naturwissenschaften*, **61**, 200–206.

56. Rao, J.R., Richter, G.J., von Sturm, F., and Weidlich, E. (1976) Performance of glucose electrodes and characteristics of different biofuel cell constructions. *Bioelectrochem. Bioenerg.*, **3**, 139–150.

57. Oncescu, V. and Erickson, D. (2011) A microfabricated low cost enzyme-free glucose fuel cell for powering low-power implantable devices. *J. Power Sources*, **196**, 9169–9175.

58. Rapoport, B.I., Kedzierski, J.T., and Sarpeshkar, R. (2012) A glucose fuel cell for implantable brain-machine interfaces. *PLoS ONE*, **7**, e38436.

59. von Sturm, F. and Richter, G. (1976) Pacemaker with biofuel cell. US Patent 3941135, Date issued Mar. 2, 1976.

60. Preidel, W., Saeger, S., von Lucadou, I., and Lager, W. (1991) An electrocatalytic glucose sensor for in-vivo application. *Biomed. Instrum. Technol.*, **25**, 215–219.

61. Ng, D. and Wolfson, S. (1974) A apple by: implantable fuel cell. US Patent 3837922A, Date issued: Sept. 24, 1974.

62. Giner, J., Marincic, L., Soeldner, J.S., and Colton, C.K. (1981) Electrochemical glucose-oxidation on a platinized platinum-electrode in krebs-ringer solution. 4. Effect of amino-acids. *J. Electrochem. Soc.*, **128**, 2106–2114.

63. Maggs, D.G., Jacob, R., Rife, F., Lange, R., Leone, P., During, M.J., Tamborlane, W.V., and Sherwin, R.S. (1995) Interstitial fluid concentrations of glycerol, glucose, and amino-acids in human quadricep muscle and adipose-tissue – evidence for significant lipolysis in skeletal-muscle. *J. Clin. Invest.*, **96**, 370–377.

64. Tato, R.E., Frank, A., and Hernanz, A. (1995) Tau protein concentrations in cerebrospinal fluid of patients with dementia of the alzheimer type. *J. Neurol Neurosurg. Psychiatry*, **59**, 280–283.

65. Meier, D.H. and Schott, K.J. (1988) Free amino acid pattern of cerebrospinal fluid in amyotrophic lateral sclerosis. *Acta Neurol. Scand.*, **77**, 50–53.

66. Pitkänen, H.T., Oja, S.S., Kemppainen, K., Seppa, J.M., and Mero, A.A. (2003) Serum amino acid concentrations in aging men and women. *Amino Acids*, **24**, 413–421.

67. Perry, T.L. and Hansen, S. (1969) Technical pitfalls leading to errors in quantitation of plasma amino acids. *Clin. Chim. Acta*, **25**, 53–58.

68. Kruse, T., Reiber, H., and Neuhoff, V. (1985) Amino acid transport across the human blood-CSF barrier: an evaluation graph for amino acid concentrations in cerebrospinal fluid. *J. Neurol. Sci.*, **70**, 129–138.

69. Humoller, F.L., Mahler, D.J., and Parker, M.M. (1966) Distribution of amino acids between plasma and spinal fluid. *Int. J. Neuropsychiatry*, **2**, 293–297.

70. Mano, N., Mao, F., and Heller, A. (2003) Characteristics of a miniature compartment-less glucose-O-2 biofuel

cell and its operation in a living plant. *J. Am. Chem. Soc.*, **125**, 6588–6594.
71. Mano, N., Mao, F., and Heller, A. (2002) A miniature biofuel cell operating in a physiological buffer. *J. Am. Chem. Soc.*, **124**, 12962–12963.
72. Drake, R.F., Kusserow, B.K., Messinger, S., and Matsuda, S. (1970) A tissue implantable fuel cell power supply. *Trans. Am. Soc. Artif. Int. Org.*, **16**, 199–205.
73. Rao, J.R., Richter, G., Weidlich, E., and von Sturm, F. (1972) Metal-oxygen and glucose-oxygen cells as power sources for implantable devices. *Phys. Med. Biol.*, **17**, 738.
74. Kloke, A., Biller, B., Kräling, U., Kerzenmacher, S., Zengerle, R., and von Stetten, F. (2011) A single layer glucose fuel cell intended as power supply coating for medical implants. *Fuel Cells*, **11**, 316–326.
75. Kerzenmacher, S., Kräling, U., Schroeder, M., Brämer, R., Zengerle, R., and von Stetten, F. (2010) Raney-platinum film electrodes for potentially implantable glucose fuel cells. Part 2: glucose-tolerant oxygen reduction cathodes. *J. Power Sources*, **195**, 6524–6531.
76. Oncescu, V. and Erickson, D. (2013) High volumetric power density, non-enzymatic, glucose fuel cells. *Sci. Rep.*, **3**, 1226.
77. Kloke, A., Köhler, C., Zengerle, R., and Kerzenmacher, S. (2012) Cyclic electrodeposition of PtCu alloy: facile fabrication of highly porous platinum electrodes. *Adv. Mater.*, **24**, 2916–2921.
78. Kloke, A., Köhler, C., Drzyzga, A., Gerwig, R., Schumann, K., Ade, M., Zengerle, R., and Kerzenmacher, S. (2013) Fabrication of highly porous platinum by cyclic electrodeposition of PtCu alloys: how do process parameters affect morphology? *J. Electrochem. Soc.*, **160**, D111–D118.
79. Kokkindis, G., Leger, J.M., and Lamy, C. (1988) Structural effects in electrocatalysis: oxidation of D-glucose on Pt (100), (110) and (111) single crystal electrodes and the effect of UpdAdlayers of Pb Tl and Bi. *J. Electroanal. Chem. Interf. Electrochem.*, **242**, 221–242.
80. Rao, M.L.B. and Drake, R.F. (1969) Studies of electrooxidation of dextrose in neutral media. *J. Electrochem. Soc.*, **116**, 334–337.
81. Bae, I.T., Xing, X., Liu, C.C., and Yeager, E. (1990) In situ fourier transform infrared reflection absorption spectroscopic studies of glucose oxidation on platinum in acid. *J. Electroanal. Chem. Interf. Electrochem.*, **284**, 335–349.
82. Bae, I.T., Yeager, E., Xing, X., and Liu, C.C. (1991) In situ infrared studies of glucose oxidation on platinum in an alkaline medium. *J. Electroanal. Chem. Interf. Electrochem.*, **309**, 131–145.
83. Sun, Y.P., Buck, H., and Mallouk, T.E. (2001) Combinatorial discovery of alloy electrocatalysts for amperometric glucose sensors. *Anal. Chem.*, **73**, 1599–1604.
84. Rao, J.R., Richter, G.J., Luft, G., and von Sturm, F. (1978) Electrochemical behavior of amino-acids and their influence on anodic-oxidation of glucose in neutral media. *Biomater. Med. Dev. Artif. Org.*, **6**, 127–149.
85. Kerzenmacher, S., Schroeder, M., Brämer, R., Zengerle, R., and von Stetten, F. (2010) Raney-platinum film electrodes for potentially implantable glucose fuel cells. Part 1: nickel-free glucose oxidation anodes. *J. Power Sources*, **195**, 6516–6523.
86. Hibbert, D.B., Weitzner, K., and Carter, P. (2001) Voltammetry of platinum in artificial perilymph solution. *J. Electrochem. Soc.*, **148**, E1–E7.
87. Zebda, A., Cosnier, S., Alcaraz, J.P., Holzinger, M., Le Goff, A., Gondran, C., Boucher, F., Giroud, F., Gorgy, K., Lamraoui, H., and Cinquin, P. (2013) Single glucose biofuel cells implanted in rats power electronic devices. *Sci. Rep.*, **3**, 1516.
88. Olivo, J., Carrara, S., and De Micheli, G. (2011) Energy harvesting and remote powering for implantable biosensors. *IEEE Sens. J.*, **11**, 1573–1586.
89. Romero, E., Warrington, R.O., and Neuman, M.R. (2009) Energy scavenging sources for biomedical sensors. *Physiol. Meas.*, **30**, R35–R62.
90. Kerzenmacher, S., Zehnle, S., Volk, T., Jansen, D., von Stetten, F., and

Zengerle, R. (2008) An efficient DC-DC converter enables operation of a cardiac pacemaker by an integrated glucose fuel cell. Proceedings of PowerMEMS 2008, Sendai, Japan, pp. 189-192.

91. Rouse, A.G., Stanslaski, S.R., Cong, P., Jensen, R.M., Afshar, P., Ullestad, D., Gupta, R., Molnar, G.F., Moran, D.W., and Denison, T.J. (2011) A chronic generalized bi-directional brain-machine interface. *J. Neural Eng.*, **8**, art # 036018.

92. Pop, V., van de Molengraft, J., Schnitzler, F., Penders, J., van Schaijk, R., and Vullers, R. (2008) Power optimization for wireless autonomous transducer solutions. Proceedings of PowerMEMS 2008, Sendai, Japan, 2008, pp. 141-144.

93. Köhler, C., Bentler, C., Kloke, A., Oudenhoven, J., Pop, V., Op het Veld, J.H.G., and Kerzenmacher, S. (2011) Towards implantable autonomous systems a low-power transducer platform powered by a glucose fuel cell. Proceedings of PowerMEMS 2011, Seoul, Republic of Korea, pp. 19-22.

94. Falk, M., Andoralov, V., Blum, Z., Sotres, J., Suyatin, D.B., Ruzgas, T., Arnebrant, T., and Shleev, S. (2012) Biofuel cell as a power source for electronic contact lenses. *Biosens. Bioelectron.*, **37**, 38–45.

95. Gonsalves, A.N.M. and Pereira, M.M. (1996) State of the art in the development of biomimetic oxidation catalysts. *J. Mol. Catal., A: Chem.*, **113**, 209–221.

96. Chen, J.H., Clark, L.T., and Cao, Y. (2005) Ultra-low voltage circuit design in the presence of variations. *IEEE Circ. Dev.*, **21**, 12–20.

97. Wang, A. and Chandrakasan, A. (2005) A 180-MV subthreshold FFT processor using a minimum energy design methodology. *IEEE J. Solid-State Circ.*, **40**, 310–319.

15
Direct-Electron-Transfer-Based Enzymatic Fuel Cells *In Vitro*, *Ex Vivo*, and *In Vivo*

Magnus Falk, Dmitry Pankratov, Zoltan Blum, and Sergey Shleev

15.1
Introduction

In this chapter, we briefly describe some historical developments made in the field of biological fuel cells (BFCs) and also discuss important design considerations taken when constructing *enzymatic* fuel cells (EFCs). BFCs constitute a subset of fuel cells (FCs) that employ biocatalysts rather than the typical nonbiological catalysts. The biocatalyst can be an organelle or a living cell, such as organelle-based or microbial FCs, which use these natural bioconverters to extract power from various fuels and oxidants. In EFCs, the redox enzymes *per se* are exploited to directly convert chemical energy into electric energy [1]. BFCs in general, and especially EFCs, are very promising devices for *in vivo* applications, since they operate under physiological conditions (neutral pH, temperatures between 25 and 50 °C, atmospheric pressure, etc.), converting the naturally present substrates into products that are tolerable to the host [2].

Since the topics BFC and EFC are rather extensive, only biodevices utilizing enzymes in direct electron transfer (DET) reactions on both the anodic and cathodic sides are considered (Figure 15.1). Firstly, a brief overview is given of oxidoreductases for which DET reactions have been unequivocally shown. Operability and stability under physiological conditions, together with biocompatibility, are indispensable requirements to make a biodevice feasible for implantation. In this chapter, biological methods of improving the activity and stability of redox enzymes for BFC applications are not considered; the chapter is instead focused on construction approaches to achieve the above requirements. Secondly, possible designs of enzymatic biodevices, including descriptions of common electrode materials, are presented. Thirdly, the performance of mainly carbohydrate/oxygen *membrane-less* and *cofactor-free* EFC is analyzed and compared, operating in *in vitro*, *in vivo*, and *ex vivo* situations. Specifically, we describe and compare the performance of mediator-, cofactor-, and membrane-less EFCs *in vitro*, *i.e.* in simple buffers and also complex solutions mimicking the composition of human physiological fluids. In addition, the capabilities of biodevices in human blood, plasma, serum, saliva, and existing *in vivo* trials of DET-based EFCs are also reviewed.

Implantable Bioelectronics, First Edition. Edited by Evgeny Katz.
© 2014 Wiley-VCH Verlag GmbH & Co. KGaA. Published 2014 by Wiley-VCH Verlag GmbH & Co. KGaA.

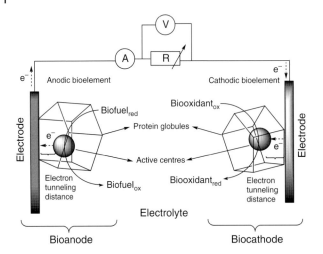

Figure 15.1 Principal scheme of a single compartment mediator- and cofactor-less direct-electron-transfer-based enzymatic fuel cell.

Finally, we compare the basic characteristics of available and tested biodevices, namely, open-circuit voltage (OCV) and operating voltage, power density, and operational stability. At the end of this work, further perspectives of EFCs in different applications are highlighted.

15.2
Oxidoreductases for Direct-Electron-Transfer-Based Biodevices

The ability of enzymes to utilize biologically derived fuels, such as carbohydrates (glucose, fructose, trehalose, and lactose), neurotransmitters (dopamine, serotonin, and adrenalines), alcohols (ethanol and methanol), and other substances (ascorbate, amino acids, etc.), along with ubiquitous molecular oxygen (O_2) and hydrogen peroxide (H_2O_2) as (bio)oxidants, makes the use of EFCs very attractive for numerous applications, especially as electric power sources for implantable devices. About 1400 oxidoreductases are known to date (www.enzyme-database.org), any of which could probably be utilized as bioelements in an EFC. However, in the majority of cases mediators are needed to shuttle electrons from the enzyme to the electrode surface, or vice versa. To a first approximation, the electron transfer (ET) rate between two redox species is determined, according to Marcus theory [3], by the driving force, the reorganization energy of the ET process, and the distance between the species. In spite of some recent reports claiming intrinsic electric conductivity of proteins [4], it is generally assumed that proteins act as electric insulators [5]. Since the catalytic sites of enzymes are often buried deeply within the protein matrix, achieving fast DET between a redox enzyme and an electrode surface is by no means trivial. Different strategies regarding the electrode design can however be employed, in order to electrically connect the catalytic site more

Table 15.1 Operational conditions for the BFCs reported herein operating under physiological conditions (see Table 15.2).

Medium	Glucose concentration (mM)	O_2 concentration (mM)	pH
In vitro devices			
Human blood	~5.0 [7]	~0.25^a (0.05–0.13 [7])^b	7.3–7.4 [7]
Human plasma	~5.6 [7]	~0.25^a	7.3–7.4 [7]
Ex vivo devices			
Human tears	~0.01 [8] to 0.6 [9]	~0.25^a (0.25 [10])^b	7.3–7.7 [9]
Human saliva	~0.10 [11a]	0.25^a (0.0025 [12])^b	6.2–7.6 [11b]
In vivo devices			
Rat (brain)	2.5–4.5 [13]	~0.05 [14]	7.1–7.3 [15]
Grape	~30^c [16, 17]	8 mM^d	2.8–4.5 [16]
Snail^e	~0.06 [18a]	~0.13 [19]	7.8–7.9 [20]
Clam^e	~0.18 [21]	0.03–0.05 [22]	7.0–8.0 [23]
Lobster^e	~0.1 [24]	~0.06 [25]	7.7–7.8 [26]

^{a}Air-saturated solution.
^{b}Condition *in vivo*.
^{c}Similar amounts of fructose.
^{d}Oxygen in air.
^{e}Large variations can occur depending on exact species, temperature, activity, and so on.
n.a. – not available.

efficiently (see below). Nevertheless, to date, less than about a hundred of the known oxidoreductases are capable of interacting with an electrode surface via a DET mechanism [6].

Of the variety of different oxidoreductases that indeed can be employed in a DET-based EFC, the bioelement to be used is mainly determined by the operating conditions of the intended device. Different implant settings are characterized by very varying compositions of the physiological fluids (Table 15.1), including amounts of fuel and oxidant, as well as pH and presence of additional compounds that can potentially affect the biodevice performance. Owing to the abundance of carbohydrates and O_2 in most implant settings, generally providing nontoxic by-products, carbohydrate-oxidizing and O_2-reducing enzymes are very attractive biocatalysts to design bioanodes and biocathodes, respectively. Also, the DET-capable oxidoreductase family is expanding as appropriately engineered redox enzymes are being developed, and with new electrode surface designs, new ways to electrically wire the biocatalysts are continuously established.

15.2.1
Anodic Bioelements

Oxidases and dehydrogenases stemming from different bacteria and fungi are well known for their efficiency in oxidizing alcohols and carbohydrates, although most often restricted to one-step incomplete oxidation process. Nevertheless, many of these redox enzymes can be, and in fact are, used to design highly efficient bioanodes of EFCs.

The most usual choice for the anodic bioelement has been the glucose-oxidizing enzyme glucose oxidase (GOx) [27]. The enzyme carries a flavin adenine dinucleotide (FAD) as a redox cofactor, as illustrated in Figure 15.2a. FAD can, in its quinone form (where the FAD is fully oxidized), accept two electrons and two protons to become $FADH_2$ (hydroquinone form), which can be oxidized to the semiquinone form (FADH), and further oxidized to return to the initial FAD (Figure 15.2c). The FAD cofactor will thus transfer high-energy electrons (Figure 15.2b). However, the FAD of GOx is buried deep within the enzyme (Figure 15.2a), which complicates the electric coupling and makes DET very

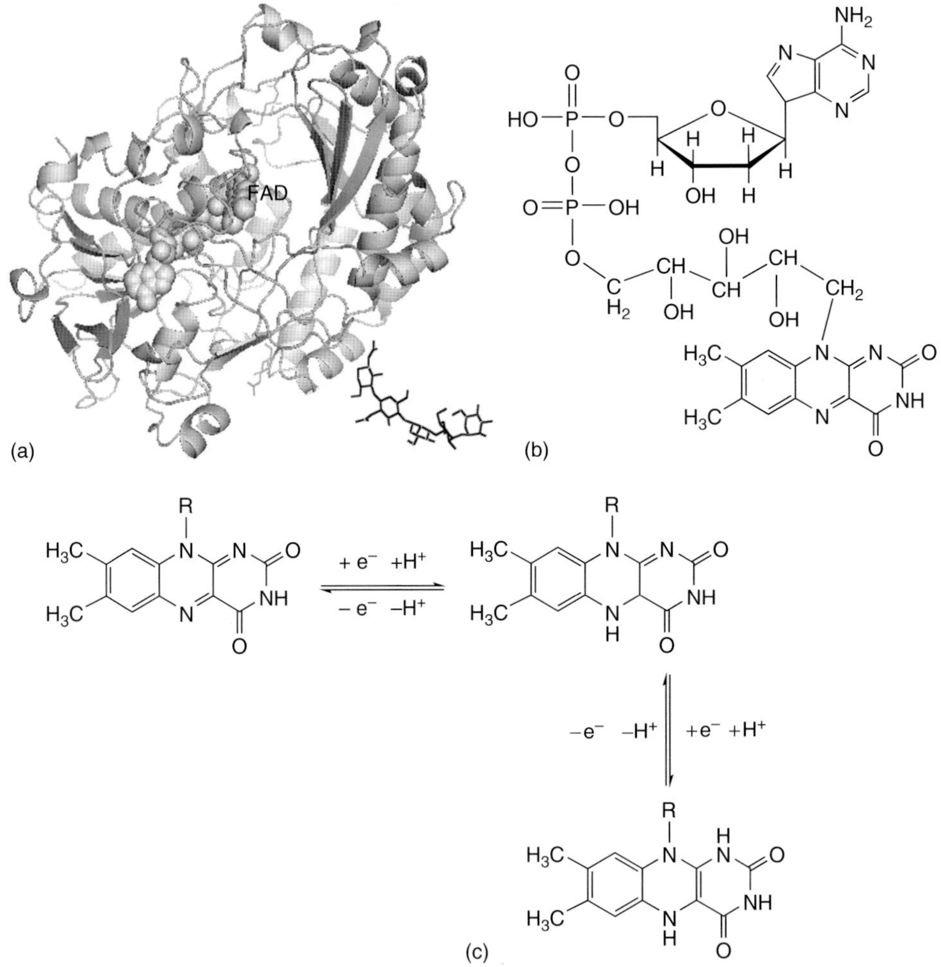

Figure 15.2 (a) Crystal structure of *Aspergillus niger* glucose oxidase (PDB file: 1CF3). (b) Structural formula of FAD. (c) Redox transformation of FAD.

sluggish in many cases. Although fast DET reactions between GOx and electrodes have been reported in many different studies, *e.g.* in Refs. [28], there is still an ongoing debate as to whether *bona fide* DET-based electrooxidation of glucose was observed, or if the process rather was mediated by some naturally occurring compound. Another complicating factor with GOx is that O_2 is the native electron acceptor of the enzyme and bioelectrocatalytic reactions will always compete with biocatalysis under aerobic conditions.

Several different dehydrogenases, which exhibit very low sensitivity toward O_2, have been utilized as anodic bioelement in glucose/O_2 biodevices. One such enzyme is cellobiose dehydrogenase (CDH) [29], which has lately received increasing attention and is utilized in DET-based EFCs [30]. CDH carries an FAD moiety that will accept two electrons from the substrate (Figure 15.3a); glucose is thereby

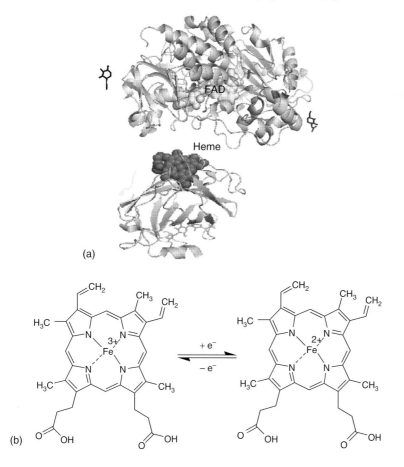

Figure 15.3 (a) Crystal structure of cellobiose dehydrogenase. The structure of the enzyme was rendered using the cytochrome and the FAD domains of *Phanerochaete chrysosporium* cellobiose dehydrogenase (PDB files: 1D7D and 1NAA, respectively). (b) Redox transformation of Heme B.

oxidized to gluconolactone, which is further spontaneously hydrolyzed to gluconic acid. The enzyme also contains a cytochrome domain, carrying the heme *b* prosthetic group with an iron center (Figure 15.3b), which can serve as a source or sink of electrons during ET, facilitating electric coupling to the electrode material [31]. Glucose is, however, not the target substrate of CDH; hence, the catalytic efficiency of the enzyme toward that particular carbohydrate is not very high. The natural substrate of CDH, as the name indicates, is the disaccharide cellobiose, but, as opposed to the case with glucose, the enzyme is also able to oxidize lactose and other carbohydrates with very high efficiency [29].

Glucose dehydrogenase (GDH) is a glucose-oxidizing enzyme that usually requires mediators in order to function in electrochemical systems (Figure 15.4a). However, DET has been shown using pyrroloquinoline quinone (PQQ)-dependent

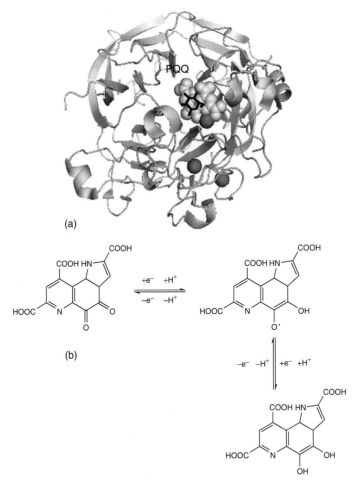

Figure 15.4 (a) Crystal structure of quinoprotein glucose dehydrogenase from *Acinetobacter calcoaceticus* (PDB file: 1CQ1). (b) Redox transformation of PQQ.

GDH, where the PQQ redox cofactor is coordinatively bound via Ca^{2+} (Mg^{2+} in some other quinoproteins) ions to the apoenzyme [32]. PQQ, similarly to FAD, is a redox cofactor, and enzymes containing PQQ are called *quinoproteins*. Similarly to FAD, three stable redox states exist for PQQ: the oxidized quinone state (PQQ), the semiquinone form (PQQH), and the reduced quinol form ($PQQH_2$) (Figure 15.4b) [33]. Similarly to CDH, PQQ-GDH performs a two-electron oxidation of carbohydrates, oxidizing, for instance, glucose to gluconic acid. Apart from glucose, PQQ-GDH can also oxidize other carbohydrates, such as cellobiose, maltose, and lactose.

Glucose is not the only possible fuel target of implantable EFCs, especially when considering biodevices operating in organisms other than vertebrates, such as plants and insects. This opens up the possibility of utilizing a large variety of enzymes, where the suitability depends on the organism in which they are intended to operate. As one example, fructose dehydrogenase (FDH) has been utilized in plants to generate power from fructose. Similarly to PQQ-GDH, FDH is a quinoprotein containing one PQQ prosthetic group and one cytochrome *c* domain, which enable DET and oxidize fructose via a two-electron mechanism [34]. The enzyme has been utilized in several different DET-based EFCs, as discussed below.

The above examples make it evident that the overall fuel economy of EFCs is intrinsically dismal. Redox enzymes, being single-step automatons, are limited in terms of columbic yield; utilizing a single enzyme to oxidize a carbohydrate, only two electrons are generally obtained from one carbohydrate molecule. As far as the EFC is concerned, the fuel is spent and the remaining energy content is wasted. On the cathode, the situation is different since not more than two electrons can be accepted by dioxygen. Strategies have however been developed in order to increase the efficiency of bioanodes by co-immobilizing different enzymes (see below). Furthermore, apart from being based on different carbohydrate-oxidizing enzymes, other DET-based EFCs have been designed to utilize alcohols, employing immobilized alcohol dehydrogenase (ADH) [35] or molecular hydrogen (H_2), exploiting hydrogenases (Hds) [36]. These BFCs are obviously not intended for implantable devices, but rather as power sources, for example for portable electronic gadgets.

15.2.2
Cathodic Bioelements

The enzymatically catalyzed reduction of bio-oxidants, mostly reduction of O_2 or H_2O_2, is based around active centers containing iron and/or copper ions. Consequently, all biocatalysts widely used in biocathodes nowadays are iron-, copper-, or iron/copper-containing redox enzymes. Many of these oxidoreductases display well-pronounced bioelectrocatalysis and almost all of them display efficient DET behavior.

The group of bioelements most suitable for biocathode design are blue multi-copper oxidases (BMCOs), such as laccase (Lc), ceruloplasmin (Cp), and bilirubin

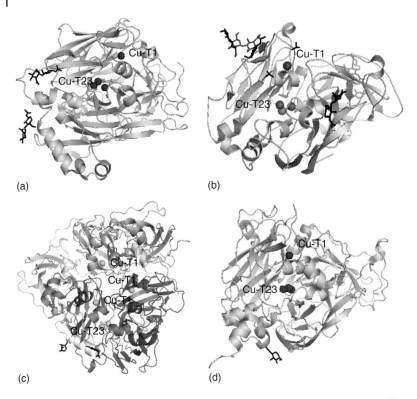

Figure 15.5 (a) Crystal structure of *Myrothecium verrucaria* bilirubin oxidase (PDB file: 2XLL). (b) Crystal structure of *Trametes versicolor* laccase (PDB file: 1GYC). (c) Crystal structure of human ceruloplasmin (PDB file: 1KCW). (d) Crystal structure of *Cucurbita pepo var. melopepo* (Zucchini) ascorbate oxidase (PDB file: 1AOZ).

and ascorbate oxidases (BOx and AOx, respectively). Hence, BMCOs have been extensively investigated as cathodic biocatalysts for DET-based biodevices [6b, 37]. Most BMCOs (except e.g. Cp) have a catalytic center with four copper ions (Figure 15.5): a T1 site (Cu-T1) containing a single copper atom, which accepts electrons from the substrate or from the electrode surface, and a T2/T3 cluster (Cu-T23) containing three copper atoms, where O_2 is reduced directly to water (H_2O) without formation of highly reactive oxygen species, such as H_2O_2, or hydroxyl and superoxide radicals.

Lcs and BOxs have similar tertiary structures, as shown in Figure 15.5a,b, and can oxidize a wide range of substrates (a one-electron process), while concomitantly reducing O_2 to H_2O (a four-electron process). High redox potential Lcs and BOxs constitute the most suitable choices as cathodic bioelements for implantable biodevices and can be exploited to build efficient DET-based O_2-reducing biocathodes. DET between the enzymes and different electrode materials, gold and carbon based, has been achieved using different immobilization strategies in attempts to appropriately orient the enzymes for efficient ET [38]. Cathodes based on BMCOs can often generate current densities in the milliamperes per square centimeter range, the efficiency being limited by O_2 diffusion to the electrode surface [39].

High redox potential Lcs usually exhibit the highest activity in acidic media, about pH 3–5, and the enzymes are inhibited in the presence of chloride ions. This makes the use of Lc in biodevices unsuitable for many implantable applications. Contrary to the Lcs, BOx usually retains a high activity at neutral pH and is less sensitive to chloride ions – circumstances that make BOx an appropriate choice for EFCs operating in human physiological fluids. However, the BOx class of enzymes have slightly lower redox potentials of the Cu-T1 compared to high redox potential fungal Lcs [40], which results in a lower voltage of the biodevices created. Moreover, deactivation of the enzyme can occur in the presence of urate, a compound present in serum [41].

As for other BMCOs, Cp (Figure 15.5c) even though showing efficient DET, lacks bioelectrocatalytic activity [42], whereas AOx (Figure 15.5d), as well as plant and bacterial Lcs, have substantially lower redox potentials [37b, 43], making the enzymes less suitable as cathodic bioelements [44]. Apart from Cu-containing enzymes, different peroxidases and oxidoreductases containing heme groups, *e.g.* horse radish and lignin peroxidases, are known to catalyze the reduction of organic and inorganic peroxides, where a two-electron reduction of the peroxide has been observed on different electrodes [6a, 45].

15.3
Design of Enzyme-Based Biodevices

Redox enzymes are in general exceptional catalysts, regularly reaching catalytic turnover numbers of $10^3 \, s^{-1}$, *cf.* the bioelements described just above, but some oxidoreductases, *e.g.* catalase, are able to operate with biocatalytic constants up to $4 \cdot 10^7 \, s^{-1}$ [46], i.e. close to the diffusion-controlled rates of redox reactions. Thus, at least in theory, enzymes could be used to create the most powerful FCs, compared to microbial or mitochondrial biodevices, and even conventional FCs [47]. The high selectivity of enzymes makes their utilization in FC applications highly advantageous by eliminating problems of cross-reactions and poisoning of the electrodes. This is important in the design of biodevices, since it allows fabrication of membrane-less single compartment BFCs, removing voltage losses that could otherwise arise. Hence, employing a DET-based design allows for significant simplifications and improvements in the construction of BFCs; no soluble compounds need to be added, mediator-induced voltage losses can be avoided, and possibly toxic mediator compounds can be excluded; also in addition, these factors simplify miniaturization and practical realization of efficient and simple biodevices. Furthermore, many oxidoreductases are active at neutral pH and at room temperature, *i.e.* under conditions at which an implanted medical device would operate. Finally, redox enzymes can potentially be produced at a very low cost.

A couple of issues should be noted when describing the design of EFCs. First, enzymes are rather large molecules, thus, reducing the actual mole fraction efficiency, which becomes a very important issue for biodevices generating current densities in the milliamperes per square centimeter range. Through calculations based on a monolayer of a 100 nm^2 enzyme, with a turnover number of 500 s^{-1}, a bioelectrocatalytic limiting current density of only about 0.1 mA cm^{-2} is estimated [2a]. In order to increase the efficiency of biodevices, additional enzyme (mono)layers need to be immobilized, while maintaining efficient electric coupling to the electrode surface [48]. Utilization of nanostructures to create expanded three-dimensional (3D) assemblies, thus increasing the available area, is an important and promising approach.

Second, of all the enzymes that exhibit DET properties, a majority contains an active site *i.e.* either relatively exposed at the surface of the protein, where the maximal practical distance is limited by the distance electrons can tunnel downhill with very high rates, *i.e.* about 2 nm [49], or buried in the protein matrix, but connected to the surface of the protein by a set of cofactors. In order to enhance ET, enzymes should therefore be appropriately oriented on the electrode surface.

Third, it should be noted, when comparing EFCs to other types of BFCs, that the efficiency of these biodevices is limited by incomplete fuel oxidation (*cf.* BFCs based on anaerobes, which can degrade the biofuel almost completely) and shorter lifetime because of limited intrinsic enzyme stability.

Overall, several key parameters can be identified in order to design efficient and stable EFCs:

- effective ET between electrode surfaces and active centers of enzymes (for this purpose proper orientation and interaction between the enzyme and the electrode surface should be obtained, providing a sufficiently short distance between the active center and the surface, *i.e.* less than about 20Å);
- high electric conductivity of electrodes ($\gg$ 200 S m^{-1} at room temperature) and high ion conductivity ($\gg$ 0.1 S m^{-1}) between biocathode and bioanode of the complete biodevice;
- high useful surface area (roughness factor $\gg$ 5) and surface to volume ratio ($\gg$ 10) along with dense monolayer loading of biocatalysts, which is usually achieved by employing 3D nanostructured electrodes;
- suitable porosity to promote enzyme stability and to facilitate efficient mass transport of both fuel(s) and oxidant(s);
- biocompatibility of biodevices for implantable applications.

Stating the obvious, there are two parallel and complementary strategies to realize these properties of EFCs, namely, the development and modification of electrode materials and improvement of electrode functionality.

15.3.1
Electrode Material

Only a limited number of materials have been used to design DET-based bioelectrodes. These materials can be divided into two main groups, namely metal- and carbon-based supports (Figure 15.6).

Carbon-based materials include spectrographic graphite (SPG), carbon nanotubes (CNTs), carbon black nanomaterials (CBNs) (including Ketjenblack (KB), Vulcan® XC-72, etc.), and their derivatives. The first DET-based, mediator- and membrane-less glucose/O_2 EFCs were designs based on SPG electrodes [30a,b] because, aside from being inexpensive, they are well-characterized porous materials [52], widely used to construct both DET-based O_2-reducing bioelectrodes [44b, 53] and carbohydrate-oxidizing biodevices [54]. CBN-modified electrodes with high roughness factors were used to demonstrate DET for several enzymes: FDH [55], Lc [55b], and BOx [56]. CBN-modified electrodes are also highly porous and have relatively high surface areas, coupled with excellent conductivity [57].

CNTs and derivatives are currently the most widely used carbonaceous supports for BFCs (Figure 15.6A–C). Both multiwalled carbon nanotubes (MWCNTs) and single-walled carbon nanotubes (SWCNTs) offer high conductivity, excellent

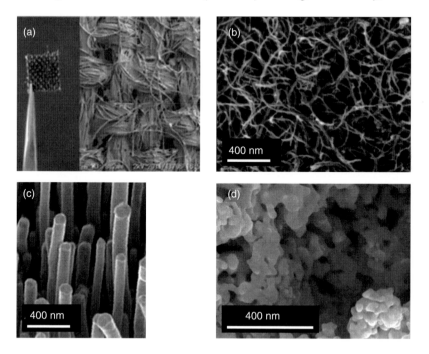

Figure 15.6 SEM images of (a) carbon fiber strip modified with CNTs and a photograph of the modified strip, (b) bucky paper surface, (c) carbon nanofiber forest surface, and (d) gold nanoparticles modified surface. (Image a and c are adapted from [50a] and [51], Copyrights (2012 and 2006), with permission from the American Chemical Society.)

chemical stability, and good mechanical strength. Moreover, available diameters are well matched to the size of redox enzymes used as biocatalysts, a fact that makes CNTs uniquely suited to facilitate the interaction between proteins and electrode surfaces. Further chemical functionalization of CNTs (surface oxidation, amination, hydroxylation, modification with thiols [58], pyrene derivatives [59], redox dyes [60], etc.) can be done to (i) improve conductivity, (ii) provide additional sites for specific enzyme attachment, and (iii) to enhance DET reactions. Highly flexible electrodes have been prepared by utilizing a carbon sheet base, modified with CNTs to facilitate the electric connection, as illustrated in Figure 15.6A [50]. To form biocompatible composites for enzyme-based electrodes, CNTs were mixed with biological materials, such as cellulose [61], chitosan [62], and silk [63]. CNTs are also used to fabricate buckypaper (BP), which is a novel conductive material for BFC applications with superior properties, such as very high surface area, porosity, chemical stability, flexibility, high mechanical strength, and, importantly, low toxicity both *in vitro* and *in vivo* [64]. BP is a plexiform film of densely packed CNTs, shown in Figure 15.6B, maintaining close contact between nanotubes due to Van der Waals forces. BP, as well as the Toray® carbon-fiber paper, is commercially available [28e, 65].

The structures described above have randomly oriented building blocks; however, DET-based BFCs with superior properties can be designed using aligned or oriented CNTs grown by chemical vapor deposition [66]. *e.g.* vertically aligned carbon nanofiber forests, as shown in Figure 15.6C, have been utilized as scaffolds for redox proteins [51] and this approach opens up new possibilities for the development of efficient and stable CNT-based EFCs.

Novel carbon materials, *e.g.* single-walled carbon nanohorns [67] and graphene [68], have also been used in BFC applications [69] and outstanding results were realized; *e.g.* a graphene-based EFC delivered nearly two times the maximum power of the SWCNT-based counterpart [70].

The usage of metals as the material for BFCs is not as diversified as carbon structures, and most metal-based EFCs are made from Au, although attempts to use platinum nanoparticles (PtNPs) have also been made [71]. Obviously, the conductivity of Au-based electrodes is high, but bare Au electrodes have a very low roughness factor (about 2 [53c]), *i.e.* too low to provide sufficient enzyme loading. Hence, modification of bare Au with AuNPs [30c, 39a] and AuNPs/CNTs [72] is usually carried out to design 3D nanostructured electrodes with much higher surface-to-volume ratio (Figure 15.6D), *i.e.* with roughness factors up to 300 [39a]. Many redox enzymes, *e.g.* CDH or Lc, are not catalytically active when absorbed on bare Au surfaces, and hence, formation of thiol monolayers [30c] or mixed monolayers based on aromatic diazonium salt derivatives and thiols [38c, 73] is essential to design catalytically active and stable CDH- and Lc-based bioanodes and biocathodes, respectively.

Despite all the positive qualities an electrode material presents, care should be taken regarding possible side reactions occurring at the support. Several auxiliary supports, namely metal or silica NPs, can show activity toward glucose oxidation [74]. Unpurified CNTs inevitably contain contaminations from the metal catalyst used during preparation [75]; *i.e.* transition metal oxide impurities can catalyze oxidation of the enzyme substrate [76].

Most major improvements in BFCs over the last 10 years have resulted from the development and application of new materials [57], and further progress in nanostructured materials is crucial for fabrication of efficient and stable DET-based EFCs.

15.3.2
Electrode Function

The limited long-term stability of EFCs, caused by the inherent short active lifetimes of many oxidoreductases, is a serious problem. The redox enzymes involved are insufficiently stable at 37 °C [65c], and enzymes in solution often have lifetimes of several hours up to a few days only. A handful of strategies for proper enzyme immobilization have been suggested to improve the stability and efficiency of biodevices, namely non-covalent linkage via molecular tethering [77], covalent binding [78], encapsulating enzymes within the voids of suitable polymers [79], nanoporous silica [80] or gold [81], and creation of silica-gel-based matrices [57, 82]. By using rational ways to immobilize the enzyme on the electrode surface, the lifetime can be increased up to the order of months [83] and further extended beyond one year through encapsulation of the enzyme in appropriately designed micelles, which physically confine the enzyme and prevent denaturation [79a, 84]. In spite of the problems that are likely to influence the mass transport of fuel and oxidant, encapsulating or entrapping of enzymes might still be necessary to make EFCs feasible for long-term use, and thus also make biodevices viable for applications intended to run in the human body.

Physiological fluids are very complex systems containing a multitude of low molecular weight compounds that can affect the performance of EFCs (Table 15.1) [2c]. Moreover, they have lower conductivity than model cell-free solutions (e.g., 1.455 and 0.775 S m^{-1} for 0 and 0.4 hematocrit, respectively [85]), with limited amounts of free biooxidants and biofuels (Table 15.1). The problem of limited biofuel availability can be addressed by designing bioanodes based on enzyme cascades, allowing a more exhaustive oxidation of fuels. Gorton *et al.* [86] showed that six electrons can be extracted from one glucose molecule at a two-enzyme anode, and attempts to realize a six enzyme-dehydrogenase system for lactate oxidation were made by the group of Minteer [87]; however, these approaches have yet to be applied *in vivo*.

In order to increase the effectiveness of biocathodes, gas-diffusion devices can be used instead of immersed biodevices [6b, 39b, 88], where O_2 is supplied to the electrode surface from the air and protons are supplied from the electrolyte (Figure 15.7). The O_2 content in air-saturated aqueous solutions is about 0.25 mM, while the biooxidant content in air is about 8 mM. Also, the O_2 diffusion coefficient in air is roughly 10 000 times larger, namely (at 1 atm) about $2 \cdot 10^{-5}$ cm^2 s^{-1} and about $2 \cdot 10^{-1}$ cm^2 s^{-1}, respectively [89]. The air-breathing floating biocathode is made of a hydrophobic diffusion layer and a hydrophilic catalytic layer, forming

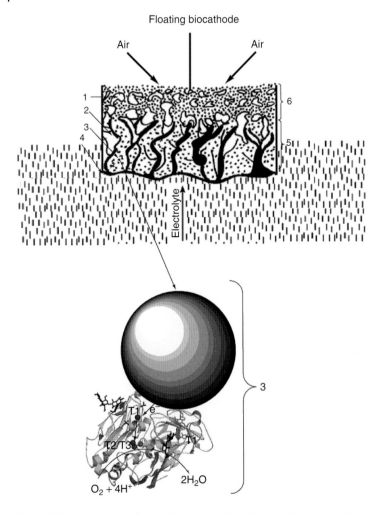

Figure 15.7 A principal scheme of an air-breathing biocathode based on *Trametes hirsuta* laccase 1-pores filled with air, 2-pores filled with electrolyte, 3-biocatalysts based on *Trametes hirsuta* laccase adsorbed on dispersed carbon materials, 4-Nafion layer, 5-active layer, 6-hydrophobic layer. (Adapted from [90a], with some additions and changes.)

the interface between the liquid electrolyte, solid electrode, and ambient air [39b, 90]. This design improves the mass transport of O_2 and prevents contact with enzyme inhibitors present in physiological fluids (Figure 15.7). Integration of a flow-through anode with an air-breathing cathode into a membrane-less BFC design was proposed [91], but, for obvious reasons the method is less than optimal for implantable biodevices.

In the case of O_2-sensitive anodic enzymes, it is necessary to use a membrane for fuel separation, which in general will increase the ohmic resistance between the anode and the cathode. Miyake *et al.* [50b] demonstrated a solution to this problem by using a hydrogel membrane with negligibly low resistance, *e.g.* 10 Ω for the membrane and 26 Ω for the cell, measured by impedance spectroscopy. Furthermore, researchers are focused on improving the ion conductivity at neutral pH by development of new polyelectrolyte membranes, although efficient devices have yet to be realized [57].

The problem of bioinertion, biocompatibility, and escaping the immune system in the case of highly evolved organisms, as well as miniaturization of EFCs, is in total a very complex issue. To resolve the matter in question, *i.e.* realizing long-term EFC operation under physiological conditions, joint efforts of scientists from different branches, *e.g.* materials science, biochemistry, physiology, immunology, and so on, is required.

15.4
Examples of Direct Electron Transfer Enzymatic Fuel Cells

A multitude of different DET-based EFCs have been designed utilizing different fuels and operating under distinctly different conditions (Tables 15.1 and 15.2). Most reported EFCs have been investigated *in vitro* in various simple buffers or complex fluids, and even though many have been designed with the intention to be used for implantable applications, very few reports of actual implantation exist. The reported DET-based EFCs are classified in this chapter according to the investigation conditions, *i.e. in vitro, in vivo,* and *ex vivo*. The third group, *ex vivo*, describes EFCs intended to operate in noninvasive contact situations, where the operational conditions lie somewhere in between *in vitro* and *in vivo*. At the end of this chapter, a summary of all the EFCs described herein (Table 15.2) and the different physiological fluids (Table 15.1) is provided.

15.4.1
Enzymatic Fuel Cells Operating *In Vitro*

A couple of reports exist of BFCs utilizing H_2 as fuel, and indeed the very first report where both the anodic and the cathodic bioelements were in DET contact with electrodes was a H_2/O_2 EFC reported in 2002 by Tarasevich *et al.* [36a]. The biodevice was designed with separated compartments with a pH 4.2 catholyte and an anolyte with a pH of 8.0, generating an extraordinarily high OCV of 1.35 V and maximum power density of 400 μW cm^{-2}. However, due to the absence of detailed information concerning the fabrication and characterization of the bioelectrodes, this paper is not cited by the scientific community, notwithstanding the impressive characteristics of the H_2/O_2 EFC presented, even when compared to current BFC developments (*cf.* H_2/O_2 EFC in Table 15.2). Three years later a mediator-, cofactor-, and membrane-less H_2/O_2 BFC was reported by Armstrong's group [36b]. The group created a membrane-less biodevice using Hd and Lc adsorbed on graphite electrodes. Moreover, the Hd-modified anode was completely unaffected by CO and only partially inhibited by O_2, owing to a proper choice of

Table 15.2 Comparison of basic characteristics of DET-based EFCs.

Fuel/Oxidant	Bioelements (anode/cathode)	Operational conditions	OCV (V)	Maximum power ($\mu W\ cm^{-2}$)	Publication
In vitro devices					
H_2/O_2	Hd/Lc	Citrate-PB pH 8 anolyte, pH 4.2 catholyte	1.35	400	Tarasevich et al. [36a]
H_2/O_2	MBH/Lc	Citrate buffer pH 5, bubbling of H_2 and air	0.97	7	Vincent et al. [36b]
$EtOH/H_2O_2$	ADH/MP-11	Acetate buffer pH 6 with 10 mM Glc and EtOH	0.27	0.18	Ramanavicius et al. [35b]
Frc/O_2	FDH/Lc	McIlvaine buffer pH 5, 200 mM Frc and O_2 sat., stirring	0.79	840	Kamitaka et al. [55b]
Frc/O_2	FDH/Lc	Citrate buffer pH 5, 200 mM Frc and air bubbled	0.66	126	Wu et al. [61]
Frc/O_2	FDH/BOx	Acetate buffer pH 6, 200 mM Frc	~0.73	660	Murata et al. [39a]
Frc/O_2	FDH/Lc	McIlvaine buffer pH 5, 200 mM Frc and O_2 sat., stirring	0.77	1800	Miyake et al. [66a]
Frc/O_2	FDH/BOx	McIlvaine buffer pH 5, 200 mM Frc in agarose hydrogel	0.7	550	Haneda et al. [50a]
Frc/O_2	FDH/BOx	McIlvaine buffer pH 5, 500 mM Frc in agarose hydrogel	2.09^a	640	Miyake et al. [50b]
Glc/O_2	CDH/Lc	Citrate buffer pH 4.5, 5 mM Glc	0.73	>5	Coman et al. [30b]
Glc/O_2	CDH/BOx	PBS pH 7.4, 5mM Glc	0.62	3	Coman et al. [30a]
Glc/O_2	CDH/BOx	Human serum	0.58	4	Coman et al. [30a]
Glc/O_2	CDH/BOx	PBS pH 7.4, 5mM Glc	0.66	3.2	Wang et al. [30c]
Glc/O_2	CDH/BOx	Human blood	0.66	2.8	Wang et al. [30c]
Glc/O_2	CDH/BOx	Rat cerebrospinal fluid	0.54	7	Andoralov et al. [2c]
Glc/O_2	PQQ-GDH/Lc	Human serum in flow cell	0.63	130	MacVittie et al. [18c]
Ex vivo devices[b]					
Glc/O_2	CDH/BOx	Human tears	0.57	3.5	Falk et al. [92]
Glc/O_2	CDH/BOx	Human saliva	0.63	2.1	Falk et al. [93]
In vivo devices					
Frc/O_2	FDH/BOx	Grape	~0.67	115	Miyake et al. [94]
Glc/O_2	CDH/BOx	Rat (brain)	0.54	2	Falk et al. [2c]
Glc/O_2	(PQQ)-GDH/Lc	Snail	0.53	30	Halmakova et al. [18a]
Glc/O_2	(PQQ)-GDH/Lc	Clam	$0.3-0.4^c$	40	Szczupak et al. [18b]
Glc/O_2	(PQQ)-GDH/Lc	Lobster	~0.6	640	MacVittie et al. [18c]

[a] Triple-layer cell.
[b] Actual *ex vivo* studies have not yet been performed.
[c] Large variations between different specimens, depending on size and physiological condition.

bioelement and construction of the enzyme electrode. When the BFC operated in a pH 5 citrate buffer with air and H_2 inlets in close proximity to the cathode and anode, respectively, an OCV of 0.97 V was registered and a maximum power density of 7 µW cm^{-2} was generated. The EFC provided stable currents for over 15 min.

The very first and, to the best of our knowledge, the only report on an ethanol (EtOH)/H_2O_2 EFC, where both bioelements were in DET contact with the electrodes, was published at the beginning of 2005 by Ramanavicius and coworkers [35b]. Moreover, this work can be considered as the first report on a membrane- and mediator-less EFC. The redox enzymes, *i.e.* ADH for the bioanode and a microperoxidase (MP-11)/GOx combination for the biocathode, were chemisorbed by means of glutaraldehyde (GA) cross-linking on carbon rod electrodes. Microperoxidase was in DET contact with the electrode, reducing the H_2O_2 produced by GOx when glucose was present in the solution, whereas ADH oxidized EtOH to acetaldehyde at the anode. The EFC was able to generate 0.18 µW cm^{-2} at 0.1 V and showed an OCV of 0.27 V in an acetate buffer, pH 6, with 10 mM of H_2O_2 and EtOH added to the solution. The EFC had an operational half-life of 2.5 days (Table 15.2).

Evidently, the intended applications for EtOH and H_2 EFCs are not as *in vivo* power-generating biodevices, but could rather be used *e.g.* for portable electronic gadgets. Instead, to allow for *in vivo* operation, different DET-based carbohydrate-oxidizing EFCs have been investigated. These could also be utilized for portable electronics as well, which indeed has been the target for some of the designs.

Several reports have been published on DET-based EFCs investigated *in vitro* using fructose as fuel, by employing the efficient enzyme FDH (Table 15.2) [39a, 50, 55b, 61, 66a]. In the first of such reports, Kano and coworkers constructed a one-compartment EFC by adsorbing FDH on a KB particle modified carbon anode and Lc on a carbon aerogel cathode [55b]. The EFC was investigated in an O_2-saturated, pH 5.0 McIlvaine buffer, containing 200 mM of fructose under stirring; an OCV of 0.79 V was observed and power densities reached 850 µW cm^{-2} at 0.41 V. Under these conditions, the biodevice displayed an operational half-life of over 12 h. Mass transfer of O_2 limited the cathode, reducing the power output to 390 µW cm^{-2} when measurements were performed without agitation. Slade and colleagues later reported on an FDH/Box-based EFC, where the enzymes were adsorbed on MWCNT-modified glassy carbon electrodes [61]. The EFC displayed relatively high stability, with a half-life of about 87 h, although the power output was lower compared to the previous work; the EFC generated 126 µW cm^{-2} at 0.35 V, with an OCV of 0.66 V. However, the measurements were conducted in air-saturated solutions without agitation, and when large currents were drawn, O_2 mass transfer limited the power.

Instead of utilizing carbon as a supporting material, Murata *et al.* [39a] employed electrodes modified with AuNPs. The authors used nanostructured electrodes to make an O_2-reducing cathode by adsorbing BOx, which did retain 90% of the current density after 48 h of continuous operation. The biocathode was combined with an FDH-based anode on AuNPs modified electrodes to assemble a membrane- and mediator-less fructose/O_2 EFC. When operated in acetate buffer, pH 6.0 containing

200 mM fructose, the EFC generated a maximum power of 0.66 mW cm^{-2} at 0.36 V. The power output was increased to 0.87 mW cm^{-2} (at 0.3 V) as a result of stirring, clearly indicating that the biocathode was diffusion limited.

Nishizawa and coworkers have reported on several FDH/O$_2$ EFCs. In 2011, the group developed a freestanding carbon nanotube forest film (see above), which was utilized for the entrapment of FDH on the bioanode and Lc on the biocathode, respectively, realizing a membrane- and mediator-less fructose/O$_2$ biodevice [66a]. When operated in stirred O$_2$-saturated McIlvaine buffer, pH 5.0, containing 200 mM fructose, an OCV of 0.77 V was registered, with a power output of 1.8 mW cm^{-2}. Eighty four percent of the power was maintained after 24 h of continuous operation. In order to demonstrate the flexible character of the EFC allowed by the CNTs ensemble, films were looped around the electric leads of an LED, where the blinking interval corresponded to the generated power, increasing the interval time by 20% only in the looped condition, as compared to merely clamping the EFC to the LED leads. The group developed the concept of flexible EFCs further, reporting on a sheet-shaped biodevice utilizing an anode made of a MWCNT-modified carbon fiber with FDH and a gas diffusion cathode made of KB-modified carbon fiber with BOx; the latter was coated with an additional layer of KB on top of the enzyme layer to create the hydrophobic layer [50a]. The two electrodes were connected by means of an agarose hydrogel made with a McIlvaine buffer, pH 5.0 containing 200 mM fructose, as shown in Figure 15.8. The cell registered an OCV of 0.7 V and a maximum power output of 550 μW cm^{-2} at 0.4 V, where the overall performance was limited by the biocathode. Later, the same group reported on a flexible fructose/O$_2$ stack EFC [50b]. The bioelectrodes were designed using CNT-decorated carbon fiber fabrics with FDH and BOx, where the EFC stack was composed of a bioanode, hydrogel sheets containing electrolyte and fructose, and an O$_2$ diffusion biocathode. The triple-layer cell was able to generate a maximum power of 0.64 mW cm^{-2} at 1.21 V with an OCV of 2.09 V, when a hydrogel sheet with McIlvaine buffer, pH 5.0, containing 500 mM fructose was used; similarly to the sheet EFC, the stack EFC was limited by the inferior performance of the biocathode. Owing to the flexible design the EFC stack could be bent and still retain high performance, and the laminar cell required no sealing frame. The stability

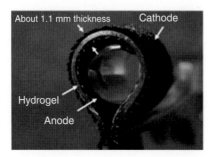 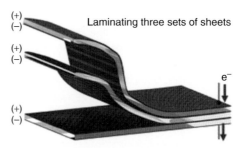

Figure 15.8 Photograph of flexible fructose/oxygen biofuel cell and schematic of triple layer cell. (Adapted from [50b], Copyright (2012), with permission from Elsevier B.V.)

of the biodevice was, however, very limited because of drying out of the hydrogel; in buffer, however, it maintained 85% of the performance for 24 h. Such flexible power sources are promising for powering wearable electronics.

The first DET-based EFC utilizing glucose as the biofuel and O_2 as the biooxidant was reported by Coman et al. [30b] as late as 2008. CDH and Lc were adsorbed on SPGE and, when the BFC was operated in a citrate buffer, pH 4.5 containing 5 mM glucose, it was able to generate over $5\,\mu W\,cm^{-2}$ at 0.5 V, displaying an OCV of 0.73 V (Table 15.2). The power output increased insignificantly by stirring the solution, indicating that the limitation was a consequence of the comparatively low performance of the CDH anode. In terms of stability, the biodevice was able to deliver power with about a 10% drop in efficiency after 5 h of continuous operation. The same group later used a similar CDH-based anode together with a Box-modified cathode [30a]. The change of bioelements, i.e. BOx instead of Lc, and CDH from ascomycete instead of basidiomycete, was needed to design a biodevice, which operated under physiological conditions, i.e. at neutral pH and in the presence of chloride ions. The EFC generated a peak power density of around $3\,\mu W\,cm^{-2}$ at a cell voltage of 0.37 V, with an OCV of 0.62 V in PBS, pH 7.4, containing 5 mM glucose. In order to demonstrate the possibility of operating the EFC in vivo, the performance was also investigated in human serum. A power maximum similar to the value obtained in buffer was observed, with a slight OCV reduction to 0.58 V. However, a second maximum of $4\,\mu W\,cm^{-2}$ at 0.19 V appeared in the power density plot. This could be attributed to the oxidation of some redox-active species, apart from glucose, present in serum, e.g. ascorbic acid or urate [30a]. As regards operational stability, a half-life of only 6 and 1.5 h was observed in buffer and human serum, respectively – clearly inadequate for long-term implantation.

A DET-based glucose/O_2 EFC with enhanced properties (compared to the biodevice described above) was designed by utilizing AuNP-modified Au electrodes [30c]. The EFC was constructed using Au macroscale electrodes modified with AuNPs, where the anode was created by cross-linking the CDH with GA on top of thiol-modified electrodes; the cathode was prepared by simple adsorption of BOx, similar to the design described by Murata et al. [39a]. The performance of the biodevice was evaluated both in PBS, pH 7.4, as well as in human blood and plasma. When the EFC was operated in 5 mM glucose containing buffer, the cell had an OCV of 0.66 V with a maximum power output of $3.3\,\mu W\,cm^{-2}$ (roughly one third of the power output using 50 mM glucose) at a cell voltage of 0.52 V. Similarly to the situation reported by Coman et al., the performance was only slightly reduced in the physiological fluids, with a power generation of roughly $3\,\mu W\,cm^{-2}$ and a slight reduction in OCV. When operated in buffer, the biodevice had a half-life of 30 h; the half-life was reduced to about 8 and 2 h in plasma and blood, respectively.

Common to all studies utilizing CDH as the anodic bioelement to design glucose/O_2 EFCs is a limitation of the overall power generation due to the low current output of the anode, caused by the inefficient oxidation of glucose by the enzyme. This is directly opposite to the situation for FDH-based biodevices, which, because of the efficient oxidation of fructose, are instead limited by the biocathode.

GOx is a significantly more efficient glucose-oxidizing enzyme than CDH, and several reports exist of DET-based EFCs utilizing GOx. Such examples include a microscale device investigated in blood [95] and a high-performance compressed MWCNT-based EFC [96]. Because of the ongoing scientific dispute concerning the performance of these biodevices, we choose not to describe GOx-based EFCs in further detail. Readers are instead referred to the literature. As another alternative to using CDH as the anodic bioelement in DET-based glucose/O_2 biodevices, PQQ-GDH can be utilized, as has been investigated *in vivo* (see below).

15.4.2
Biodevices Operating *In Vivo*

Although few records of *bona fide* implanted EFCs exist, the last couple of years have seen several examples of EFCs operating *in vivo* in animals (vertebrates, insects, molluscs, and crustaceans), as well as plants. The first report of an implanted DET-based EFC in an organism, a grape, was published in 2011 by Miyake *et al.* [94], although the very first implanted EFC was reported by Heller and his coworkers earlier in 2003 [97]; however, that biodevice was based on redox mediators. Grapes contain rather high amounts of glucose, over 30 mM, and have an acidic pH (roughly between pH 2.8 and 4.5) [16]. In order to circumvent the limited amount of O_2 usually available in living organisms (Table 15.1), Miyake and coworkers designed a needle bioanode that they combined with a gas diffusion biocathode, as shown in Figure 15.9a. The bioanode was formed by letting FDH adsorb on CNPs (KB)-modified electrodes. The gas diffusion biocathode was constructed by adsorbing BOx on KB-modified carbon paper, additionally coated with carbon particles to create the hydrophobic layer. By using an ion-conducting agarose hydrogel as an inner matrix, the bioanode and biocathode were assembled. When the FDH-based bioanode was inserted into a grape, a maximum power of 115 $\mu W\,cm^{-2}$ at 0.34 V was generated (Table 15.2). A similar design using a (mediator-based) bioanode with GDH was employed to generate power from glucose in the blood of a rabbit ear *in vivo*. Excluding separator membranes in the design allowed for significant simplification in the construction, enabling biofuels to be accessed from living organisms by the bioanode, while the biocathode avoided contact with potential inhibitors present in physiological fluids utilizing the abundant O_2 available in the air.

Very recently, Shleev and coworkers reported on a microscale EFC that was implanted in the brain of a living rat [2c]. The EFC design was used in earlier investigations by the group, in human tears (see below) [92], where Au wire electrodes with a diameter of 100 μm were modified with AuNPs and further modified with enzymes, utilizing a simple DET-based approach without employing membranes or separators. CDH was immobilized on one electrode to form the bioanode and BOx to another to from the biocathode. The use of such a simple design in the construction of the BFC allowed for miniaturization and easy fabrication of the biodevice. The biomodified microelectrodes were attached to micromanipulators and inserted directly into the brain tissue, as displayed in

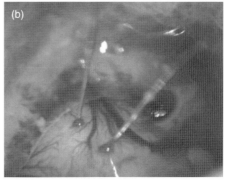
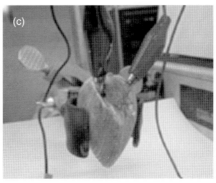

Figure 15.9 Photographs of biofuel cells operating *in vivo* in grape (a), rat brain (b), and clam (c). (Adapted from [94, 2c, and 18b, respectively], and, Copyrights (2011, 2013, and 2012), with permission from Elsevier B.V., Wiley-VCH, and The Royal Society of Chemistry, respectively.)

Figure 15.9b, where the EFC generated a maximum power of $2\,\mu W\,cm^{-2}$ at a cell voltage of 0.4 V, with an OCV of 0.54 V. When the same cell was investigated *in vitro* in cerebrospinal fluid, the power increased to roughly $7\,\mu W\,cm^{-2}$. The apparent drop in power density *in vivo* compared to the *in vitro* situation can be explained by the lower amount of glucose and O_2 available in the tissue than in air-saturated cerebrospinal fluid (Table 15.1). As regards stability, the voltage was roughly halved in 2 h when a constant load was applied. However, a small manipulation of the position of the electrodes brought back the voltage to the initial level. This clearly illustrates the limitation of an implanted EFC operating in a living organism, being influenced by the diffusional properties of the surrounding tissue. When several EFCs were connected in series in the same rat, no increase in voltage was observed [2c], since the animal body is a single-compartment electrolyte system.

Owing to the properties of blood and the advanced immune system of the hosts, operation of EFCs in vertebrates adds further difficulties. However, apart from vertebrates and plants, BFCs could also be envisioned to operate in simpler organisms, such as insects, molluscs, and crustaceans. Such EFCs would mainly have environmental or other applications. The circulatory fluid of molluscs consists of hemolymph instead of blood, which supplies the organisms with nutrients and O_2 in an open system [98]. The hemolymph resident immune system is not nearly as sophisticated as that of vertebrates, facilitating long-term implantation. The research group headed by Prof. Katz have investigated this possibility, with three consecutive reports of glucose/O_2 EFCs operating in molluscs and crustaceans [18]. The electrodes were fabricated using a DET-based design composed of compressed CNTs (see above), with Lc and PQQ-dependent GDH immobilized by cross-linking. As discussed above, the DET approach simplifies the construction of the EFC and excludes the use of possibly toxic mediator compounds, especially important for *in vivo* operation. When the EFC was implanted in snails and clams, the set-up used for investigating in clams shown in Figure 15.9c, a maximum power of 30 μW cm^{-2} at a cell voltage of 0.39 V [18a] and 40 μW cm^{-2} at a cell voltage of 0.17 V [18b] was obtained. A rapid voltage drop was observed when a constant load was applied to the EFCs, halving the voltage in less than 15 min. This can be attributed to inefficient mass transport in the tissue, similar to what was observed by Shleev and coworkers in the brain tissue. When the electrodes were disconnected and the organism was allowed a certain relaxation time, the power obtained initially could yet again be generated. When several EFCs were connected in series and in parallel in different clams, a significant increase in voltage and in current could be obtained. Similarly, when two EFCs were connected in series in two different lobsters, an OCV of 1.2 V was obtained and enough power was generated to power an electric watch [18c]. Connecting several EFCs in series in the same organism did not produce any appreciable increase in voltage. For a single EFC, an OCV of up to 0.6 V with a maximum power output of 0.64 mW cm^{-2} at a cell voltage of 0.28 V was registered. Apart from different *in vivo* trials, the EFC was also investigated in a flow cell system in human serum, where an OCV of about 0.63 V was observed, with a maximum power output of roughly 130 μW cm^{-2} at a cell voltage of 0.31 V [18c].

15.4.3
Enzymatic Fuel Cells Operating *Ex Vivo*

In this section, a special niche for application of BFCs in general, and EFCs in particular, namely, biodevices operating "*ex vivo*," is emphasized, with focus on attachable, adhesive, or floating biodevices operating in tears, saliva, and sweat, *i.e.* operations which somehow lie in between *in vivo* and *in vitro* situations [93]. This might be a very promising field of application for EFCs, which could be utilized to power biomedical devices in noninvasive contact situations, *e.g.* skin and oral electronic patches [99], as well as "smart" electronic contact lenses [100]. At these conditions, most of the *in vivo* shortcomings are non-issues, *e.g.* immune response, encapsulation, and many more.

A specific example of a possible application of BFCs operating *ex vivo* is a power supply for an electronic lens. High-performance bionic contact lenses for sensing purposes have been designed by a couple of different research groups [100], and even simple single-pixel displays have been incorporated into contact lenses [101]

(Figure 15.10a). Instead of powering such devices with inductive links or RF circuits, EFCs present a very attractive alternative as power supplies [92, 93]. A principal scheme of a "smart" electronic contact lens with an embedded DET EFC, acting as a green self-sufficient power source, is presented in Figure 15.10c. The EFC is able to utilize glucose or other biofuels, such as ascorbate and dopamine, readily available in tears, supplying electrical power for the components of the self-contained gadget, *e.g.* electronic elements (microchip) and display.

Miniature EFCs can potentially be produced at a low cost and without complicated designs, especially when utilizing enzymes immobilized directly on electrode surfaces without any mediators, and the otherwise major stability issue would not be as significant for biodevices intended to operate on the order of days to weeks, up to possibly months. In order to investigate this niche for BFC applications, a microscale glucose/O_2 EFC based on an AuNP-modified electrode was fabricated and tested in human lachrymal liquid, *i.e.* basal tears [92], *cf.* the EFC described above for *in vivo* investigations [2c]. The EFC, when operating in human tears, showed an OCV of 0.57 V, whereas a maximum power output of 3.5 $\mu W\,cm^{-2}$ was realized at an operating voltage of only 0.2 V, a much lower voltage compared to the *in vivo* studies (see above). At 0.5 V the EFC had a power output of only 0.8 $\mu W\,cm^{-2}$. The maximum at the lower cell voltage can be attributed to other compounds being oxidized at the anode, such as ascorbic acid and dopamine, which are both present in tears at very significant concentrations, *i.e.* with reported values of up to roughly 0.6 mM [102] and 2 mM [103], respectively. The glucose concentration in tears is also much lower than in blood, *i.e.* as low as 13 μM [8], whereas the biooxidant concentration in lachrymal liquid is very high, since human tears are O_2 saturated during waking hours (Table 15.1) [10]. All these factors make the performance of the EFC seriously limited by the bioanode.

Very recently, transparent and flexible EFCs have been designed in our laboratory – suitable systems for real *ex vivo* studies of self-powered contact lenses (Figure 15.10b). The devices were assembled on a biocompatible transparent polymer, initially metallized with a thin layer of titanium (5 nm), and afterwards with Au (10 nm). Then the transparent and flexible Au electrodes were biomodified with CDH and BOx to form the bioanodes and biocathodes of a glucose/O_2 EFC, following previously developed protocols for immobilization of both redox enzymes on Au surfaces [30c, 92]. Initial tests of separate transparent and flexible biocathodes and bioanodes using a macroscle electrochemical cell operating in a complex buffer mimicking the composition of human tears (a scratch on the top right corner of the electrode in Figure 15.10b corresponds to the point of attachment of a Au alligator clip for *in vitro* studies of the biodevice) showed almost identical performance of these bioelectrodes compared to opaque rigid biodevices under the same conditions. To design practically applicable biodevices, a "floating" biocathode (Figure 15.7, see above) could be incorporated into the transparent and flexible electrode, *e.g.* by employing the nonmetallized (opposite) side of the Au electrode. This work has already been initiated using composite nanobiomaterials and biopolymers, such as poly(hydroxyethyl methacrylate) (PHEMA) [104], currently in use in regular contact lenses for vision correction. To maximize the biodevice performance and

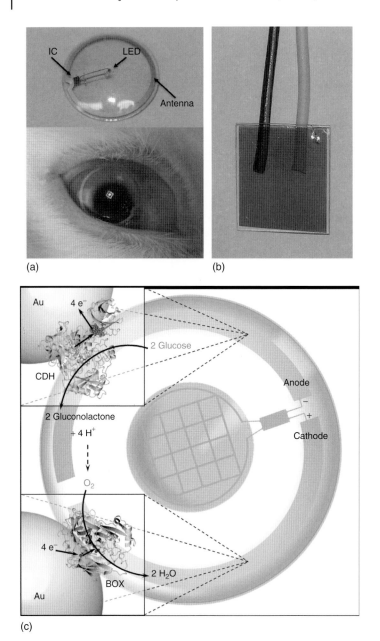

Figure 15.10 (a) Photographs from tests of the contact lens single-pixel display on a live rabbit. (Reprinted from [101], Copyright (2011), with permission from the IOP Publishing Ltd.) (b) A photograph of a transparent, flexible, and biocompatible gold bioelectrode. (c) A principal scheme of an electronic contact lens with an embedded glucose/O_2 enzymatic fuel cell, acting as a green self-sufficient power source. (Reprinted from [92], Copyright (2012), with permission from Elsevier B.V.)

to simultaneously avoid possible risks of eye damage due to exhaustion of basal tear O_2, the final self-powered biodevice would use biofuels from human lachrymal liquid, whereas molecular oxygen could be consumed directly from the air by the air-breathing BOx-based biocathode [105].

Apart from investigations in human tears and complex buffers mimicking the composition of human lachrymal liquid, a similar but macroscale EFC was also investigated in human resting saliva, with an OCV of 0.63 V and a maximum power output of 2.1 $\mu W\,cm^{-2}$ at an operating voltage of only 0.16 V [93]. The shape of the power output plot resembled the plot shapes from investigations on human lachrymal liquid, and could be similarly explained by the low glucose concentration in saliva (Table 15.1), *i.e.* as low as about 100 μM [11], as well as the presence of other compounds that can be oxidized by the Au support at low electrode potential.

15.4.4 Summary

A summary of all the BFCs described herein is shown in Table 15.2. It should be noted that a direct comparison of the performance of the reported EFCs is less than straightforward, considering the different operational conditions between different experiments (stirring, substrate concentration, etc.), as well as different assessment protocols regarding characterization and performance of biodevices. With regard to the stability of a particular EFC, one thing to take note of is that stability results might be interpreted as adequate, when in fact the enzyme is continuously denatured or desorbed from the electrode surface. This can happen when the power generated by the EFC is limited by substrate mass transport, and not by the amount of active enzyme. Because of this fact, stability data was excluded from the summary.

For an overview of the different conditions in different implantable situations, a short summary regarding the appropriate amounts of fuel and oxidant available, *i.e.* in different human physiological fluids such as blood, plasma, saliva, and tears, as well as in plants and different molluscs, is given in Table 15.1. It should be emphasized that the available amounts of fuel and oxidant *in vivo* are not as high as it is usually suggested. The concentration of free O_2 is almost 5 times lower in blood and 10 times lower in saliva, compared to air-saturated electrolytes, since most of the biooxidant molecules are bound to hemoglobin in the human vascular system [7] or to hemocyanine in some crustaceans, as well as consumed during different ongoing enzymatic reactions occurring in the human oral cavity [12]. In addition, a large variation in reported concentrations can be found in the literature because of uncertainties in measurement techniques, differences in collection procedures, and inadequate detection limits in previously used methods. A large variation can also exist locally in living organisms simply due to biological variability.

When biodevices are based on highly efficient anodic bioelements, such as FDH, the overall power output is determined by the power of the biocathode. Agitation and saturation with O_2 both increase the performance, and when possible, employing an air-diffusion design improves the efficiency even more. When less efficient anodic enzymes are utilized, such as CDH, the overall power output is instead limited by the anode. This is very significant in *ex vivo* situations, where a very limited amount of fuel is available. When biodevices are actually implanted *in vivo*, a strong mass transport dependence on both fuel and O_2 is usually observed.

15.5
Outlook

Integrated self-sustained bioelectronic devices, *i.e.* self-powered and wireless biodevices, will doubtlessly become increasingly important, not only in terms of basic scientific development but also as application vehicles in medicine, high-tech industry, military endeavors, and biocomputing. Substantial investments are placed to resolve the unfortunate conflict between the main parameters, *i.e.* voltage and power required versus appropriate scaling, of the major parts of a potentially implantable biodevice, namely the electric energy source and the working and controlling units, that at present hampers the practical realization of such devices. The by far largest component in contemporary commercially available implantable devices is the power source. We believe that membrane-less and mediator-free EFCs will replace the batteries commonly used, and will thus drastically decrease both the size and weight of the devices, given that the critical components of the biobattery, the biofuel and biooxidant, are readily extractable from animal and human tissues. Recent nanotechnological achievements have resulted in a marked decrease in both size and power consumption of transmitter/transducer systems, both RF and optical signal based. In parallel, also drawing on the phenomenal nanotechnological expansion, a distinct increase of the basic parameters of EFCs has been accomplished. Taking these recent technological feats together, it is only safe to assume that the combined product – nano/microscale self-contained biodevices (nano/microbiorobotic systems) – may be constructed in the near future. The theoretical performance limits of EFCs will depend on each particular fuel/oxidant couple, determining the maximal current density and voltage of a given biodevice. For a glucose/O_2 EFC operating *in vitro* the calculated limits are about a few hundred microamperes per square centimeter and 1.2 V, derived using diffusion coefficients of biocompounds in quintessential buffers and the potential difference between the gluconolactone/glucose and O_2/H_2O redox couples, respectively. A 3D EFC with a useful volume of 1 mm^3, operating at the performance limit (delivering a few microwatts of electric power) could supply enough energy even for microscale sensors, as well as control units based on modern microchips. However, the parameters of EFCs achieved so far, especially with regard to the operating voltage (Table 15.2), are still far from the thermodynamic limits. Even if transistors operating at a few tens of millivolts are described in the literature, 0.4–0.5 V is a crucial minimum voltage to drive modern semiconductor-based circuits, and hence, to allow the design of useful (bio)electronic devices. As a further complication, the operational stability of DET-based EFCs already fabricated and tested has been found wanting as regards practical biomedical applications. Both issues, operating voltage and operational stability, can be significantly improved by proper choice of bioelements, *i.e.* redox enzymes with required redox potentials and augmented stability in immobilized situations. Even though it is not discussed in this chapter, rational redesign of oxidoreductases by genetic modification and directed evolution could provide a route for practically applicable implantable EFCs.

References

1. Ivanov, I., Vidakovic-Koch, T., and Sundmacher, K. (2010) *Energies*, **3**, 803–846.
2. (a) Barton, S.C., Gallaway, J., and Atanassov, P. (2004) *Chem. Rev.*, **104**, 4867–4886. (b) Heller, A. (2004) *Phys. Chem. Chem. Phys.*, **6**, 209–216. (c) Falk, M., Narvaez Villarrubia, C., Atanassov, P., and Shleev, S. (2013) *ChemPhysChem*, **14**, 2045–2058.
3. Marcus, R.A. (1956) *J. Chem. Phys.*, **24**, 966–978.
4. Cahen, D. (2013) *AIP Conf. Proc.*, **1512**, 1326.
5. Guo, L.H. and Hill, H.A.O. (1991) *Adv. Inorg. Chem.*, **36**, 341–375.
6. (a) Christenson, A., Dimcheva, N., Ferapontova, E., Gorton, L., Ruzgas, T., Stoica, L., Shleev, S., Haltrich, D., Thorneley, R., and Aust, S. (2004) *Electroanalysis*, **16**, 1074–1092. (b) Shleev, S., Tkac, J., Christenson, A., Ruzgas, T., Yaropolov, A.I., Whittaker, J.W., and Gorton, L. (2005) *Biosens. Bioelectron.*, **20**, 2517–2554. (c) Ramanavicius, A. and Ramanaviciene, A. (2009) *Fuel Cells*, **9**, 25–36.
7. Guyton, A.C. and Hall, J.E. (2006) *Textbook of Medical Physiology*, 11th edn, Elsevier Saunders.
8. Taormina, C.R., Baca, J.T., Asher, S.A., Grabowski, J.J., and Finegold, D.N. (2007) *J. Am. Soc. Mass Spectrom.*, **18**, 332–336.
9. Berman, E.R. (1991) *Biochemistry of the Eye*, Plenum Press.
10. Efron, N. (2002) *Contact Lens Practice*, 1st edn, Butterworth-Heinemann.
11. (a) Jurysta, C., Bulur, N., Oguzhan, B., Satman, I., Yilmaz, T.M., Malaisse, W.J., and Sener, A. (2009) *J. Biomed. Biotechnol.*, **2009**, 430426. (b) Edgar, M., Dawes, C., and O'Mullane, D. (2004) *Saliva and Oral Health*, 3rd edn, British Dental Association, London.
12. Cohen, F., Burdairon, G., Rouelle, F., and Chemla, M. (1989) *J. Electroanal. Chem.*, **1**, 523–527.
13. Mak, W., Cheng, T.S., Chan, K.H., Cheung, R.T.F., and Ho, S.L. (2005) *Hong Kong Med. J.*, **11**, 457–462.
14. Bazzu, G., Puggioni, G.G.M., Dedola, S., Calia, G., Rocchitta, G., Migheli, R., Desole, M.S., Lowry, J.P., O'Neill, R.D., and Serra, P.A. (2009) *Anal. Chem.*, **81**, 2235–2241.
15. Chesler, M. (2003) *Physiol. Rev.*, **83**, 1183–1221.
16. Kliewer, W.M. (1967) *Am. J. Enol. Vitic.*, **18**, 87–96.
17. Palmer, J.K. and Brandes, W.B. (1974) *J. Agric. Food Chem.*, **22**, 709–712.
18. (a) Halamkova, L., Halamek, J., Bocharova, V., Szczupak, A., Alfonta, L., and Katz, E. (2012) *J. Am. Chem. Soc.*, **134**, 5040–5043. (b) Szczupak, A., Halamek, J., Halamkova, L., Bocharova, V., Alfonta, L., and Katz, E. (2012) *Energy Environ. Sci.*, **116**(15), 4457–4464. (c) MacVittie, K., Halamek, J., Halamkova, L., Southcott, M., Jemison, W.D., Lobel, R., and Katz, E. (2013) *Energy Environ. Sci.*, **6**, 81–86.
19. Mikkelsen, F.F. and Weber, R.E. (1992) *Physiol. Zool.*, **65**, 1057–1073.
20. Barnhart, M.C. (1986) *J. Comput. Phys. B*, **156**, 347–354.
21. Kraeuter, J.N. and Castagna, M. (2001) *Biology of the Hard Clam*, 1st edn, Elsevier Science B.V, Amsterdam.
22. Tran, D., Boudou, A., and Massabuau, J.-C. (2000) *Can. J. Zool.*, **78**, 2027–2036.
23. Bayne, B.L. (1976) *Marine Mussels: Their Ecology and Physiology*, 1st edn, Cambridge University Press, New York.

24. Chang, E.S. (2005) *Integr. Comp. Biol.*, **45**, 43–50.
25. Taylor, E.W. and Whiteley, N.M. (1989) *J. Exp. Biol.*, **144**, 417–436.
26. Dove, A.D.M., Allam, B., Powers, J.J., and Sokolowski, M.S. (2005) *J. Shellfish Res.*, **24**, 761–765.
27. Wong, C.M., Wong, K.H., and Chen, X.D. (2008) *Appl. Microbiol. Biotechnol.*, **78**, 927–938.
28. (a) Hu, F., Chen, S., Wang, C., Yuan, R., Chai, Y., Xiang, Y., and Wang, C. (2011) *J. Mol. Catal. B: Enzym.*, **72**, 298–304. (b) Wang, Y., Liu, L., Li, M., Xu, S., and Gao, F. (2011) *Biosens. Bioelectron.*, **30**, 107–111. (c) Courjean, O., Gao, F., and Mano, N. (2009) *Angew. Chem. Int. Ed.*, **48**, 5897–5899. (d) Holland, J.T., Lau, C., Brozik, S., Atanassov, P., and Banta, S. (2011) *J. Am. Chem. Soc.*, **133**, 19262–19265. (e) Ivnitski, D., Branch, B., Atanassov, P., and Apblett, C. (2006) *Electrochem. Commun.*, **8**, 1204–1210.
29. Ludwig, R., Harreither, W., Tasca, F., and Gorton, L. (2010) *ChemPhysChem*, **11**, 2674–2697.
30. (a) Coman, V., Ludwig, R., Harreither, W., Haltrich, D., Gorton, L., Ruzgas, T., and Shleev, S. (2010) *Fuel Cells*, **10**, 9–16. (b) Coman, V., Vaz-Dominguez, C., Ludwig, R., Harreither, W., Haltrich, D., De Lacey, A.L., Ruzgas, T., Gorton, L., and Shleev, S. (2008) *Phys. Chem. Chem. Phys.*, **10**, 6093–6096. (c) Wang, X., Falk, M., Ortiz, R., Matsumura, H., Bobacka, J., Ludwig, R., Bergelin, M., Gorton, L., and Shleev, S. (2012) *Biosens. Bioelectron.*, **31**, 219–225.
31. Zamocky, M., Ludwig, R., Peterbauer, C., Hallberg, B.M., Divne, C., Nicholls, P., and Haltrich, D. (2006) *Curr. Protein Pept. Sci.*, **7**, 255–280.
32. (a) Tanne, C., Goebel, G., and Lisdat, F. (2010) *Biosens. Bioelectron.*, **26**, 530–535. (b) Ivnitski, D., Atanassov, P., and Apblett, C. (2007) *Electroanalysis*, **19**, 1562–1568. (c) Flexer, V., Durand, F., Tsujimura, S., and Mano, N. (2011) *Anal. Chem.*, **83**, 5721–5727. (d) Okuda, J. and Sode, K. (2004) *Biochem. Biophys. Res. Commun.*, **314**, 793–797.
33. Oubrie, A. (2003) *Biochim. Biophys. Acta*, **1647**, 143–151.
34. (a) Ameyama, M., Shinagawa, E., Matsushita, K., and Adachi, O. (1981) *J. Bacteriol.*, **145**, 814–823. (b) Ikeda, T., Matsushita, F., and Senda, M. (1991) *Biosens. Bioelectron.*, **6**, 299–304. (c) Tominaga, M., Shirakihara, C., and Taniguchi, I. (2007) *J. Electroanal. Chem.*, **610**, 1–8. (d) Tkac, J., Svitel, J., Vostiar, I., Navratil, M., and Gemeiner, P. (2009) *Bioelectrochemistry*, **76**, 53–62.
35. (a) Ikeda, T., Miyaoka, S., Matsushita, F., Kobayashi, D., and Senda, M. (1992) *Chem. Lett.*, **111**, 847–850. (b) Ramanavicius, A., Kausaite, A., and Ramanaviciene, A. (2005) *Biosens. Bioelectron.*, **20**, 1962–1967.
36. (a) Tarasevich, M.R., Bogdanovskaya, V.A., Zagudaeva, N.M., and Kapustin, A.V. (2002) *Russ. J. Electrochem.*, **38**, 335. (b) Vincent, K.A., Cracknell, J.A., Lenz, O., Zebger, I., Friedrich, B., and Armstrong, F.A. (2005) *Proc. Natl. Acad. Sci. U.S.A.*, **102**, 16951–16954.
37. (a) Solomon, E.I., Sundaram, U.M., and Machonkin, T.E. (1996) *Chem. Rev.*, **96**, 2563–2605. (b) Sakurai, T. and Kataoka, K. (2007) *Chem. Rec.*, **7**, 220–229. (c) Morozova, O.V., Shumakovich, G.P., Gorbacheva, M.A., Shleev, S.V., and Yaropolov, A.I. (2007) *Biochemistry*, **72**, 1136–1150.
38. (a) Blanford, C.F., Heath, R.S., and Armstrong, F.A. (2007) *Chem. Commun.*, 1710–1712. (b) dos Santos, L., Climent, V., Blanford, C.F., and Armstrong, F.A. (2010) *Phys. Chem. Chem. Phys.*, **12**, 13962–13974. (c) Pita, M., Gutierrez-Sanchez, C., Olea, D., Velez, M., Garcia-Diego, C., Shleev, S., Fernandez, V.M., and De Lacey, A.L. (2011) *J. Phys. Chem. C*, **115**, 13420–13428.
39. (a) Murata, K., Kajiya, K., Nakamura, N., and Ohno, H. (2009) *Energy Environ. Sci.*, **2**, 1280–1285. (b) Kontani, R., Tsujimura, S., and Kano, K. (2009) *Bioelectrochemistry*, **76**, 10–13.
40. Christenson, A., Shleev, S., Mano, N., Heller, A., and Gorton, L. (2006) *Biochim. Biophys. Acta*, **1757**, 1634–1641.

41. (a) Kang, C., Shin, H., and Heller, A. (2006) *Bioelectrochemistry*, **68**, 22–26. (b) Shin, H., Kang, C., and Heller, A. (2007) *Electroanalysis*, **19**, 638–643.
42. (a) Yaropolov, A.I., Kharybin, A.N., Emneus, J., Marko-Varga, G., and Gorton, L. (1996) *Bioelectrochem. Bioenerg.*, **40**, 49–57. (b) Haberska, K., Vaz-Dominguez, C., De Lacey, A.L., Dagys, M., Reimann, C.T., and Shleev, S. (2009) *Bioelectrochemistry*, **76**, 34–41.
43. (a) Sakurai, T. and Kataoka, K. (2007) *Cell. Mol. Life Sci.*, **64**, 2642–2656. (b) Quintanar, L., Stoj, C., Taylor, A.B., Hart, P.J., Kosman, D.J., and Solomon, E.I. (2007) *Acc. Chem. Res.*, **40**, 445–452.
44. (a) Miura, Y., Tsujimura, S., Kamitaka, Y., Kurose, S., Kataoka, K., Sakurai, T., and Kano, K. (2007) *Chem. Lett.*, **36**, 132–133. (b) Shleev, S., Wang, Y., Gorbacheva, M., Christenson, A., Haltrich, D., Ludwig, R., Ruzgas, T., and Gorton, L. (2008) *Electroanalysis*, **20**, 963–969.
45. (a) Gorton, L., Lindgren, A., Larsson, T., Munteanu, F.D., Ruzgas, T., and Gazaryan, I. (1999) *Anal. Chim. Acta*, **400**, 91–108. (b) Ferapontova, E. and Gorton, L. (2002) *Bioelectrochemistry*, **55**, 83–87. (c) Andreu, R., Ferapontova, E.E., Gorton, L., and Calvente, J.J. (2007) *J. Phys. Chem. B*, **111**, 469–477.
46. Garrett, R.H. and Grisham, C.M. (2008) *Biochemistry*, 4th edn, Cengage Learning, Inc., Boston, MA, p. 1058.
47. Oman, H. (1999) *MRS Bull.*, **24**, 33–39.
48. (a) Dronov, R., Kurth, D.G., Moehwald, H., Scheller, F.W., and Lisdat, F. (2007) *Electrochim. Acta*, **53**, 1107–1113. (b) Dronov, R., Kurth, D.G., Mohwald, H., Scheller, F.W., and Lisdat, F. (2008) *Angew. Chem. Int. Ed.*, **47**, 3000–3003. (c) Dronov, R., Kurth, D.G., Mohwald, H., Spricigo, R., Leimkuhler, S., Wollenberger, U., Rajagopalan, K.V., Scheller, F.W., and Lisdat, F. (2008) *J. Am. Chem. Soc.*, **130**, 1122–1123. (d) Spricigo, R., Dronov, R., Rajagopalan, K.V., Lisdat, F., Leimkuehler, S., Scheller, F.W., and Wollenberger, U. (2008) *Soft Matter*, **4**, 972–978.
49. Page, C.C., Moser, C.C., Chen, X., and Dutton, P.L. (1999) *Nature*, **402**, 47–52.
50. (a) Haneda, K., Yoshino, S., Ofuji, T., Miyake, T., and Nishizawa, M. (2012) *Electrochim. Acta*, **82**, 175–178. (b) Miyake, T., Haneda, K., Yoshino, S., and Nishizawa, M. (2013) *Biosens. Bioelectron.*, **40**, 45–49.
51. Baker, S.E., Colavita, P.E., Tse, K.-Y., and Hamers, R.J. (2006) *Chem. Mater.*, **18**, 4415–4422.
52. Jaegfeldt, H., Torstensson, A.B.C., Gorton, L.G.O., and Johansson, G. (1981) *Anal. Chem.*, **53**, 1979–1982.
53. (a) Shleev, S., El Kasmi, A., Ruzgas, T., and Gorton, L. (2004) *Electrochem. Commun.*, **6**, 934–939. (b) Shleev, S., Jarosz-Wilkolazka, A., Khalunina, A., Morozova, O., Yaropolov, A., Ruzgas, T., and Gorton, L. (2005) *Bioelectrochemistry*, **67**, 115–124. (c) Ramirez, P., Mano, N., Andreu, R., Ruzgas, T., Heller, A., Gorton, L., and Shleev, S. (2008) *Biochim. Biophys. Acta*, **1777**, 1364–1369.
54. (a) Stoica, L., Ruzgas, T., Ludwig, R., Haltrich, D., and Gorton, L. (2006) *Langmuir*, **22**, 10801–10806. (b) Stoica, L., Ludwig, R., Haltrich, D., and Gorton, L. (2006) *Anal. Chem.*, **78**, 393–398.
55. (a) Kamitaka, Y., Tsujimura, S., and Kano, K. (2007) *Chem. Lett.*, **36**, 218–219. (b) Kamitaka, Y., Tsujimura, S., Setoyama, N., Kajino, T., and Kano, K. (2007) *Phys. Chem. Chem. Phys.*, **9**, 1793–1801.
56. (a) Tominaga, M., Otani, M., Kishikawa, M., and Taniguchi, I. (2006) *Chem. Lett.*, **35**, 1174–1175. (b) Habrioux, A., Napporn, T., Servat, K., Tingry, S., and Kokoh, K.B. (2010) *Electrochim. Acta*, **55**, 7701–7705.
57. Minteer, S.D., Atanassov, P., Luckarift, H.R., and Johnson, G.R. (2012) *Mater. Today*, **15**, 166–173.
58. Rahman, M.M., Umar, A., and Sawada, K. (2009) *Adv. Sci. Lett.*, **2**, 28–34.
59. Pang, H.L., Liu, J., Hu, D., Zhang, X.H., and Chen, J.H. (2010) *Electrochim. Acta*, **55**, 6611–6616.
60. (a) Li, X., Zhang, L., Su, L., Ohsaka, T., and Mao, L. (2009) *Fuel Cells*, **9**, 85–91. (b) Li, X.C., Zhou, H.J., Yu, P., Su,

L., Ohsaka, T., and Mao, L.Q. (2008) *Electrochem. Commun.*, **10**, 851–854.
61. Wu, X., Zhao, F., Varcoe, J.R., Thumser, A.E., Avignone-Rossa, C., and Slade, R.C.T. (2009) *Biosens. Bioelectron.*, **25**, 326–331.
62. Park, H.J., Won, K., Lee, S.Y., Kim, J.H., Kim, W.J., Lee, D.S., and Yoon, H.H. (2011) *Mol. Cryst. Liq. Cryst.*, **539**, 238–246.
63. (a) Kang, M., Chen, P., and Jin, H.J. (2009) *Curr. Appl. Phys.*, **9**, S95–S97. (b) Liu, J., Zhang, X.H., Pang, H.L., Liu, B., Zou, Q., and Chen, J.H. (2012) *Biosens. Bioelectron.*, **31**, 170–175.
64. Bellucci, S., Chiaretti, M., Cucina, A., Carru, G.A., and Chiaretti, A.I. (2009) *Nanomedicine*, **4**, 531–540.
65. (a) Hussein, L., Feng, Y.J., Aonso-Vante, N., Urban, G., and Kruger, M. (2011) *Electrochim. Acta*, **56**, 7659–7665. (b) Hussein, L., Rubenwolf, S., von Stetten, F., Urban, G., Zengerle, R., Krueger, M., and Kerzenmacher, S. (2011) *Biosens. Bioelectron.*, **26**, 4133–4138. (c) Pankratov, D., Zeifman, Y., Morozova, O., Shumakovich, G., Vasil'eva, I., Shleev, S., Popov, V., and Yaropolov, A. (2013) *Electroanalaysis*, **25**, 1143–1149.
66. (a) Miyake, T., Yoshino, S., Yamada, T., Hata, K., and Nishizawa, M. (2011) *J. Am. Chem. Soc.*, **133**, 5129–5134. (b) Alonso-Lomillo, M.A., Rudiger, O., Maroto-Valiente, A., Velez, M., Rodriguez-Ramos, I., Munoz, F.J., Fernandez, V.M., and De Lacey, A.L. (2007) *Nano Lett.*, **7**, 1603–1608. (c) Kim, J., Parkey, J., Rhodes, C., and Gonzalez-Martin, A. (2009) *J. Solid State Electrochem.*, **13**, 1043–1050.
67. (a) Wen, D., Deng, L., Zhou, M., Guo, S.J., Shang, L., Xu, G.B., and Dong, S.J. (2010) *Biosens. Bioelectron.*, **25**, 1544–1547. (b) Wen, D., Xu, X.L., and Dong, S.J. (2011) *Energy Environ. Sci.*, **4**, 1358–1363.
68. Zheng, W., Zhao, H.Y., Zhang, J.X., Zhou, H.M., Xu, X.X., Zheng, Y.F., Wang, Y.B., Cheng, Y., and Jang, B.Z. (2010) *Electrochem. Commun.*, **12**, 869–871.
69. Tamaki, T. (2012) *Top. Catal.*, **55**, 1162–1180.
70. Liu, C., Alwarappan, S., Chen, Z.F., Kong, X.X., and Li, C.Z. (2010) *Biosens. Bioelectron.*, **25**, 1829–1833.
71. Li, Y., Chen, S.M., Chen, W.C., Li, Y.S., Ali, M.A., and AlHemaid, F.M.A. (2011) *Int. J. Electrochem. Sci.*, **6**, 6398–6409.
72. Naruse, J., Hoa, L.Q., Sugano, Y., Ikeuchi, T., Yoshikawa, H., Saito, M., and Tamiya, E. (2011) *Biosens. Bioelectron.*, **30**, 204–210.
73. Vaz-Dominguez, C., Pita, M., de Lacey, A.L., Shleev, S., and Cuesta, A. (2012) *J. Phys. Chem. C*, **116**, 16532–16540.
74. (a) Vidakovic-Koch, T., Ivanov, I., Falk, M., Shleev, S., Ruzgas, T., and Sundmacher, K. (2011) *Electroanalaysis*, **23**, 927–930. (b) Choi, Y.K., Wang, G., Nayfeh, M.H., and Yau, S.T. (2009) *Biosens. Bioelectron.*, **24**, 3103–3107.
75. Ge, C.C., Lao, F., Li, W., Li, Y.F., Chen, C.Y., Qiu, Y., Mao, X.Y., Li, B., Chai, Z.F., and Zhao, Y.L. (2008) *Anal. Chem.*, **80**, 9426–9434.
76. (a) Kaur, B., Prathap, M.U.A., and Srivastava, R. (2012) *ChemPlusChem*, **77**, 1119–1127. (b) Anu Prathap, M.U., Kaur, B., and Srivastava, R. (2012) *J. Colloid Interface Sci.*, **381**, 143–151.
77. (a) Ramasamy, R.P., Luckarift, H.R., Ivnitski, D.M., Atanassov, P.B., and Johnson, G.R. (2010) *Chem. Commun.*, **46**, 6045–6047. (b) Chen, R.J., Zhang, Y.G., Wang, D.W., and Dai, H.J. (2001) *J. Am. Chem. Soc.*, **123**, 3838–3839.
78. (a) Beneyton, T., Wijaya, I.P.M., Ben, S.C., Griffiths, A.D., and Taly, V. (2013) *Chem. Commun.*, **49**, 1094–1096. (b) Gutierrez-Sanchez, C., Pita, M., Vaz-Dominguez, C., Shleev, S., and De Lacey, A.L. (2012) *J. Am. Chem. Soc.*, **134**, 17212–17220.
79. (a) Akers, N.L., Moore, C.M., and Minteer, S.D. (2005) *Electrochim. Acta*, **50**, 2521–2525. (b) Ghanem, A. and Ghaly, A. (2004) *J. Appl. Polym. Sci.*, **91**, 861–866.
80. Wang, Y.J. and Caruso, F. (2004) *Chem. Commun.*, 1528–1529.
81. Salaj-Kosla, U., Poller, S., Beyl, Y., Scanlon, M.D., Beloshapkin, S., Shleev, S., Schuhmann, W., and Magner, E. (2012) *Electrochem. Commun.*, **16**, 92–95.

82. Sarma, A.K., Vatsyayan, P., Goswami, P., and Minteer, S.D. (2009) *Biosens. Bioelectron.*, **24**, 2313–2322.
83. Kim, J., Jia, H., and Wang, P. (2006) *Biotechnol. Adv.*, **24**, 296–308.
84. Moore, C.M., Akers, N.L., Hill, A.D., Johnson, Z.C., and Minteer, S.D. (2004) *Biomacromolecules*, **5**, 1241–1247.
85. Yoon, G. (2010) *World Academy of Science, Engineering and Technology*, **60**, 640–643. http://www.waset.org/journals/waset/v60/v60-120.pdf
86. Shao, M.L., Zafar, M.N., Sygmund, C., Guschin, D.A., Ludwig, R., Peterbauer, C.K., Schuhmann, W., and Gorton, L. (2013) *Biosens. Bioelectron.*, **40**, 308–314.
87. Sokic-Lazic, D., de Andrade, A.R., and Minteer, S.D. (2011) *Electrochim. Acta*, **56**, 10772–10775.
88. (a) Babcock, B., Tupper, A.J., Clark, D., Fabian, T., and O'Hayre, R. (2010) *J. Fuel Cell Sci. Tech.*, **7**, 021017. (b) Gupta, G., Lau, C., Rajendran, V., Colon, F., Branch, B., Ivnitski, D., and Atanassov, P. (2011) *Electrochem. Commun.*, **13**, 247–249. (c) Gellett, W., Schumacher, J., Kesmez, M., Le, D., and Minteer, S.D. (2010) *J. Electrochem. Soc.*, **157**, B557–B562.
89. Denny, M.W. (1995) *Air and Water: The Biology and Physics of Life's Media*, Reprint edn, Princeton University Press, Princeton, NJ.
90. (a) Shleev, S., Shumakovich, G., Morozova, O., and Yaropolov, A. (2010) *Fuel Cells*, **10**, 726–733. (b) Ciniciato, G.P.M.K., Lau, C., Cochrane, A., Sibbett, S.S., Gonzalez, E.R., and Atanassov, P. (2012) *Electrochim. Acta*, **82**, 208–213. (c) Brocato, S., Lau, C., and Atanassov, P. (2012) *Electrochim. Acta*, **61**, 44–49.
91. Rincon, R.A., Lau, C., Luckarift, H.R., Garcia, K.E., Adkins, E., Johnson, G.R., and Atanassov, P. (2011) *Biosens. Bioelectron.*, **27**, 132–136.
92. Falk, M., Andoralov, V., Blum, Z., Sotres, J., Suyatin, D.B., Ruzgas, T., Arnebrant, T., and Shleev, S. (2012) *Biosens. Bioelectron.*, **37**, 38–45.
93. Falk, M., Blum, Z., and Shleev, S. (2012) *Electrochim. Acta*, **82**, 191–202.
94. Miyake, T., Haneda, K., Nagai, N., Yatagawa, Y., Onami, H., Yoshino, S., Abe, T., and Nishizawa, M. (2011) *Energy Environ. Sci.*, **4**, 5008–5012.
95. Pan, C., Fang, Y., Wu, H., Ahmad, M., Luo, Z., Li, Q., Xie, J., Yan, X., Wu, L., Wang, Z.L., and Zhu, J. (2010) *Adv. Mater.*, **22**, 5388–5392.
96. Zebda, A., Gondran, C., Le Goff, A., Holzinger, M., Cinquin, P., and Cosnier, S. (2011) *Nat. Commun.*, **2**, 1–6.
97. Mano, N., Mao, F., and Heller, A. (2003) *J. Am. Chem. Soc.*, **125**, 6588–6594.
98. Gilbert, L.I. (2012) *Insect Molecular Biology and Biochemistry*, 1st edn, Elsevier Academic Press, San Diego, CA.
99. Duun, S., Haahr, R.G., Hansen, O., Birkelund, K., and Thomsen, E.V. (2010) *J. Micromech. Microeng.*, **20**, 075020/1–075020/15.
100. (a) Liao, Y.-T., Yao, H., Lingley, A., Parviz, B., and Otis, B.P. (2012) *IEEE J. Solid-State Circuits*, **47**, 335–344. (b) Chu, M.X., Miyajima, K., Takahashi, D., Arakawa, T., Sano, K., Sawada, S.-I., Kudo, H., Iwasaki, Y., Akiyoshi, K., Mochizuki, M., and Mitsubayashi, K. (2011) *Talanta*, **83**, 960–965. (c) Yao, H., Shum, A.J., Cowan, M., Lahdesmaki, I., and Parviz, B.A. (2011) *Biosens. Bioelectron.*, **26**, 3290–3296.
101. Lingley, A.R., Ali, M., Liao, Y., Mirjalili, R., Klonner, M., Sopanen, M., Suihkonen, S., Shen, T., Otis, B.P., Lipsanen, H., and Parviz, A. (2011) *J. Micromech. Microeng.*, **21**, 125014/1–125014/8.
102. (a) Choy, C.K.M., Cho, P., Chung, W.Y., and Benzie, I.F.F. (2001) *Invest. Ophthalmol. Vis. Sci.*, **42**, 3130–3134. (b) Gogia, R., Richer, S.P., and Rose, R.C. (1998) *Curr. Eye Res.*, **17**, 257–263.
103. Agarwal, S., Agarwal, A., Apple, D.J., Buratto, L., and Alió, J.L. (2002) *Textbook of Ophthalmology*, 1st edn, Jaypee Brothers Medical Publishers, p. 3000.
104. Bose, R.K. and Lau, K.K.S. (2010) *Biomacromolecules*, **11**, 2116–2122.

105. Villarrubia, C.N., García, S.O., Shleev, S., and Atanassov, P. (2013) Bioelectrochemistry 2013. 12th Topical Meeting of the International Society of Electrochemistry and XXII International Symposium on Bioelectrochemistry and Bioenergetics of the Bioelectrochemical Society, Bochum, Germany, March 17–21, 2013.

16
Enzymatic Fuel Cells: From Design to Implantation in Mammals

Serge Cosnier, Alan Le Goff, and Michael Holzinger

16.1
Introduction

Battery-powered implantable devices have been in use for 50 years, following the implantation of the first successful cardiac pacemaker in 1960. Since then, a variety of implantable battery-powered devices have been developed and introduced for the treatment of conditions ranging from neurological disorders to hearing loss. The development of lithium batteries in the late 1960s led to better and smaller devices exhibiting a high reliability combined with a longevity exceeding 5 years for the most recent designs. The latter continue to be the unique commercial tool for supplying power to electronic medical implants such as pacemakers.

Sealed batteries are thus adequate for pacemakers (of about 10 ml) with leads that consume about $20\,\mu W$. Moreover, the packaging and electronic design of pacemakers has witnessed marked progress in the last decades. As a consequence, the power requirement of pacemakers has been significantly reduced over the past years and Ohm and Danilovic reported that $1\,\mu W$ is a reasonable upper estimate of the required power of modern pacemakers [1]. Nevertheless, it was commonly accepted that a pacemaker consumed an average of $10\,\mu W$. In parallel to pacemakers, sealed batteries may be also employed for larger cardiac resynchronization therapy devices with leads that consume $40\,\mu W$ when pacing both ventricles. However, these devices are insufficient for more demanding applications such as leadless pacemakers where the maximum capsule volume allowed is 0.7 ml or for neurostimulators where more power is needed.

Although energy demand for artificial heart (several Watts) is beyond the current outlook for future implantable batteries [2], the development of a new generation of artificial implanted organs that require only between several hundred microwatts and several tens of milliwatts strongly depends on the emergence of new sources of energy. Yet, these putative artificial implanted organs may constitute an attractive solution to terminal failures of organs such as pancreas, urinary sphincters, or even kidneys. For instance, an artificial pancreas constituted by an implantable insulin pump combined with an implanted glucose sensor assuming the continuous glucose monitoring could offer a much more efficient insulin delivery, and hence

Implantable Bioelectronics, First Edition. Edited by Evgeny Katz.
© 2014 Wiley-VCH Verlag GmbH & Co. KGaA. Published 2014 by Wiley-VCH Verlag GmbH & Co. KGaA.

treatment of diabetes. However, its power requirement exceeds the power provided by sealed batteries. Potential applications in the medical field with actuators that could be combined with sensors may open up new ways of treatment for chronic diseases such as diabetes. They could also address physical deficiencies created by surgical removal of organs during the treatment of cancer. Such robots, unfortunately, consume more power than what conventional batteries can provide. For instance, manual artificial urinary sphincters are implanted each year in 10 000 new patients suffering from incontinence after radical prostatectomy. Such artificial sphincters are driven by the patient himself through a pump inserted in his scrotum, which he has to press to allow urination [3]. Robotization of this implanted system would provide much more comfort and ease of use and theoretically should require only about 200–300 µW [4]. More futuristic applications include research on implanted artificial kidneys [5], which can be envisioned if a permanent source of power generating around 20 mW necessary for the osmotic work of kidneys in human beings can be conceived.

As a consequence, numerous efforts are afoot to develop alternative power supply systems that are capable of operating independently over prolonged periods of time without the need for external recharging or refueling [6]. Different alternatives such as mechanical and thermoelectric generators that utilize temperature differences within the human body or vibrations or body movements are being explored in order to power implanted devices with energy scavenged from the human body. However, systems that utilize temperature differences within the human body or vibrations or body movements to scavenge power are currently limited to few microwatts per milliliter because of the erratic and low frequency of body movements or heart contractions as well as the weak difference in temperature inside the body for Seebeck thermoelectric effect [7].

The ever-increasing depletion of fossil fuels and the need for clean methods of producing electricity have stimulated the emergence of new sources of sustainable and renewable energy without greenhouse gas emissions or environmental pollution. Among these clean alternative sources, energy production via electrochemical means was seriously considered. One of the main alternative sources is fuel cells that convert chemical energy into electrical energy. This concept was extrapolated to harvest energy from human body fluids [8]. In their pioneering work in the 1970s, Drake and colleagues envisioned the production of electric power through the oxidation of glucose and the simultaneous reduction of oxygen by fuel cells using noble metals as catalysts [9]. This revolutionary approach went against the dogma of the insertion of a limited amount of energy (sealed batteries), proposing to consume substrates present *"in situ"* (glucose and oxygen) for producing this energy. Taking into account that these substrates are constantly renewed by human metabolism, this approach opened fabulous perspectives as these systems may theoretically generate electricity indefinitely. Moreover, implanted medical devices require to be installed in easy-to-reach places in the body to replace their battery whereas powering them by abiotic fuel cell means they could be installed in more favorable positions to carry out their functions. In addition, fuel cells, theoretically, would eliminate the need for periodic surgeries to install new batteries or considerably decrease their frequency. However, the low activity of the abiotic catalysts or their unfavorable kinetic at the specific "*in vivo*" conditions, namely, low temperature and neutral pH lead to weak cell voltage and hence low power output density. Unfortunately, these original implanted systems have not kept their promises and have thus almost fallen into oblivion. Nevertheless, some work based on fuel cells that use platinum alloys or activated carbon as catalyst or single-layer design based on single wafers has recently reported [8, 10]. However, their power remains relatively weak, namely, a power fuel cell density of $2\,\mu W\,cm^{-2}$ and a volumetric power density of $16\,\mu W\,cm^{-3}$.

The drastic improvement in the catalytic reactions of glucose and oxygen degradation in physiological conditions through the use of biological catalysts, mainly enzymes, has recently restarted this extremely promising axis of research [11]. Development of biofuel cells based on the consumption of glucose and oxygen are mainly focused on the use of glucose oxidase (GOx) and glucose dehydrogenase at the anode for the oxidation of glucose. Concerning oxygen reduction at the cathode, it was almost exclusively catalyzed by laccase or bilirubin oxidase (BOD). The "*in vitro*" performance of these biofuel cells approaching 1 V for open-circuit voltage and achieving significant power densities ($1-1.54\,mW\,cm^{-2}$) makes them a promising source of power for implanted devices [12].

As a consequence, considerable attention has recently been paid to the possibility of implantation of biofuel cells in the human body. In order to demonstrate the feasibility of the concept of implantable batteries, the first tests were performed with various living organisms such as rats [13], rabbits [14], cockroaches [15], clams [16], snail [17], or lobsters [18]. It should be noted, however, that apart from the first implanted biofuel cell in a rat, the examples reported so far were distant from the mammals since they involved insects, mollusks, or arthropods. Moreover, taking into account the size of the living target, the biofuel cell was sometimes partially implanted; only the tips of the electrodes were in contact with the living organism as reported by Miyake *et al.*, [14] who inserted a needle anode in the vein of a rabbit's ear and used an air-breathing biocathode. Another example is the deposition of the bioelectrodes onto the surgically exposed cremaster tissue of a rat [19]. Nevertheless, these implanted biofuel cells have demonstrated that it is possible to power electronic systems from biological fluids of living organisms. Thus, three biofuel cells implanted in clams and connected in parallel can activated, via an electronic circuitry, an electric motor inducing the rotation of a rotor [16]. In the same vein, two biofuel cells implanted in lobsters and connected in series were able to power an electrical watch [18]. In parallel, the first example of a biofuel cell implanted in mammals and capable of activating electronic devices was recently described [20]. A biofuel cell was thus implanted in the abdominal cavity of rats, which, when combined with a voltage boost converter, can flash a light-emitting diode (LED) or activate a digital thermometer. As previously reported for the first biofuel cell implanted into a living organism, it must be emphasized that this biofuel cell was completely inserted in the rat and not partially in contact with its physiological fluids. In addition, it should be noted that this is the first example of a biofuel cell operating in freely moving animals. Finally, Crespilho's group described recently the first example of a micro biofuel cell implanted into the jugular vein of a rat [21]. The latter produced 9.5 nW or 95 µW cm^{-2} from rat blood. However, it should be noted that the enzyme coatings were directly in contact with blood and hence exposed to the reaction of the rat's immune system.

Although these experiments demonstrate the validity of this strategy, many hurdles are to be overcome for the development of implanted biofuel cells. The main obstacles to circumvent are the sterilization of the biofuel cell, its biocompatibility, lifetime, and low energy efficiency *in vivo*.

In order to eliminate all forms of microbial life, including fungi, bacteria, viruses, and also bacterial spores that may contaminate biofuel cells, the latter must be sterilized before implantation. However, methods widely used for the sterilization of medical devices consist in heat by autoclave treatment; radiations such as electron beams, X-rays, or γ-rays; and chemical sterilization by ethylene oxide, sodium hypochlorite, and oxidizing agents such as hydrogen peroxide and ozone. However, these methods of sterilization should strongly impact the activity of enzymes and therefore greatly reduce the performance of biofuel cells. The packaging of biofuel cells could be designed to withstand such treatment. Another alternative consists in using sterilized components except enzymes that could be purified by filtration and then to fabricate biofuel cells in a sterile environment just prior to implantation.

Biocompatibility of biofuel cells itself is another challenge. First, its location within the body is a subject of controversy. Many research groups claim that their biofuel cells may be implanted in the blood vessels to harvest energy from blood. Effectively, the oxygen concentration must be higher in blood so that the blood flow ensures a steady supply of glucose and oxygen. Some projects, indeed, are focused on electricity production from the human bloodstream via a tiny turbine installed in a blood vessel or nanopiezoelectronics based on zinc oxide nanowires that generate electricity when subjected to mechanical stress. However, this approach may slow down the bloodstream or induce an overwork on the heart because of more resistance in the blood flow. One of the biggest concerns with the presence of devices such as a turbine or biofuel cells in a blood vessel is the possibility of blood clots resulting from the generated turbulence. If such clots occur, they could prove deadly as they move through the bloodstream. Moreover, any object present in a vessel creates a potential source of thrombosis and embolism. Perfectly "flat" objects, such as valves, stents, or vascular prostheses, can be tolerated, but this is only because their geometry is designed to avoid any turbulence and because they are coated with materials that are known to limit the risk of thrombosis. However, in most cases, the patients have to undergo life-long anticoagulation treatments, such as aspirin, with an induced morbidity that has to be considered. Safe introduction of a non-flat biofuel cell in an artery would almost certainly lead to thrombosis in a matter of days. In addition, it is very well documented that hemocompatibility is much more demanding than biocompatibility. In view of these remarks, we consider that estimation of the increase in performance that could be expected by introducing biofuel cells in the blood would not be very useful from a clinical viewpoint because of the clinically unacceptable risks following such an implantation. Our group has thus favored the implantation of biofuel cells in the retroperitoneal space because it is a zone where the difference in the concentration of glucose and oxygen between the blood and the extracellular space in the body is likely to be the lowest. This is because the peritoneum is a highly vascularized zone, and therefore provides a short distance between any point in the retroperitoneal space and a blood vessel that is rich in glucose and oxygen.

Concerning the biofuel cell, each bioelectrode should be enveloped by a membrane allowing only the diffusion of small molecules such as glucose or oxygen. This membrane may thus retain at the electrode proteins and all nano-objects currently used in the elaboration of biofuel cells such as carbon nanotubes, nanobeads, or graphene layers, which are known to be potentially dangerous. Moreover, a biocompatible bag must wrap the biofuel cell to avoid inflammatory reactions in the body. Polyester materials such as Dacron® are excellent examples of successful synthetic material used within the human body that could be employed for bagging of implanted biofuel cells. Dacron is completely compatible with the body so that rejection and calcification do not occur. In addition, the durability of these materials exceeds that of the human life span.

The lifetime of the biofuel cells is another significant parameter that determines the success of biofuel cells. The literature is replete with information on the lifetime of biosensors based on enzyme electrodes, with examples indicating up to 80% retention of electro-enzymatic activity after 1 year [22]. However, this information generally reflects the stability of proteins in storage or in intermittent operation but not in continuous operation after implantation in a living organism. In addition, the two most commonly used enzymes (laccase and BOD) at the cathode for the reduction of oxygen could be inhibited by urate and chloride ions present in biological fluids. In addition, the majority of enzymatic biofuel cells consuming glucose are based on the direct or indirect electrical wiring of GOx. The efficiency of this connection determines the extent of the side reaction producing H_2O_2. Hydrogen peroxide can denature GOx and hence decrease the lifetime of the biofuel cell. Consequently, the enzyme immobilization method and the design of microenvironment constitute critical steps in the biofuel cell elaboration.

Regarding the performance of the biofuel cells after implantation, the latter are obviously limited by the concentration of substrates, namely, 5 mM and 45 µM for glucose and oxygen, respectively. One possibility to boost the performance of biofuel cells would be to associate several biofuel cells in series or in parallel, as recently demonstrated by the group of Katz [18]. Another possibility is to increase the surface area or volume of these biofuel cells on condition that they do not exhaust the local concentration of glucose and oxygen. Finally, obtaining an effective direct electron transfer (DET) between enzymes and electrodes, particularly at the anode, would optimize the voltage supplied by the biofuel cells and hence their power.

Several criteria have to be considered before implantable biofuel cells can be considered as real alternatives to existing batteries. Besides the effective power output of implanted biofuel cells capable of energizing the designated device, the biocompatibility determining the impact on the patient's life quality with such a device has to be taken into account.

16.2
Design of Implantable Bioelectrodes of Glucose Biofuel Cells

The performance of a biofuel cell is first of all related to the wiring efficiency of enzymes or other catalyst to the conducting substrate or matrix for both, the bioanode and the biocathode. This is directly correlated to the choice of the (bio) catalyst. The three reported examples of implanted biofuel cells inside a rat have quite different approaches even when GOx was used in all cases for the bioanode.

First, the different approaches use carbon-based materials with different degrees of porosity or roughness. In fact, control over porosity of the material is one important parameter to maximize the electroactive area per volume unit of the implanted glucose biofuel cell (GBFC). This allows optimizing the surface concentration of biocatalysts in order to efficiently oxidize or reduce low concentrated substrates, that is, glucose and oxygen. In this matter, an increased degree of porosity was obtained using graphite, carbon micro- to nano-fibers, and 10 nm-diameter carbon nanotubes. The main difference between the few examples of biofuel cells implanted in mammals is the size of the GBFC. While microelectrodes have the advantage to be easily inserted either directly through the skin or to complicate accessible locations in the body such as brain or veins, the major drawback is the low power output of such micro devices. Furthermore, stability issues increase because of the low amounts of catalysts deposited on the microelectrodes. On the contrary, the second approach is the use of macroporous electrodes that have the advantages of delivering high power outputs while using higher amounts of enzymes that can represent advantageous biocatalyst resources for long-term implantation. In this matter, the first implantations in rats were based on graphite electrodes that were surgically implanted in the rat's retroperitoneal space. The volumes of these bioelectrodes were in the order of 130 µl and the amount of enzymes was in the order of tens of milligrams. A second generation of such macroelectrodes has employed a carbon nanotube matrix that considerably increases electrical conductivity, electroactive surface, and electron transfers between enzymes, mediators, and electrode. For comparison, graphite pellets exhibit a specific surface of around $10 \, m^2 \, g^{-1}$ while Brunauer–Emmett–Teller (BET) measurements revealed a specific surface in the order of $200 \, m^2 \, g^{-1}$ for multiwalled carbon nanotube (MWCNT) electrodes. These MWCNT-based pellets were formed by compression of MWCNT and enzyme mixtures, while ensuring excellent electrical conductivity in the order of $3000 \, S \, m^{-1}$ [20]. More importantly, MWCNTs have shown highly efficient DET properties toward many types of redox enzymes such as copper oxidases employed at the biocathode [23]. On the contrary, graphite electrodes always require the use of additional redox mediators to achieve bioelectrocatalysis at both electrodes. This leads to an important increase of the open-circuit voltage of the biofuel cell by replacing graphite with MWCNTs. The graphite-based bioelectrodes could only be able to operate at an OCV of 0.25 V [13] while MWCNT bioelectrodes achieved an OCV of 0.95 V [12]. It is noteworthy that the OCV has to be as high as possible for the integration of a GBFC in a real electronic circuit to supply biomedical devices.

Among the important parameters to maximize the GBFC performances, the nature of the wired enzymes, the choice between mediated and DET, and the nature of the sealing membrane are of particular interest.

At the biocathode, the nature of the copper enzyme involved in the four proton/four electron oxygen reduction determines the onset potential and electrocatalytic current densities. Tyrosinase was used in the first implanted biofuel cells for its ability to operate in physiological conditions [13]. The enzymes were indirectly wired on graphite electrodes by using quinone as a redox partner. The biocathodes exhibited an OCP of 0.15 V, corresponding to the redox potential of ubiquinone, and current densities of $0.22\,\mathrm{mA\,cm^{-2}}$. Using MWCNT electrodes, the possibility of DET between tyrosinase and MWCNT and the ability of this T3 copper enzyme to achieve catalytic oxygen reduction has been demonstrated [23b]. However, despite the fact that tyrosinases have the ability to operate at physiological pH, the electrocatalytic reduction of O_2 is less efficient compared to the direct oxygen reduction achieved by laccases or BOD immobilized on MWCNTs [12a, 24]. In particular, laccases exhibit lower overvoltages, that is, 0.58 V versus SCE, for oxygen reduction and higher catalytic activity [12c]. Therefore, even if the optimal pH for this enzyme is around 5, novel implanted GBFC based on MWCNT matrices with wired laccases are often favored. As shown in Figure 16.1, laccases allow the biocathode to operate at high OCP values of 0.58 V compared to tyrosinases operating at 0.34 V for DET and 0.15 V for mediated electron transfer (MET). The choice of laccase as the biocatalyst at the biocathode allows achieving, *in vitro*, maximum current densities up to $3.25\,\mathrm{mA\,cm^{-2}}$ in oxygen-saturated solutions [12c].

Another important parameter in the design of MWCNT electrodes is the nature of the membrane employed for the separation between the MWCNT matrix and the physiological medium. Since diffusion in highly porous matrix is crucial to maximizing the GBFC performances, different types of membranes for the diffusion of glucose and oxygen were investigated [12c]. Figure 16.1d shows the effect of different membranes on the *in vitro* electrocatalytic current. At the biocathode, nafion has led to higher performances, compared to a polypyrrole/polystyrenesulfonate (PPy/PSS) membrane, toward the diffusion of oxygen in the biocathode. In contrast, nafion lowers the diffusion of glucose at the bioanode where the PPy/PSS membrane resulted in higher performances of the bioanode.

At the bioanode, GOx is still the privileged enzyme employed for glucose oxidation since it has many advantages such as long-term stability, facile availability, and high specificity. Glucose dehydrogenases are also used in other implanted GBFCs, taking advantage of their inactivity toward oxygen. Among the important issues when GOx is used for electrocatalytic glucose oxidation, its electrical wiring is one critical aspect when optimizing the GBFC performances [25]. It has been shown that the cofactor FAD of the GOx is relatively deeply confined in the protein, at approximately 13 Å from the surface of the enzyme. For these reasons, DET between the FAD cofactor and the electrode is difficult to observe [26]. When it is observed, it is difficult to address a true DET from electrochemical experiments. One other issue for the wiring of GOx is the competition between the two ways (oxygen reduction and DET by the electrode) for the regeneration of the prosthetic

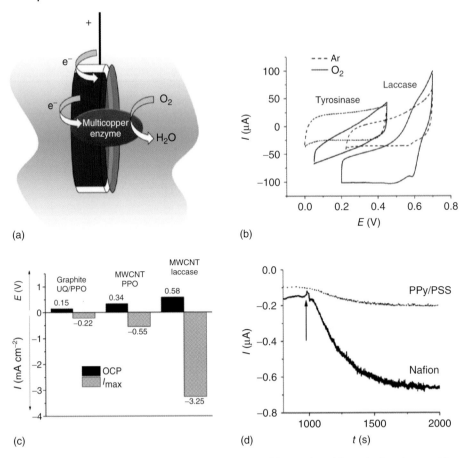

Figure 16.1 (a) Schematic representation of a copper-oxidase-based biocathode; (b) cyclic voltammetry performed for a tyrosinase- and a laccase-based MWCNT biocathode (0.2 mV s^{-1}, oxygen-saturated PBS pH 7); (c) comparison of OCP and I_{max} for tyrosinase- and laccase-based biocathode using graphite or MWCNT matrix; and (d) chronoamperometric measurements comparison between of laccase-based biocathodes sealed with a polypyrrole/polystyrenesulfonate (PPy/PSS) and a nafion membrane.

site of GOx (FADH$_2$) [12b, 27]. To address the oxygen reduction issue and the side production of hydrogen peroxide, catalase is used as an enzymatic partner at the bioanode. Catalase not only removes the enzymatically produced H$_2$O$_2$ but also locally removes oxygen from the bioanode in combination with the reduction of O$_2$ by GOx in the presence of glucose. This local oxygen depletion allows GOx to operate at low overpotentials without any competition with oxygen reduction. OCV values as low as −0.35 V was recorded for oxidation of glucose at the bioanode [12b]. However, inefficient DET between GOx and the electrode was further investigated by comparing with a bioanode working with a redox mediator [12c, 26]. In this case, the redox mediator is able to closely approach the deeply confined FAD cofactor and increase the yield of wired GOx inside the MWCNT matrix. Much higher catalytic activity and stability were observed in this case. Figure 16.2b shows the comparison between the performances obtained at a mediator-less and a naphthoquinone-mediated bioanode. While the OCV is lower in the case of the mediator-less bioanode, a sevenfold increase of catalytic current was obtained by using a redox mediator. Similarly to the biocathode, the sealing membrane has to allow efficient diffusion of glucose. In this case, the PPy/PSS film has shown superior performances over nafion (Figure 16.2a). Figure 16.2c,d shows the nanostructured surface of the MWCNT electrode before and after deposition of a thin layer of a PPy/PSS film.

From the design of each bioelectrode, we developed high-power GBFCs based on MWCNT electrodes. The first reported mediator-less GBFC, which was limited by the DET of the FAD-GOx, delivered a maximum power density of $1.6\,\text{mW}\,\text{ml}^{-1}$.

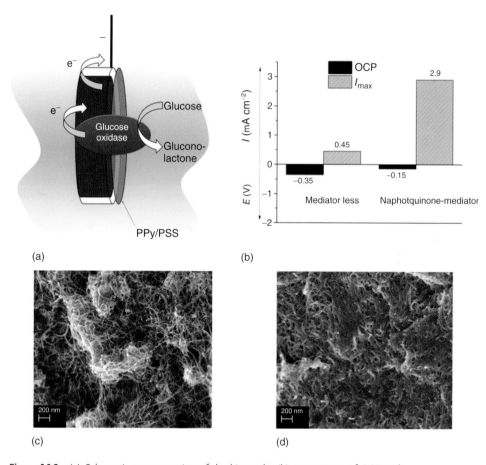

Figure 16.2 (a) Schematic representation of the bioanode; (b) comparison of OCP and I_{max} for a glucose-oxidase-based bioanode; (c) Surface of the MWCNT before and (d) after electrodeposition of a PPy/PSS membrane.

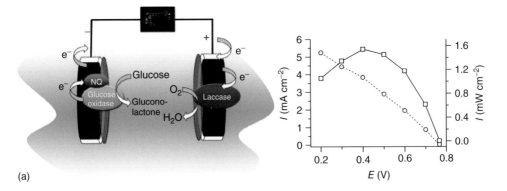

Figure 16.3 (a) Schematic representation of the GBFC and (b) *in vitro* performances of the GBFC in 50 mM glucose under oxygen.

The naphthoquinone-mediated bioanode increases the total power output of the biofuel cell to approximately $2\,mW\,ml^{-1}$ in 50 mM glucose (Figure 16.3). While the bioanode limits the overall GBFC performances in the mediator-less GBFCs, the laccase-based biocathode becomes the limiting electrode in the case of the naphthoquinone-based GBFC. *In vitro* storage stability over months was obtained for the different GBFCS. The naphthoquinone GBFC configuration is also able to constantly deliver hundreds of microwatts under long-term discharge.

16.3
Packaging of Implanted Biofuel Cells

In order to prevent diffusion of the biocatalysts, mediators (if used), or carbon matrix to the physiological liquids and furthermore allow optimal diffusion of glucose and oxygen throughout the bioelectrodes, the packaging and confinement of these bioelectrodes represents a particular challenge. Furthermore, the whole packaging should have minimum size and optimal biocompatibility. The first packaging design for a completely implanted biofuel cell consisted in the sealing of each bioelectrode in a dialysis bag. Both bagged electrodes were then placed in a second, external dialysis bag and finally sutured inside an *exPTFE* coating for biocompatibility, leading to an implant of around 10 ml [13] (Figure 16.4).

Even when this setup showed good appropriateness, improvements were targeted in terms of size of the implant and the diffusion of glucose and oxygen to the electrodes. The second generation packaging consisted in the coating of each bioelectrode with a dialysis membrane and their face-to-face placement in a perforated silicone tube (Figure 16.5). This not only allowed mechanical stabilization of the bioelectrodes but also fixed the distance between these electrodes. Further packaging was similar to the first-generation design except that instead of the exPTFE tissue a Dacron® sleeve was used for improved biocompatibility and wettability, enabling improved diffusion of the body fluids through this membrane. The final

16.3 Packaging of Implanted Biofuel Cells

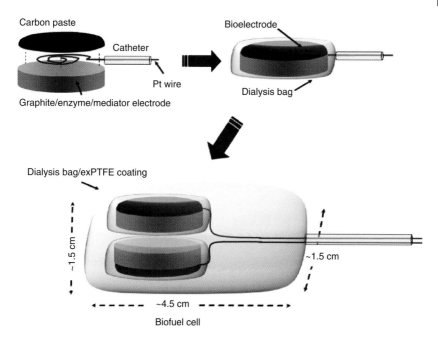

Figure 16.4 Packaging of the first implanted biofuel cell.

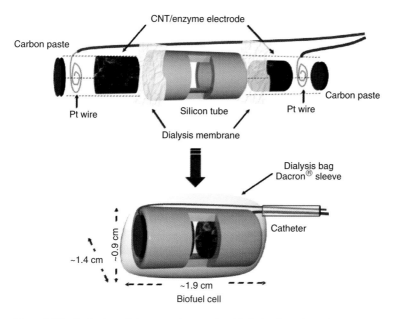

Figure 16.5 Packaging of the second-generation biofuel cell implanted in a rat.

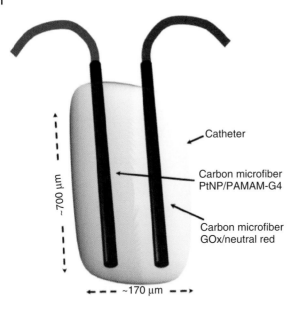

Figure 16.6 Scheme of the two microelectrodes implanted in a vein of a rat.

implant had a volume of 2.4 ml [20]. Both of these implants did not provoke any inflammations even after several months in the rat and a highly vascularized tissue covered these implants.

Using modified carbon-fiber-based microelectrodes, the packaging volume was reduced further because these microelectrodes were inserted into the superior vena cava through the left jugular vein using a polyethylene jugular catheter (Figure 16.6) [21]. Since the experiments were foreseen to be carried out within 24 h, no further precautions concerning inflammation were taken. Moreover, in this case, the use of a platinum-based cathode instead of an enzyme-based electrode has the important limitation to be nonselective toward reduction of oxygen.

16.4
Surgery

Such precautions, however, are necessary when long-term stability of the fuel cells is studied. Furthermore, the position of the implanted biofuel cell in the body is quite important because the present body liquids should contain enough glucose and oxygen. Furthermore, rejection risks for the implant should be minimal at these locations. Another criterion is the life quality of the living being with such an implant, which should not be altered after the surgical implantation. The first implanted biofuel cell was surgically inserted into the retroperitoneal space of a rat, where sufficient glucose and O_2 are present in the extracellular fluid. The cables were then tunneled to the rat's neck to enable the connection to an external circuit

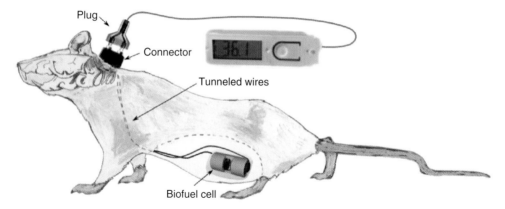

Figure 16.7 Sketch of a rat with an implanted biofuel cell in the retroperitoneal space where the cables were tunneled to the neck and soldered to a connector, which is then fixed to the rat's skull (artwork realized by Judith Cosnier).

for performance measurements. Lots of improvements were made concerning the wire outlet for the second generation of implanted biofuel cells. By keeping the same procedure for the implantation of tunneling of the wires, these wires were finally soldered to a female microconnector and fixed to the rat's skull by acrylic cement, which facilitated the connection to the external circuit (Figure 16.7). This allowed free movement of the rat even during measurements (see video at [28]) and therefore increased drastically the life quality of the rat. Concerning the implantation of a biofuel cell in the vein of a rat, the animal was kept under anesthesia not only during surgery but also during the performance measurements [21].

16.5
Implanted Biofuel Cell Performances

All examples of implanted biofuel cells in mammals have shown clear differences in performances compared to equivalent *in vitro* experiments. This is not surprising because, for example, the oxygen concentration in body liquids ($\sim 50\,\mu M$) is hard to control under laboratory conditions and therefore, such experiments are performed in solutions exposed to air, which leads to an average concentration of dissolved oxygen of around $200\,\mu M$. Furthermore, real physiological fluids such as extracellular fluids or blood are complex electrolytes: (i) diffusion is difficult to control; (ii) concentrations of glucose or oxygen can vary during the course of the experiments; (iii) the presence of inhibitors such as chlorides can decrease GBFC performances; and (iv) inflammatory responses or fouling of the electrode can occur during mid-term experiments. For all these reasons, steady-state values are difficult to obtain during electrochemical measurements. Table 16.1 briefly summarizes the biofuel cell performances *in vitro* and *in vivo*.

The reported maximum power densities of all examples of implanted biofuel cell in mammals are difficult to compare because of the size or volume of the different bioelectrodes and the different types of measurements used to determine the performances. In brief, the power output is only estimated during different periods of time depending on the setup of the experiments. By taking the performances of the first implanted biofuel cell in a rat as reference, clear improvements were obtained within a few years of development where the obtained power densities increased by more than one order of magnitude from 6.5 [13] to 95 [21] and 193.5 µW cm^{-2} [20]. However, these values have to be considered as normalized indicators since only the surface, exposed to the body fluids, is taken into account even when the bioelectrodes have a porous 3D design. For this, volume power densities were provided in [13] and [20] (24.4 and 161 µW ml^{-1}, respectively). Furthermore, following the purpose of later application of such implanted biofuel cells to supply implanted electronic devices, the effective power obtained is therefore to be evaluated.

Modern pacemakers operate at around 10 µW [1] whereas artificial sphincters need 200 µW [4]. Other implantable devices such as neurostimulators (10–100 nW), sensors (100 nW to 10 µW), or drug delivery devices (1–10 mW) can be envisioned to be powered by a biofuel cell whereas artificial organs ($\geq$1 W) represent the greatest challenge.

Considering the actually obtained power from the described implanted biofuel cells and furthermore the delivered power during a certain time period giving the converted energy, the real efficiency of each biofuel cell can be better evaluated. The first biofuel cell based on enzyme-graphite pellets could supply constantly 2 µW during almost 3 h, thereby converting almost 20 mJ of energy. The CNT-enzyme-pellet-based biofuel cell could flash five times an LED within 3–4 min producing around 8 mJ during this time. Continuous discharge during periods longer than 10 min was unfortunately not reported. Nevertheless, this biofuel cell converted 100 times more energy in 3–4 min than the microelectrode-based biofuel cell within 24 h. Even though we can get more insight about which biofuel cell principle is more promising for the power supply of implanted electronic devices, these values do not take into account the advantages of miniaturized microsystems.

Table 16.1 Comparison of biofuel cell performances *in vitro* and *in vivo*.

Electrode type	Physiological liquid	Performance *in vitro*	Performance *in vivo*	References
Microelectrodes	Blood	200 µW cm^{-2} at 0.25 V (OCV: 0.4 V)	100 µW cm^{-2} at 0.08 V (OCV: 125 mV)	[21]
CNT/enzyme electrodes	Extracellular liquid	1.25 mW cm^{-2} at 0.57 V (OCV: 0.95 V)	193.5 µW cm^{-2} at 0.23 V (OCV: 0.57 V)	[20]
Graphite/mediator/ enzyme electrodes	Extracellular liquid	48 µW cm^{-2} at 0.2 V (OCV: 0.32 V)	6.5 µW cm^{-2} at 0.13 V (OCV: 0.27 V)	[29]

Table 16.2 Performances of implanted glucose biofuels inside a rat.

Bioelectrodes	GBFC location	Surface power density ($\mu W\, cm^{-2}$)	Electrode surface (cm^2)	Measured power (μW)	Period of measurement (time) (discharge power)	Generated energy (mJ)	References
Graphite/mediator/ enzyme pellets	Retroperitoneal space	5	1.33	6.5	2.75 h (2 μW)	19.8	[13]
CNT/enzyme pellets	Retroperitoneal space	193.5	0.19	38.7	3–4 min (38.7 μW)	~8.1	[20]
Carbon microfibers/ mediator/GOx and PtNP	Rat jugular vein	~100	1×10^{-4}	~0.001	24 h (1 nW)	0.08	[21]

Table 16.2 summarizes the performances of the up-to-date reported biofuel cells generating power out of a mammal's body liquids (extracellular liquid or blood).

From this table, it can be seen that principally some medical devices can already be supplied by an implanted biofuel cell. However, in terms of life time, even the most successful biofuel cells are still not competitive with conventional lithium ion batteries. In this context, lots of efforts are still to be invested not only in biofuel cell design, performance, and life time but also in new technologies in terms of power consumption and management [30].

References

1. Ohm, O.-J. and Danilovic, D. (1997) *Pacing Clin. Electrophysiol.*, **20**, 2–9.
2. Dowling, R.D., Gray, L.A. Jr., Etoch, S.W., Laks, H., Marelli, D., Samuels, L., Entwistle, J., Couper, G., Vlahakes, G.J., and Frazier, O.H. (2003) *Ann. Thoracic Surg.*, **75**, S93–S99.
3. Kim, S.P., Sarmast, Z., Daignault, S., Faerber, G.J., McGuire, E.J., and Latini, J.M. (2008) *J. Urol.*, **179**, 1912–1916.
4. Lamraoui, H., Mozer, P., Robain, G., Bonvilain, A., Basrour, S., Gumery, P.Y., and Cinquin, P. (2008) Design and implementation of an automated artificial urinary sphincter. Presented at pHealth2008.
5. Davenport, A., Gura, V., Ronco, C., Beizai, M., Ezon, C., and Rambod, E. (2007) *Lancet*, **370**, 2005–2010.
6. (a) Roundy, S. (2005) *J. Intell. Mater. Syst. Struct.*, **16**, 809–823. (b) Woias, P. (2005) *VDE Mikrosystemtechnik-Kongress 2005*, VDE-Verlag, Freiburg, pp. 405–410. (c) Görge, G., Kirstein, M., and Erbel, R. (2001) *Herz*, **26**, 64–68. (d) Watkins, C., Shen, B., and Venkatasubramanian, R. (2005) Low-grade-heat energy harvesting using superlattice thermoelectrics for applications in implantable medical devices and sensors. Presented at 24th International Conference on Thermoelectrics, ICT 2005, June 19–23, 2005.
7. Kang, J.Y. (2006) *Encyclopedia of Medical Devices and Instrumentation*, Vol. p, John Wiley & Sons, Inc., Hoboken.
8. Kerzenmacher, S., Ducrée, J., Zengerle, R., and von Stetten, F. (2008) *J. Power. Sources*, **182**, 1–17.

9. Drake, R.F., Kusserow, B.K., Messinger, S., and Matsuda, S. (1970) *Trans. Am. Soc. Artif. Int. Organs*, **16**, 199–205.
10. (a) Kloke, A., Biller, B., Kräling, U., Kerzenmacher, S., Zengerle, R., and von Stetten, F. (2011) *Fuel Cells*, **11**, 316–326. (b) Kerzenmacher, S., Kräling, U., Metz, T., Zengerle, R., and von Stetten, F. (2011) *J. Power. Sources*, **196**, 1264–1272. (c) Oncescu, V. and Erickson, D. (2013) *Sci. Rep.*, **3**, 1226.
11. (a) Schröder, U. (2012) *Angew. Chem. Int. Ed.*, **51**, 7370–7372. (b) Meredith, M.T. and Minteer, S.D. (2012) *Annu. Rev. Anal. Chem.*, **5**, 157–179.
12. (a) Gao, F., Viry, L., Maugey, M., Poulin, P., and Mano, N. (2010) *Nat. Commun.*, **1**, 2. (b) Zebda, A., Gondran, C., Le Goff, A., Holzinger, M., Cinquin, P., and Cosnier, S. (2011) *Nat. Commun.*, **2**, 370. Reuillard, B., Le Goff, A., Agnès, C., Holzinger, M., Zebda, A., Gondran, C., Elouarzaki, K., and Cosnier, S. (2013) *Phys. Chem. Chem. Phys.*, **15**, 4892–4896.
13. Cinquin, P., Gondran, C., Giroud, F., Mazabrard, S., Pellissier, A., Boucher, F., Alcaraz, J.-P., Gorgy, K., Lenouvel, F., Mathé, S., Porcu, P., and Cosnier, S. (2010) *PLoS ONE*, **5**, e10476.
14. Miyake, T., Haneda, K., Nagai, N., Yatagawa, Y., Onami, H., Yoshino, S., Abe, T., and Nishizawa, M. (2011) *Energy Environ. Sci.*, **4**, 5008–5012.
15. Rasmussen, M., Ritzmann, R.E., Lee, I., Pollack, A.J., and Scherson, D. (2012) *J. Am. Chem. Soc.*, **134**, 1458–1460.
16. Szczupak, A., Halámek, J., Halámková, L., Bocharova, V., Alfonta, L., and Katz, E. (2012) *Energy Environ. Sci.*, **5**, 8891–8895.
17. Halámková, L., Halámek, J., Bocharova, V., Szczupak, A., Alfonta, L., and Katz, E. (2012) *J. Am. Chem. Soc.*, **134**, 5040–5043.
18. MacVittie, K., Halamek, J., Halamkova, L., Southcott, M., Jemison, W.D., Lobel, R., and Katz, E. (2013) *Energy Environ. Sci.*, **6**, 81–86.
19. Castorena-Gonzalez, J.A., Foote, C., MacVittie, K., Halámek, J., Halámková, L., Martinez-Lemus, L.A., and Katz, E. (2013) *Electroanalysis*, **25**, 1579–1584.
20. Zebda, A., Cosnier, S., Alcaraz, J.-P., Holzinger, M., Le Goff, A., Gondran, C., Boucher, F., Giroud, F., Gorgy, K., Lamraoui, H., and Cinquin, P. (2013) *Sci. Rep.*, **3**, 1516.
21. Sales, F.C.P.F., Iost, R.M., Martins, M.V.A., Almeida, M.C., and Crespilho, F.N. (2013) *Lab Chip*, **13**, 468–474.
22. (a) Cosnier, S., Fombon, J.J., Labbé, P., and Limosin, D. (1999) *Sens. Actuators B*, **59**, 134–139. (b) Minteer, S.D., Liaw, B.Y., and Cooney, M.J. (2007) *Curr. Opin. Biotechnol.*, **18**, 228–234.
23. (a) Bourourou, M., Elouarzaki, K., Lalaoui, N., Agnès, C., Le Goff, A., Holzinger, M., Maaref, A., and Cosnier, S. (2013) *Chem. Eur. J.*, **19**, 6718–6723. (b) Reuillard, B., Le Goff, A., Agnès, C., Zebda, A., Holzinger, M., and Cosnier, S. (2012) *Electrochem. Commun.*, **20**, 19–22.
24. Little, S.J., Ralph, S.F., Mano, N., Chen, J., and Wallace, G.G. (2011) *Chem. Commun.*, **47**, 8886–8888.
25. Kavanagh, P. and Leech, D. (2013) *Phys. Chem. Chem. Phys.*, **15**, 4859–4869.
26. Goran, J.M., Mantilla, S.M., and Stevenson, K.J. (2013) *Anal. Chem.*, **85**, 1571–1581.
27. Vaze, A., Hussain, N., Tang, C., Leech, D., and Rusling, J. (2009) *Electrochem. Commun.*, **11**, 2004–2007.
28. http://www.nature.com/srep/2013/130322/srep01516/extref/srep01516-s1.avi.
29. Giroud, F., Gondran, C., Gorgy, K., Vivier, V., and Cosnier, S. (2012) *Electrochim. Acta*, **85**, 278–282.
30. Southcott, M., MacVittie, K., Halamek, J., Halamkova, L., Jemison, W.D., Lobel, R., and Katz, E. (2013) *Phys. Chem. Chem. Phys.*, **15**, 6278–6283.

17
Implanted Biofuel Cells Operating *In Vivo*
Evgeny Katz

17.1
Implanted Biofuel Cells

Harvesting electrical power from a human body [1–6] or from animals [7, 8] for various microelectronic, biomedical, and biosensor applications is a challenging aim that can be achieved using different physical or chemical methods [9–15]. Power generated *in situ* from biological sources can increase the autonomy of various biomedical implanted devices (e.g., cardiac pacemakers, cardioverter-defibrillators, cochlear implants, etc.) [2], prosthetic devices [16, 17], and implantable biosensors [18], allowing their operation without implanted batteries. Energy harvested from living species can be used far beyond medical applications, being useful for powering various portable microelectronic, computerized, and sensor devices utilized in communication and transportation systems, environmental, animal, and homeland security monitoring, and so on. Physical methods of the energy harvesting from living species can be based on transducers utilizing mechanical energy [5], such as muscle stretching [10], arm/leg swings [2], walking/running [1, 2, 9], heart beats [19, 20], blood flow [21], gas flow upon respiration [2, 11, 20], and so on. Different thermoelectric [2], piezoelectric [2], and electrical potential gradient effects [22] can be also used for the energy harvesting from a living body. It should be noted, however, that all physical energy conversion methods are based on complex machinery and represent an engineering rather than a biological approach. They are highly dependent on the human/animal physical activity and the environment conditions. For example, power generation by physical motion (e.g., walking/running) can be interrupted at sleeping time and methods based on the temperature gradient between the body and air cannot work at high air temperature. On the other hand, bioelectrochemical methods of the energy conversion based on biofuel cells [23] powered by organic "fuel" and producing electrical energy through biological catalysis are much more natural and bio-inspired. They are based on enzymatic chemical conversion processes that are responsible for power generation in living systems. Obviously, biochemical processes are active even without visible physical activity, for example, when the living species are sleeping and not moving. One of the major biological

Implantable Bioelectronics, First Edition. Edited by Evgeny Katz.
© 2014 Wiley-VCH Verlag GmbH & Co. KGaA. Published 2014 by Wiley-VCH Verlag GmbH & Co. KGaA.

power sources–glucose–is always available at almost constant levels. Therefore, when biological species are considered for power generation, one should look at biomolecular processes rather than creating artificial machinery, which might be good for robotic applications but not natural in the biological systems.

Biofuel cells [23] that extract electrical energy by oxidizing biomolecules have been developed on the basis of the biocatalytic activity of microbial cells [24–27], enzymes [28–36], or biological organelles (e.g., mitochondria) [37]. While microbial biofuel cells are usually constructed as large-scale biological reactors generating substantial electrical power (for example, for a household or wastewater treatment) [27, 38, 39], enzyme biofuel cells are mostly considered as micro [40–42] or even nano power [12] sources for implantable biomedical devices. Despite the fact that implantable biofuel cells operating *in vivo* have been suggested a long time ago [43], they are still exotic and challenging to design. While *potentially implantable* biofuel cells have been discussed, in reality most of the experiments were performed in model solutions [44–46] or at best in human serum or blood *in vitro* [47, 48]. Significant breakthrough has been achieved in this area in the last 2 years with several papers reporting on implanted biofuel cells that operate *in vivo* in living animals: insects [49], mollusks (snail (Figure 17.1a) and clams (Figure 17.1b)) [50, 51], lobsters (Figure 17.1c) [52] , rats (Figure 17.2) [53–57], and rabbits [58]. The implanted biofuel cells utilized physiologically produced "fuel" and oxygen dissolved in biofluids for biocatalytic generation of electrical power by catalyst-modified electrodes. The "fuel" was usually represented by glucose in freely circulating hemolymph of invertebrates (snails, clams, and lobsters) [50–52] and in blood vessels of vertebrate animals (rats and rabbits) [53–58]. In the case of insects another physiologically produced sugar, trehalose, was utilized as the "fuel" for extracting electrons [49]. Biocatalytic anodes modified with special enzymes for oxidizing trehalose have been developed to allow operation of biofuel

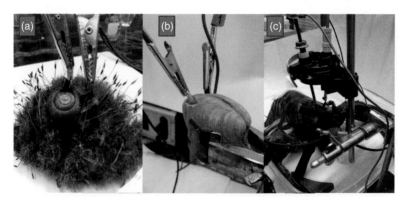

Figure 17.1 Implanted biofuel cells for operating *in vivo* in invertebrates: (a) snail (*Neohelix albolabris*) [50], (b) clam (*Mercenaria mercenaria*) [51], and (c) lobster (*Homarus americanus*) [52]. (Part "a" is adapted from Ref. [50] with permission, copyright American Chemical Society, 2012; parts "b" and "c" are adapted from Refs. [51, 52], reproduced with permission of the Royal Society of Chemistry.)

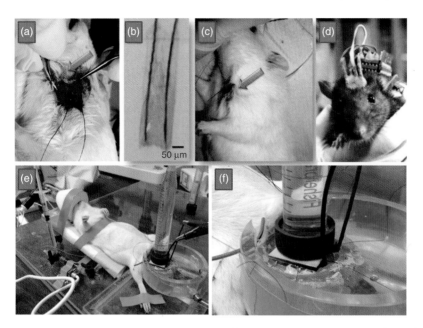

Figure 17.2 Implanted biofuel cells operating *in vivo* in rats: (a) a photograph of the catheter implanted into the jugular vein of rat (*Rattus norvegicus*) for use as a glucose/O_2 biofuel cell [54]; (b) an optical microscope image of the enzyme-modified electrodes inside the catheter [54]; (c) the catheter was surgically introduced at the ventral surface of the living rat; [54] (d) a rat (Wistar) with a biofuel cell implanted in the abdominal cavity and the electrical outlet on the head connected to a voltage boost converter for increasing the output voltage [56], (e) a rat (Sprague-Dawley) with the cremaster tissue surgically exposed to enzyme-modified biocatalytic electrodes [57]; and (f) the cremaster tissue exposed to the biocatalytic electrodes (close-up view) [57]. (Parts "a–c" are adapted from Ref. [54] with permission of the Royal Society of Chemistry; part "d" is adapted from Impact Lab (http://www.impactlab.net/2010/09/17/new-biofuel-cell-powered-by-rats-body-fluids/)with the permission, corresponding Ref. [56]; parts "e–f" are adapted from Ref. [57], with permission.)

cells implanted in insects [49, 59]. The catalytic oxidation of the biofuel (in most of the cases glucose) can be achieved using inorganic catalytic species in the so-called "abiotic" implantable fuel cells [60, 61]. The advantages of abiotic catalytic electrodes might be in their simple sterilization and extended operational lifetime due to higher than enzyme–electrode stability. This research sub-direction can result in relatively simple technological advances, and the developed abiotic fuel cells could be combined with other implantable energy-harvesting devices resulting in hybrid systems. However, enzyme-modified electrodes represent the mainstream research direction because of better biocompatibility and in general because they represent a bio-inspired approach.

Although simple methods for electrical contacting of redox enzymes, including the use of non-immobilized cofactors and mediators, are possible in biofuel cells operating *in vitro* [22, 23], the implanted electrodes should exclude any non-immobilized species for their operation *in vivo*. This can be achieved by co-immobilization of enzymes and electron-transporting species, for example, by entrapment of enzymes in redox polymers shuttling electrons between the enzyme active centers and conducting supports [41]. Direct non-mediated electron transport would be an advantage for constructing biofuel cells: [62] it can exclude potentially bio-incompatible redox mediators and result in higher voltage generated by a biofuel cell because the interfacial electron transfer would proceed directly at the enzyme active center without intermediate steps, which always result in a voltage decrease. However, direct non-mediated electrical "wiring" of enzymes on electrodes is always a challenging problem [63]. The direct electron transfer between enzyme active centers and electrode conducting supports is usually much less efficient than mediated electron transport unless very sophisticated multimolecular ensembles are architectured on electrode surfaces [64, 65]. The methods based on these ensembles require very complex steps in the modified surface preparation and, although they are interesting for basic research in biomolecular electrochemistry, they are not practically applicable in real biofuel cells. In order to achieve simple-in-preparation and robust-in-operation bioelectrocatalytic electrodes providing efficient non-mediated electrical "wiring" for immobilized enzymes, nano-structured buckypaper composed of compressed multiwalled carbon nanotubes (CNTs) has been suggested recently [66–70]. In general, CNTs in various configurations and particularly the buckypaper can be used for different mediated and non-mediated electrical wiring of enzymes; however, the non-mediated mode is particularly attractive for *in vivo* applications [70].

Selection of enzymes associated with the biocatalytic electrodes, including those made of CNTs, is a very critical issue for the implantable biofuel cell. The cathodic reaction of oxygen reduction is usually biocatalyzed by laccase (E.C. 1.10.3.2) [50–52, 55–57] or bilirubin oxidase (E.C. 1.3.3.5) [49, 58]; however, some other enzymes, for example, polyphenol oxidase (E.C. 1.14.18.1) [53], cytochrome oxidase (E.C. 1.9.3.1) [23], or inorganic catalysts, such as Pt-nanoparticles [54], are also applicable. The most important feature of the catalytic reduction of oxygen is the formation of water as the product upon a four-electron transfer process biocatalyzed by enzymes on the cathode. Although in biofuel cells operating *in vitro* the formation of hydrogen peroxide upon oxygen reduction is possible, or even hydrogen peroxide can be used as an electron acceptor instead of oxygen [23], in the implanted biofuel cells operating *in vivo* hydrogen peroxide must be excluded as the toxic bioincompatible substance. The oxygen-reducing enzymes might operate at different optimal pH values, but the most important difference of these enzymes is the potential generated on the cathode. For example, cytochrome oxidase operating in the complex with cytochrome c can generate a cathode potential of about 0 V (vs Ag/AgCl reference electrode), which is limited by the heme potential in cytochrome c mediating the electron transfer [71]. The use of laccase or bilirubin oxidase can significantly shift the oxygen reduction process to positive potentials, reaching about 0.6 V [50], thus increasing the voltage generated by the biofuel cells. Depending on the used oxygen reducing enzyme and the material selected for the conducting support, the electron transfer can proceed with [49, 53] or without [50–52, 55–58] redox mediators facilitating the electron transport. However, once the electron transfer between the enzyme and the conducting support is achieved, the cathodic reaction proceeds in almost the same way for all oxygen-reducing enzymes.

Enzyme selection for the anodic reaction is trickier. The most frequently used enzyme for glucose oxidation is glucose oxidase (GOx, E.C. 1.1.3.4) [72], which is used in most of glucose biosensors [73] and glucose-powered biofuel cells [28]. Therefore, it is not surprising that GOx was also used in the implanted biofuel cells operating *in vivo* [53, 54], including those based on CNTs as the conducting support (Figure 17.3a) [56]. However, it should be remembered that O_2-dependent oxidases (particularly GOx) [72] generate H_2O_2 in the presence of oxygen, which is toxic when it is produced in the implanted biofuel cell. Also, the enzyme reaction with oxygen competes with the electron transfer to the electrode, thus inhibiting current generation. Decomposition of H_2O_2 using catalase can eliminate toxic material from the biological environment [53], but it will not improve the current yield on the anode. Indeed, the electrons transported from GOx to oxygen will be lost for the current generation. Usually GOx-bioelectrocatalytic electrodes operate *in vitro* under anaerobic conditions when O_2 is removed from solutions by bubbling with Ar or N_2 gas. When these electrodes are used as anodes in biofuel cells, they are usually separated from the oxygen-containing cathodic solution with a membrane

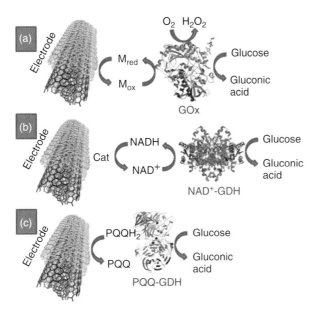

Figure 17.3 Glucose oxidation by different enzymes associated with CNTs as the conducting support: (a) glucose oxidase (GOx) operating in the presence of O_2 with the help of an electron transfer mediator; (b) NAD^+-dependent glucose dehydrogenase (GDH) operating in the presence of the soluble NAD^+ cofactor and a catalyst facilitating oxidation of NADH on the electrode surface; and (c) PQQ-dependent glucose dehydrogenase (PQQ-GDH) operating with the direct non-mediated electron transfer to CNTs.

[28]. Eventually, a GOx-based biocatalytic anode operating in the presence of oxygen is possible, but it requires a very sophisticated multimolecular ensemble to achieve kinetically preferential electron transfer to the electrode instead of oxygen, thus allowing a membrane-less biofuel cell [71]. Simpler GOx/mediator assemblies efficiently operating in oxygen-containing solutions and used in membrane-less biofuel cells have also been reported based on redox polymers with high density of redox centers allowing fast electron transport from the flavin adenine dinucleotide (FAD)-center of GOx to the conducting support and effectively competing with the electron transfer to oxygen [74]. Actually, limited amount of O_2 in biofluids and its slow transport because of high viscosity of the biofluids makes easier current generation by the GOx-anode operating *in vivo*.

Another choice of the glucose-oxidizing enzymes is NAD^+-dependent glucose dehydrogenase (GDH; E.C. 1.1.1.47) (Figure 17.3b). However, this enzyme, as well as all other NAD^+-dependent dehydrogenases, requires NAD^+ cofactor in a solution [66], that is, not permitted in a membrane-less implantable biofuel cell. Co-immobilization of NAD^+ cofactors on electrode surfaces always results in highly sophisticated procedures [75], and thus cannot be a good solution for robust implantable biofuel cells. In order to overcome this problem, the implanted anode functionalized with NAD^+-dependent glucose dehydrogenase and operated in the presence of soluble NAD^+ was separated from the biological environment with a membrane [55, 58]. It should be also noted that electrochemical oxidation of the biocatalytically produced NADH requires special catalysts (e.g., methylene green) associated with the anode surface [55] (Figure 17.3b). Therefore, the potential generated on the anode is controlled by the redox potential of the catalyst, which is significantly more positive than the thermodynamic potential of $NADH/NAD^+$. This decreases the voltage output of the biofuel cell. On the other hand, NAD^+-dependent enzymes can be organized in a multistep/multienzyme catalytic system for deep oxidation of the biofuel [76, 77]. For example, three NAD^+-dependent dehydrogenase enzymes, alcohol dehydrogenase, aldehyde dehydrogenase, and formate dehydrogenase, combined in a biocatalytic cascade result in complete oxidation of methanol to carbon dioxide [76]. Similar processes can be realized in implantable biofuel cells where the oxidation reactions mimic metabolic pathways [77]. This can significantly increase the current and power generated by a biofuel cell [77]. Therefore, by compromising on the voltage and allowing a membrane separating the anodic solution from the biofluids, one can increase the number of electrons extracted from the biofuel, thus increasing the current drawn from the cell.

In order to exclude the problems associated with the NAD^+- and O_2-dependent enzymes, pyrroloquinoline quinone (PQQ)-dependent GDH (PQQ-GDH; E.C. 1.1.5.2) has been selected for the biocatalytic oxidation of glucose at the implanted anode [50–52, 57], Figure 17.3c. This enzyme does not require soluble cofactors and its reaction does not interfere with oxygen. However, this enzyme is much less studied for biofuel cell applications and its direct electrical communication with the CNT-functionalized electrodes was demonstrated only recently [30]. In the implanted biofuel cells the PQQ-GDH on the anode and laccase on the cathode were linked to the CNTs using a heterobifunctional cross-linker, 1-pyrenebutanoic acid succinimidyl ester (PBSE), which provides covalent binding with amino groups of protein lysine residues through formation of amide bonds and interacts with CNTs via π–π stacking of polyaromatic pyrenyl moiety [50–52, 57] (Figure 17.4a). It should be noted that this linker results in random orientation of the enzyme molecules on the conducting support because of the large number of amino groups differently positioned in the protein structure [78]. When this immobilization approach is used on a flat electrode surface, it does not result in the direct non-mediated wiring for the majority of the bound enzyme molecules because most of the enzyme active centers are not in proximity to the conducting support [78]. It was hoped that in case of the 3D-nano-structured buckypaper the enzyme immobilization with the random orientation would allow direct electron transfer for many active centers that can find a nearby conducting CNT. Actually, this statement was overoptimistic since only about 6% of the immobilized enzymes were found to be electrochemically active in this electrode configuration. Indeed, the distances separating the CNTs in the buckypaper are much longer than the size of the enzyme molecules (Figure 17.4b); thus, only a fraction of the immobilized enzyme molecules can be in the position favorable for the direct electron transfer between the active centers and CNTs [50].

Upon recording cyclic voltammograms for the PQQ-GDH-electrode, the anodic current corresponding to the bioelectrocatalytic glucose oxidation was developed at the potentials more positive than −0.1 V (vs Ag/AgCl) (Figure 17.5a). The cyclic voltammograms obtained for the laccase-modified electrode demonstrated the cathodic bioelectrocatalytic current corresponding to the oxygen reduction starting at 0.6 V [50] (Figure 17.5b), thus allowing about 700 mV potential difference between the anodic and cathodic reactions. The biocatalytic electrodes were implanted in a

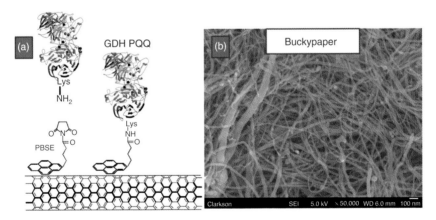

Figure 17.4 (a) Immobilization of the PQQ-GDH on CNTs with the help of the heterobifunctional linker 1-pyrenebutanoic acid succinimidyl ester (PBSE), which provides covalent binding with amino groups of protein lysine residues through formation of amide bonds and interacts with CNTs via π–π stacking of polyaromatic pyrenyl moiety. (b) SEM image of the buckypaper used as the conducting support for the enzyme-modified electrodes.

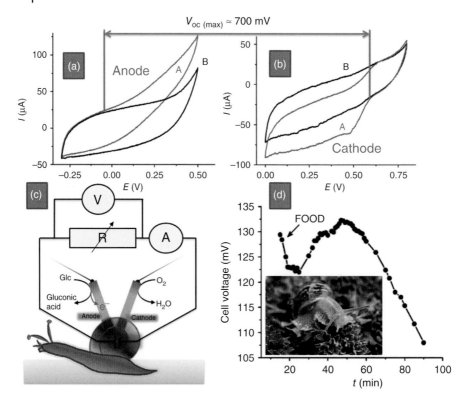

Figure 17.5 A biofuel cell implanted in a snail: [50] (a) cyclic voltammograms of the PQQ-GDH-CNTs-anode: curves A and B were obtained in the presence and absence of 20 mM glucose, respectively. (b) Cyclic voltammograms of the laccase-CNTs-cathode: curves A and B were obtained in the presence and absence of O_2, respectively. All cyclic voltammograms were obtained *in vitro* in a buffer solution, pH 7.4, scan rate $1\,mV\,s^{-1}$. (c) The electrical circuitry including a biofuel cell implanted in a snail and connected to a variable load resistance. (d) Variation of the voltage produced by the biofuel cell *in vivo* in real time upon feeding the snail. (Parts "a–c" are adapted from Ref. [50] with permission, copyright American Chemical Society, 2012.)

snail (*Neohelix albolabris*) (Figure 17.5c) [50], and current–voltage characteristics of the biofuel cell were obtained for the variable load resistance. The current produced by the implanted biocatalytic electrodes was rapidly decreasing because of the slow glucose diffusion through the viscous biological medium; however, it was restored when the electrodes were disconnected and glucose diffused to their surfaces. The most impressive was the observation that the electrical output produced by the implanted biofuel cell increases upon feeding the snail, thus confirming that it is indeed proportional to the physiological concentration of glucose in hemolymph (snail's analog of blood) (Figure 17.5d).

Keeping in mind that implanted biofuel cells are aiming at the powering of electronic (particularly biomedical) devices, one can realize that there is mismatch between the voltage produced by biofuel cells, which is usually about 0.5 V being thermodynamically limited by the redox potentials of the biological fuel (usually glucose) and oxygen, and the voltage required by electronic systems, which is, for most of devices, about 2–3 V. It should be noted that modern microelectronic devices may require very small power for their operation; however, the input voltage is still in the range of several volts [79]. Despite the fact that many researchers demonstrated power release from biofuel cells, which may be potentially enough for activating microelectronic devices, real interfacing of electronics with implanted biofuel cells was demonstrated only in a few recently published papers [51, 52, 56]. It should be noted that most of the papers on biofuel cells do not discuss the problem of their interfacing with electronics, while being concentrated on resolving internal problems of the biofuel cells, such as current efficiency, stability, and so on [22–37]. Two approaches have been applied to resolve the problem of the low voltage produced by the biofuel cells: (i) assembling biofuel cells in series electrically, thus increasing the total output voltage [51, 80–82] and (ii) collecting produced electrical energy in capacitors/charge pumps for the burst release in short pulses [56, 83–86].

The first approach can be illustrated by the biofuel cells implanted in living lobsters (*Homarus americanus*) [52]. A single pair of the biocatalytic electrodes (PQQ-GDH-anode and laccase-cathode on the buckypaper conductive support) were implanted in a lobster tissue, being in the contact with hemolymph containing glucose and oxygen. The biofuel cell was connected to a variable load resistance allowing the voltage and current measurements (Figure 17.6a). In a typical experiment, open-circuit voltage, V_{oc}, and short circuit current, I_{sc}, achieved in the biofuel cell *in vivo* were about 550 mV and about 1 mA, respectively. The maximum power, P_{max}, produced in the typical experiment by the implanted biofuel cell at the optimum resistance of 500 Ω was about 0.16 mW. While the generated power might be enough for activating some low-power microelectronic devices, the voltage output was obviously below the threshold required by electronics. In order to increase the voltage two pairs of the biocatalytic electrodes were implanted in a lobster's tissue and connected in series (Figure 17.6b). The result was quite disappointing – the series connection of two biofuel cells showed only a minor (50–100 mV) increase in the produced voltage, which is much less than the expected double voltage. This negative experimental result was easily explained by the low electrical resistance of the lobster's body tissue, R_T, (the resistance between the electrodes implanted in the lobster's back at a distance of 2 cm was about 180 kΩ), which formed a low impedance path (jumper) between the anodes and cathodes (Figure 17.6c), thus preventing the desired series operation of the two fuel cells. To resolve the problem, two biofuel cells were implanted in two different lobsters and then connected in series (Figure 17.6d). In this configuration, the voltage output was doubled, as expected, and enough for activating an electrical watch used as an example electronic load with the minimum required voltage of about 1 V (Figure 17.7). While connecting two live lobsters with the implanted biofuel cell is still possible, further increase of the voltage by adding more lobsters to the circuit is impractical. Therefore, the next set of experiments moved to the *in vitro* model system. Again, as it was demonstrated for two biofuel cells implanted in the same lobster, two pairs of

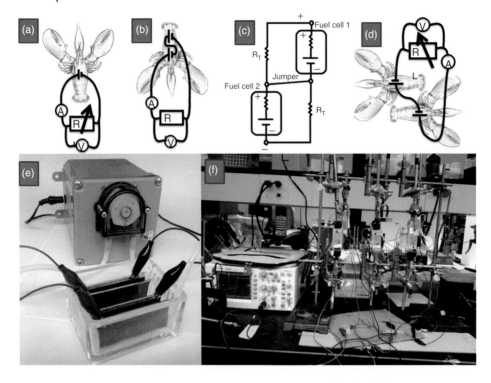

Figure 17.6 (a–d) The cartoons showing schematically different wirings of the biocatalytic electrodes implanted in lobsters: [52] (a) a single pair of the biocatalytic cathode–anode implanted in the lobster; (b) two pairs of the biocatalytic cathodes–anodes implanted into *the same* lobster and connected in series; (c) the electrical circuitry equivalent to the wiring scheme shown in (b); and (d) two pairs of the biocatalytic cathodes–anodes implanted into two *different* lobsters and connected in series. (e) A flow-biofuel cell with two pairs of the biocatalytic electrodes immersed in the same solution and connected in series. (f) The setup composed of five *separate* flow-biofuel cells used for powering the pacemaker (three biofuel cells are well visible in the front row, while two other biofuel cells are only partially visible in the back row) [52]. (Parts "a–d" and "f" are adapted from Ref. [52], reproduced with permission of the Royal Society of Chemistry.)

biocatalytic electrodes immersed in the same solution, do not allow increasing the output voltage when the electrodes are connected in series (Figure 17.6e). Therefore, five *separate* flow biofuel cells filled with human serum solution mimicking human blood circulatory system were connected in series (Figure 17.6f), resulting in voltage output increase up to about 3 V, which was already enough for activating most of microelectronic devices. Note that the biofuel cells were connected to the *separate* flow pathways in order to prevent electrical shortcut connections between the electrodes. In order to illustrate the interfacing of the biofuel flow-cells with an implantable biomedical device, a pacemaker [87] (Affinity DR 5330L, St. Jude Medical) was selected as an example [52]. A sealed pacemaker was cut, opened,

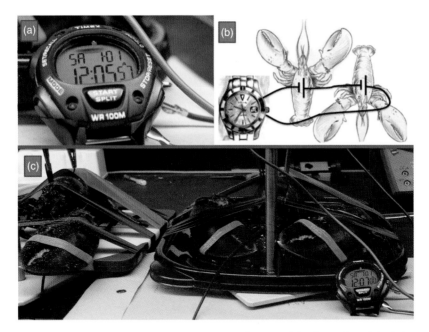

Figure 17.7 The biofuel cells composed of two pairs of the biocatalytic cathodes–anodes implanted in two lobsters wired in series and used for powering an electronic watch [52]. (a) The operating watch powered by the implanted biofuel cells; (b) the wiring scheme; and (c) the photo of the setup. (The figure is adapted from Ref. [52], reproduced with permission of the Royal Society of Chemistry.)

and the internal battery was removed, allowing for wiring of the pacemaker to the external biofuel cell (Figure 17.8A). When the pacemaker was connected to the biofuel flow-battery (Figure 17.6f) it started to generate pulses as expected. The pulses being registered by the oscilloscope demonstrated voltage spikes of about 5 V with the duration of about 0.6 ms separated by time-gaps of about 1 s, which correspond to the normal performance of the pacemaker (Figure 17.8B) [88]. It is particularly important that the shape of the generated pulses was characteristic of the normal behavior of the pacemaker [88]. The present result, being the very first for activating a pacemaker by a biofuel cell mimicking human blood circulation system, is still very disappointing and impractical – the power was extracted from five *independently* working biofuel cells connected to the *separate* fluidic channels. In other words, if this approach is translated to medical applications, five people with implanted biofuel cell, all together connected with wires, can activate only one implanted pacemaker, which is absolute nonsense. Note that five biofuel cells cannot increase the voltage if they are implanted in the same body, as it was shown for lobsters and flow cells. Obviously, this approach cannot be practically useful.

The second approach based on electronic interface devices such as charge pumps and other forms of DC–DC converters has already been applied for activating a

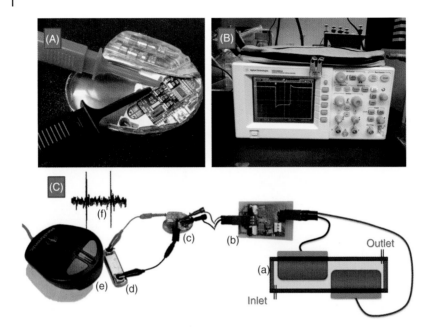

Figure 17.8 (A) Open pacemaker – a close view showing the microscheme and wiring leads connected to the external power source. Note the empty space (left part of the device) from which the original battery was removed. (B) Oscilloscope measuring the electrical pulses produced by the pacemaker activated by the biofuel cell. (C) Experimental setup which includes: (a) the flow biofuel cell with the inlet/outlet connected to a peristaltic pump (not shown in the scheme), (b) the charge pump – DC–DC interface circuit, (c) Affinity DR 5330L, St. Jude Medical, pacemaker, (d) Medtronic Reveal XT, Model 9529, implantable loop recorder (ILR), (e) sensor device for the Medtronic CareLink Programmer, Model 2090, and (f) registered pulses generated by the pacemaker when it is powered by the biofuel cell. (Part "c" is adapted from Ref. [89], reproduced with permission of the Royal Society of Chemistry.)

wireless transmitting electronic device, however, using non-implantable enzyme-based [86] or microbial biofuel cells [83]. Application of an interface device to increase the voltage is rather well known [90]; however, it should be remembered that the voltage increase is achieved at the expense of the current consumed by the charge pump, thus placing additional demand on the current output of the biofuel cell. Implantable microsize electrical energy generators connected to an electrical interface can be effectively used for activating microelectronic devices operating in the short-pulses regime, while the time between pulses is used for the accumulation of energy [20]. However, an implantable biofuel cell connected to a charge pump has never been used for the continuous powering of an electronic device. To satisfy the high current demand for the operation of the charge pump, large biocatalytic electrodes (buckypaper with geometrical area of $6\,cm^2$) modified with PQQ-GDH on the anode and laccase on the cathode were used in a biofuel cell filled with human serum solution and operating under flow condition [89]. The biofuel cell was connected to a variable load resistance and polarization was measured demonstrating the open-circuit voltage, V_{oc}, about 470 mV and short circuitry current, I_{sc}, about 5 mA. The biofuel cell mimicking an implantable device was connected to the charge pump and DC–DC converter interface circuit, which was further connected to a pacemaker (Figure 17.8C). In order to analyze the pacemaker performance, the pacemaker output leads were connected to the implantable loop recorder (ILR) – a subcutaneous electrocardiographic (ECG) monitoring device. In the present setup, it was used as a medically useful analyzer of the electrical pulses produced by the pacemaker receiving the power from the biofuel cell. The loop recorder output was wirelessly read by the sensor device of the Medtronic CareLink Programmer, Model 2090, typically used for the programming of pacemakers and loop recorders after implantation and for the remote control of their operation [91]. The electrical pulses produced by the pacemaker and registered by the ILR were very similar when the pacemaker was powered by a standard battery and by implantable biofuel cell, thus confirming the correct pacemaker operation while receiving power from the external biofuel cell through the charge pump and DC–DC converter interface circuit. This approach is already practically applicable in future biomedical applications (note that the pacemaker was powered by a *single* implantable biofuel cell). Still major problems are awaiting additional research and engineering. The biocatalytic electrodes presently used in the fluidic system operating *in vitro* are too large to be implanted in a human body; thus, current efficiency should be increased to allow smaller electrodes. Taking into account that in the present design of the biocatalytic electrodes only about 6% of the immobilized enzymes are electrically wired to the conducting support [50], one can expect that upon increasing the efficiency of the electron transport between the enzyme active sites and the electrode support, the current density can be increased by more than 10-fold, thus resulting in the electrode miniaturization down to about $0.5\,cm^2$, which is already reasonable for the implanted electrodes. The biofuel cells used for activating the pacemaker can operate for hours, at best for several days, while the standard batteries operating as electrical power supplies for pacemakers provide power for at least 10 years [87, 88]. The stability issue for the implanted biofuel cells is a critic problem for their future biomedical applications. The implanted biofuel cells will be competitive with the presently used batteries only if they can operate *in vivo* more than 10 years. However, this long-time operation is only needed for biomedical applications when the biofuel cells are implanted in a human body. If the implantable biofuel cells are aiming at operation in animals, their stability for much shorter time might be sufficient for some technological applications.

Another important issue is the current value produced by the implanted biofuel cells. When the biocatalytic electrodes are implanted in small animals (rats, rabbits) [53–58] or even insects [49] their size is rather small (some electrodes have micro- or even nano-size dimensions). For very small electrodes, even if their operation is very efficient and the generated current density is large, the total current output is small and might be insufficient for continuous operation of microelectronic devices interfaced through the charge pump. Figure 17.9 shows the polarization functions of various implanted/implantable biofuel cells as well as the demand curve of the charge pump used for the pacemaker activation. The low voltage generated by the

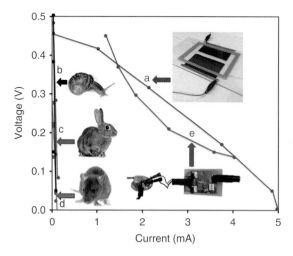

Figure 17.9 Polarization curves of the biofuel cells measured on the variable resistances – voltage and current produced by the cell as the function of the Ohmic resistance load: (a) an implantable biofuel cell operating *in vitro* in a flow device filled with a human serum solution [89], (b) a biofuel cell implanted in a snail and operating *in vivo*, (c) a biofuel cell implanted in a rabbit and operating *in vivo* [58], (d) a biofuel cell implanted in a rat and operating *in vivo* [53], and (e) *I–V* curve characterizing operation of a charge pump loaded with a pacemaker [89]. Curves "c" and "d" were recalculated from the data available in the original publications.

biofuel cell can be increased with the help of the charge pump only if the current supplied by the cell to the charge pump is above a certain threshold value. In the present plot it is reflected by positioning the polarization curve of the biofuel cell above the demand $I-V$ curve of the charge pump [89]. It is very clear that only the large-size electrodes [89] operating *in vitro* satisfy the current demand of the charge pump, while the biofuel cells implanted in snails [50], rabbits [58], and rats [53] do not provide enough current to activate the charge pump for the continuous powering of the pacemaker. This problem can be partially resolved by using more efficient and larger size biocatalytic electrodes as well as using more efficient charge pumps (note that the charge pump used in the present study [89] had only 60% efficiency). However, really small implanted biofuel cells, for example, operating in insects, will never satisfy the conditions required for the *continuous* operation of microelectronic devices. Still, they might be very useful for devices activated for a short time with significant time-gaps between their active states, thus allowing accumulation of the electrical energy for the burst release.

In conclusion, we note that the recent rapid progress in the design of implanted biofuel cells operating *in vivo* promises in the future various medical electronic implants powered by implanted biofuel cells and resulting in bionic human–machine hybrids [92]. Aside from biomedical applications, one can foresee bioelectronic self-powered "cyborgs," which can autonomously operate using power from biological sources and used for environmental monitoring, homeland security, and military applications.

Acknowledgment

This research was supported by the National Science Foundation (Award CBET-1066397) and by the Semiconductor Research Corporation (Award 2008-RJ-1839G).

References

1. Rome, L.C., Flynn, L., Goldman, E.M., and Yoo, T.D. (2005) *Science*, **309**, 1725–1728.
2. Sue, C.Y. and Tsai, N.C. (2012) *Appl. Energy*, **93**, 390–403.
3. Dai, D. and Liu, J. (2012) *IEEE Trans. Consum. Electron.*, **58**, 767–774.
4. Yun, J., Patel, S.N., Reynolds, M.S., and Abowd, G.D. (2011) *IEEE Trans. Mob. Comput.*, **10**, 669–683.
5. Riemer, R. and Shapiro, A. (2011) *J. NeuroEng. Rehabil.*, **8**, art. # 22.
6. Jia, D., Liu, J., and Zhou, Y. (2009) *Phys. Lett. A*, **373**, 1305–1309.
7. Reissman, T., MacCurdy, R.B., and Garcia, E. (2011) *Proc. SPIE*, **7977**, art. # 797702.
8. Nadimi, E.S., Blanes-Vidal, V., Jørgensen, R.N., and Christensen, S. (2011) *Comput. Electron. Agric.*, **75**, 238–242.
9. Donelan, J.M., Li, Q., Naing, V., Hoffer, J.A., Weber, D.J., and Kuo, A.D. (2008) *Science*, **319**, 807–810.
10. Yang, R., Qin, Y., Li, C., Zhu, G., and Wang, Z.L. (2009) *Nano Lett.*, **9**, 1201–1205.
11. Sun, C., Shi, J., Bayerl, D.J., and Wang, X. (2011) *Energy Environ. Sci.*, **4**, 4508–4512.
12. Paradiso, J.A. and Starner, T. (2005) *IEEE Pervasive Comput.*, **4**, 18–27.
13. Starner, T. and Paradiso, J.A. (2004) in *Low-Power Electronics Design* (ed C. Piguet) Chapter 45, CRC Press, Boca Raton, FL, pp. 1–35.
14. Hansen, B.J., Liu, Y., Yang, R., and Wang, Z.L. (2010) *ACS Nano*, **4**, 3647–3652.
15. Olivo, J., Carrara, S., and De Micheli, G. (2011) *Sci. Adv. Mater.*, **3**, 420–425.
16. Almouahed, S., Gouriou, M., Hamitouche, C., Stindel, E., and Roux, C. (2011) *IEEE/ASME Trans. Mechatron.*, **16**, 799–807.
17. Rasouli, M. and Phee, L.S.J. (2010) *Expert Rev. Med. Devices*, **7**, 693–709.
18. Vaddiraju, S., Tomazos, I., Burgess, D.J., Jain, F.C., and Papadimitrakopoulos, F. (2010) *Biosens. Bioelectron.*, **25**, 1553–1565.
19. Zurbuchen, A., Pfenniger, A., Stahel, A., Stoeck, C.T., Vandenberghe, S., Koch, V.M., and Vogel, R. (2013) *Ann. Biomed. Eng.*, **41**, 131–141.
20. Li, Z., Zhu, G., Yang, R., Wang, A.C., and Wang, Z.L. (2010) *Adv. Mater.*, **22**, 2534–2537.
21. Deterre, M., Lefeuvre, E., and Dufour-Gergam, E. (2012) *Smart Mater. Struct.*, **21** art. # 085004.
22. Mercier, P.P., Lysaght, A.C., Bandyopadhyay, S., Chandrakasan, A.P., and Stankovic, K.M. (2012) *Nat. Biotechnol.*, **30**, 1240–1245.
23. Davis, F. and Higson, S.P.J. (2007) *Biosens. Bioelectron.*, **22**, 1224–1235.
24. Scholz, F. and Schröder, U. (2003) *Nat. Biotechnol.*, **21**, 1151–1152.
25. Osman, M.H., Shah, A.A., and Walsh, F.C. (2010) *Biosens. Bioelectron.*, **26**, 953–963.
26. Logan, B.E. (2008) *Microbial Fuel Cells*, John Wiley & Sons, Inc., Hoboken, NJ.
27. Fornero, J.J., Rosenbaum, M., and Angenent, L.T. (2010) *Electroanalysis*, **22**, 832–843.
28. Cracknell, J.A., Vincent, K.A., and Armstrong, F.A. (2008) *Chem. Rev.*, **108**, 2439–2461.
29. Yu, E.H. and Scott, K. (2010) *Energies*, **3**, 23–42.
30. Meredith, M.T. and Minteer, S.D. (2012) *Annu. Rev. Anal. Chem.*, **5**, 157–179.
31. Leech, D., Kavanagh, P., and Schuhmann, W. (2012) *Electrochim. Acta*, **84**, 223–234.
32. Moehlenbrock, M.J. and Minteer, S.D. (2008) *Chem. Soc. Rev.*, **37**, 1188–1196.

33. Rincón, R.A., Lau, C., Luckarift, H.R., Garcia, K.E., Adkins, E., Johnson, G.R., and Atanassov, P. (2011) *Biosens. Bioelectron.*, **27**, 132–136.
34. Zhou, M. and Dong, S. (2011) *Acc. Chem. Res.*, **44**, 1232–1243.
35. Zhang, L., Zhou, M., Wen, D., Bai, L., Lou, B., and Dong, S. (2012) *Biosens. Bioelectron.*, **35**, 155–159.
36. Zhou, M. and Wang, J. (2012) *Electroanalysis*, **24**, 197–209.
37. Bhatnagar, D., Xu, S.A., Fischer, C., Arechederra, R.L., and Minteer, S.D. (2011) *Phys. Chem. Chem. Phys.*, **13**, 86–92.
38. Kalyuzhnyi, S. (2008) *Pure Appl. Chem.*, **80**, 2115–2124.
39. Oh, S.T., Kim, J.R., Premier, G.C., Lee, T.H., Kim, C., and Sloan, W.T. (2010) *Biotechnol. Adv.*, **28**, 871–881.
40. Barton, S.C., Gallaway, J., and Atanassov, P. (2004) *Chem. Rev.*, **104**, 4867–4886.
41. Heller, A. (2004) *Phys. Chem. Chem. Phys.*, **6**, 209–216.
42. Song, Y., Penmasta, V., and Wang, C. (2011) in *Biofuel's Engineering Process Technology* (ed M.A. Dos Santos Bernardes) Chapter 28, InTech, Rijeka, pp. 657–684.
43. Drake, R.F., Kusserow, B.K., Messinger, S., and Matsuda, S. (1970) *Trans. Am. Soc. Artif. Intern. Organs*, **16**, 199–205.
44. Rapoport, B.I., Kedzierski, J.T., and Sarpeshkar, R. (2012) *PLoS ONE*, **7**, art. # e38436.
45. Sun, M.G., Justin, G.A., Roche, P.A., Zhao, J., Wessel, B.L., Zhang, Y.Z., and Sclabassi, R.J. (2006) *IEEE Eng. Med. Biol. Mag.*, **25**, 39–46.
46. Mano, N., Mao, F., and Heller, A. (2002) *J. Am. Chem. Soc.*, **124**, 12962–12963.
47. Coman, V., Ludwig, R., Harreither, W., Haltrich, D., Gorton, L., Ruzgas, T., and Shleev, S. (2010) *Fuel Cells*, **10**, 9–16.
48. Pan, C., Fang, Y., Wu, H., Ahmad, M., Luo, Z., Li, Q., Xie, J., Yan, X., Wu, L., Wang, Z.L., and Zhu, J. (2010) *Adv. Mater.*, **22**, 5388–5392.
49. Rasmussen, M., Ritzmann, R.E., Lee, I., Pollack, A.J., and Scherson, D. (2012) *J. Am. Chem. Soc.*, **134**, 1458–1460.
50. Halámková, L., Halámek, J., Bocharova, V., Szczupak, A., Alfonta, L., and Katz, E. (2012) *J. Am. Chem. Soc.*, **134**, 5040–5043.
51. Szczupak, A., Halámek, J., Halámková, L., Bocharova, V., Alfonta, L., and Katz, E. (2012) *Energy Environ. Sci.*, **5**, 8891–8895.
52. MacVittie, K., Halámek, J., Halámková, L., Southcott, M., Jemison, W.D., Lobel, R., and Katz, E. (2013) *Energy Environ. Sci.*, **6**, 81–86.
53. Cinquin, P., Gondran, C., Giroud, F., Mazabrard, S., Pellissier, A., Boucher, F., Alcaraz, J.P., Gorgy, K., Lenouvel, F., Mathé, S., Porcu, P., and Cosnier, S. (2010) *PLoS ONE*, **5**, art. # e10476.
54. Sales, F.C., Iost, R.M., Martins, M.V., Almeida, M.C., and Crespilho, F.N. (2013) *Lab Chip*, **13**, 468–474.
55. Cheng, H., Yu, P., Lu, X., Lin, Y., Ohsaka, T., and Mao, L. (2013) *Analyst*, **138**, 179–185.
56. Zebda, A., Cosnier, S., Alcaraz, J.P., Holzinger, M., Le Goff, A., Gondran, C., Boucher, F., Giroud, F., Gorgy, K., Lamraoui, H., and Cinquin, P. (2013) *Sci. Rep.*, **3**, art. # 1516.
57. Castorena-Gonzalez, J.A., Foote, C., MacVittie, K., Halámek, J., Halámková, L., Martinez-Lemus, L.A., and Katz, E. (2013) *Electroanalysis*, **25**, 1579–1584.
58. Miyake, T., Haneda, K., Nagai, N., Yatagawa, Y., Onami, H., Yoshino, S., Abe, T., and Nishizawa, M. (2011) *Energy Environ. Sci.*, **4**, 5008–5012.
59. Pothukuchy, A., Mano, N., Georgiou, G., and Heller, A. (2006) *Biosens. Bioelectron.*, **22**, 678–684.
60. Oncescu, V. and Erickson, D. (2011) *J. Power. Sources*, **196**, 9169–9175.
61. Kloke, A., Biller, B., Kräling, U., Kerzenmacher, S., Zengerle, R., and von Stetten, F. (2011) *Fuel Cells*, **11**, 316–326.
62. Falk, M., Blum, Z., and Shleev, S. (2012) *Electrochim. Acta*, **82**, 191–202.
63. Ghindilis, A.L., Atanasov, P., and Wilkins, E. (1997) *Electroanalysis*, **9**, 661–674.
64. Katz, E., Sheeney-Ichia, L., and Willner, I. (2004) *Angew. Chem. Int. Ed.*, **43**, 3292–3300.
65. Zayats, M., Katz, E., and Willner, I. (2002) *J. Am. Chem. Soc.*, **124**, 2120–2121.

66. Narváez Villarrubia, C.W., Rinćon, R.A., Radhakrishnan, V.K., Davis, V., and Atanassov, P. (2011) *ACS Appl. Mater. Interfaces*, **3**, 2402–2409.
67. Zebda, A., Gondran, C., Le Goff, A., Holzinger, M., Cinquin, P., and Cosnier, S. (2011) *Nat. Commun.*, **2**, art. # 370.
68. Hussein, L., Rubenwolf, S., von Stetten, F., Urban, G., Zengerle, R., Krueger, M., and Kerzenmacher, S. (2011) *Biosens. Bioelectron.*, **26**, 4133–4138.
69. Strack, G., Luckarift, H.R., Nichols, R., Cozart, K., Katz, E., and Johnson, G.R. (2011) *Chem. Commun.*, **47**, 7662–7664.
70. Holzinger, M., Le Goff, A., and Cosnier, S. (2012) *Electrochim. Acta*, **82**, 179–190.
71. Katz, E., Willner, I., and Kotlyar, A.B. (1999) *J. Electroanal. Chem.*, **479**, 64–68.
72. Wilson, R. and Turner, A.P.F. (1992) *Biosens. Bioelectron.*, **7**, 165–185.
73. Wang, J. (2008) *Chem. Rev.*, **108**, 814–825.
74. Mano, N., Mao, F., and Heller, A. (2004) *ChemBioChem.*, **5**, 1703–1705.
75. Zayats, M., Katz, E., and Willner, I. (2002) *J. Am. Chem. Soc.*, **124**, 14724–14735.
76. Kar, P., Wen, H., Li, H.Z., Minteer, S.D., and Barton, S.C. (2011) *J. Electrochem. Soc.*, **158**, B580–B586.
77. Moehlenbrock, M.J., Toby, T.K., Waheed, A., and Minteer, S.D. (2010) *J. Am. Chem. Soc.*, **132**, 6288–6289.
78. Katz, E. (1994) *Electroanal. Chem.*, **365**, 157–164.
79. Razavi, B. (2013) *Fundamentals of Microelectronics*, Wiley Inter-Science.
80. Ieropoulos, I., Greenman, J., and Melhuish, C. (2010) *Bioelectrochemistry*, **78**, 44–50.
81. Ieropoulos, I., Greenman, J., and Melhuish, C. (2008) *Int. J. Energy Res.*, **32**, 1228–1240.
82. Aelterman, P., Rabaey, K., Pham, H.T., Boon, N., and Verstraete, W. (2006) *Environ. Sci. Technol.*, **40**, 3388–3394.
83. Meehan, A., Gao, H., and Lewandowski, Z. (2011) *IEEE Trans. Power Electron.*, **26**, 176–181.
84. Ieropoulos, I., Greenman, J., Melhuish, C., and Horsfield, I. (2010) Proceedings of the Alife XII Conference, Odense, Denmark, pp. 733–740.
85. Hanashi, T., Yamazaki, T., Tsugawa, W., Ferri, S., Nakayama, D., Tomiyama, M., Ikebukuro, K., and Sode, K. (2009) *Biosens. Bioelectron.*, **24**, 1837–1842.
86. Hanashi, T., Yamazaki, T., Tsugawa, W., Ikebukuro, K., and Sode, K. (2011) *J. Diabetes Sci. Technol.*, **5**, 1030–1035.
87. Ellenbogen, K.A., Kay, G.N., Lau, C.P., and Wilkoff, B.L. (2011) *Clinical Cardiac Pacing, Defibrillation, and Resynchronization Therapy*, 4th edn, Elsevier, Philadelphia, PA.
88. Barold, S.S., Stroobandt, R.X., and Sinnaeve, A.F. (2010) *Cardiac Pacemakers and Resynchronization Step-by-Step*, Wiley-Blackwell, Chichester, p. 23.
89. Southcott, M., MacVittie, K., Halámek, J., Halámková, L., Jemison, W.D., Lobel, R., and Katz, E. (2013) *Phys. Chem. Chem. Phys.*, **15**, 6278–6283.
90. Pan, F. and Samaddar, T. (2006) *Charge Pump Circuit Design*, McGraw-Hill Professional, New York, NY.
91. Kusumoto, F. and Goldschlager, N. (2010) *Clin. Cardiol.*, **33**, 10–17.
92. Johnson, F.E. and Virgo, K.S. (eds) (2006) *The Bionic Human: Health Promotion for People with Implanted Prosthetic Devices*, Humana Press, Totowa, NJ.

18
Biomedical Implantable Systems – History, Design, and Trends
Wen H. Ko and Philip X.-L. Feng

18.1
Introduction

Implantable systems are a technique of biomedical instrumentation for sensing body information or soliciting body reaction within a living organism from a location outside of the body through a wireless communication link. In a broad sense, the system can be surgically implanted inside the body; or located inside a cavity – intestines, mouth, and so on; or attached on the body surface. It is human nature to want to explore nature beyond the boundary of the body. From the body surface outward one may explore family, community, Earth, other planets, and outer space; in the same way, one may wish to explore inside the body from organ systems, organ parts, and brain function to muscles, nerves, cells, cell components, DNA, and molecules. As shown in Figure 18.1, if the unit circle is the body boundary that one can sense, outwardly, one may explore from "1" to ∞; one can also explore inwardly from "1" to $1/n$ to $1/n^2$ to near 0. Every point R_o outside the unit circle $R_o(r, \theta)$ can be mapped to a point R_1 $(1/r, \theta)$ inside the circle. There are infinite things to explore inside our body as there are infinite spaces to explore outside of our body. However, the inward search relates to body health and to one's well-being, providing a strong motivation for research in biology, physiology, and medicine. Implantable systems are tools for exploring living organisms around us and body functions inside us. Implantable systems look inward from the body to organs, tissues, cells, and subcellular molecules including proteins and DNA/RNA. Implantable systems are the tools for our inward exploration just as spaceships are the tools for exploration of outer space. The science, technology, methodologies, and instruments for both endeavors are similar but great differences exist in scale, size, and cost.

Implantable systems represent a technique of biomedical instrumentation; they are the tools for sensing body information or soliciting body reaction within a living organism from a location outside of the body through a wireless communication link for our inward exploration.

The essential parts of all implantable systems are shown in Figure 18.2: the oval is the body and the rectangular block is the external equipment. The body is linked

Implantable Bioelectronics, First Edition. Edited by Evgeny Katz.
© 2014 Wiley-VCH Verlag GmbH & Co. KGaA. Published 2014 by Wiley-VCH Verlag GmbH & Co. KGaA.

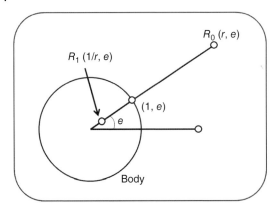

Figure 18.1 The body is the unit circle; every point R_0 outside can be mapped to a point R_1 inside the circle.

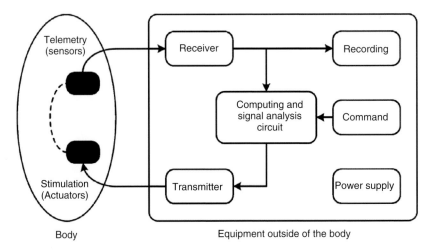

Figure 18.2 Conceptual diagram of an implantable system.

to the external equipment by wireless links for telemetry (sensing) or stimulation (actuation).

The building blocks of implanted systems, shown schematically in Figure 18.3, can function for telesensing (telemetry) or remote stimulation (tele-actuation). When both are linked together, it is a closed-loop control system. The system may be all electronic or may have parts of the implant that are mechanical, chemical, or biological. All implantable systems have the same essential building blocks irrespective of whether the implant device is for sensing or actuating or closed-loop control. For implant telemetry systems, the sensors convert the biologic parameters into electrical signals that are processed by the interface electronics and then transmitted by the radio frequency (RF) (or other wireless means such as ultrasound, optical, etc.) link to an external receiver and the recording or processing

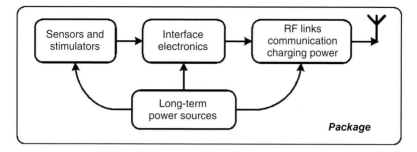

Figure 18.3 The essential building blocks of the implantable system for the body.

facilities. For the implant actuation systems, the external command unit sends command signals through the wireless link, which is then processed by the interface electronics to drive the actuators that solicit certain actions in the body. For the implantable electronics system, the common actuators are the electrical stimulators that send electrical currents to the muscles or nerves through stimulating electrodes that solicit the desired body reactions. For implant control systems, the telemetry unit and the stimulating unit are linked to control specific body function to a desired level, with the telemetry part acting as the feedback unit of a conventional control system. All these systems need to be packaged for biocompatibility before implant insertion (e.g., through surgical operations or ingestion).

The sensor/stimulator, the electronics and the wireless communication link, and power supply of an implantable system may be located totally within the body or on the surface layer of the organism. The location may be in the body or intracavity (implantable system); within the intestines or mouth, (in-dwelling system); or may be attached on the external surfaces of the body, may be subcutaneous, or on the organ surface (attachable system). The implant electronic systems are valuable instruments for life sciences research, as well as health care. Telemetry systems are used for monitoring and diagnosis such as ECG measurement and blood pressure monitoring. Stimulation systems are used for prostheses, therapy, and treatment, such as cochlear implants, heart pacemakers, and nerve signal blockage for pain suppression. The closed-loop systems are used for paralyzed arm prostheses, automated health care, and prevention, such as the variable rate heart pacemakers that can adjust the heart rate according to the oxygenation level in the blood or the activity of the body [1].

18.2
History: Review of Implant Systems

Although radio transmission of analog signals has been known since 1884 [2] and frequency-modulation (FM) radio links were used to transmit pneumograms in 1948 [3], development of biomedical implant electronic techniques did not really get started until the transistor was discovered in 1948 and made available after 1954. Especially after the Sputnik (1957) excitement, the comparatively small size and power consumption of this new device as compared with vacuum tubes made possible the construction of practical telemetry transmitters for biomedical implant measurements. Since then, the developments of implant systems run parallel with the advances in solid-state devices and microelectronics. The advances of electronic devices from germanium to silicon transistors, from bipolar to MOS (metal-oxide-semiconductor) transistors, and from millimeter scale integrated circuits (ICs) to 0.2 µm scale large-scale integrated circuit (LSIC) and application-specific integrated circuit (ASIC) have further enhanced implant systems into a field of flourishing research in biomedical instrumentation. The curiosity, motivation, and desire to explore the invisible internal events within the body were there and the types of biomedical information that were being looked for were there, from the very beginning. It was because of the limitations of the tools – the instrumentation and the technology – that the history of biomedical implant systems advanced in synchrony with the technology in three stages as summarized below.

18.2.1
Historical Review of Early Implant Systems, 1950–1970

Before 1954, vacuum tube radio instruments were used for military communication and space exploration in telemetry and stimulation. O. H. Schmitt [4] designed a vacuum tube AM- or FM-modulated RF oscillator coupled to a germanium varistor (diode) to stimulate single nerve fibers with variable pulse amplitude and duration, as shown in Figure 18.4.

Similarly, electrical stimulation was tried for cochlear hearing in 1930s [5, 6] and cardiac pacemakers developed in early 1930s [7]. The electrical effect on muscle and nerve was also explored in various experiments in the 1880s by Carlo Matteucci. He actually discovered the muscle action potential (EMG) and observed electrical stimulation of muscle reaction [8].

18.2.1.1 Historical Review of Early Implant *Telemetry* Systems, 1950–1970
With the discovery of transistors and diodes, signal processing and transmission skills were greatly improved, and the size and power consumption reduced as the implantable systems started to take shape and developed into practical and useful tools. It started with the biological tracking area, and advanced to implant studies of biomedical signals. The tracking instrument was used for studies of flying birds, underwater fish, and roaming animals. In flying birds, it was used to track (i) homing behavior of pigeons; (ii) the behavior of ruffed grouse [9] in northeastern Minnesota; and (iii) eagles, gulls, mallards (*Anas platyrhynchos*), and bobwhites (*Colinus virginianus*). In underwater telemetry, RF transmitters were inserted in dead fish, which were then swallowed whole by dolphin or other large fish (Caceres' book, p. 214, [10]). In roaming animals, tracking was used to study the behavior of grizzly bears [11] and elk. The tracking transmitter was either the simple blocking oscillator shown in Figure 18.5 or a more complicated circuit for

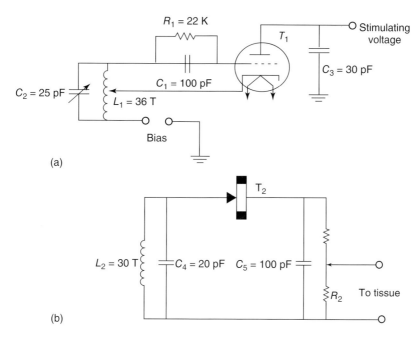

Figure 18.4 (a) The T_1 vacuum tube RF oscillator, controlled by bias and C_2, generates the stimulation signal, which is transmitted to (b), the insulated receiving LC circuit and rectified by a germanium diode to form the stimulating pulses with amplitude adjusted by R_2 and R_3.

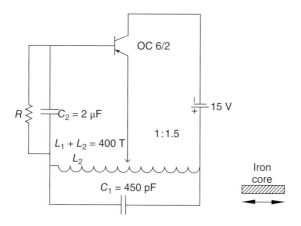

Figure 18.5 The one-transistor Hartley oscillator circuit working as a blocking oscillator.

larger distances (Caceres book p. 155, and p. 208). They were either fixed-duration pulse generators or were modulated by sensor signals of temperature- sensitive R and C, or pressure-sensitive, R, L, and C, or position-sensitive L and C.

For biological and medical information telemetry, the circuit in Figure 18.5 could be incorporated with (not sure what this means: "use?") sensors to telemeter biomedical information from inside the body. The sensors were added elements or existing components. For example, R can be temperature or pressure sensor; L, can be a pressure, displacement, or position sensor; and C can be temperature, pressure, and displacement sensors. Biopotential sensors can be electrodes of various materials. This simple circuit can be packaged with a battery, antenna coil, and sensors in a small (0.8 cm diameter × 5 cm) cylinder and become an endoradiosonde [12]. Sensors for pH were a mixture of polyacrylic acid and poly(vinyl alcohol), which changes volume to 60% when the pH changes from 3 to 8 and can be used with the endoradiosonde to study the pH and contraction of the intestines [13]. A similar telemetry design, shown in Figure 18.6, was used to study the contraction of the gut and used for diagnosis of problems in the digestive track [14] in England. Figure 18.7 shows intestinal pressure waves from patients, using the telemetry unit of Figure 18.6.

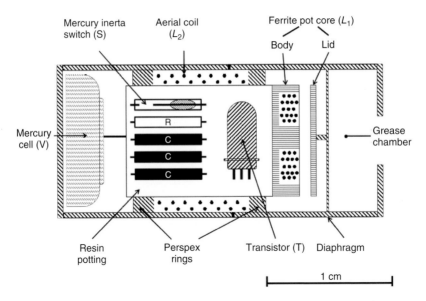

Figure 18.6 The wireless telemetry unit used for the study of the digestive track.

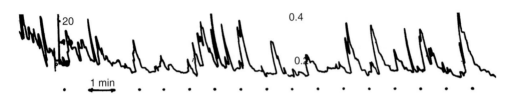

Figure 18.7 Pressure waves in the area of the left caecal valve. On the extreme left is small intestinal activity; on the right is caecal activity.

Endoradiosondes with R or C temperature sensors were used to monitor/measure the temperature of tortoises and dolphins by feeding them food stuffed with telemetry units. It is a mystery how penguins can incubate their eggs in Antarctic winter at a temperature −77 °F and how the embryos can develop. These questions were answered by Eklund and Charlton [15] when they monitored the temperature of incubating eggs by placing the transmitter in an empty penguin egg shell, and then injecting albumen into the sealed shell. They placed the transmitter egg with other normal eggs for the penguin to hatch. It was found that the average temperature of the incubating egg was 9.5° less than the average penguin body temperature of 106 °F.

Simple active transmitters with various sensors were used to explore many biomedical investigations in animals and patients [16, 17]. Passive transmitters of very small size (2 mm diameter × 1 mm thick) were used to measure the pressure rise in rabbit eyes influenced by sensory input changes. [19]. Figure 18.8 shows the principle of the Mackay–Marg tonometer [20] for eye pressure measurement in connection with glaucoma testing. It is being used in recent blood pressure monitoring devices and fluid circuit measurements [20].

In the early stage of development, most work was pursued by field biologists using simple electronic circuits and germanium transistors. From the late 1950s to the early 1960s Mackay, Noller, Wolff, Zworykin, and others developed active radiotelemetry units for use in the gastrointestinal tract and other cavities of the body. Subcutaneous and deep-body implantation of telemetry units were initiated by Essler, Ko, Mackay, Cole, Young, and others to measure physiologic information in animals as well in humans. For a historical review of these and other areas of biomedical telemetry see [16, 18, 21, 22]. Many application ideas were explored. They paved the way for the recent development of RFID, heart pacemakers, injectable radio capsules, pressure tonometers, industrial RF control, and biomedical instrumentation.

In the period between 1960 and 1970, besides improved performance of germanium and silicon junction transistors, the tunnel diode was invented in August

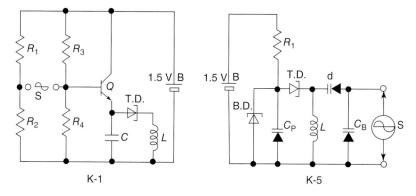

Figure 18.8 Two of the six tunnel diode low-power telemetry circuits developed in Ko's group.

1958 by Leo Esaki. It reduced power consumption and made RF generation and modulation simple. At the same time, the launch of Sputnik on 4 October 1957 marked the dawn of the space age. Electrical engineers devoted time to develop telemetry and stimulation technology for space biology and medical problems on earth. Ko and his group received the first engineering grant with no medical doctors as principle investigator, from NIH Institute of General Medical Sciences to study the implantable systems in 1967. A series of six telemetry circuits were designed. Figure 18.8 shows the K-1 to K-5 devices. K-1 used a transistor to drop the battery voltage to 0.2 V for the tunnel diode and the K-5 used a diode varactor (variable capacitor) to FM modulate the carrier with input signal "s."

K-5 and K-6 devices were packaged with Hysol® epoxy and medical-grade silicone for short-term (<30 days) implants; flat pack boxes made of metal and Macor® (a machinable ceramic) and glass capsules were used for packaging of longer term implants. The volumes were in the range of 800 mm^3, and the power levels were from milliwatts to microwatts. They were intended for biomedical monitoring and diagnosis, including (i) physiology research using small animal models; (ii) system biology study; (iii) patient monitoring in the hospital; (iv) response and progress of medical treatment; and (v) monitoring vital signs of patients for preventive care and early warning. Figure 18.9 shows two of the telemetry units that were used for many of these applications from 1961 to 1975 [1, 23, 24]. The world's first pocket

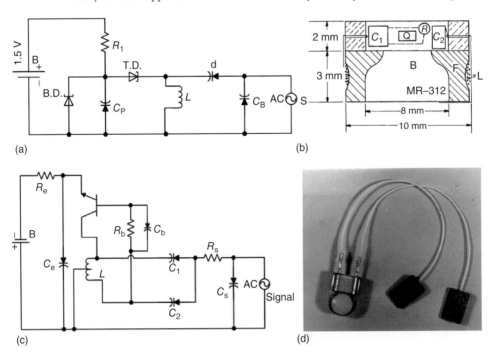

Figure 18.9 (a) K-5 tunnel diode telemetry unit, (b) packaged K-5 telemetry device, (c) K-6 circuit, and (d) packaged K-6 implantable FM telemetry transmitter for voice, EKG, and EMK.

size mobile microphone was demonstrated in 1964 with a PZT microphone and a MR312 mercury battery [23, 27]. The K-6 circuit was integrated into a monolithic chip; the photo and performance of the chip as well as the packaged implant unit are shown in Figure 18.10 [25].

Bedsides the telemetry units, the group worked on microsensors, an integrated pressure sensor, and the earliest active integrated sensor that was tested in 1971, as shown in Figure 18.11 [26].

A piezoelectric heart energy harvester was designed for long-term EKG telemetry monitoring in dogs. The resonant mode was used to increase the power converted. Sixteen milliwatts was obtained from the dog's heart for months. The device is shown in Figure 18.12, where (a) is the structure of the PZT beam; the movement of the heart excites the PZT beam to oscillation. The AC voltage of the PZT beam is rectified by the doubler circuit and the energy is stored in (c) for output. Sixteen microwatt power was harvested from the dog's heart [27].

The Ko group also studied all technical problems of implant systems including (i) RF powering and design [28]; (ii) fabricated micropower (<1 μW) pulsed carrier transmitter circuits, in multi-chip pack, housed in Macor (a machinable ceramic) packaging and distributed to many biomedical researchers [29]; (iii) multichannels cortex stimulation circuits [30] and active electrodes for EEG monitoring [31] for

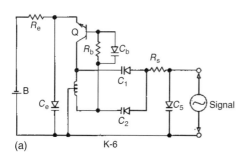

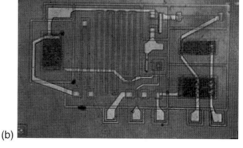

Typical values of K-6 transmitter characteristics

Characteristic	Value	
Operating frequency	90–130	MHz, tunable
Current drain	300–500	μA at 1.3 V
Modulation sensitivity	10–20	kHz mV^{-1} rms at 400 Hz sinus
Noise level	5–25	μV rms
RF signal strength	15–30	μV at 2.5 m distance with 1/2 wavelength receiving antenna
Normal operating range	3–10	m in free air
Operating voltage	1.1–3.5	V
Transmitter volume	3.0	cm^3
Transmitter mass	4.9	grams

Figure 18.10 The K-6 circuit (a) was integrated in a chip (b) and packaged, same as K-5, in epoxy with a silicone surface layer (c). Characteristics of K-6 circuit for continuous broad band input (d). (It was the world's smallest implantable telemetry unit in 1965; it weighed 1.5 g, with battery and sensor.)

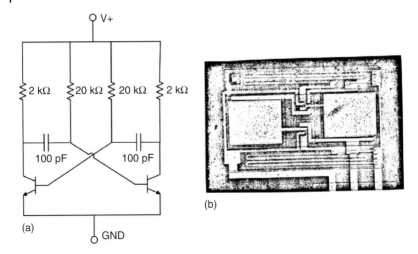

Figure 18.11 The integrated piezoresistive pressure sensor. The two base resistors are piezoresistive (World's first integrated sensor, 1971; made in class-10 000 clean room, Photograph (b) shows all the dust). (a) the circuit and (b) photograph of the chip.

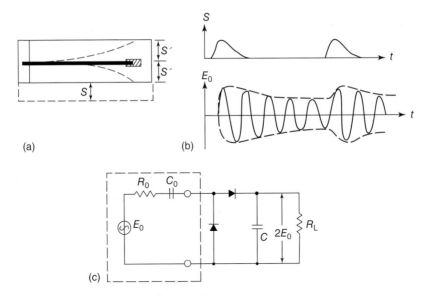

Figure 18.12 The heart energy harvester: (a) structure, (b) waveform, and (c) circuit [27].

brain stimulation; (iv) biomedical sensors – pressure, strain, pH, pO_2 sensors, and ion-sensitive field effect transistors (FETs) [32]; (v) body reaction and implant packaging [33, 34]; (vi) taped-on telemetry transmitters [35]; (vii) body surface potential mapping [36]; (viii) RF radiation from a small loop inside a passive medium (the body) [37, 38]; (ix) implant evaluation of a nuclear battery [39]; and (x) intracranial pressure and temperature telemetry monitoring systems [40]. In this period, the telemetry systems were developed in many fields including biology tracking and behavior studies. A collection of the biotelemetry work for tracking can be found in [41].

18.2.1.2 Historical Review of Early Implant *Stimulation* Systems, 1950–1970

The peripheral and central nervous system in the body can be stimulated by small electrical currents to generate electrical nerve signals or to block nerve signal transmission, thus inhibiting related functions, such as pain blocking [42]. Tissues can also be stimulated to contract and to generate EMG signals. The theory and application of *functional electrical stimulation* were developed at the Engineering Design Center, Case Institute of Technology, in 1960–1970 [43]. A typical single-channel electrical stimulation circuit is shown in Figure 18.13. Here, the K-5 unit is a feedback circuit used to check the magnitude of the electrode current delivered to the body. Multiple-channel stimulator units were developed for hand control applications for tetraplegia patients [30, 44]. Many research groups tried various applications of the functional stimulation; examples include (i) cardiac pacemaker and defibrillation devices; (ii) an electronic pain suppressor; (iii) implantable middle ear and cochlear hearing aids; (iv) visual prosthesis (with multiple electrodes attached to the forehead or the visual cortex surface); (v) diaphragm pacing for respiration; (vi) epileptic seizure control; (vii) hand and aim control for patients with spinal cord injury; (viii) leg and foot control for walking; and (ix) brain stimulation for emotion control of monkeys and other animal models [44, 45].

18.2.1.3 Historical Review of Early Implant *Control* Systems, 1950–1970

Many implant control systems were studied by adding a feedback telemetry circuit to monitor the response to the stimulation for adjusting the strength of command to reach a desired level of body action. Examples include the on-demand pacemaker with heart rate monitor feedback or the EMG-controlled hand prosthesis that telemeters back the level of muscle response to the command unit to adjust the

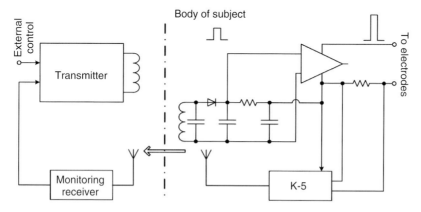

Figure 18.13 The single-channel stimulation circuit with K-5 feedback.

stimulation level for the desired consciously controlled motion. In the early stage of implantable systems many control systems were studied; examples include (i) paralyzed limb control for upper and lower extremities [46]; (ii) drug infusion control; (iii) organ function regulation, such as heart pacing and lung pacing; (iv) biological research; (v) pain suppression; and (vi) drop foot and other muscle control systems. Demonstrative projects were tried, but the practical use of these systems took many years of developmental work. Figure 18.14 illustrates the Arm Aid system proposed in 1960 at the Electronics Design Center (EDC), Case Institute of Technology. In this system, the brain generates the commands and controls through the shoulder muscles; the implanted or surface-mounted telemetry unit on the shoulder sends EMG signals to the external control unit, which in turn generates stimulation signals that control the forearm and the fingers; the visual feedback signal closes the control loop. These early experiments were used to study the feasibility of prosthesis control, brain-controlled machines, and brain-controlled environment (only recently the Case Arm Aid System has been approved by the FDA for use on patients to gain some use of the hand for daily activity [44]).

Figures 18.15–18.17 are three historical photographs demonstrating the realization of the dream of using the brain to control outside machines and other

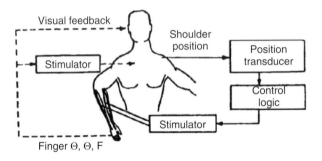

Figure 18.14 Illustration of the conceptual arm aid system of 1960.

Figure 18.15 The EMS signal from Dr. Vodovnik's shoulder was operating the machine, as Dr. J. Reswick was observing.

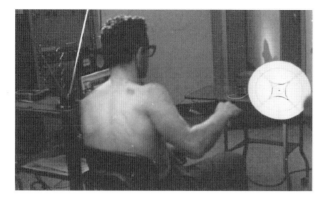

Figure 18.16 The EMG signal from the implanted transmitter is controlling the speed of the disk and motor.

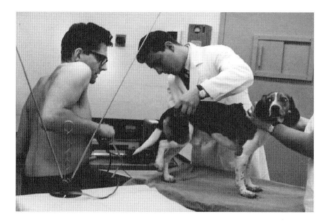

Figure 18.17 Human EMG implant device overrides and controls the leg motion of a dog (1970). (The other person in the picture is Dr. Ron Lorig, an M.D.-Ph.D. student in the laboratory.)

bodies' motion through implant telemetry and stimulation systems (1970). They are included here to preserve them in the literature. In these experiments, a K-6 telemetry unit was implanted under the skin on the shoulder of Dr Lojze Vodovnik [47], an outstanding visiting scholar from Czechoslovakia to the Engineering Design Center of the Case Institute of Technology in the 1960s. In these systems, the brain generates the commands to control the shoulder muscle; the implanted telemetry unit sends EMG signals of this muscle to an FM receiver nearby, and the receiver output is the command signal for control. In Figure 18.15, the command signal was used to select codes from a machine to activate preselected actions to aid a paralyzed patient. In Figure 18.16, the command signal was used to control the speed of the motor, and in Figure 18.17, an electrical stimulation unit was implanted in a dog's rear leg muscle. The EMG signal of Dr. Vodovnik was used to override and control the up and down movement of the dog's leg. These early experiments demonstrated the feasibility of brain-controlled machines, or brain-controlled environment [48, 49].

Besides the stimulation of peripheral nerves for prosthesis, the stimulation of brain and spinal cord were tried and studied. Multiple-channel stimulation of the central nerve system (CNS) for disease control, emotion control, and for hearing and visual prosthesis [30] were explored [43, 50].

18.2.2
Historical Review of Implant Systems – 1970–1990

The period 1950–1970 is one of discovery and exploration for implant systems; many new technologies were studied, and many new applications were investigated. The field blossomed. From 1970 to 1990, the implant systems matured. Field biologists used telemetry systems with various sensors and camera for the tracking and behavior study of wild species. A collection of biotelemetry works for tracking can be found in [41].

By 1980, the precision technologies for mass production of microelectronics and ICs were mature. The micromachining technology was adopted for the fabrication of micromechanical elements, and chemical, physical, biological sensors and actuators. These sensors and actuators can be incorporated with telemetry and stimulation systems for new exploration and research. Researchers in this field recognized the need for a specialized journal in this new field of transducers, and that is when the *Sensors and Actuators* was born in 1980. The first transducer conference was held in Boston as a part of the Materials Conference in 1981. When transducers were embedded with LSIC and computing circuits, another industrial revolution was inaugurated. During the period 1986–1988, U.C. Berkeley and MIT fabricated micromotors and published papers on them, nearly at the same time. In November 1987, the National Science Foundation of USA called a meeting at Hyannis, Massachusetts, to discuss possible research in micro-robots, sensors, and actuators. By 1988, the term *Microelectromechanical system* or *MEMS* was created, which opened up a new field of engineering and industrial research. In the same year, the IEEE and ASME joint publication, the *Journal of Microelectromechanical Systems* (JMEMS) brought out its first issue. With these technological advances the implant systems incorporating the micro sensors (physical and biochemical) become valuable biomedical research and health care instruments. However, the cost is still high; most progresses were made in research, while the implantable system is maturing gradually. Some of the systems got to be used in medical care, such as the heart pacemaker/monitor and defibrillator, as well as small external hearing aids and ingestive capsules, and matured to become common medical devices.

18.2.3
Historical Review of Implant Systems – 1990–2012

The great advances of the last three decades include the manufacturing of very large scale integrated circuits (VLSI) and ASICs, the miniaturization of physical, chemical, and biological transducers (sensors, actuators, and electrodes), especially MEMS and nano-electromechanical systems (NEMS). These advances have cleared many barriers for practical implantable microsystems in medical care and life sciences research. The pioneering techniques, of short-range MEMS wireless remote monitoring and control, have advanced from biomedical research and aerospace applications to preventive care, and to industrial, environmental, and social communication fields. New biomedical implantable systems have been explored in all directions, such as the cochlear hearing aids, visual prostheses, tough sensors, drug infusion systems, artificial organs, and surgical robots. Figure 18.18

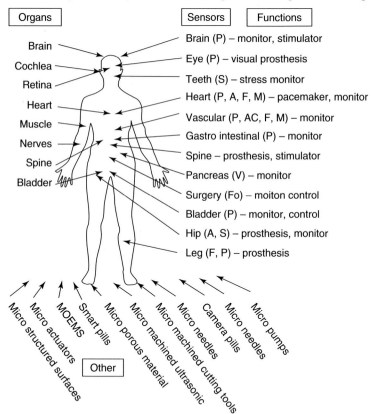

Figure 18.18 Microsensors and implant systems used for medical care. Functions – monitor, stimulation, and control; sensors – P = pressure, A = acceleration, Fl = flow, AC = AC conductance, Ma = magnetic, S = strain, Fo = force, M = multiple, V = viscosity; other – other medical devices, and MOEMS = opto-microelectromechanical systems.

illustrates some of the microsensors and implant systems used for medical care and research now.

The literature in this field is vast. A comprehensive literature review can be found in [50, 51]. No detailed and focused review on implantable MEMS systems from 1990 to the present is included in this chapter. A sample list of implantable systems developed from 1970–2012, which have made considerable impacts, are briefly summarized below.

1) Implantable electronic devices: Telemetry, stimulation, and control systems.
   a. *Limb prostheses*: hand, arm, and leg motion control, drop foot stimulator.
   b. *Sensory prostheses*: middle ear and cochlear hearing aids, visual prostheses, touch sensors, artificial taste and smell devices.
   c. *Assistive/artificial organs*: such as cardiovascular assistive devices, heart pacemaker, respiratory pacer, urinary control system, artificial vocal cord.
   d. *Drug infusion devices*: insulin release device, cancer treatment drug infusion devices, pain release drug infusion devices.
   e. *Monitoring organ function or disease progression*: cancer growth, transplant organs.
2) In-dwelling devices: systems in-dwell in the digestive channel or other body cavities such as ingestive capsules, devices implanted in the bladder, oral cavity, or ear channel.
3) CNS (brain) to machine interface research: With large electrode arrays implanted in the brain or on the dura surface for telemetry and stimulation, in order to treat diseases and to establish communication links with the brain to develop CNS and computer communication networks.
4) Surface-attached devices for mobile baseline monitoring and early warning of patients' vital signs.
5) Body energy-harvesting and various wireless powering and distribution techniques.
6) Part of medical robots: for surgical operations and routine work in patient care.

18.3
Design of Implant Systems

The implant systems are wireless communication systems used in the biomedical field. There are several ways to achieve wireless communication, such as by RF, sonic or ultrasonic, and optical means. The design principles are the same; only the range of carrier frequency is changed. Wireless communication systems are used in industries and environmental studies and in these cases they are called *embedded systems*. Only RF implant systems for biomedical applications are discussed in this chapter.

18.3.1
Basic Considerations and Characteristics of RF MEMS Implantable Systems

18.3.1.1 Legal Considerations of the Radio Frequency (RF), Field Strength, and Power Levels

The wireless operation of implantable systems can be accomplished by modulating a high frequency radiating carrier that can transmit signals through the body to a receiving station nearby or vice versa. The carrier frequency used can be in the range of ultrasound, RF, and infrared (IR) frequencies. The most commonly used frequencies are in the RF range, from kilohertz (kHz) to megahertz (MHz) to gigahertz (GHz). There are government regulations and international standards to guide the proper use of frequency and radiated carrier power level. These basic considerations are (i) that the carrier signal is noninterfering to other users and (ii) the assurance that the system is not harmful to the body. In the United States, the regulations are detailed in the electronic Code of Federal Regulations (eCFR) of the Federal Communication Commission (FCC), titled 47 Telecommunication, Parts 15, 18, 90, and 95, and IEEE code C-95-2005 [66]. These regulations are specific, detailed, and complicated. The following is a greatly simplified sample of the FCC regulations. The designers and the users of implantable systems should seek professional advice on selecting the operation frequency and power level. In many cases, an FCC license will be required.

The IEEE C95-2005 is the IEEE standard for safety levels with respect to human exposure to RF electromagnetic fields from 3 kHz to 300 GHz. It *provides recommendations to minimize aversive or painful electro-stimulation in the frequency range of 3 kHz to 5 MHz and to protect against adverse heating in the frequency range of 100 kHz to 300 GHz. In the transition region between 100 KHz to 5 MHz.* Protection against both electrostimulation and thermal effects is provided through two separate sets of limits. Below 100 kHz only the electrostimulation limits apply, above 5 MHz only the thermal limits apply, and both sets of limits apply in the transition region. In the transition region, the limits based on electrostimulation will generally be more limiting for low duty cycle exposures, while the thermal-based limits will be more limiting for continuous wave fields (IEEE C-95 Section 4.1 basic restrictions). The eCFR of the FCC allocates the use of RF frequencies and regulates RF devices, equipment, and operators' licenses.

In the United States, implantable system frequencies may have the following options: (i) The FCC-assigned frequency bands for medical telemetry service: 608–614, 1390–1395, 1429–1429.5, 1427-1429, and 1429.5–1431.5 MHz; as well as the frequency bands for medical radio-communication service in 401–406 MHz (eCFR title 47 part 95- Subpart H and E). (ii) Frequency bands assigned to industry, science, and medicine (ISM bands), as listed in Table 18.1 ([52], Title 47, Part 95, Parts 15, and part 18). (iii) Amateur RF bands. (iv) Frequencies that satisfy the low-power, noninterference provision.

Operation of the equipment according to the ISM radio bands within the following safety, search, and rescue frequency bands is prohibited: 490–510 kHz,

Table 18.1 ISM frequencies and tolerance ([52] part 18 Section 18.301).

Frequency (MHz)	6.78	13.56	27.12	40.68	915	2450	5800	24125	6125	12250
Tolerance (kHz)	15.0	7.0	163	20	13 M	50 M	75 M	125 M	250 M	500 M

2170–2194 kHz, 8354–8374 kHz, 121.4–121.6 MHz, 156.7–156.9 MHz, and 242.8–243.2 MHz.

The regulations on power level and maximum field strength outside of ISM band are given in IEEE C95. There will be no summary given in this chapter. Because the information contained is complex and will be updated soon, there will be changes made as technology advances and more RF devices are used. Furthermore, the operating power level and the maximum field strength involve personal safety and interference to other's communication. The readers are referred to the updated IEEE C-95 or to consult specialists on these topics in selecting operation power level and field strength.

18.3.1.2 Biocompatibility and Protection of the Biomedical Implant Systems

The body is a fragile but harsh environment for implantable electronics. The implantable systems should not seriously affect or harm the tissues surrounding the implant as well as the body, that is, the device/system has to be biocompatible. At the same time, the implanted system also needs protection from the tissue's activity and damage from corrosive, conducting body fluid. Biocompatibility demands proper design of the implantable systems in size, weight, functional electronics, and packaging to minimize electrical, mechanical, thermal, and optical interference to the normal activity of surrounding tissues and the body as well as to eliminate any chemical and biological toxicity and contamination to the tissue and the body.

For physical biocompatibility, the size and weight should be small so that the implant will not affect the normal activity of the body. It is generally accepted that the volume and weight of implantable devices or systems should not be more than 2% by volume and weight of the host organ or body to avoid discomfort or loading stress on the organ or body in their normal activity and movement. Besides the small size, the density of the implant should closely match that of the tissues so that when the body moves, the implant should not cause undue stress to the tissues surrounding the device. For thermal considerations, the implant should not have any hot spots that are 2 °C higher than the normal temperature of the tissue. The surface of the packaged implant should be smooth, and not have any sharp parts that would cause localized stress or irritation to the tissues. Some means to stabilize the position of the implant should be provided to prevent the implant from migrating from the designated site to other parts of the body [34, 33].

When the implantable system is surgically inserted in the body, there will be biological and chemical reactions occurring at the implant location as well as in the body. For biocompatibility, the implant should not have any toxic materials leaking out of the package and the package should be designed to minimize biological reaction to the implant and the surgical operation. The small size, the smooth surface, the use of biocompatible packaging materials and techniques, as well as the proper sterilization and surgical operation can reduce biological reactions to the implant. For good implantable systems, inflammatory reactions should be minimized and the thickness of the connecting tissues should be kept thin, in the micrometer range.

For the protection of the implantable device/system and in order for it to operate properly throughout its designated lifetime, the implant should be designed and packaged to achieve the following: (i) To prevent the electrical damage from the corrosive conducting body fluid to the electronic circuits and devices. The leakage resistance between all parts of the implantable system should be maintained high (with sheet resistance on the order of 100 MΩ/square), to ensure proper operation of the system. This can be achieved by hermetic packaging using compact metal, ceramic and glass boxes and hermetic feed through, or alternatively by non-hermetic packaging with polymeric materials and thin film vapor barriers [53, 54]; (ii) To protect the system from mechanical damage that may be caused by improper handling during the implant operation and after implantation; and (iii) To withstand the harsh chemical and biological environment throughout its lifetime without degradation. Many electronic devices are sensitive to light; therefore, some form of shielding would be desired during the testing or preparation stages.

In the biocompatibility area, there are many unsolved problems related to the body reaction and packaging. For example, the correlation between the thickness of the connecting tissue surrounding the implants and the package as well as implant processes is not yet fully understood. The non-hermetic packaging technology is not fully developed. The technology needs to be better understood, and the techniques to stabilize and integrate the chronic implant with surrounding tissue need to be further developed. The body environment is dynamic and has conductive and corrosive body fluids; the body also generates antibodies that attack foreign materials including implants. The surrounding tissue also may change its properties with time. Sometimes, implants may be rejected by the body. The causes and the process of implant rejection need to be further studied [55, 33].

18.3.1.3 Characteristics of Biological and Medical Signals

Biological and medical signals have different characteristics when compared with industrial, engineering, and scientific signals in the requirements for accuracy, sensitivity, reliability, and signal bandwidth. When measuring biomedical signals, accuracy may be less important than sensitivity. For biomedical signals, such as blood pressure, heart rate, respiration rate, and so on, their normal ranges and useful characteristics may vary between individuals. Even with the same person, the signal may vary by a few percentages from time to time and may be affected by many body and mental activities and general health status. The quantitative effects of these personal activities on the biomedical signals are not precisely understood at present. Therefore, for measurement instruments designed for these signals now, the required accuracy is generally not as high as in other scientific research. One percent full-scale accuracy is generally sufficient, although the sensitivity or resolution may need to be very high because the signals are weak. The input electrical signals are usually in the millivolt (mV) to microvolt (μV) to nanovolt (nV) ranges.

Most biomedical measurement instruments will be used by nontechnical personnel under strenuous circumstances and high reliability, ease of operation, and robust structure of these instruments may be required beyond the levels set for general instruments used in an engineering laboratory.

Since biomedical signals are related to the molecular manipulations in the body and not to RF processes, the frequency bandwidth required for biomedical signals are generally lower than that for electronic signals. In general, the bandwidth of biomedical signals is below 10–30 kHz, and many clinical instruments will have signal bandwidths limited to just below 50/60 Hz – the AC power frequency – to avoid the strong interference of AC power lines. However, signal amplitude may be in the range from millivolt to microvolt levels for the organ and body signals, and may extend to nanovolts or below for cellular and subcellular signals.

18.3.2
Design Considerations of Implantable Systems

When designing electronics for implantable or surface attached biomedical systems, besides the basic and common considerations related to biocompatibility, reliability, and ease of operation as discussed before, there are fundamental considerations of size and weight, the body reaction, and the package used to protect the implantable system. When translated to engineering design of the electronic system the major considerations are (i) use of micropower components and circuits to reduce power consumption including micropower active devices, micropower electronic circuits (use ASIC whenever possible), and micro sensors and micro actuators; (ii) selecting the smallest power supply suited to the application. The system may use (a) the smallest reliable power supply that can last the required life time; (b) a passive wireless power unit; or (c) a rechargeable thin battery and wireless charge circuit to charge the battery and to turn on/off the system; and (d) energy scavenging from inside the human body; (iii) packaging the system to minimize harmful body reactions and having adequate protection to the system in the body environment as well minimizing volume and weight [1]. *All medical devices or instruments should be designed with **reliability and safety** as the prime considerations above all cost and technical factors.*

There are a few interesting electronic design approaches that were developed for micropower electronics for implantable MEMS systems in the early development stage of implantable systems before the LSIC was developed. They may still be of value in the present when we look for surface-attached instruments for individualized mobile monitoring and early warning systems. These approaches are (i) using micropower devices at low voltage, low current such as tunnel diodes and sub threshold MOS transistors [67, 68]; (ii) integrating sensor/actuator with circuits using micropower ASIC; (iii) using multifunctional electronic circuits; (iv) sharing battery current by cascade electronic circuits; (iv) splitting lithium battery voltage to two or three low voltage supplies for the electronic circuits. Some examples are given in Section 18.3.3 to explain these approaches.

18.3.3
Micropower Electronic Design Approaches and Samples

As discussed in Sections 18.3.1 and 18.3.2, the implant system design has different considerations than other RF MEMS systems. The implantable systems are operating in a very harsh environment the implantable systems are operating in a very harsh environment. Therefore, the material used for the implant should be nontoxic and properly packaged to protect the tissue, the implant, and the host body from any harmful reactions.

Implant operations are expensive and time consuming and the implant device takes a long time to reach stable operation after it is implanted. Therefore, when designing implantable systems, reliability and safety is more important than many quality factors such as broadband response and high accuracy.

Biological rhythms generally repeat in fractions of a second to many seconds per cycle. Therefore, the bandwidths of biomedical signals are usually in the range from 0.01 Hz to 10 kHz.

For long-term implants with lifetimes from months to years, the basic requirements are as follows: (i) The volume and weight of the packaged implant device shall be smaller and light so that the device will not affect the normal activity of the host. It is generally agreed that 2% of the host body size and weight are the upper limit. (ii) Biocompatibility – the implant shall not contain any toxic or harmful materials and the package shall be able to protect the body and the implant from any harm. (iii) The implant system – implant and the external unit – shall be safe, reliable, and easy to operate by nontechnical biomedical personnel [69].

When designing long lifetime implant electronic systems to meet the size, weight, and other requirements, all the building blocks in Figure 18.1 need to be considered. The largest and heaviest blocks are the chronic power supply and the biocompatible package. They have been the most difficult problems, from the early stages of the field to the present.

With the advances of LSICs and MEMS/NEMS, the interface electronics and the RF link can be designed to achieve micropower consumption that leads to minimal total size and weight. Several approaches were developed in the early stages that may still be of interest today. A few examples are given below:

1) Use of micropower electronic devices such as tunnel diodes [56], low current/voltage transistors, and subthreshold operation of transistors.
2) Integrating R, C, and sensors with all the signal-processing circuits into a single ASIC chip.
3) Use of low-duty-cycle (nanosecond duration) pulse frequency or pulse position modulated RF approach to greatly reduce the power consumption by a factor equal to the duty cycle, which can be on the order of 10^{-2}–10^{-6} [56].

4) Splitting the power supply voltage to share the same current for several devices/circuits, such as using two tunnel diodes to share the same battery current with a low voltage transistor, as shown in Figure 18.19 where the 1.35 V battery is shared by a transistor and two tunnel diodes (as subcarrier oscillators). The two-channel telemetry unit consumed less than 1 milliwatt power.

5) Multiple functions of one active device. Figure 18.20 shows the single antenna, two-frequency (one for RF powering and one for signal transmission) circuit. Normally Q2 and Q3 are off, the Q1 FET functions as a diode to rectify the RF power at the frequency tuned by L and (C1, C2); when there is transmitting signal from the device, Q2 and Q3 are on, and Q1 and Q3 function as oscillators and RF transmitters at the frequency of L and C2 [57]. Two or three signal bands with very different frequency bands can share a single amplifier by using filters to separate the amplified signals to each band, such as that used in the early stages of radio receivers.

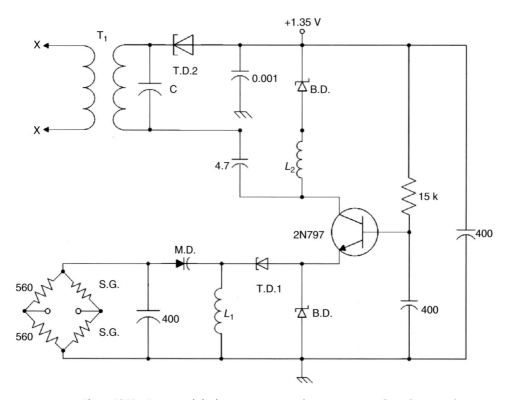

Figure 18.19 Two tunnel diodes are in series with a transistor to share the 1.35 V battery current.

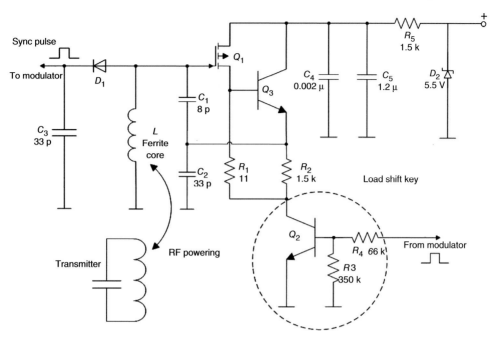

Figure 18.20 The single-antenna two-frequencies RF-link circuit [57].

18.3.4
Power Supply Design

On the selection of power supplies for the systems, batteries are a common choice. There are several new developments in high energy density batteries. Among them the thin film lithium (primarily rechargeable) batteries are frequently used. The selection considerations should include (i) the volume and weight per unit energy stored (current × time × voltage), (ii) low charging time, (iii) small internal resistance or large discharge pulse current, (iv) large recharge cycles during the lifetime, and (v) sufficient storage and operation temperatures to allow the battery to be sterilized by steam or other common methods.

When rechargeable batteries are used or the battery-less passive wireless powered approach is selected, the RF power or charging circuit can use the following: (i) Single frequency time sharing for powering and signal transmission with shared antenna (see Figure 18.18 and reference). (ii) Two frequencies for powering and signal transmission, each with its own antenna and RF electronic circuits. Wireless charging of small batteries or super-capacitor approaches is popular. Wireless charging can use (i) the magnetic field of a near-field RF radiation, (ii) low frequency (kHz) time-varying magnetic induction field, or (iii) sonic or ultrasonic vibration with sonic to electrical convertor transducers, such as PZT. The wireless charging with on–off control would be the desired approach whenever it is used.

All RF and wireless powering/charging methods have very low power transfer efficiency (10^{-2}–10^{-6}) and drop off quickly with increased spacing and misalignment between primary and secondary units. When the wireless power source is used to power deep implants, the field intensity on the body surface may be much greater than that at the implant system site. Careful consideration should be made to protect the skin from being burned or harmed. In all these RF charging and powering situations, the surface and tissue power loss density should be kept below the legal limits for body exposure (for example, the tissue temperature rise should be smaller than 20 °C). When wireless systems are implanted in moving subjects, the sharp drop off of the received power due to misalignment and the change of coupling coefficients can be a big problem. The use of all-direction powering antenna or all-direction receiving detectors would be needed. A three-direction RF power receiving coil is shown in Figure 18.21. High-efficiency, high-power RF powering may be achieved by using an external ferrite core to focus the external powering magnetic field to maintain alignment with one of the receiving coils as the body moves around [1].

Implantable body energy harvesting and distribution of harvested energy throughout the body are interesting fields of research and challenge. The safe long-term nuclear power sources would also be interesting to explore [39]. For a discussion on implantable power selection, the reader is referred to the literature on thin-film lithium rechargeable batteries [58].

18.3.5
System Integration and Micro-Packaging

For large-scale implant systems with many channels for telemetering and stimulation, the system integration and component layout need careful consideration. The main concerns are unwanted coupling and interference; near uniform density of components; location of various sensors separately from the electronic circuits; and antenna location to reduce RF loss and frequency drift caused by nearby components. These are similar to the layout problems of non-implantable electronic system design.

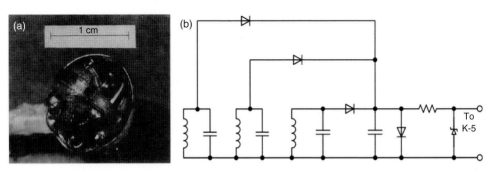

Figure 18.21 The three-dimensional RF power-receiving coil structure and circuit diagram. (a) Structure of a 3D RF power unit and (b) circuit diagram of the RF power unit.

For the packaging of implant systems, as discussed under general design considerations in Section 18.3.3, the package should be biocompatible to protect the host body and to protect the implant unit by (i) protection of the implant from water vapor and ionic chemicals to cause low resistance leakage and component failure; (ii) protection of the implant from mechanical damage during the implant operation and in the body; and (iii) ensuring the proper functioning of the implant over the desired lifetime.

The conventional implant packages use hermetically sealed boxes made of nontoxic metal, ceramic, and glass. They can have very long lifetimes, but generally are large, heavy, expensive, and have long turnaround time. The volume and weight of hermetic packages are much larger than the unpackaged MEMS implantable systems using MEMS sensors/actuators and ASIC. Furthermore, the sensors that need to communicate with the body's environment cannot be boxed and the feed-through and interconnecting leads are not reliable. New micropackaging technology is being developed for microimplant systems [59, 60]. Micro package technology are being developed to provide small volume, light weight, low cost packaging of MEMS implant systems for short and long term implant devices, [61].

18.4
Present Challenges

The potential of microfabricated implant systems is great. They have enabled us to reach into the body to monitor the body's operation and that of the internal organs, as well as to influence them when needed by electrical stimulation or chemical intervention. To reach and realize the potential of implant systems there are still great challenges that we are faced with at the present!

Although MEMS/NEMS transducers and IC technologies have reduced the size and power consumption of sensory and electronic systems by several orders of magnitude since 2000, the integration of transducer and LS-ASIC to further reduce the size, weight, and power consumption of implant systems is the real challenge now [71]. The power consumption per telemetry channel for 1 kHz bandwidth is about 1–5 µW for the advanced implantable systems at present. If it is possible that MEMS/NEMS technology can be developed to reduce the per telemetry channel power by two orders of magnitude to the 10 nW level using electronic devices that require supply voltages in the range of 0.2–1.0 V, then the energy storage component and the implantable systems package can have their volume reduced from the larger than 1000 mm^3 now to 10 mm^3. Many future applications of injected implant systems on or in body organs could then be realized. The remaining challenge would be in micropackaging and a wireless micropower supply network within the body [70].

In the micropackaging area, technologies need to be developed to provide micropackages that add small package weight (0.1 g) and volume (mm^3) to the system, with an implant lifetime from months to 10 years, while still satisfying all the requirements of good packages. The non-hermetic packages that use polymeric materials such as epoxy, silicone, parylene, and ceramic thin films as coating materials are being actively pursued [60]. When properly developed, a micropackage may be made with very small volume and weight and can be processed quickly at low cost, but the lifetime is expected to be shorter, ranging from several months [62] up to 2–10 years [59, 60, 63].

In the power supply area, present micropower electronic design efforts to reduce the power requirements for signal processing and wireless communication are aimed to reach sub-microwatts per channel consumption (0.1–0.01 µW per channel). However, in the future there will be several or many implant systems in one body and so the individualized power supply approach may need to be changed to some form of wireless power network. In this case, the source of energy from external or body energy harvesting or from extracting bioenergy in the body (body fluid fuel cells, pH difference, body potential difference in the cochlea, etc.) can be wirelessly distributed to power all the implant systems while all the information from different implant systems would also be collected and transmitted together. Some research is being initiated in these areas in US universities.

These theories and technologies need to be developed and practical tools and processes need to be proved reliably and verified consistently. Great challenges are upon us.

18.5
Future Trends

From 1960 to 2010, implantable electronics systems have made tremendous progress thanks to the relentless miniaturization of transducers and electronics and the efforts of implant instrumentation. The pioneering techniques of short-range MEMS wireless remote monitoring and control have propagated from biomedical research and aerospace applications into clinical health care and to industrial, environmental, and social communication fields. Short-range RF links (powering and signal transmission) are everywhere in our daily life now. On the basis of our experience and understanding of the challenges in implantable systems for biomedical research and health care, we suggest that the future trends of implantable, in-dwelling, and attachable systems for biomedical applications may include the following:

1) Nanoscale transducers and electronics that can further miniaturize the system volume and weight, and enable minimally invasive, higher performance implantable systems, for research on organs, tissues, and cells. Many of these would be used for clinical care of these parts.
2) Individualized medical health care will be made possible, where normal baseline health information can be recorded and monitored through RF links to healthcare institutions while the person assumes normal activities at home or in the office. Deviations from the individualized norm would initiate preventive actions to avoid serious medical problems from developing. The

system needs to be reliable, easy, and comfortable to use, and available at low cost.

3) Home and online (mobile) monitoring and treatment systems will be available. When trends A and B are accomplished, the personalized monitoring and warning system can be extended to mobile locations such as during travel or away from office. Medical information collection and remotely controlled treatment can be accomplished through cell phones or other mobile communication systems.

4) On-organ monitoring and therapeutic devices realized. When the implantable device is small and reliable, on-organ monitoring and therapeutic devices may be developed, such as cardiac pacemakers and defibrillators on the heart, and powered by the energy harvested from the heart's motion [64, 65]. Similarly, on the liver, on the kidney, on the bladder, on the brain, and other on-organ function monitoring and therapeutic devices may be developed by implant system engineers teaming with healthcare researchers.

5) Larger scale multiple channel (1000 and up) brain–machine interconnected networks for biology and physiology research, medical diagnosis, prosthesis, and treatment of body malfunction may be developed.

6) Artificial tissues and organs such as blood vessels, muscle tendons, and artificial kidney, liver, pancreas, lung, heart, and digestive channels may be developed by multidisciplinary teams including scientists, engineers, and biomedical and clinical researchers.

7) Large-scale multichannel monitoring and control systems as well as brain–computer interfaced RF MEMS networks for biomedical as well as for engineering, industrial, social, environmental, and energy operation and research may be studied.

8) Bioelectrical energy and body energy harvest may be used to power wireless power networks to power all implant or attached systems regularly and continuously without care from individual patients.

Acknowledgments

The authors are grateful for the partial support from the NIH grant from NIBIB #RE B014442A. The editorial assistance of Dr. Christ Zorman and Dr. Linda Ko-Ferrigno and editorial help of Dr. Peng Wang are gratefully acknowledged.

References

1. Ko, W.H. and Neuman, M.R. (1967) Implant biotelemetry and microelectronics. Science, 156(773), 351–360.
2. Prescott, G.B. (1884) Bell's Electric Speaking Telephone: Its Invention, Construction, Application, D. Appleton & Company, New York, Reprinted as a Google e book, ASIN: B0021A0150.
3. Fuller, L. and Gordon, T.M. (1948) The radio inductograph—a device for recording physiological activity in unrestrained animal. Science, 108, 287.

4. Schmitt, O.H. (1948) A radio frequency coupled tissue stimulator. *Science*, **107**, 432.
5. Hallpike, C.S. and Hartridge, H. (1937) Electrical stimulation of the human cochlea. *Nature*, **139**, 192.
6. Kellaway, P. (1946) The mechanism of the electrophonic effect. *J Neurophysiol.*, **9**, 23–32.
7. Hyman, A.S. (1932) Resuscitation of the stopped heart by cardiac therapy: II. Experimental use of an artificial pacemaker. *Arch. Intern. Med.*, **50**, 283.
8. Geddes, L.A. and Hoff, H.E. (1971) The discovery of bioelectricity and current electricity The Galvani-Volta controversy. *IEEE Spectrum*, **8**(12), 38–46.
9. Schladweiler, P. (1965) Movements and activities of ruffed grouse (Bonasa umbellus L.) during the summer period. MS thesis. University of Minnesota, St. Paul
10. Mackay, R.S. (1964) The application of physical transducers to intra-cavity pressure measurement, with special reference to tonometry. *Med. Electron. Biol. Eng.*, **2**, 3–19.
11. Craighead, F.C. Jr., and Craighead, J.J. (1965) Tracking grizzly bears. *BioScience*, **15**(2), 88–92.
12. Mackay, R.S. and Jacobson, B. (1957) Endoradiosonde. *Nature*, **179**, 1240.
13. Jacobson, B. and Mackay, R.S. (1957) A pH endoradiosonde. *Lancet*, **1**, 224.
14. Connell, A.M. and Rowlands, E.N. (1960) New methods for diagnosis and research—Wireless telemetering from the digestive tract. *Gut*, **1**, 266–272.
15. Eklund, C.R. and Charlton, F.E. (1959) Measuring the temperatures of incubating penguin eggs. *Am. Sci.*, **47**(1), 80–86.
16. Caceres, C.A. (ed.) (1967) *Bio-Medical Telemetry*, Academic Press, New York, June 3 1960.
17. Mackay, R.S. (1959) Radio telemetry from within the human body. *IRE Inst. Trans. Med. Electron*, **ME-6**, 100–105.
18. Mackay, R.S., Burton, L.B., and Clark, R.H. (1964) Dolphin telemetry. *Science*, **145**(3629), 296.
19. Collin, C. (1966) A study of the mechanism of sencery evoke pressure response in the rabbit eyes. Doctoral thesis. University of California, Berkeley, and Mackay Book, p. 19, Fig. 1.12.
20. MacKay, R.S., Marg, S.E., and Raymond, O. (1960) Automatic tonometer with exact theory: various biological applications. *Science*, **131**(3414), 1668–1669.
21. Slater, L. (ed) (1963) *Bio-Telemetry*, Pergamon, New York.
22. Slater, L. (1966) *Survey on Implant Biotelemetry*, American Institute of Biological Sciences.
23. Ko, W.H. (1960) Tunnel diode FM wireless microphone. *Electronics*.
24. Ko, W.H. (1965) Progress in miniaturized telemetry. *Bioscience*, **15**, 118–120.
25. Ko, W.H., Conrad, D., Yon, E.T., and Hynecek, J. (1971) A general purpose micropower monolithic telemetry system. 9th International Conference on Medical and Biological Engineering, Melbourne, Australia.
26. Ko, W.H., Blaser, E., and Yon, E.T. (1971) A monolithic biomedical pressure sensor utilizing the piezoresistive effect. 9th International Conference on Medical and Biological Engineering, Melbourne, Australia.
27. Ko,W.H. (1966) Piezoelectric energy converter for electronic implants. 19th Annual Conference of the Society for Engineering in Medicine and Biology, San Francisco, CA, p. 67.
28. Ko, W.H., Liang, S.P., and Fung, C.D. (1977) Design of radio frequency powered coils for implant instruments. *Med. Biol. Eng. Comput.*, **15**, 634–640.
29. Ko, W.H. and Lin, W.C. (1968) A study of microwatt power pulsed carrier transmitter circuits. *Med. Biol. Eng.*, **6**, 309–317.
30. Lin, W.C., Ruffing, F., and Ko, W.H. (1972) Feasibility study of engineering problems in multi-electrode visual cortex stimulation system. *Med. Biol. Eng.*, **10**, 365–375.
31. Ko, W.H. and Hynecek, J. (1974) Implant evaluation of a nuclear power source-betacel battery. *IEEE Trans. Biomed. Eng.*, **21**(3), 238–241.
32. Cheung, P.W., Ko, W.H., Fung, C.D., and Wong, A.S. (1978) Theory, fabrication, testing, and chemical response of ion selective field effect transistor

devices, in *Theory, Design, and Biomedical Applications of Solid State Chemical Sensors*, CRC Press, West Palm Beach, FL.
33. Ko, W.H. and Spear, T.M. (1983) Packaging materials and techniques for implantable instruments. *Eng. Med. Biol.*, **2**, 24–38.
34. Ko, W.H. (1995) Packaging of microfabricated devices and systems. *Mater. Chem. Phys.*, **42**, 169–175.
35. Ko, W.H., Yon, E.T., Mabrouk, S., and Hynecek, J. (1971) Tapped-on heart rate and EKG telemetry transmitters. *J. Assoc. Adv. Med. Instrum.*, **5**(5), 268–272.
36. Ko, W.H., Bertgmann, B.P., and Plonsey, R. (1977) Data acquisition system for body surface potential mapping. *J. Bioeng.*, **2**, 33–46.
37. Ko, W.H., Plonsey, R., and Kang, S.R. (1972) The radiation from an electrically small circular wire loop implanted in a dissipative homogeneous spherical medium. *Ann. Biomed. Eng.*, **1**(2), 135–146.
38. Ko, W.H. and Kang, S.R. (1974) The radiation from AN electrically small off centered loop in a dissipative homogeneous spherical medium. *Ann. Biomed. Eng.*, **2**, 321–325.
39. Ko, W.H. and Hynecek, J. (1974) *Biomedical Electrode Technology*, Academic Press, New York, pp. 169–181.
40. Leung, A., Ko, W.H., Lorig, R.J., and Cheng, E.M. (1976) Design of a two-channel intracranial pressure and temperature telemetry system. 20th Annual Conference on Engineering in Medicine and Biology, p. 420.
41. Amlaner, C.G. and MacDonald, D.M. (eds) (1986) *A Handbook on Biotelemetry and Radio Tracking*, Pergamum Press, Oxford and New York.
42. Shealy, C., Silgalis, E., and Ko, W.H. (1968) Differential blocking of peripheral nerve fibers. 21st Annual Conference on Engineering in Medicine and Biology, Section 17.3, Houston, TX, November 1968.
43. Hambrecht, F.T. and Reswick, J.B. (eds) (1977) *Functional Electrical Stimulation: Applications in Neural Prostheses*, Marcel Dekker, New York.
44. Peckham, P.H., Keith, M., and Kilgore, K. (1993) Restoration of upper extremity unction in tetraplegia. *IEEE Trans. Rehabil. Eng.*, **1**(1), 8–11.
45. Peckham, P.H., Poon, C.W., Ko, W.H., Marsolais, E.B., and Rosen, J.J. (1981) Multichannel implantable stimulator for control of paralyzed muscle. *IEEE Trans. Biomed. Eng.*, **BME-28**(7), 530–536.
46. Crago, P.E., Mortimer, J.T., and Peckham, P.H. (1980) Closed-loop control of force during electrical stimulation of muscle. *IEEE Trans. Biomed. Eng.*, **27**(6), 306–312.
47. Miklavčič, D., Kotnik, T., and Serša, G. (eds) (2003) *Lojze Vodovnik –Zbrana Dela (Collected Works)*, University of LjubljanaImprint International d.o.o.
48. Grotz, R.C., Yon, E.T., Long, C., and Ko, W.H. (1964) Intramuscular FM radio transmitter of muscle potential. Proceedings of the Congress of Physical Medicine and Rehabilitation, Chicago, IL.
49. Lorig, R.J., Vodovnik, L., Reswick, J., et al. (1967) An implantable system for the myo-electric stimulation of skeletal muscle using radio frequency links. 20th Annual Conference on Engineering in Medicine and Biology, Section 28.6, Boston, MA.
50. Receveur, R.A.M., Lindemans, F.W., and de Rooij, N.F. (2007) TOPICAL REVIEW-microsystem technologies for implantable applications. *J. Micromech. Microeng.*, **17**, R50–R80.
51. Poellabauer, C. (2011) Pervasive Health, http://www.cse.nd.edu/~cpoellab/teaching/cse40816/Lecture6.pdf.
52. eCFR (2011) Electronic Code of Federal Regulations Part 15, Part 18, Part 90 and Part 95. Federal Communication Commission.
53. Lachhman, S. (2011) Evaluation and accelerated lifetime studies of medical grade poly-dimethylsiloxane as an encapsulation material. MS thesis. EECS Department, Case Western Reserve University, Cleveland, OH.
54. Wong, C.P. (1998) Polymers for encapsulation: materials processes and reliability. *Chip Scale Rev.*, **2**(1), 30.

55. Ko, W.H., Neuman, M.R., and Lin, K.Y. (1969) in *Body Reaction of Implant Packaging Materials*, Vol. 10 (ed. L. Stuart), Plenum Press, New York, pp. 55–65.
56. Ko, W.H., Yon, E.T., and Hynecek, J. (1970) Micropower pulse modulated telemetry transmitter. Proceedings of the 23rd Annual Conference on Engineering in Medicine and Biology, Washington, DC, p.119.
57. Hynecek, J. and Ko, W.H. (1975) Single frequency RF powered telemetry system. Abstract from 28th ACEMB–Fairmont Hotel, New Orleans, LA, p. 97.
58. Ko, W.H. (1979) Power sources for implant telemetry and stimulation systems. Proceedings of an International Conference on Telemetry and Radio Tracking in Biology and Medicine, Oxford, UK, March 1979. Also in Amlaner, C.G. (1980) in *A Handbook on Biotelemetry and Radio Tracking* (ed. D.M. MacDonald), Pergamum Press, Oxford and New York.
59. Lachhman, S., Zorman, C.A., and Ko, W.H. (2012) Multi-layered polydimethylsiloxane as a non-hermetic packaging material for medical MEMS. 34th Annual International Conference of the IEEE Engineering in Medicine and Biology Society, San Diego, CA.
60. Xie, X.Z., Rieth, L., Tathireddy, P., and Solzbacher, F. (2011) Long-term in-vivo investigation of Parylene-C as encapsulation material for neural interfaces. *Procedia Eng.*, **25**, 483–486.
61. Wang, P., Lachhman, S.B., Sun, D., Majerus, S.J.A., Damaser, M.S., Zorman, C.A., Feng, P.X.-L., and Ko, W.H. (2013). Non-hermetic micropackage for chronic implantable systems. Presented at the IMAPS 46th Int. Symp. on Microelectronics Orlando, FL, USA.
62. Bu, L.P., Cong P., Kuo, H.I., Ye, X.S., and Ko W.H. (2009) Micro package of short term wireless implantable microfabricated systems. Proceedings of the IEEE EMBC Annual Conference, Vol. 31(1), pp. 6395–6399.
63. Xie, X., Rieth, L., Merugu, S., Tathireddy, P., and Solzbacher, F. (2012) Plasma-assisted atomic layer deposition of Al2O3 and parylene C bi-layer encapsulation for chronic implantable electronics. *Appl. Phys. Lett.*, **101**, 093702.
64. Feng, P.X.L. and Ko, W.H. (2010) Bio-implantable piezoelectric microsystems for heart health monitoring and self-powered cardiac arrhythmia management. Pending Disclosure Case No. 2012–2124.
65. Zhang, R., Chen, Y., Ko, W.H., Rosenbaum, D.S., Yu, X., and Feng, P.X.L. (2011) Ex vivo monitoring of rat heart wall motion using piezoelectric cantilevers. Proceedings of 5th Joint Conference of the 65th IEEE International Frequency Control Symposium (IFCS 2011) and 25th European Frequency and Time Forum (EFTF 2011), San Francisco, CA, pp. 868–873.
66. IEEE Standards Association (2005) IEEE C 95. *IEEE Standard for Safety Levels with Respect to Human Exposure to Radio Frequency Electromagnetic Fields 3 kHz to 300 GHz*, IEEE Standards.
67. Ko, W.H. (1961) Designing tunnel diode oscillators. *Electronics*.
68. Ko, W.H. (1962) Tunnel diode oscillator delivers RF and audio. *Electronics*, **35**(41), 56.
69. Ko, W.H. et al. (1995) Packaging of microfabricated devices and systems. International Conference on Electronic and Materials, Hsinchu, Taiwan, ROC, December 19–22, 1994, Paper published in *Materials Chemistry and Physics*, Elsevier, Vol. 42 (3), pp. 169–175.
70. Nichols, M.F. (1994) The challenges for hermetic encapsulation of implanted devices – a review. *Biomed. Eng.*, **22**(1), 39–67.
71. Narasimhan, S., Chiel, H.J., and Bhunia, S.K. (2011) Ultra low-power and robust digital signal processing hardware for implantable neural interface microsystems. *IEEE Trans. Biomed. Circuits Syst. (TBioCAS)*, **5**(2), 169–178.

19
Brain–Computer Interfaces: Ethical and Policy Considerations
Ellen M. McGee

19.1
Introduction

The time of the cyborg has arrived. Revolutions in semiconductor device miniaturization, bioelectronics, and applied neural control technologies are enabling scientists to craft machine-assisted minds, science fiction's "cyborgs [1]." A paper published in 1999 [2] sought to draw attention to the advances in prosthetic devices, to the myriad of artificial implants, and to the early developments of this technology in cochlear and retinal implants. The concern, then and now, is to draw attention to the ethical issues arising from these innovations. Since that time, advances have occurred at a breathtaking pace. Scientists, researchers, and engineers, using differing methodologies, are pursuing the possibilities of direct interfaces between brain and machine.

Technological innovations as such are neither good nor evil; it is the uses devised for them that create moral implications. As there can be ethical problems inherent in the proper human use of technologies, and as brain chips are a very likely future technology, it is prudent to formulate policies and regulations that will mitigate their ill effects before the technologies are widespread. Unlike genetic technologies, which have received widespread scrutiny, brain–machine interfaces (BMIs) have received little social or ethical analysis. However, since genetic enhancements are inherently limited by biology and the single location of an individual, whereas hybrids of humans and machines are not so restricted, the potential of this technology to change and significantly affect humans is potentially far greater than that of genetic enhancements. Today, intense interest is focused on the development of drugs to enhance memory; yet, these drugs merely promise an improvement of normal memory, not the encyclopedic recall of a computer-enhanced mind combined with the ability to share information at a distance. The potential of brain chips in transforming humanity is astounding. BMIs, also referred to as brain–computer interfaces (BCIs), or neuromotor prostheses (NMP), are technological interventions that establish direct communication pathways between the brain and an external device. The interface interprets signals from an array of neurons and uses computer chips and programs to translate the signals into a desired action. As this technology

Implantable Bioelectronics, First Edition. Edited by Evgeny Katz.
© 2014 Wiley-VCH Verlag GmbH & Co. KGaA. Published 2014 by Wiley-VCH Verlag GmbH & Co. KGaA.

is progressively developed and implemented, humanity will increasingly have an intimate relationship with machines and become cybernetic organisms, humans who are intrinsically coupled to bioelectronic devices.

This chapter describes advances in hybrid BMIs, offers some likely hypotheses concerning future developments, reflects on the implications of combining cloning and transplanted brain chips, and suggests some potential methods of regulating these technologies. It proceeds from the standpoint of neuroethics.

19.2
Neuroethics

Neuroethics is a term used to describe the study of the ethical and social implications of new technologies from neuroscience, including consideration of how neurotechnology can affect humanity. Inasmuch as neuroethics is a subfield of bioethics, it has adopted the principles and rules commonly utilized in the field of bioethics and applied them to the issues arising in the neuroethical domain. Thus, as in bioethics, the primary working principles employed are nonmaleficence, beneficence, justice, and respect for autonomy [3]. Nonmaleficence is concerned with the responsibility not to intentionally harm another, beneficence is the requirement that if one can do good one has an obligation to do so, justice refers to the fair allocation of scarce resources, and respect for autonomy emphasizes the duty to value the self-direction of persons. Thus, anxieties arising from safety and efficacy, which fall under both maleficence and beneficence, need to be addressed, as well as issues of fairness and justice. In particular, the ramifications of BCIs for privacy and autonomy must be examined. Insofar as possible, regulatory provisions to mitigate deleterious outcomes should be implemented. Indeed, the broader question of whether humanity should proceed toward this future needs to be raised, debated, and answered.

19.3
Brain–Computer Interfaces

Two kinds of interfaces can be identified – those that input to the neural base and those that output or record electrical brain signals. Interfaces that input to the neural base include clinical devices that aim to restore function to body systems.

Input interfaces are comprised of three varieties that are presently undergoing research: noninvasive, partially invasive, and invasive. A report issued in 2006 noted that there are 175 American, 69 European, and 54 Asian laboratories pursuing BCIs [4]. In North America, the majority of these laboratories are pursuing invasive BCIs, while those pursued in Europe emphasize noninvasive modalities, and the development of biologically inspired robots.

19.4
Noninvasive Interfaces

Noninvasive neural interfaces record brain activity from an external device mounted on the scalp. Recording of electrocorticographic activity from the cortical surface has been used to create games that read alpha and beta waves, and to allow patients, after extensive training, to detect, modify, and use a computer to direct a cursor on a screen or to control lights, TV, and stereo sound [5]. Electroencephalography (EEG) recording is the most widely studied noninvasive BCI; in addition, magnetoencephalography and functional magnetic resonance imaging are employed. These noninvasive methods suffer from poor signaling resolution due to interference from the skull, and the intensive and demanding training needed to operate the technology. Nevertheless, several commercial models, from companies such as Emotive and Neurosky, are available to control gaming systems, to provide educational applications, and to investigate medical applications [6]. The company InteraXon has created a suite of brain training games, and has introduced a device called *MUSE* which measures brainwaves, and "allows you to control games, reduce stress, improve memory and concentration, and eventually to control devices directly with your mind" [7]. The first commercial effort at a computer interface designed for patients with locked-in syndrome, the Intendix, lets users input text using only their brains; another application lets users create paintings.

19.5
Partially Invasive Interfaces

Other than noninvasive interfaces, slightly better results have been achieved using partially invasive BCIs. In this type, the device is implanted inside the head, but on the surface of the brain rather than inserted into the brain. These have the advantage of producing better signals than noninvasive devices, but, unfortunately, also have the potential to form scar tissue in the brain. Electrocorticography (ECoG) is the most frequently used partially invasive technique. It uses the same technology as noninvasive EEG, but the electrodes are placed outside the dura. In one application, researchers Eric Leuthardt and Daniel Moran [8] from Washington University in St. Louis have enabled a teenage boy to play Space Invaders using his ECoG implant.

19.6
Invasive Interfaces

The third type of BCI is surgically implanted directly into the gray matter of the brain. These devices produce the highest quality signals, and as materials are developed which permit long-term implantation without scar tissue buildup and without degradation of the signal, they will become the preferred method of enhancement because of their superior qualities.

19.7
Development of Brain–Computer Interfaces

Devices produced to restore functions to disabled limbs, hearts, and brains have facilitated the development of BCIs. The field has progressed forward in stages, marked by advances in prosthetic technology and computer science. Initial use of prosthetics, used to restore function to disabled limbs, has evolved from the use of crutches and peg legs to the development of devices that employ microprocessors becoming "bionic" joints, artificial knees, and smart legs. Research has now progressed to a "biohybrid" limb actually using brain signals to directly control a prosthesis [9].

At the same time, revolutions in bioengineering and implant technology have resulted in millions of people using implants (pectoral, testicular, chin, calf, hair, hormonal, dental, and breast) and millions more using cardiac pacemaker and cardiac assist devices. In developing these implants, safer interfaces between neural tissues and the substrate micro probes have been produced.

Cochlear implants, which successfully make hearing possible for totally deaf individuals, have been in use since the 1980s, and have over 220 000 users worldwide. Prosthetic vision, studied since the 1960s, is in an earlier stage of development; it employs a diversity of visual stimulating implants: retinal, cortical, optic nerve, and biohybrid. One type, the retinal visual prosthesis, delivers direct electrical stimulation to those cells that carry visual information to the brain from a tiny camera; data is then wirelessly transmitted to a microelectronic prosthesis. Another type of vision prosthesis, the cortical implant, directly stimulates the visual cortex, and can be either surface type implants, such as the Dobelle implant, or penetrating ones. An artificial vision system, announced in January 2000, enables the blind, using a cortical implant, to navigate independently, to "read" letters, and through a special electronic interface to "watch" television, use a computer, and access the Internet [10]. In 2013, an FDA panel recommended approval of Argus II, a retinal prosthesis system designed to restore vision to the blind [11]. In the future, the use of nanoscale electronic components should result in the production of high-quality vision.

Applied neural control is also used for the bladder, to help Parkinson's patients control tremors, to treat epilepsy, and to mitigate intractable depression. Deep brain stimulation (DBS) utilizes "brain-pacers" that connect wires from a device implanted in the chest to targeted areas of the brain to control unwanted symptoms. There are more than 80 000 people already implanted with these devices, treating, in addition to Parkinson's disease, depression and epilepsy, dystonia, essential tremor, pain conditions, and obsessive-compulsive disorders. A second generation of these devices will be capable of responding to brain activity, shutting off when not needed, and increasing action when required. The Rehabilitation Nano Chip (ReNaChip) is bidirectional – it is involved in monitoring and regulating both electrical input and output [12]. The development of this capacity, where output interfaces record electrical signals from the brain, foreshadows the emergence of new forms of connection to humans. Neural activity can and will be monitored, interpreted, and directed. These devices will permit the decoding of human intentions. Initially, the intentions studied are those that presage limb movement, allowing actions to be initiated by paralyzed patients. The Modular Prosthesis Limb (MPL) project is planning to test a mind-controlled arm [13]. In June 2011 scientists published results of experiments that demonstrated creation of an implant for the hippocampal system that restored lost memory function in rats. In addition, they were able to strengthen memory capacity in normal rats [14].

These neuroprosthetics represent a step toward the emergence of more comprehensive BCIs, which involve the connection of the brain with a computer system. BCIs, where the interface is surgically implanted in the brain, provide for greater energy efficiency and will eliminate the need for TVs, newspapers, GPS units, and cell phones or other separate devices [15]. Because they are more energy efficient, have greater bandwidth, and are invisible, it is probable that these devices will be preferred to partially invasive or noninvasive interfaces.

Kevin Warwick, through projects entitled Cyborg 1 and 2, has conducted experiments involving the insertion of an active microchip into the nerves of his left arm in order to link his nervous system directly to a computer. Initially in 1998, the device served to simply open doors or turn on lights, and heaters. In 2002, with the implantation of a more complex neural system into both Kevin and his wife, the first purely electronic communication experiment between the nervous systems of two humans was achieved [16].

For the most part, however, and because of the strictures for research on humans, animals are the first research subjects, which are then closely followed by trials on human patients. In 2002, a computer-directed rat was unveiled at the State University of New York, amid suggestions that it could search for people after an earthquake, and detect explosives or fulfill other dangerous tasks [17]. In research on monkeys, a primate at Duke University was implanted with a BCI, and it succeeded in controlling, with its thoughts, a robot walking on a treadmill, continents away in Japan [18]. Brain implants have been used to detect activity in the parietal reach region of monkeys' brains, the area where higher level thoughts are initiated. This enables researchers to actually assess the degree of enthusiasm of monkeys and involves decoding higher brain functions. Such work enables understanding not only the goals and intentions of individuals, but also their moods and motivation [19]. Initial work, on linking the brain directly with both local and remote manipulators, has been demonstrated by neuroscientists at Duke University who trained a monkey to control a mechanical arm just by thinking [20]. Miguel Nicolelis, whose research has focused on decoding the brain activity of rats and owl monkeys, reported in October of 2011 that monkeys were enabled "to interpret the signals fed to their brains as a kind of artificial tactile sensation that allowed them to identify the 'texture' of virtual objects [21]." Nicolelis promises to empower by 2014 a paralyzed individual wearing a robotic exoskeleton, which will be operated by electrodes implanted in the brain, to walk up to 20 steps and kick a soccer ball [22]. In pursuit of this goal Nicolelis recently announced that his laboratory had wired the brains of two rats together, allowing signals from one rat's brain to solve a problem for the other rat [23]. These efforts to decode the brain's information processing system were jump-started by the discovery of Miguel Nicolelis and John Chapin that electrodes with flexible tips did not damage the brain and that enough information to recognize commands could be produced by decoding only a small number of the neurons in a brain.[1]

The first research to treat a human with a brain implant was led by investigators at Emory University in 1998. A "locked in" patient was enabled to communicate by using his thoughts to move a cursor [24]. This work resulted in the formation of a company to develop these interfaces for clinical use in patients. A device, called *Braingate*, allows the paralyzed to control a computer through a neural interface, and has been tested successfully on severely paralyzed patients since 2004 [25]. The first subject, a quadriplegic 25 year old, was effectively implanted with a brain chip that enabled him to check e-mail, play computer games, control a television, and turn lights on and off by thought alone; he succeeded in employing an artificial hand directed by his thinking alone. Braingate2, now in clinical trials, is expected to operate with a wireless interface, which has already been tested on non-human primates: "Sensors attached to the neurons in your brain would be implanted as with the original Braingate technology. Now, however, power, and control would be supplied by a radio frequency signal (RF) into the brain [26]."

The United States Defense Advanced Research Projects Agency (DARPA) provides support for BCI research by allocating over $24 million to support the projects of six laboratories [27]. These projects seek to manipulate airplanes and robots through the mind alone. The United States National Aeronautic and Space Administration's (NASA) *Extension of the Human Senses Group* (EHS) "focuses on developing alternative human-machine interfaces by replacing traditional interfaces (keyboards, mice, joysticks, microphones) with bio-electric control and augmentation technologies [28]." The goal of these efforts is the Cyber Soldier.

Both government-sponsored entities in Europe, Asia, and the United States, and commercial projects are involved in furthering these attempts. The commercial promise of this technology has stimulated more than 300 private companies to investigate such devices, seeking cures for conditions caused by defective nerves, brains, and spinal columns. Once clinical uses are implemented, nonclinical enhancement uses will follow. The developmental midpoint of these projects will involve humans, implanted with these devices, who will be constantly linked to each other, to their own perfect memory systems, and to the Internet. The boundaries between self and the other will be undermined; the community will be thoroughly part of the individual, the self will no longer have the possibility of being an isolated entity, nor will traditional concepts of privacy and autonomy be maintained. BMIs will permit humans not only to be constantly linked to the Internet and to cyber think, that is allow for invisible communication with others, but will also permit technology to lift information directly from the brain, even to direct the thoughts and actions of technology-enabled entities. BCIs, where a chip is implanted in the brain, will facilitate a tremendous augmentation of human capacities and a radical enhancement of the human ability to remember

1) See Ref. [27], Zimmer 2004.

and reason. Upcoming BMIs may also enable the individual to achieve immortality through cloning and brain downloading; the thoughts, memories, and emotions of an individual may be able to be downloaded and stored. Humans are already familiar with a life that is constantly online and connected to the Internet. The wearable, personal information structures used by Thad Starner [29] and the "Wear Cam" of Steve Mann [30] are early harbingers of technologies that combine wireless communication with information systems and allow the augmentation and enhancement of experiences, memories, and networking. Gordon Bell and a team at Microsoft Research, for example, have been working on creating a digital archive of his entire existence, of every aspect of his life, including all images, sounds, memories, and experiences [31]. Memoto is being marketed as the world's smallest wearable camera; it can record every instance of life and have it stored and organized [32].

In September of 2012, The *Journal of Neural Engineering* reported on "a device that improves brain function internally, by fine-tuning communication among neurons [33]." The device, a brain prosthesis, restored memory functions in monkeys, and improved decision-making.

At Johns Hopkins, surgeons have implanted a DBS device that has stimulated the growth of the hippocampus in several Alzheimer's patients [34]. The Intel Corporation is proposing that consumers will adopt brain implants to control a myriad of gadgets by 2020, dispensing with the need for I phones and keyboards to access the web [35]. The development of a wireless interface scheme, which was developed by Brown University's ArtoNurmikko, promises to facilitate the adoption of BCIs, because multiple chips can be implanted in the brain, allowing access to more neurons and permitting complicated thoughts to progress to action [36].

In a risky procedure aimed at restoring hearing for two deaf women, a penetrating device was inserted directly into the brain stem [37]. The hippocampus functions to "encode" experiences; it stores new memories, so that memories can be stored elsewhere as long-term memories. The team, working on replacing these functions of the hippocampus, has copied its behavior, rather than waiting to understand its intricacies. This shift from trying to map detailed neural function and instead exploit the user's ability to learn is expected to increase the pace of development.

Researchers at Caltech have succeeded in using implanted electrodes to detect activity in monkeys' "parietal reach region," where higher level thoughts such as "get the key and use it" are generated [38]. These results are promising for the development of neural prostheses that would enable users to move mechanical devices with thoughts, and monitor "not only patients' goals, what they want to reach for, but also their mood and motivation [39]." Researchers at Washington University, using an implanted brain grid, have found that patients can imagine moving and thus control a cursor with thought alone; their goal is to create a BMI for long-term use [40].

Toward that goal Brown University researchers have reported on the use of a durable, fully implantable, and rechargeable wireless brain sensor. It enabled the observation, recording, and analysis of signals from a multitude of neurons in the brain [41]. In the future, as nanotechnology develops and is utilized, extremely small, injectable brain-controlling devices should enable implantation of long-lasting and easily inserted devices [42].

19.8
Therapy/Enhancement

A distinction, the therapy/enhancement distinction, is commonly made between interventions which are therapeutic in their intent, used to treat disease or disability, and interventions to enhance or improve on normal species function, or to bestow entirely new capacities, nonhealth related improvements – the therapy/enhancement distinction [43]. This seemingly obvious distinction can be highly problematical, because in reality there is no bright line separating therapy and enhancement (how poor does my memory need to be to justify a remedial brain chip?); normal is a difficult concept to specify regarding many human capacities (for example, moral sensitivity), and within a societal context the norm itself may change (as the average IQ score has been rising, the so-called Flynn effect) [44]. Nevertheless, as regards brain chip implants, it seems apparent that there are real distinctions between using the technology for therapy and for enhancement. When used for therapy, implantable brain chips are seemingly noncontroversial, enabling those who are paralyzed, or naturally less cognitively endowed, to achieve on a more equitable level. The issues that arise with therapeutic use of implanted brains chips primarily involve questions of equity, access, and the costs of implementing this technology. However, these concerns will be complicated by the technology's ability to initiate a constantly changing standard of normalcy. This will be made all the more difficult to restrain, as the derivative of change will be positive – thus providing strong feedback leading to increasingly greater expectations.

More problematical technical, ethical, and social questions are raised by this technology's potential for enhancement and control of humans. Brain implants used to provide vision to the blind are seen as highly desirable devices. Extending their use, in order to provide night vision, X-ray vision, and long-range zoom capacities to the normally sighted, raises considerably different issues. Enhancement in and of itself is not necessarily objectionable; vaccines, in vitro fertilization, breast augmentation surgery, Viagra (used as a recreational agent, rather than to alleviate erectile dysfunction), are all instances of readily accepted and widely sought enhancement technologies. However, BMIs will put new forms of stress on privacy, autonomy, and justice, and more importantly, on what it means to be human. BMIs will enable humans to be constantly logged onto the Internet, and this augmented human–system interaction will assist not only those with failing memory, but might even bestow fluency in a new language, enable "recognition" of previously unmet individuals, and provide nearly instantaneous access to encyclopedic databases. BCIs promise to change the capacities of humans to such a degree that they become fundamentally different. In particular, noteworthy ethical concerns are raised by three future possibilities: (i) the prospect of using these technologies to improve and augment human capabilities; (ii) the prospect of achieving a type of immortality through cloning of an individual and implanting the clone with a chip that contains the uploaded memories, emotions, and knowledge of the clone's source; and (iii) the chance that humankind, as we know it, may eventually be phased out or become just a step in guided evolution. Humanity itself, at least those (former) members of *Homo sapiens* who have access to the technology, will be substantially different.

19.9 Ethical Issues

Ethical concerns raised by BCIs include apprehensions about safety, risk, and informed consent, issues of manufacturing and scientific responsibility, anxieties about the psychological impacts of enhancing human nature, worries about possible usage in children, concerns about increasing the divide between the rich and the poor, and most troublesome, issues of privacy and autonomy. As is the case in any technology assessment, it is unlikely that we can reliably predict all effects. Nevertheless, the potential for harm must be considered.

Safety issues include both short- and long-term risks. Short-term risks include those associated with surgery. Of greater concern are possible changes in cognition, mood, and personality, or the risks of incorrect placement of electrodes. Long-term risks are more challenging to evaluate but certainly may include immune reactions to foreign substances. These types of concerns are already present in many surgical procedures and existing malpractice law should cover these cases.

Non-implanted technologies are already being used to track children and those with dementia [45]. At present, chips have been implanted in scores of household pets, and the company, Applied Digital Solutions, originally suggested marketing its chip, the "Guardian Angel," for children and those with dementia [46]. Similarly, the emergence of rising numbers of "security cameras" has provided the ability to track people along the streets of several large urban cities. Software is under development to automatically track individuals using images from these cameras. When similar technologies are implanted, significantly novel difficulties arise. The differences are (i) that once implanted there might never be a time or place when the individual could not be tracked, (ii) remote stimulation of the brain could be used to cause behavior that the individual might not even be conscious of, and (iii) if there are regions where the individual could not be tracked/controlled, then the individual could be "trained" so that they stay out of such areas. With implantable devices, messages and information could be transmitted to the brain, actions could be initiated by remote control, and information could be transmitted both to and from the brain. Remote control of rats has already been demonstrated [47] and remote control of humans is presumably equally feasible. When chips are connected directly into a brain, signals from another human or a machine could directly impact the individual, raising the potential for power over the subject. Chips could easily be used as a way to monitor the movements of people and to transmit personal information. As noted earlier, some neural researchers believe that present experiments indicate that patient's moods as well as their motivations could be monitored. [2] While it is often possible with careful observation, augmented with physiological monitoring and interviews, to discern a subject's mood and motivations, the ability to do so remotely and potentially on a massive scale fundamentally changes the notions of privacy. One can think of a direct "Nielsen" style rating of a politician's speech, or interview – where the participants might not be able to say, "No, I don't want to participate."

The kinds of warranties users should receive and the liability responsibilities of manufacturers for defective equipment need to be addressed. Manufacturers should also make provisions to facilitate upgrades, because users presumably would want neither to undergo multiple operations nor to possess obsolete technology. Further, manufacturers must understand and devise programs for teaching users how to implement the new systems. Other practical problems with ethical ramifications include whether there will be a competitive market in such systems and whether there will be industry-wide standards for the devices. To approach these questions, we need data on the usefulness of the implants to individual recipients and on whether all of us will benefit equally.

Questions of justice also arise. In a culture such as ours with different levels of care available on the basis of ability to pay, it is plausible to suppose that the technology will be available as enhancement only to those who can afford a substantial investment, and that this will further widen the gap between the haves and the have-nots. A major anxiety should be the social impact of implementing a technology that widens the divisions not only between individuals and genders, but also between the rich and the poor. As enhancements become more widespread, enhancement becomes the norm, and there is increasing social pressure to avail oneself of the "benefit." Thus, even those who initially shrink from the surgery may find it a necessity. As a society, then, we need to think carefully about the wisdom of leaving development and dissemination of this technology to market forces.

19.10
Brain Chips and Cloning

Among the most complex of the ethical uncertainties is that raised by the prospect of combining brain chips and cloning. Cloning involves the asexual reproduction of genetically identical individuals. If the nuclear DNA from a woman were put into her own egg, the resulting child would be a complete replication of the woman [48]. Presumably, it will become possible in the near future to clone an individual, and bring it to birth.

Insofar as the self is identified with a particular body, the clone duplicates the self. Cloning the self would ensure a certain type of immortality – genetic. However, insofar as the self is other than the genetic body, fully replicating the

---

2) See Ref. [19] Begley 2004: Bw.

self requires more than biological identity. In 30 years, it may be possible to store the data representing all of a human being's sensory experiences in a storage device implanted in the body. This data could be collected by biological probes receiving electrical impulses, and would enable a user to recreate experiences, or even to transfer (transplant) memories from one brain to another. Another technique for achieving this goal is to implant a chip behind the eye in order to record all of a person's thoughts, sensations, and experiences. British Telecom's Artificial Life Team is working on this device, called *Soul Catcher 2025*, and it is estimated that it will be ready for use in 2025. Dr. Winter's claim is that "by combining this information with a record of the person's genes, we could recreate a person physically, emotionally, and spiritually [49]." In actuality, it would probably be necessary to implant multiple chips in order to capture all the sensory data that is sent to the brain. Gordon Bell's project for Microsoft's Media Presence Group is recording a lifetime's worth of articles, books, cards, CDs, letters, memos, papers, photos, pictures, presentations, home movies, videotaped lectures, and voice recordings and storing them digitally, along with creating software to selectively replay this information [50] Under the auspices of the Department of Defense, DARPA proposed, but subsequently ended (probably due to public concerns about privacy), the LifeLog project to create a comprehensive database of an individual human's life [51, 52]. The introduction of Google Glass allows for augmented reality at voice command; the wearer sees information relayed from a web browser and a tiny camera displayed on a tiny screen slightly above the wearer's eyes [53]. Google glasses increase public concern about the loss of personal privacy and the collection of data without permission.

It is an open question whether BMI technology will in the near future facilitate uploading our memories to a chip. Some researchers argue that as we develop capacities to scan the brain, research that is ongoing, we will learn to scan the brain in order to download it, thus not even needing to store the raw data as it is generated. It is theoretically possible that we will map the locations and interconnections of neurons and synapses, and eventually be able to transfer an analog of the brain and its memory to a digital–analog computer [54]. Indeed Ray Kurzweil claims, "By the end of this century, I don't think there will be a clear distinction between human and machine [55]." Futurologist Ian Pearson of BT posits that at some point it should be possible to make a "full duplex mind link between man and machine [56]." Thought transmission between humans will then be achievable, backup copies of our brains could be made, and a global network will be part of our consciousness. One result of uploading minds is that immortality would be assured since uploaded minds would not age; in addition, such humans could travel at the speed of light, have enhanced memory and knowledge capabilities, and communicate from mind to mind.

In this eventuality, psychological continuity of personal identity could be immortalized in a series of cloned selves, bestowing immortality, and raising anew philosophical questions regarding personal identity. If all that is required for the persistence of personal identity is the sustaining of memory and physical continuity [57], then the clone with a previous or ongoing individual's memories uploaded to a chip implanted and activated would be the same person ongoing in time. Arguments minimizing cloning's effects and claiming that cloning is unlikely to affect a person's sense of self or identity [58] become irrelevant when the clone receives all the memories and experiences of a previous individual. In this case, concerns about the loss of an open future for the clone and the impacts upon autonomy and freedom are warranted. Certainly, the cloned individual's individuality and uniqueness could be overwhelmed to such an extent that the new individual might simply be the ongoing previous individual now experiencing a new history; a clone's independent learning might even be suppressed to facilitate this. The extent to which the clone's identity would be impacted by the implant would depend upon the age of implantation, and the control exerted over the new memories of the host clone. In considering the question of whether such an implant would produce an extension of the same person or a duplicate, one disquieting question is: how many can exist at the same time? If only one, then it would be an extension, if more than one, a duplication. It is, in some real sense, the same person, and not the same person, just as I am not the same today as I was yesterday, because things have happened in the meantime and this changes who I am. What the ability to transfer memories does is to enable this evolution of "self" across a much longer time than a single body might normally exist, possibly forever.

A multitude of other questions emerge when contemplating this eventuality. When would the chip be implanted? Or enabled? What would it be like to be an already aware individual with an ongoing history imprisoned in a child's body? Would the cloned, implanted entity feel like a unique person? Who would the clone be? These are similar to questions about brain transplants and personal identity considered in a dialog between Derek Parfit and Godfrey Vesey in 1974, in which Parfit argued that there "isn't anything more to personal identity than what you call psychological continuity in a one-one case [59]."

The question "what is man?" has no definitive answer. Yet, mind is surely the most salient feature of *Homo sapiens*. Once memories can be transferred from one brain to another or perhaps even several others – even to a computer or other species – questions regarding personal identity, the nature of memory, and the meaning of memory will be even more insistent. Neuroscientists, who hold that the mind is essentially computational and that consciousness is an emergent property of complicated information patterns, are comfortable with the notion of uploading the mind to a chip. For, they are essentially materialists or physicalists who posit mind as the result of physical processes, and challenge the view of dualists who believe that mind is nonphysical, spiritual. Derek Parfit proposes that "on the Reductionist View, each person's existence just involves the existence of a brain and body, the doing of certain deeds, the thinking of certain thoughts, the occurrence of certain experiences, and so on [60]."

At the moment, human intelligence is superior to that of machines, at least in terms of general intelligence. But, as machines improve, they will successfully compete with humans, and given sufficient time, surpass humans. Several researchers in a variety of articles and books have projected the imminent superiority of artificial intelligence [61]. Intelligent machines could then supersede mere humans.

We are re-evolving artificial minds at ten million times the original speed of human evolution, exponentially growing robot complexity. Currently, a guppylike thousand MIPS and hundreds of megabytes of memory enable our robots to build dense, almost photorealistic 3D maps of their surroundings and navigate intelligently. Within three decades, fourth-generation universal robots with a humanlike 100 million MIPS will be able to abstract and generalize – perhaps replace us [62].

On the basis of the speed with which computers are gaining processing power, already existent input/output technologies and the research potential for understanding the principles of operation of the human brain and copying its workings (either through computational neuroscience or emulating "the scanned brain on a computer by running a fine-grained simulation of its neural network ") [61], artificial intelligence and super-smart robots may well be developed within a few decades. Unless humanity embraces cyber technology, its hegemony may eventually yield to intelligent machines. If cybernetic technology is guided to allow the development or evolution of a human that is not merely human – cyborg – the supremacy of humans could be ensured, at least until the time when the machines evolve the next generation without our assistance.[3]

Taking these possible, even probable, developments into accounts reveals that ethical analysis needs to focus on the question of the good for man. Questions of this nature are more likely to be debated by theorists who address the question of what is the good or the best life for human beings. In contrast to those who espouse justificatory neutrality, "perfectionists" hold that some conceptions of human flourishing are superior to others [63]. If the state accords adequate priority to individual freedom, then the government need not be neutral on the question of the good for humans. This position is particularly applicable to technologies that entail a comprehensive transformation of humanity. There is no way to consider whether or not to regulate, ban, or adopt a permissive attitude toward radical human enhancement without considering the question of what is the human. How do humans differ from nonhumans? How do humans differ from the future's "posthumans?" What does it mean to be human? Would it be better to be more than human? Or more precisely, other than human? Presently, being human certainly means flourishing within a biological substrate, being a certain kind of animal, and not just a disembodied mind. Even if there are a multiplicity of goods, and different conceptions of how to achieve those goods, the underlying reality is that they are human goods. If the entity realizing those goods is no longer "human" but rather "posthuman," what is one to make of the good for that entity? Surely, in this eventuality assessment requires more than risk/benefit calculations. The question of whether such technological development should be considered acceptable at all entails a new normative framework. Enhancement BCIs pose the central and most challenging question the legal system will be confronted with: What are the barriers for intervening in human nature and for a person's right to alter his or her bodily functions or nature beyond natural barriers? The fundamental problem is that our

---

3) See Ref. [54], Kurzweil 2002 :99.

notion of normalcy or naturalness, which may form barriers to intervention in human features and capacities, gets blurred. The legal discourse on how to regulate human enhancement technologies depends on cultural and ethical discourses on human nature. Therefore, it is necessary that a general discussion inform the regulatory and legal discussion about where to draw the line.

There are culturally diverse approaches to this technology. In the United States, there is less unease about future developments; optimism and utilitarian considerations dominate the discussion, as in the report "Converging Technologies for Improving Human Performance" on nanotechnology, biotechnology, information technology, and cognitive science [64]. Here, the reflections on nanotechnology and "converging technologies" applaud the enhancement of human performance and look forward to the visionary enhancements of brain implants. However, this optimistic evaluation is not universal; European assessments are often more guarded, proceeding from the standpoint of the "precautionary principle." This principle holds that "when an activity raises threats of harm to human health or the environment, precautionary measures should be taken even if some cause and effect relationships are not fully established scientifically [65]." Nations and world societies need to apply not only a traditional risk/benefit analysis to assess the risks to society as a whole, and to the nature of humanity and its self-evolution from the development of these technologies, but also to engage in substantive discussions on the nature of the good for man. Global discourse leading to guidelines about the forms of development preferred for humankind must be structured and sound principles, rules, and sets of law must be put into practice. For the most part, existing regulatory bodies, such as the U.S. Food and Drug Administration (FDA), can only consider standards such as safety and efficacy. Such standards are inadequate to consider the social and policy questions of these enhancement devices. In order to implement such a discourse, new national and international bodies need to be created. Development and realization of technology are not inevitable; reflection and restriction are possible. In dealing with radical human enhancement technologies, wisdom is required. The question of what constitutes a meaningful future for humanity must take center stage. Therefore, a new system needs to be instituted for the rationalization of public policy in dealing with radical human enhancement technologies. Innovative global forums need to be fashioned to consider and deliberate these social and policy questions. From these discussions, new international regulations for enhancement technologies should be implemented to protect future generations.

19.11
Regulatory Procedures

Because of the enormous potential impact of cybernetics, this area of scientific development must be closely linked to ethical guidelines. Developments in cybernetics should be openly debated with accountability to the public, especially for investment of public funds. Advances that can impact society so intensely require public scrutiny. Procedures should be established for evaluating safety and efficacy, and for consideration of the need for equitable distribution of the benefits of this new technology. Some call for outright bans, citing the "precautionary principle" on the development of technologies that have the potential to significantly change human nature. Using language that speaks of hubris, and the "giveness" of humanness, they challenge developments that would threaten the "dignity of the naturally human way of activity [66]." George J. Annas, in proposing the "human species protection treaty," has recommend that human–machine cyborgs be banned along with cloning, genetic engineering, and artificial organs [67]. However, since Scientific progress promises to fulfill our desire for improvement; banning that progress is unrealistic and probably futile. In liberal democracies, the actions of persons are generally not interfered with unless they cause harm to others, and in very limited cases, to self. Nevertheless, between the options of unfettered freedom and outright prohibition lies the province of regulation.

Now is the time to consider whether, and if so, how, to regulate, rather than ban, the enhancement uses of this technology. Presently, before a medical device can be marketed in the United States it must meet the requirements of the FDA. Implantable brain chips would be listed as Class III devices since they are implanted, and may present a potential risk of illness or injury. Clinical trials will be used to establish the efficacy of the device and its safety. As with the development of the Activa Dystonia Therapy System, which was approved in April of 2003 for treatment of a movement disorder, the expedited development of therapeutic uses of brain implants may proceed under a humanitarian device exemption process. The development of brain implant technology in the United States is, then, already subject to a layer of governmental scrutiny. Whether this scrutiny is adequate to the task of reviewing brain implants as used for enhancements is questionable; even required post market safety reviews of devices are rarely done [68]. Moreover, the focus of FDA review is the establishment of indications for use, methods of safe placement, risks to subjects of surgery and anesthesia, and compilation of adverse events, particularly for those requiring device removals. No system exists for consideration of the extraordinary social and policy questions raised by these devices when used for enhancement. This level of scrutiny should be added, and considered the equivalent of an environmental impact statement.

Such deliberation is all the more significant because implantable brain chips can be a positive and transformative step in the evolution of humans. The differences in the kinds of humans this future will create need national and international consideration. Although the United States often approaches new technological developments from within an optimistic, ameliorist framework that privileges capitalist innovation and scientific freedom, many areas with social impact are regulated. Unfettered and unregulated scientific activity is a myth; existing regulations govern research on humans, on active infectious agents such as smallpox and Ebola, encryption software, and even research on marijuana.[4] Nations can and should deliberate and pass laws governing technology. The history

---

4) Regulations do not do away with violations any more than laws against murder eliminate murder.

of the guidelines formulated for recombinant DNA research in the United States suggests possible avenues for regulation, and the usefulness of creating a parallel system to the human subjects review process.

Controls need to be pursued on the national and international levels, with self-regulation by the scientific community leading the way. Self-regulation by those involved in brain implant technology should be pursued, in much the way that the Asilomar meeting created guidelines for recombinant DNA research [69]. The Asilomar meeting resulted in a moratorium, one possible method for ensuring that deliberation occurs before innovation. In the present case – development of enhanced humans through brain implants – there is adequate time to formulate guidelines without the use of a moratorium, if the proper forums for reflection can be created or accessed; however, this window will close more rapidly than is the case for strictly biological developments. In the United States, reestablishment of an Office of Technology Assessment would facilitate legislative examination of the complex issues surrounding brain implants. Other committees, such as those at the National Academy of Sciences, the National Academy of Engineering, and the Institute of Medicine could conceivably investigate the safety, social, and political implications of cybernetic technologies, but a nonpartisan agency serving in an advisory role to Congress has much to recommend it. Such an agency could commission an analysis of this future technology and provide forums for discussion of its possibilities and perils. Some mechanism such as that used to regulate recombinant DNA research, including the establishment of research guidelines, the holding of public hearings, and the formation of an agency to review protocols, is required [70]. Recommending and establishing standards to regulate the enhancement possibilities of cybernetic technology in order to ensure safety, efficacy, privacy, consent, and justice should not prove as difficult as attempts to regulate reproductive technologies that entail more deeply held value conflicts.

However, inasmuch as the complexity of enhancement techniques makes providing general policy challenging, a new regulatory body needs to be created to deal with the myriad of knotty issues. Such an agency could provide for a public review process at the national level to facilitate the consideration of risk and benefit. With regard to brain implants, the societal and transformative effects of uploading memories and implanting in third parties are concerns that need extensive consideration.

19.12
Principles and Standards for Adoption

A discussion of standards for enhancement technologies is long overdue. A principle, which needs to be considered for possible adoption, is that the risk/benefit ratio applied in evaluating the safety and effectiveness of enhancement techniques should be higher than that required for therapeutic interventions. Criterion for the use of this technology in the normal individual for purposes of enhancement should be placed at a higher level than the norms for review of devices to heal the sick.

Thus, concerns about risks of infection and brain damage are more pressing when the individual is healthy. In the deliberations concerning brain chips, this safety assessment will be complicated by the fact that researchers have demonstrated that for the monkeys that have been trained to control robot arms, the brain itself actually changes with more neurons emerging so that "the brain is assimilating the robot. It's creating a representation of it in different areas of the motor cortex."[5] Presumably, human brains will be similarly plastic and change by using an interface. This also raises the issue of removing the device – or even if removing it will revert the individual to the state before the device was implanted. A safety issue with even wider implications involves achieving control over remembering. Forgetting seems to provide benefits, and brain implant technology will need to be studied for its effects on our ability to deal with a painful past. Another area requiring study will be the effect that constant connection to others would have on our attention abilities, and on our needs for isolation [71].

Likely and reasonable initial standards would ensure that enhancement uses of interface technology include provisions for (i) reversibility in the event of adverse events, or changed preferences; (ii) informed consent; and (iii) limited access for initial studies. The requirement of reversibility would guarantee that if unforeseen risks develop there is a possibility of avoiding permanent problems; it is not altogether clear that superior, even perfect memory, would be a good, nor is high intelligence always beneficial. The requirement of informed consent would restrict usage, for enhancement, to adults with decisional capacity. Usage for prisoners, children, the military, and citizenry of despotic regimes would be forbidden, since their capacity for voluntary informed consent is limited. This provision would eliminate the risks that a clone could be precluded by brain chip implantation in early childhood from the development of personal individuality. Utilizing these principles, and initially restricting implantation to a small and controlled group, would secure time for evaluation of these technologies before widespread implementation. These requirements, although they may serve to secure the safety and autonomy of individuals, are insufficient to deal with more encompassing concerns, which include the relationship between these superhumans and the inferior species that is unenhanced. Nations and world societies have a stake in assessing the costs to society as a whole from the introduction of novel technologies. Fukuyama's concerns about a "future world in which ... human homogeneity splinters ... into competing human biological kinds," [72] although addressed to genetic technologies, is applicable also to BMIs, and raises the prospect of diminishing tolerance and democracy.

Nor are national laws and regulation sufficient to effectively monitor and control technologies that will cross national boundaries. It is necessary and should be possible to create international regulatory bodies for biotechnology developments. As Amitai Etzioni has pointed out, many transnational authorities and structures already exist – the World Trade Organization, the Bank for International Settlements, the World Health organization's new powers after SARS, agreements on

---

5) See Ref. [27], Zimmer. 2004.

biodiversity, and on pollution [73]. The legitimacy of such endeavors relies on the agreements of established governments and is often associated with the United Nations. International documents, such as the Universal Declaration on Human Genome and Human Rights (The United Nations Educational, Scientific, and Cultural Organization, UNESCO), 1997, and the World Medical Association's Resolution on Cloning (1997), have begun the task of addressing enhancement issues, and demonstrate an emerging willingness to regulate biomedical technologies globally.

It is ambiguous whether patent laws cover the manipulation of nature, and thus controversy on this topic has been provoked. Because of the "patent first, ask questions later" approach of the US patent law and the demise of the "moral utility" doctrine, effective regulation will necessitate action on the part of Congress [74]. Presently, neither morality nor public-policy issues are within the purview of the United States Patent Office. This is not the case in Europe because the International Agreement on Trade-Related Aspects of Intellectual Property Rights [75], which the United States did accept, contains a provision "which requires that members provide patents for inventions in all fields of technology with one significant caveat: 'Members may exclude from patentability inventions … [where such exclusion] is necessary to protect ordre [sic] public or morality, including to protect human, animal, or plant life or health …'" [6] This exclusion from patentability would diminish the economic value of research and innovation, although it would not prohibit it.

Some considerations, which could make regulations restricting such enhancement technologies difficult, include rights under the Constitution and the lack of any present law. According to Michael H. Shapiro, "a First Amendment argument" can be made "for a right to use intellectual enhancement resources without government interference – as well as the right to refuse such resources [76]." A statement from a conference held in May 2006 at Stanford Law School affirms that, in opposition to those who propose a global treaty on human-enhancement technologies, a case can be made that the right to enhance oneself is a human right [77]. At that same conference, Henry T. Greely "surveyed the right to enhancement within the current U.S. legal system through the lens of constitutional law, statutory regulation, and private claims." He noted that when an action is not prohibited by law in the United States, it is permitted [78]. Presently, there is no agency or method in the United States or internationally for regulating these technologies.

The scenario envisioned in this chapter's analysis, where a clone is created and implanted with a brain chip containing all of a previous individual's thoughts and memories, could be effectively regulated by national and international bans on human reproductive cloning. However, the effort to pass effective legislation does not appear promising. Presently in the United States, although there appears to be a scientific self-imposed ban, and there is a ban on government-funded projects, there is no law outlawing human cloning. Of the 193 countries in the world, 26 have passed laws or moratoriums or have implicitly banned human cloning

---

6) Bagley, *supra* Ref. [75], at 531.

to replicate an individual [79]. The effort to pass a worldwide ban on human reproductive cloning through an international treaty under the auspices of the United Nations was abandoned in November 2004 [80]. Nevertheless, the United Nations proposed a recommendation that each country pass laws banning human reproductive cloning. In this, as in any efforts to regulate BMIs, the covenants of the United Nations lack any effective mechanism for enforcement; the system of international law is voluntarist and depends upon the consent of each state. Despite this, the United Nations could and should take the lead in deliberation and policy recommendations concerning BMIs. On this issue, although the concerns are real and significant, the international stakes are not so high as to preclude agreement and international regulation.

The globalization of bioethics through international commissions has been proceeding by providing directives to encourage legislation in particular nation states. International law is traditionally fairly weak, depending on the consent of the member states. In the United States, there has been a tremendous decline in support for U.N. multilateral treaty systems [81]. These treaty systems are the foundation of international cooperation, and there needs to be hope of developing international regulation in this field.

It may be that progress toward that end will arise from consideration of nanotechnology's impact on development of these interfaces; nanotechnology will be essential in facilitating such progress. The Congress of the United States has already acted to a limited extent in this field [82]. Act 189, called the *21st Century Nano-Technology Research and Development Act*, calls for ensuring that ethical, legal, environmental, and other appropriate societal concerns, including the potential use of nanotechnology in enhancing human intelligence and in developing artificial intelligence that exceeds human capacity are considered during the development of nanotechnology.

Internationally, the United Nations agency charged with the responsibility for science and technology is UNESCO. Since 1993, UNESCO's International Bioethics Committee, a permanent committee, has been considering advances in science and technology in light of ethical concerns. The Intergovernmental Bioethics Committee represents the views of governments. Further input is received from the World Commission on the Ethics of Scientific Knowledge and Technology (COMEST), which advises UNESCO on matters concerning the ethics of science and technology. Inasmuch as forthcoming developments in BMIs will impact globally, member states of the United Nations can plausibly seek to develop international guidelines. The global adoption of the Universal Declaration on Bioethics and Human Rights provides a commitment to international agreement on principles that should aid in considering a response to potential brain–machine implants: consent, respect for personal integrity, privacy, confidentiality, equality, justice, and sharing of benefits [83]. The Declaration calls for society as a whole to be engaged in pluralistic public debate and to seek to protect future generations.

Thus, a framework for public dialog already exists within UNESCO. The need is apparent inasmuch as the ethical and social challenges that arise in this area are presently unregulated. Immense benefits can be expected from bioelectronics and implanted devices; possible harms need to be examined and precluded. Time is running out; the efforts to regulate this technology should begin and begin soon.

References

1. This Article expands on previous papers, including (a) McGee, E.M. and Maguire, G.Q. Jr., (1999) Implantable brain chips? Time for debate. *Hastings Cent. Rep.*, **29** (1), 7; (b) McGee, E. and Maguire, G.Q. Jr., (2007) Becoming borg to become immortal: regulating brain implant technologies. *Cambridge Q. Healthcare Ethics*, **16** (3), 291–303; (c) McGee, E.M. (2007) Should there be a law? Brain chips: ethical and policy issues. *Hilary Term*, **24** (1), 81–97; (d) Cooley, T.M. (2010) Law review; toward regulating human enhancement technologies. *AJOB Neurosci.*, **1** (2), 49–50; (e) Gordijn, B. and Chadwick, R. (eds) (2009) Bioelectronics and implanted devices, in *Medical Enhancement and Posthumanity*, Springer.
2. McGee, E.M. and Maguire, G.Q. (1999) Implantable brain chip? Time for debate. *Hastings Cent. Rep.*, **29** (1), 7–13.
3. Beauchamp, T. and Childress, J. (1979) *Principles of Biomedical Ethics*, Oxford University Press, New York.
4. Berger, T., Chapin, J. et al. (2007) WTEC Panel Report on Brain Computer Interfaces Research, http://www.dtic.mil/dtic/tr/fulltext/u2/a478887.pdf (accessed 18 March 2013).
5. Donoghue, J. (2006) A Link between mind and machine that can turn thought into movement. *Nature*, **442**, 13.
6. Emotiv http://www.emotiv.com; Neurosky. http://www.neurosky.com (accessed 13 March 2013).
7. Interaxon http://www.interaxon.ca (accessed 13 March 2013).
8. Leuthardt, E.C., Schalk, G., Wolpaw, J.R., Ojemann, J.G., and Moran, D.W. (2004) A brain–computer interface using electrocorticographic signals in humans. *J. Neural Eng.*, **1**, 63–71.
9. Aaron, R.K. et al. (2006) Horizons in prosthesis development for the restoration of limb function. *J. Am. Acad. Orthop. Surg.*, **14** (10), S198–S204.
10. Artificial Vision System for the Blind Announced by the Dobelle Institute Jan. 18 2000, http://www.sciencedaily.com/releases/2000/01/000118065202.htm (accessed 18 November 2013).
11. Second Sight FDA Panel Recommends FDA Approval for Second Sight's Argus® II Retinal Prosthesis System, http://2-sight.eu/landing-spot-fda-panel (accessed 6 February 2013).
12. Halley, D. (2010) Computer Chip Implant to Program Brain Activity, Treat Parkinson's, Singularity Hub, http://singularityhub.com/2010/07/21/computer-chip-implant-to-program-brain-activity-treat-parkinsons/ (accessed 20 July 2011).
13. Drummond, K. (2010) Human Trials Next for Darpa's Mind-Controlled Artificial Arm, Wired, http://www.wired.com/dangerroom/2010/07/human-trials-ahead-for-darpas-mind-controlled-artificial-arm/ (accessed 20 July 2011).
14. Berger, T. et al. (2011) A cortical neural prosthesis for restoring and enhancing memory. *J. Neural Eng.*, **8-11**, 046017 http://iopscience.iop.org/1741-2552/8/4/046017/pdf/1741-2552_8_4_046017.pdf (accessed 19 September 2011).
15. Maguire, G. (1999) Transforming Humanity: Toward the Implantable Walkstation, http://www.scribd.com/doc/27151750/1/Transforming-Humanity-Toward-the-Implantable-Walkstation (accessed 20 July 2011).
16. Warwick, K. Te Next Step Towards True Cyborgs, http://www.kevinwarwick.com/Cyborg2.htm (accessed 19 January 2013).

17. Talwar, S., Xu, S., Hawley, E. *et al* (2002) Behavioral neuroscience: rat navigation guided by remote control. *Nature*, **417** (6884), 37–38.
18. Greenemeier, L. (2008) Monkey Think, Robot Do, Scientific American, http://www.scientificamerican.com/article.cfm?id=monkey-think-robot-do (accessed 27 February 2011).
19. Begley, S. (2004) Prosthetics Operated by Brain Activity Move Closer to Reality. Wall Street Journal, http://online.wsj.com/article/0,,SB108931910809458926,00.html (accessed 22 July 2011).
20. Lemonick M. (2003) Robo-Monkey's Reward. Time (Oct. 27), pp. 45–47.
21. Guizzo, E. (2011) Monkeys Use Brain Interface to Move And Feel Virtual Objects. IEEE Spectrum, http://spectrum.ieee.org/automaton/robotics/medical-robots/monkeys-use-bidirectional-brain-machine-interface-to-feel-virtual-objects (accessed 20 January 2013).
22. Miller, G. (2013) The Wildly Ambitious Quest to Build a Mind-Controlled Exoskeleton by 2014, http://www.wired.com/wiredscience/2013/02/robotic-exoskeleton/ (accessed 12 March 2013).
23. Miller, G. (2013) Rodent Mind Meld: Scientists Wire Two Rats' Brains Together, http://www.wired.com/wiredscience/2013/02/rodent-mind-meld/ (accessed 12 March 2013).
24. Headlam, B. (2000) The Mind that Moves Objects. New York Times Magazine, pp. 63–64.
25. Hooper, S. (2004) Brain Chip Offers Hope for Paralyzed. CNN.com, http://edition.cnn.com/2004/TECH/10/20/explorers.braingate/ (accessed 20 July 2011).
26. Saenz, A. (2009) Braingate2: Your Mind Just Went Wireless. Singularity Hub, http://singularityhub.com/2009/06/17/braingate2-your-mind-just-went-wireless/ (accessed 22 July 2011).
27. Zimmer, C. (2004) Mind Over Machine. Popular Science, pp. 47–48.
28. Dino, J. (2008) Extension of the Human Senses, Ames Technology Capability and Facilities, http://www.nasa.gov/centers/ames/research/technology-onepagers/human_senses.html (accessed 22 July 2011).
29. Boran, M. (2011) Thad Starner: Wearable Computing For Smarter Living, http://newtechpost.com/2011/03/16/thad-starner-wearable-computing-for-smarter-living (accessed 21 July 2011).
30. Mann, S. (1997) Wearable computing: a first step toward personal imaging. *Computer*, **30** (2), 25–32. http://wearcam.org/ieeecomputer/r2025.htm (accessed 20 July 2011).
31. Bell, G. and Gemmell, J. (2007) A Digital Life. Scientific American, http://www.scientificamerican.com/article.cfm?id=a-digital-life (accessed 18 July 2011).
32. Memoto http://memoto.com (accessed 21 January 2013).
33. Carey, B. (2012) Brain Implant Improves Thinking in Monkeys, First Such Demonstration in Primates. The New York Times (Sept. 14).
34. Rubio, J. (2012) US Begins Testing Brain Implants with Hopes of Slowing Alzheimer's The Verge, December 7 2012, http://mobile.theverge.com/2012/12/7/3740988/us-deep-brain-stimulation-implant (accessed 18 January 2013).
35. Hsu, J. (2009) Intel Wants Brain Implants in Its Customers' Heads by 2020, POPSI, http://www.popsci.com/technology/article/2009-11/intel-wants-brain-implants-consumers-heads-2020 (accessed 18 January 2013).
36. Patel, P. (2009) The Brain-Machine Interface, Unplugged. IEEE Spectrum, http://spectrum.ieee.org/biomedical/devices/the-brainmachine-interface-unplugged (accessed 11 January 2013).
37. Graham-Rowe D. (2013) The First Brain Prosthesis –an Artificial Hippocampus- is in Development. New Scientist, http://www.newscientist.com/article.ns?id=dn3488 (accessed May, 2004).
38. Graham-Rowe, D. (2004) Brain implants 'read' monkey minds. *New Sci.*, **305**,

258 http://www.newscientist.com/article.ns?id=dn6127 (accessed 10 August 2004).

39. Begley S. (2004) Prosthetics Operated by Brain Activity Move Closer to Reality. The Wall Street Science Journal, p. B1.
40. Implanted Brain Grid Reads Minds (2004) www.betterhumans.com/News/news.aspx?articleID-2004-06-10-2 (accessed 21 August 2004).
41. Borton, D., Yin, M. *et al.* (2013) An implantable wirless neural interface for recording cortical circuit dynamics in moving primates. *J. Neural Eng.*, **10** (2), 026010.
42. Lindsay, G. (2012) The Future of Brain-Computer Interface; A Glimpse into the Nanomembrane-Filled Crystal Ball, http://www.neurdon.com/2012/11/06/the-future-of-brain-computer-interface-a-glimpse-into-the-nanomembrane-filled-crystal-ball/ (accessed 18 March 2013).
43. Parens, E. (1998) in *Enhancing Human Traits: Ethical and Social Implications* (ed. E. Parens), Georgetown University Press, Washington, DC, pp. 1–28.
44. Wikipedia Flynn Effect, Wikipedia the Free Encyclopedia. http://en.wikipedia.org/wiki/Flynn_effect (accessed 17 April 2005).
45. ManageWare Introducing the Original GPS Personal Locator for Children and "911" Call-for-Help Watch! www.gpschildlocatorwatch.com (accessed 22 July 2004).
46. Digital Angel Corporation http://www.digitalangelcorp.com (accessed 10 January 2004).
47. Chapin, J.K., Moxon, K.A., Markowitz, R.S., and Nicolelis, M.A.L. (1999) Real-time control of a robot arm using simultaneously recorded neurons in the motor cortex. *Nat. Neurosci.*, **2**, 664–670.
48. Elliot, D. (1998) Uniqueness, individuality, and human cloning. *J. Appl. Philos.*, **15** (3), 218.
49. Archives Society of Alberta, (1996) Newsletter, Vol. 16, Issue 2, www.archivesalberta.org/vol16_2/brain.htm (accessed 7 January 2005).
50. Gemmel, J, Bell, G, Lueder, R, Drucker, S, and Wong C. (2002) My Life Bits: Fulfilling the Memex Vision, Microsoft Research http://research.microsoft.com/~jgemmell/pubs/MyLifeBitsMM02.pdf (accessed 7 January 2005).
51. DARPA http://www.darpa.mil/ipto/Solicitations/PIP_03-30.html (accessed 28 January 2005).
52. Schachtman, N. (2004) Pentagon Kills Lifelog Project. Wired News (Feb. 4), http://www.wired.com/news/privacy/0,1848,62158,00.html (accessed 19 September 2013).
53. Google Glass. http://www.google.com/glass/start/what-it-does/ (accessed 13 September 2013).
54. Kurzweil, R. (2002) in *Understanding Artificial Intelligence* (ed. S. Fritz), Warner Books, Inc., New York, pp. 90–99.
55. McCullag, D. and Kurzweil, R. (2000) Rooting for the Machine. Wired News (Nov. 3), http://www.wired.com/news/technology/0,1282,39967,00.html (accessed 31 January 2005).
56. Pearson, I. (2013) Future Branches on the Human Tree, http://www.futurehumanevolution.com/future-branches-on-the-human-tree (accessed 19 November 2013).
57. Olson, E.T. (2002) Personal identity, in *The Stanford Encyclopedia of Philosophy* (ed. E.N. Zalta), Fall 2002 edn http://plato.stanford.edu/, The Metaphysics Research Lab Center for the Study of Language and Information Stanford University, Stanford, CA 94305-4115 (accessed 17 April 2005).
58. Brock, D.W. (2002) Human cloning and our sense of self. *Science*, **296**, 314–317.
59. Parfit, D. and Vesey, G. (2000) in *Philosophy and Contemporary Issues*, 8th edn (eds J. Burr and M. Goldinger), Prentice Hall, Upper Saddle River, NJ, p. 440.
60. Parfit, D. (1986) *Reasons and Persons*, Oxford University Press, Oxford, p. 211.
61. (a) Bostrom, N. (2000) When machines outsmart humans. *Futures*, **35** (7), 759–764 http://www.nickbostrom.com/2050/outsmart.html (accessed 16 April 2005); (b) Kurzweil, R. The Age of Intelligent Machines, 1990 MIT Press
62. Moravec, R. (2000) Robots, Re-Evolving Mind. Kurzweil AI Net,

http://www.kurzweilai.net/meme/ frame.html?main=/articles/art0145.html (accessed 17 April 2005).

63. Wall, S. (2007) Perfectionism in moral and political philosophy. *Stanford Encyclopedia of Philosophy*, http://plato.stanford.edu/entries/perfectionism-moral, The Metaphysics Research Lab, Stanford, CA, (accessed 3 February 2010)

64. Roco, M. and Bainbridge, W. (eds) (2001) *Converging Technologies for Improving Human Performance: Nanotechnology, Biotechnology, Information Technology and Cognitive Science*http://www.wtec.org/ConvergingTechnologies (accessed 13 September 2013).

65. Science and Environmental Health Network's The Wingspread Statement on the Precautionary Principle January 1998, *http://www.sehn.org/state.html*, The Metaphysics Research Lab, Stanford, CA, (accessed 18 March 2013).

66. Kass, L. (2003) *Beyond Therapy: Biotechnology and the Pursuit of Happiness*, Regan Books, New York, p. 292.

67. Annas, G.J. (2000) The man on the moon, immortality and other millennial myths: the prospects and perils of genetic engineering. *Emory Law J.*, **49** (3, Summer), 778–780.

68. Kerber, R. (2004) Globe Staff. 22% of Medical Device Follow-up Studies Left Undone. The Boston Globe (Dec. 24).

69. Barinaga, M. (2003) Asilomar revisited: lessons for today? *Science*, **287** (5458), 1584–1585. *http://www.biotech-info.net/asilomar_revisited.html* (accessed 3 February 2005)..

70. Fredrickson, D.S. (2001) *The Recombinant DNA Controversy: A Memoir*, American Society Microbiology Press.

71. Dertougos M. (1999) Brain Implants a Lousy Idea, July/August 1999, *www.lcs.mit.edu/about/brainimplants.html* (accessed 28 January).

72. Fukuyama F. (2002) Sensible Restrictions. March 19, 2002 *http://reason.com/archives/2002/03/19/sensible-restrictions* (accessed 18 November 2013).

73. Etzioni, A. (2004) *From Empire to Community. A New Approach to International Relations*, Palgrave Macmillan, New York.

74. Bagley, M.A. (2003) Patent first, ask questions later: morality and biotechnology in patent law. *William Mary Law Rev.*, **45**, 469–473.

75. GATT (1994) General Agreement on Tariffs and Trade – Multilateral Trade Negotiations of The Uruguay Round: Agreement on Trade Related Aspects of Intellectual Property Right, Including Trade in Counterfeit Goods, Annex. 1C, April 15, 1994, 33 I.L.M. 81.

76. Shapiro, M.H. (2002) Does technological enhancement of human traits threaten human equality and democracy? *San Diego Law Rev.*, **39**, 769–834.

77. SLS (2006) Human Enhancement Technologies and Human Rights, Stanford Law School-Details, May 2006, *http://www.law.stanford.edu/calendar/details/115/Human%20Enhancement%20Technologies%20and%20Human%20Rights/* (accessed 22 May 2013).

78. Guon, J. (0000) Human Enhancement, Rights, and the Venture Beyond Humanity, Institute on Biotechnology and the Human Future, at 2, *http://www.thehumanfuture.org/commentaries/human_enhancement/humanenhancement_commentary_guon01.html* (accessed 22 May 2007).

79. Matthews, Kirstin Overview of World Human Cloning Policies, Connexions. August 3, 2007 *http://cnx.org/content/m14834/latest/* (accessed 18 November 2013).

80. Pearson, H. (2004) UN Ditches Cloning Ban. news@nature.com (Nov. 22), *http://www.nature.com/news/2004/041122/pf/041122-2_pf.html* (accessed 5 February 2005).

81. See Patricia, J. (2005) Isolationism. Washington Spectator (May 1), *http://www.washingtonspectator.com/articles/20050501treaties_1.cfm* (accessed 19 September 2013).

82. US Congress, S. 189, 108th Cong. § 2 (codified as 15 U.S.C. § 7501 (2003)).

83. UNESCO (2005) Universal Declaration on Bioethics and Human Rights 75, October 19, 2005, *http://portal.unesco.org/en/ev.php-URL_ID=31058&URL_DO=DO_TOPIC&URL_SECTION=201.html* (accessed 18 March 2013).

20
Conclusions and Perspectives
Evgeny Katz

Rapid progress in science and engineering has made possible and commercially available many implantable biomedical devices (e.g., pacemakers [1], deep-brain neurostimulators [2], spinal cord stimulators [3], etc.) improving the natural operation or substituting the missing function of a human body. The development of sophisticated implantable devices (e.g., autonomously operating insulin pumps) [4] can improve the quality of life and contribute to the novel concept of personalized medicine [5]. More sophisticated implantable devices, such as an artificial eye [6], are currently being designed and studied in research laboratories and will come into medical practice in the near future. Long-term implants for different biomedical applications present specific engineering challenges related to the minimization of energy consumption, physical miniaturization, and stable performance optimization. The successful integration of machines with biological systems requires energy sources harvesting power directly from physiological processes [7] to unify the energy supply for the biological and electronic/mechanical parts of the integrated system. The design of implanted biofuel cells [8] operating *in vivo* promises for future use various medical electronic implants powered by implanted biofuel cells and resulting in bionic human–machine hybrids [9]. Aside from biomedical applications, one can foresee bioelectronic self-powered "cyborgs" based on various animals, which can operate autonomously using power from biological sources and which are used for environmental monitoring, homeland security, and military applications.

Projecting into the future, one can foresee devices extending the human functions to additional unnatural possibilities. A "sixth sense" [10] could be implemented, for example, extending human vision to a broader range of electromagnetic spectrum allowing infrared and ultraviolet vision. Different implantable prosthetic devices can enhance human physical power by integrating a human body with a machine yielding "cyborgs," presently not only existing in science fiction but also discussed seriously in scientific publications [11, 12]. Direct brain–computer interfacing [13] will allow not only operation of prosthetic devices controlled by brain, which is already possible at the present level of technology, but also provide connections to the Internet, ultimately resulting in a futuristic human-cybernetic ("humanetic") society, where all individuals can be

Implantable Bioelectronics, First Edition. Edited by Evgeny Katz.
© 2014 Wiley-VCH Verlag GmbH & Co. KGaA. Published 2014 by Wiley-VCH Verlag GmbH & Co. KGaA.

directly wired to a computer network, connecting the whole human society in one super-brain. This sounds too futuristic and potentially scary, but it is not technically impossible and is already visible at the present level of science and technology. This trend in the development of implantable bioelectronics certainly requires plenty of research in microelectronics, power sources, biotechnology, medicine, cognitive science, computer science, materials science, and so on. Even more important will be the understanding of cultural, ethical, and social problems originating from the technological developments [14]. When considering human cyborgs with some machinery parts integrated with a human body, another option – robots with human brains – should be taken into account. Ultimately, the brain function could be also partially or completely given to electronic devices. This technologically possible perspective could promise something similar to personal immortality, but, on the other hand, will result in significant social and cultural changes, which will not be easily adopted. Perhaps, research and technology will not go so far and cyborgs will remain in science fiction stories. Still, significant advances will be obviously achieved in medicine, saving lives and improving life quality for many people. Moreover, the interfacing of human neurons with machines can result in novel robotic systems where robots will be directly controlled by a human operator [15].

The present book is a small step from the present to the future, where science fiction predictions become reality. It is also our goal that this reality will be beneficial rather than scary for mankind.

References

1. Hayes, D.L., Asirvatham, S.J., and Friedman, P.A. (eds) (2013) *Cardiac Pacing, Defibrillation and Resynchronization: A Clinical Approach*, Wiley-Blackwell.
2. Marks, W.J. Jr., (ed.) (2010) *Deep Brain Stimulation Management*, Cambridge University Press, Cambridge.
3. Kreis, P. and Fishman, S. (2009) *Spinal Cord Stimulation Implantation: Percutaneous Implantation Techniques*, Oxford University Press.
4. Garmo, A., Hornsten, A., and Leksell, J. (2013) *Diabet. Med.*, **30**, 717–723.
5. Ruano, G., Bronzino, J.D., and Peterson, D.R. (eds) (2014) *Personalized Medicine: Principles and Practices*, CRC Press ISBN: 10: 1439874646.
6. Rasmussen, M.L.R., Prause, J.U., and Toft, P.B. (2010) *Acta Ophthalmol.*, **88**, 41.
7. Sue, C.Y. and Tsai, N.C. (2012) *Appl. Energy*, **93**, 390–403.
8. Katz, E. and MacVittie, K. (2013) *Energy Environ. Sci.* **6**, 2791–2803.
9. Johnson, F.E. and Virgo, K.S. (eds) (2006) *The Bionic Human: Health Promotion for People with Implanted Prosthetic Devices*, Humana Press, Totowa, NJ.
10. Abbott, A. (2006) *Nature*, **442**, 125–127.
11. Norman, D. (2001) *Commun. ACM*, **44**, 36–37.
12. Warwick, K. (2004) *I, Cyborg*, University of Illinois Press.
13. Rao, R.P.N. (2013) *Brain-Computer Interfacing*, Cambridge University Press, Cambridge.
14. Warwick, K. (2010) *Stud. Ethics Law Technol.*, **4**, 1–18.
15. Warwick, K., Xydas, D., Nasuto, S.J., Becerra, V.M., Hammond, M.W., Downes, J., Marshall, S., and Whalley, B.J. (2010) *Defence Sci. J.*, **60**, 5–14.

Index

a

abiotic (nonenzymatic) implantable biofuel cells
- abiotic catalysts 297
- applications 308
- challenges and future trends 309
- depletion design fuel cell 297
- fuel cell designs and power densities, comparison
-- depletion design 299–300
-- glucose fuel cell 301, 302–304
-- Raney-platinum film electrodes 300
-- reactant separation concepts 304
-- single-layer fuel cell design 300, 301
- glucose oxidation, electrocatalysts
-- platinum-gold alloys 292
-- silicon nanoparticles 292
- history
-- abiotic catalysts 286
-- batteries 285
-- cardiac pacemaker 285, 286
-- standard free energy of formation 285–286
- long-term operation, factors affecting
-- anode poisoning 305
-- endogenous amino acids, adsorption 305, 306
-- glucose oxidation, decreased 306–307
-- platinum electrodes, deactivation 304–305
-- Raney-platinum cathode 305
- mixed reactants, strategies
-- abiotic catalysts 297
-- depletion design fuel cell 297
-- platinum cathodes 297, 298, 299
-- single-layer fuel cell 297, 298
-- stacked assembly fuel cell 297
- oxygen reduction, electrocatalysts
-- graphene, use 294
-- manganese oxide and metal macrocycles 294
-- platinum 293–294
- principles
-- abiotic glucose fuel cell 287
-- Butler-Volmer equation 289
-- complete fuel cell, polarization behavior 289–290
-- conversion efficiency 290–291
-- fuel cell performance 292, 293
-- high-throughput material screening 291–292
-- maximum power point (MPP) 290
-- Nernst equation 288–289
-- passive load resistors 291
-- polarization plot or load curve 289, 291
-- standard free energy 287–288
-- standard hydrogen electrode (SHE) 288
-- standard redox potential 287
- separator membranes
-- anion exchange membranes 294
-- depletion design 294
-- hydrogels 294
- site of implantation
-- air-breathing fuel cell design 295
-- blood stream implantation 295
-- brain-machine interfaces 295
-- choice of implantation site 295, 296
-- physiological fluids 296
-- tissue implantation 295
- state-of the-art 307–308

active pixel sensor (APS)
- CMOS image sensor 199, 201
- photodiode and equivalent circuits 200,
- pixel array 200
- 3T-APS circuits 200, 201

active power 71

Index

ADCs. *See* analog-to-digital converters (ADCs)
analog signal processing, subthreshold circuits
- band-limiting filters 68
- reduction, static power consumption 69
- Sub-V_t analog circuit, gammatone filters 69, 70

analog-to-digital converters (ADCs) 62, 69 70–71
anionic magnetic nanoparticles 18–19
anodic bioelements
- bacteria and fungi 317
- CDH 319
- crystal structure, *Aspergillus niger* glucose oxidase 318
- FAD 318
- quinoprotein glucose dehydrogenase 320
- quinoproteins 321

antenna and rectification power module
- AC/DC rectifier 255–256
- low dropout (LDO) regulator 255, 256
- planar rectangular coil 255
- Texas Instrument® TRF7960 255
- uses 255
- Verichip® Corp 255

application-specific integrated circuit (ASIC) 250, 384
architecture-level optimizations, low-power data processing
- "approximate processing" methods 78–79
- computation task optimal apportioning, analog and digital blocks 76–77

area, power reduction methodologies
- clock gating and supply voltage gating 179
- sampling frequency 180
- size and energy consumption 179
- subthreshold *vs.* super-threshold operation 181–183
- voltage scaling 180

artificial intelligence (AI) 121–122
artificial retina
- equivalent circuits and layout 205–206
- fabricated chip 206, 207
- infrared light 206
- light-sensing function 207, 208
- log sensor, pixel circuits 206, 207
- micro camera system 209–210
- microchip architecture 207, 208
- microchip, timing diagram 208–209
- MPDA 205
- principle 203–204
- retinal prostheses 204–205
- stimulator 209

ASIC. *See* application-specific integrated circuit (ASIC)

b

BCIs. *See* brain–computer interfaces (BCIs)
BFCs. *See* biological fuel cells (BFCs)
biocatalysts 15
biochemical–electrode integration, neural growth 99
biocompatibility, definition 267
bioelectrical energy and body energy harvest 407
bioelectronic device materials, implantable
- biocompatibility, definition 267
- biofouling 268
- biofunctionality 268
- biosafety 267–268
- challenges 267
- evaluation, guidelines 266,
- gold (Au) and platinum (Pt) 268
- indium tin oxide (ITO) 268–269
- polymeric materials 269
- polymeric materials, insulating or encapsulating 268
- properties 267
- silver–silver chloride electrode (Ag/AgCl) 268

biofouling 268
biofunctionality 268–265
biological brains, robot body
- brain cells networks 117
- control element 118
- culture size 119–120
- electrochemical activity 118
- electrodes 118
- input–output response map 118
- MEA 117–118
- monitoring activity 120
- motor signal 119
- neurons 116
- performance 118 119
- physical appearance 116
- physical robot body 118
- plasticity process 119
- sensory input 120
- weeled robot body and brain 118 119

biological fuel cells (BFCs) 315
biological response test 266–267
biomedical devices, in vivo analysis
- ASIC 250,
- diabetes care devices market 260–261
- electrochemical cell-on-a-chip 249
- implantable bio-MEMSs 249–250
- implantable device, architecture 251–253

- implantable front-end architecture 254–259
- innovative biomedical device 250–251
- power and communication requirements 250
- radio frequency (RF) 250
- technological solutions 249

biomedical implantable systems
- active transmitters 387
- biological and medical information telemetry 386
- body functions 381
- building blocks 382, 383
- closed-loop systems 383
- conceptual diagram 381, 382
- early control 1950–1970, 391–394
- early stimulation 1950,–1970 391
- electronic circuits and germanium transistors 387
- electronic devices 384
- endoradiosondes 387
- heart energy harvester 389, 390
- historical review 1970–1990, 394
- historical review 1990–2012, 395–396
- integrated piezoresistive pressure sensor 389, 390
- Mackay–Marg tonometer 387, 381
- microfabricated 405
- micropackaging area 405
- and microsensors, medical care 395
- monitoring and diagnosis 388
- non-hermetic packages 405
- one-transistor Hartley oscillator circuit 384, 385
- piezoelectric heart energy harvester 389
- pioneering techniques 395
- pressure waves 386
- radio transmission 383–384
- sensor/stimulator 383
- sensory and electronic 405
- size and power consumption 384
- telemetry systems 383
- tracking instrument 384
- T1 vacuum tube RF oscillator 384, 385
- unit circle 381, 382
- vacuum tube radio instruments 384
- wireless communication link 381
- wireless micropower supply network 405
- wireless telemetry unit 386

biomedicine, magnetically labeled cells applications
- MNPs-mediated cell delivery and tissue engineering 21–23
- MRI imaging, MNPs-labeled cells 21

bio-MEMSs. *See* Bio-micro-electromechanical systems (bio-MEMSs)
bio-micro-electromechanical systems (bio-MEMSs) 249–250
biosafety 267–268
biosensors, biofunctionality and longevity 275
biosorbents 15
blood–material interaction assays 277, 278
blue multicopper oxidases (BMCOs) 321–322
BMCOs. *See* blue multicopper oxidases (BMCOs)

brain chips and cloning
- biological identity 420–421
- BMI technology 421
- enhanced memory 421
- enhancement technologies 424
- ethical uncertainties 420
- global discourse 424
- Google Glass 421
- human enhancement technologies 424
- human intelligence 422
- innovative global forums 424
- intelligent machines 422
- legal system 423
- neural network 412–423
- personal identity 421
- physical processes 422
- processing power 423

brain–computer interfaces (BCIs)
- artificial tactile sensation 415
- bioengineering and implant technology 414
- brain chip implantation 427
- brain chips 411
- chips and cloning 420–424
- computer cursor 126–127
- EEG 126
- electrodes number 126
- energy efficiency 415
- enhancement technologies 428
- ethical concerns 419–420
- fast interface method 127
- fMRI 127
- genetic enhancements 411
- globalization, bioethics 429
- human reproductive cloning 429
- hybrid BMIs 412
- information processing system 413–416
- injectable brain-controlling devices 417–418
- input interfaces 412
- invasive interfaces 413

brain–computer interfaces (BCIs) (contd.)
– medical domain 126
– neural activity 414–415
– neural control 414
– neuroethics 412
– neuroprosthetics 415
– noninvasive interfaces 413
– partially invasive interfaces 413
– regulatory procedures 424–426
– technological innovations 411
– therapy/enhancement 418–419
– wireless brain sensor 417
– wireless interface scheme 417
brain-implantable CMOS imaging device
– chemical substance 214
– electrodes 215
– LEDs 215
– micrometers 214
– neural activity 214
– pixel size 212
– planar and needle type 212, 213
– polyimide substrate 212, 213
– slip-ring connector 212, 213
brain implants
– BCI 123
– humans 124
– microelectrode array 123–124
– nervous system, human 124
– neural signals 123
– number crunching 124
– signals 124
– telegraphic communication 125
– therapeutic purposes 125–126
– ultrasonic sensory input 124–125
brain micromotion 98–99
Butler-Volmer equation 289
BVT Technologies® 258

c
case arm aid system 392
cathodic bioelements
– bioelectrocatalysis 321
– four-electron process 322
– *Myrothecium verrucaria* bilirubin oxidase 322
– redox potential Lcs 323
cationic magnetic liposomes (CML)
– CML-labeled cells 23
– preparation 19
cationic magnetic nanoparticles 19
CDH. See cellobiose dehydrogenase (CDH)
cellobiose dehydrogenase (CDH) 319,
central nervous system (CNS)
– astroglial scar 97, 98

– biochemical–electrode integration, neural growth 99
– cellular response, oreign implant 97, 98
– glial cells. See glial cells
– infammation reduction, brain micromotion 98–99
– recording electrodes. See recording electrodes, brain
charge-coupled device (CCD) 195
CML. See cationic magnetic liposomes (CML)
CMRR. See common-mode rejection ratio (CMRR)
CNS. See central nervous system (CNS)
common-mode rejection ratio (CMRR) 239
complementary metal-oxide semiconductor (CMOS)
– advantages and challenges, biomedical devices 196–197
– APS. See active pixel sensor (APS)
– artificial retina. See artificial retina
– biomedical devices 195, 196
– brain-implantable. See brain-implantable CMOS imaging device
– CCD 195
– features, image sensors 195, 196
– log sensor. See log sensor
– photosensors 198–199
– power supply 197
– pulse width modulation sensor 202–203
– sensor structure and characteristics 197
– SPAD sensors 203
computer network 435–436
conceptual arm aid system 1960, 392
content addressable memory (CAM) 178
cross-hierarchy design, implantable electronics
– architecture-level optimizations. See architecture-level optimizations, low-power data processing
– bioelectronic implant 65–66
– energy-efficient memory. See energy-efficient memory design
– long-term implants 65
– low-power signal processing. See low-power signal processing, circuit design
– wireless communication power delivery. See wireless communication power delivery
cuff electrodes 90
cyborgs
– BCI 126–127
– biological brains, robot body. See biological brains, robot body
– biology and technology 115
– biotechnological link 115

- brain implants 115
- DBS 120–122
- description 116
- RFID implants 128–130
- robots 115
- subdermal magnetic implants. *See* subdermal magnetic implants

d

DACs. *See* digital-to-analog converters (DACs)
DBS. *See* deep brain stimulation (DBS)
deep brain stimulation (DBS)
- AI tools 121–122
- BCIs 120
- biological mechanisms 100
- device 121
- electrode design and stimulation 100–101
- L-dopa 121
- PD 120–121
- research designs 101
- stimulator 121
- surgical treatment 121
DET-based EFCs
- biodevices operating *in vivo* 334–336
- bioelectronic devices 340
- biological variability 339
- compartment mediator- and cofactor-less direct-electron-transfer-based 315, 316
- enzyme-based biodevices 323–329
- genetic modification 341
- human physiological fluids 339
- nano/microbiorobotic systems 340
- operating *ex vivo* 336–339
- operating *in vitro* 329–334
- oxidoreductases 316–323
- power generation 333
- thermodynamic limits 340
- uncertainties, measurement techniques 339
diabetes care devices market
- cutting-edge multidisciplinary research 261
- juvenile diabetes 260
- Type 1 diabetes 260
- Type 2 diabetes 260
- in vitro diagnostics (IVDs) 260
- World Diabetes Market Analysis 261
digital signal processing (DSP)
- electrodes 162
- loop neural system 160
- low-power. *See* low-power DSP
- power consumption 66
- wireless communication 160–161
digital-to-analog converters (DACs) 70–71

- disadvantages 236
- output current 234, 235
direct electron transfer (DET) reactions 315
discrete wavelet transform (DWT) 181
driving and control modules
- data extraction circuitry 256
- generation circuitry 256
- potentiostat amplifier 256–257
DSP. *See* digital signal processing (DSP)

e

eCFR. *See* electronic code of federal regulations (eCFR)
EDC. *See* electronics design center (EDC)
electrical modulation, human nervous system
- cholinergic anti-inflammatory pathway 104–105
- DBS 100–101
- nerve regeneration 101–102
- pain. *See* pain modulation, human nervous system
- vagus nerve and stimulation 103–104
electromyograms (EMGs) 27
electronic code of federal regulations (eCFR) 397
electronics design center (EDC) 392
EMGs. *See* electromyograms (EMGs)
energy-efficient memory design
- STT-MRAM. *See* spin transfer torque MRAM (STT-MRAM)
- subthreshold SRAM 79–80
energy harvesting techniques
- and power densities 47,
- power-harvesting antenna 47–48
enhancement technologies 424
environmental monitoring 435
envisaged subcutaneous device, architecture
- antenna and rectification power module 255–256
- driving and control modules 256–257
- electrodes 255
- modulation/data processing module 255
- power-on-reset module 254–255
- sensor conditioning module 255
- sensor control module 255
- signal conditioning and communication modules 257–258
enzymatic fuel cells
- abdominal cavity, rats 349–350
- artificial pancreas and urinary sphincters 347–348
- battery-powered implantable devices 347
- biocompatibility 350–351
- Dacron® 351

enzymatic fuel cells (*contd.*)
- elaboration 351
- GBFCs. *See* glucose biofuel cells (GBFCs)
- glucose and oxygen degradation, catalytic reactions 349
- greenhouse gas emissions 348
- implantation, performance 351
- lifetime 351
- living organisms 349
- low activity, abiotic catalysts 349
- packaging
  - – exPTFE coating 356, 357
  - – microelectrodes, rat 358–356
  - – second generation, rat 356, 357 358
- performances
  - – glucose biofuels 361
  - – *in vitro* and *in vivo* 359–360
  - – physiological fluids 359
- periodic surgeries 348–349
- power supply systems 348
- production, electric power 348
- sealed batteries 347
- sterilization, medical devices 350
- surgery 358–359
enzyme-based biodevices
- DET properties 324
- EFCs 324
- electrode function 327–329
- electrode material 325–327
- microbial/mitochondrial biodevices 323
- redox enzyme 323
extracellular labeling, cells 20
extraneural electrodes
- cuff 90
- FINE 90

f

FAD. *See* flavin adenine dinucleotide (FAD)
far-field electromagnetic wireless communication 76–83, 82
fast Fourier transform (FFT) 171
FCC. *See* Federal Communication Commission (FCC)
Federal Communication Commission (FCC) 397
feedback telemetry circuit 391
FETs. *See* field effect transistors (FETs)
field effect transistors (FETs) 390
FINE. *See* flat interface nerve electrode (FINE)
FinFETs. *See* fin-shaped field effect transistors (FinFETs)
fin-shaped field effect transistors (FinFETs)
- device structure 75–76
- SCE 74

flat interface nerve electrode (FINE) 90
flavin adenine dinucleotide (FAD) 318
forward data link, wireless communication
- constant amplitude modulation 51–52
- efficient ASK-PW demodulator 52, 53
functional magnetic resonance imaging (fMRI), 127

g

GBFCs. *See* glucose biofuel cells (GBFCs)
glial cells
- astrocytes 94
- microglia 93–94
- PNS 88
glucose biofuel cells (GBFCs)
- bioelectrode design, *in vitro* performances 355, 356
- copper enzyme 353, 354
- glucose oxidation 353, 354–355
- MWCNT. *See* multiwalled carbon nanotube (MWCNT)
- parameters 353
- porosity, material 352
- PPy/PSS film 355
- stability issues 352

h

hardware implementation
- digital signal-processing block 172
- FFT 171
- implantable systems 171
- neural interface system 171–172
- wavelet module 173
high molecular weight kininogen (HMWK) 274
home and online (mobile) monitoring 407
homeland security and military applications 435
hover input device 141–142
human–computer interaction (HCI) 135
human EMG implant device 392, 393
humanitarian device exemption process. 425
human physical power 435
hydrogels 294

i

immobilization methods 15
implantable bioelectronics
- biofuel cells 3
- brain-machine interface 3, 4
- cyborgs 3, 4
- electronic elements 1
- flexible supports 1, 2
- nervous systems 3

- neural signals 2
- reworking, physiology 33
- robotic hand 1, 2
- "sense-and-act" system 2
- substrates insertion, eye development. *See Zophobas morio* beetle, eye development process
- technological solutions 5

implantable biomedical devices 435

implantable device, architecture
- amperometric electrodes 252, 253
- conception 252
- front-end architecture 252, 253
- glucose monitoring 251
- implantable prototype 252, 253
- RFID tag 253
- true/false applications 251

implantable front-end architecture, in vivo detection biosensor applications
- artificial pancreas 254
- bidirectional communication 254
- BVT Technologies® 258
- commercial 130 nm technology 258
- cyclic voltammetries (CVs) 258–259
- envisaged subcutaneous device, architecture 254–258
- full-custom ICs 258
- glucose monitoring 254
- implementation and results 258–259
- pervasive monitoring 254
- placement of sensor 254
- powering and communications 254
- recovered power 258
- safety operating area (SOA), concept 258
- Texas Instrument® TRF7960 258
- threshold level 259

implantable intracardiac probe
- ADCs 61–62
- architecture, analog front end 60
- cardiac arrhythmia 59
- electrode design 60, 61

implantable loop recorder (ILR) 375

implantation, response
- biosensors, biofunctionality and longevity 275
- blood coagulation and factors 273–274
- blood–material interaction assays 277, 278
- coagulation factors, intrinsic/extrinsic pathway 275–276
- complement activation pathways 276, 277
- high molecular weight kininogen (HMWK) 274
- hydrogels 275

- non-thrombogenic polymer-modified device 277, 278
- protein adsorption process 274–275

implanted biofuel cells operating in vivo
- biocatalytic cathodes 371, 373
- biocatalytic electrodes 371, 372, 374 376
- bioelectrochemical methods 363
- biofuel cells 364
- biomolecular processes 364
- buckypaper conductive support 371
- catalytic oxidation of the biofuel 365
- cyclic voltammograms 369, 370
- cytochrome oxidase 366
- direct electrical communication 368
- direct non-mediated electron transport 366
- electrical power 363
- enzyme selection 367
- glucose biosensors 367
- glucose oxidation 367
- glucose-oxidizing enzymes 368
- human blood circulation 373
- human blood circulatory system 372
- immobilization, PQQ-GDH 369
- invertebrates 364
- medical electronic implants 376
- microelectronic devices 371
- non-immobilized cofactors and mediators 365
- oxidation reactions mimic metabolic pathways 368
- physical/chemical methods 363
- physical energy conversion 363
- polarization functions 375
- rats 364, 365
- transducers 363
- variable load resistance and polarization 374–375

implanted devices
- interactive tasks 134
- medical domain 133
- mobile 133
- output device 134
- prototype device 134, 135
- user interfaces. *See* implanted user interfaces

implanted user interfaces
- advantages 135
- artificial skin 148–149
- audio device 144–145
- communication and synchronization 137
- components 147
- devices 138–139
- evaluation 137–138
- experimenter 139

implanted user interfaces (*contd.*)
- HCI   135
- hover input device   141–142
- humans   135
- infection   151
- input   136
- insert devices   139–140
- LED output   147–148
- limb specimen   139
- location   150–151
- medical devices   152
- medical implications   152
- outcomes   150
- output device   142–144
- parameters   151
- participants   150
- powering device   145–146
- power supply   137
- procedure   139
- prototype device   147
- sensor controls   147
- skin, wireless communication   147
- task and procedure   149
- technical evaluation   152–153
- touch input device   140–141
- tracking ball   151
- wireless communication device   146

implanted wireless biotelemetry
- back-telemetry link   225
- biological systems   221
- cardioverter defibrillators and MEAs   221
- cochlear implant and bionic eye   221
- communication scheme   224–225
- development   222
- dual band   223
- external units   223, 224
- implantable devices   221
- in-body medical systems   223
- inductive link, forward data   225–226
- instrumentation amplifier. *See* instrumentation amplifier
- low-power device   221–222
- MEAs and interface electronics   232–233
- multichannel neural recording systems. *See* multichannel neural recording systems
- patients   223, 224
- recording front-ends
- – CMRR   239
- – ECG, EEG and EMG   238
- – equivalent circuit   238
- – signal quality   238
- – thermal noise power spectral density   238
- stimulation
- – active charge balance circuits   237
- – blocking capacitors, electrodes   236–237
- – charge-balancing strategies   237
- – constant value   236
- – current/voltage pulses   233–234
- – DAC, output current   234, 235
- – disadvantage, DACs   236
- – dynamic current matching   238
- – neural stimulator devices   233
- – voltage headroom and linearized voltage-controlled resistors   234, 235, 236
- technologies   222, 224
- wideband telemetry link   228, 229
- wireless endoscope. *See* wireless endoscopy
- wireless power link. *See* Wireless power link

indium tin oxide (ITO)   268–269
innovative biomedical device
- academic research projects   251
- entire value chain, approach   251
- human skin, system implanted   250–251
- true/false alarm system   251

insect flight control
- muscles extracellular stimulation, elicit turns   32–33
- neurostimulation, wing oscillations initiation   30–31
- robust stimulation schemes   30

instrumentation amplifier
- chopping-stabilized amplifiers   241–222
- electrode–tissue interface   239–240
- OTA   240, 241
- pseudoresistor   240

intracellular labeling, cells
- anionic magnetic nanoparticles   18–19
- cationic magnetic nanoparticles   19

intrafascicular electrodes
- LIFE   91
- TIME   91, 92

invasive interfaces   413
invasive intracortical electrodes
- Michigan   95–96
- microwire   96–97
- Utah   97

in vitro biocompatibility assessment   266
in vitro diagnostics (IVDs)   260

l

large-scale integrated circuit (LSIC)   384
layer-by-layer (LbL) magnetic functionalization, microbial cells
- procedure   10
- strategy   9–10
- transmission electron microscopic images   10–11, 12

LEDs. *See* light-emitting diodes (LEDs)
LIFEs. *See* longitudinal implanted intrafascicular electrodes (LIFEs)
light-emitting diodes (LEDs) 32
locomotive implant and implantable cardiac probe
– device operation, bloodstream 57
– experimental setup, MHD propulsion 59
– fluid propulsion methods 56
– implantable intracardiac probe. *See* Implantable intracardiac probe
– simulated performance, MHD propulsion 58
log sensor
– output signal 202
– pixel circuits 201–202
longitudinal implanted intrafascicular electrodes (LIFEs) 91
low dropout (LDO) regulator 255, 256
low-power DSP
– active power 71
– dynamic voltage and frequency scaling 72
– FinFETs, ultralow voltage subthreshold circuits 74–76
– minimum energy subthreshold operation 73–74
– standby mode power reduction 73
– V_{DD} scaling and parallel 71–72
low-power signal processing, circuit design
– analog signal, subthreshold circuits 68–69
– analog-to-digital conversion 69, 70–71
– bioelectronic sensor interface 67–68
– data-processing components 67
– DSP 71–74
LSIC. *See* large-scale integrated circuit (LSIC)

m

magnetically functionalized cells
– labeling, mammal (human). *See* magnetic labeling, mammal (human)
– magnetic microbial cells. *See* magnetic microbial cells
– MNPs, biomedicine 7
magnetic ferrofluids 8–9
magnetic labeling, mammal (human)
– applications, biomedicine 20–23
– extracellular. *See* extracellular labeling, cells
– intracellular. *See* intracellular labeling, cells
magnetic microbial cells
– biosorbents and biocatalysts 15
– direct deposition, MNPs 8–9
– polymer-mediated deposition. *See* polymer-mediated deposition, MNPs
– remotely controlled organisms 16–17

– whole-cell biosensors and microfluidic devices 15–16
magnetic nanoparticles (MNPs)
– direct deposition, microbial cells 8–9
– MNPs-mediated cell delivery and tissue engineering 21–23
– polymer-mediated deposition, microbial cells 9–15
magnetic resonance imaging (MRI)
– MNPs-labeled cells 21
magnetohydrodynamics (MHD) propulsion
– experimental setup 59
– operation 57
– simulated performance 58
– thrust force 58
maximum power point (MPP) 290
MEAs. *See* microelectrode arrays (MEAs)
MEMS. *See* micro-electromechanical systems (MEMS)
metamorphosis 33, 36
MHD propulsion. *See* magnetohydrodynamics (MHD) propulsion
michigan electrodes 95–96
microelectrode arrays (MEAs)
– electrical activity 229
– and interface electronics 232–233
– weak signals 230
micro-electromechanical systems (MEMS) 67
microelectronic devices and micromachines, development 265
microfluidic devices 15–16
micromachining technology 394
micro photodiode array (MPDA) 205
micropower electronic design
– biological rhythms 401
– lifetime implant systems 401
– single-antenna two-frequencies RF-link circuit 402, 403
– stages, radio receivers 402
micro sensors (physical and biochemical) 394
microwire electrodes 96–97
miniaturized biomedical implantable devices
– battery-less 46
– cardiac pacemaker 46
– energy harvesting 47–48
– implantable system 48–49
– locomotive implant and cardiac probe. *See* locomotive implant and implantable cardiac probe
– medical care 45
– neurostimulation 45–46
– replacement, battery 46

miniaturized biomedical implantable devices (contd.)
– RF power harvesting. See radio frequency (RF) power harvesting
– wireless communication. See wireless communication
mltielectrode array (MEA)   117–118
Modular Prosthesis Limb (MPL) project   415
MPP. See maximum power point (MPP)
MRI. See magnetic resonance imaging (MRI)
multichannel neural recording systems
– EEG recording   228, 229
– electrical activity   229
– medical applications   229
– on-chip spike detection and sorting   229–230
– state-of-art   230, 231
– UWB transmitter   230
multiwalled carbon nanotube (MWCNT)
– compression, electrical conductivity   352
– graphite electrodes   352
– PPy/PSS membrane   355
– and tyrosinase   353, 354
MWCNT. See multiwalled carbon nanotube (MWCNT)

n

nanobiosensor or nanosensor
– cantilever array sensors   247
– definition   247
– nanotube/nanowire sensors   247
nano-electromechanical systems (NEMS)   395
nano-enabled implantable device, in vivo glucose monitoring
– biomedical devices for in vivo analysis   249–261
– nanomedicine   248–249
– nanotechnology   247–248
nanomedicine
– definition   248
– diagnostic devices, development   249
– emerging nanomedical techniques   248
– multidisciplinary research groups   248–249
nanoscale transducers and electronics   406
nanotechnology
– bio and health market   248
– bio-nanotechnologies   248
– breakthroughs   248
– GENNESYS White Paper   248
– "knowledge triangle"   247
– nanobiosensor or nanosensor   247

– nanomachining (top-down technology)   248
– nanosynthesis (bottom-up technology)   248
– nanotech-enabled products   247, 248
near-field electromagnetic wireless communication   82
NEMS. See nano-electromechanical systems (NEMS)
Nernst equation   288–289
nerve regeneration, electrical modulation
– biological mechanisms   102
– electrode stimulation   102
– peripheral nerve injury and repair   101
neural interfaces   158–160
– CNS. See central nervous system (CNS)
– electrical modulation. See electrical modulation, human nervous system
– infrared sensor   106
– mechanical environment   106
– PNS. See peripheral nervous system (PNS)
neural signal
– analysis   162–163
– burst-level vocabulary   167
– multichannel vocabulary, behavior-specific patterns   167–169
– output packet generation   169–170
– spike characterization and sorting   166–167
– spike detection   164–166
– spike-level vocabulary   163–164
neuroethics   412
noninvasive electrodes   89–90
noninvasive interfaces   413

o

on-organ monitoring and therapeutic devices   407
operational transconductance amplifier (OTA)   240, 241
OTA. See operational transconductance amplifier (OTA)

p

pain modulation, human nervous system
– CNS stimulation   103
– outcomes   103
– PNS stimulation   102–103
Parkinson's disease (PD)   120–121
partially invasive interfaces   413
peripheral nervous system (PNS)
– electrode designs   87
– extraneural electrodes   90
– functional afferent and efferent pathways   88–89

- glial cells 88
- intrafascicular electrodes 91, 92
- noninvasive electrodes 89–90
- regeneration-based electrodes 92–93
- sensory and motor axons 93
photodiode (PD)
- avalanche mode 199
- current-voltage curve 198, 199
- description 198
- solar cell mode 198
photosensors, PD 198
PNS. *See* peripheral nervous system (PNS)
poly(2-hydroxyethyl methacrylate) (pHEMA) 269
poly(hydroxyethyl methacrylate) (PHEMA) 337
poly(vinyl alcohol) (PVA) 269
polydimethylsiloxane (PDMS) 269
polyethylene glycol (PEG) 269
polylactic acid (PLA) 269
polymeric materials
- biocompatible 269
- insulating or encapsulating 268
- properties and applications 271
polymer-mediated deposition, MNPs
- LbL-magnetic functionalization, microbial cells 9–11
- single-step magnetic functionalization, microbial cells 11, 13–15
polymethyl methacrylate (PMMA) 269
polytetrafluoroethylene (PTFE) 269
powering device 145–146
power supply design 403–404
pulse width modulation sensor 202–203

r
radio frequency (RF) 250
radio frequency identification device (RFID) implants
- antenna and microchip 129
- device transmits 128
- human rights 129–130
- security and privacy 129
- US Food and Drug Administration 129
radio frequency identification (RFID) tag 253
radio frequency (RF) MEMS implantable systems
- biocompatibility and protection 398–399
- carrier frequency 397
- characteristics, biological and medical signals 399–400
- design considerations of implantable systems 400–401
- eCFR 397

- ISM frequencies and tolerance 397, 398
- ISM radio bands 397
radio frequency (RF) power harvesting
- battery-less implantable device 49
- low-power controller and auxiliary circuits 50–51
- matching network 49
- rectifier 49–50
- regulator and bandgap reference 50
recording electrodes, brain
- extracortical 95
- invasive intracortical 95–97
- noninvasive 94–95
regeneration-based electrodes 92–93
Rehabilitation Nano Chip (ReNaChip) 414
reverse data link, wireless communication
- backscatter link 54
- configurable load, backscatter link and Smith chart representation 55
- full-duplex communication 54
RFID. *See* radio frequency identification (RFID) tag
RF power harvesting. *See* radio frequency (RF) power harvesting

s
safety operating area (SOA) 258
SAR. *See* successive approximation register (SAR)
screening tests 266
SHE. *See* standard hydrogen electrode (SHE)
signal conditioning and communication modules
- codifications and modulations 257
- current-to-voltage conversion 257
- DC modulation, PMOS 258
- digital processing 257
- threshold level 257
- threshold voltages 257
single-photon avalanche diode (SPAD) 203
single-step polymer-mediated magnetic functionalization, microbial cells
- advantage 15
- PAH-stabilized MNPs 13
- thin-sectioned intact and MNPs-functionalized microalgae cells 13–14
SOA. *See* Safety operating area (SOA)
spike detection
- coefficients 165
- reconstruction quality 166
- wavelet transform 164–165
spin transfer torque MRAM (STT-MRAM) 80–81

standard hydrogen electrode (SHE)   288
STT-MRAM. *See* spin transfer torque MRAM (STT-MRAM)
subdermal magnetic implants
– durability   127
– electromagnet   127
– mechanoreceptors   128
– Morse signals   128
– ultrasonic range information   128
subthreshold SRAM
– on-current to off-current ratio   79, 80
– sizing constraints   79
– variability   80
subthreshold *vs.* super-threshold operation
– DWT   181
– dynamic power   181
– lifting wavelet transform   181–182
– signal-processing algorithm   181–183
– voltage scaling   182–183
successive approximation register (SAR)   70
superpara magnetic iron oxide nanoparticles (SPIONs). *See* magnetic nanoparticles (MNPs)
surface composition
– biocompatible polymeric materials   269
– bioelectronic device surfaces   271
– biological surface modification strategies   270
– blood coagulation, Virchow's triangle   271, 272
– blood–material interaction   272
– contact angle measurements   272
– cross-linking   272
– device surface texture   273
– elicit platelet adhesion and aggregation   271
– hydrogels   270–271, 272
– nonbiological surface modification strategies   270
– polymers properties and applications   271
– PTFE surfaces   273
system integration and micro-packaging   404–405

t

technology transfer   248, 261
tetherless insect flight control. *See* insect flight control
Texas Instrument® TRF7960   258
texture, device surface   273
TIME. *See* transverse intrafascicular multichannel electrode (TIME)
tissue engineering and MNPs-mediated cell delivery
– CML-labeled cells   23
– magnetically facilitated formation, multicellular layers   21–22
– positioning and concentration, cells   21
touch input device   140–141
transverse intrafascicular multichannel electrode (TIME)   91, 92

u

ultralow power and robust on-chip digital signal processing
– area, power reduction methodologies   179–180
– bioimplantable devices   155, 156
– biological signal   155
– biomedical systems   188
– design flow   188–191
– design parameters   156
– DSP   157
– hardware implementation   171–173
– implantable microsystems   155
– low-power circuits   157
– microsystems   157–158
– neural interfaces. *See* neural interfaces
– neural signal. *See* neural signal
– on-chip DSP   160–162
– PEs   185–186
– process variations   183–185
– signal-processing system   186
– spike reconstruction quality   187–188
– subthreshold *vs.* super-threshold operation   181–183
– super-threshold design   185
– supply voltages   186
– vocabulary module   177–179
– wavelet module   173–177
– wireless communication protocols   190, 191
untethered insect interfaces
– beetle hybrid system   27, 29
– eight-channel system   29
– implantable bioelectronics, insects. *See* implantable bioelectronics
– tetherless insect flight control. *See* insect flight control
– wireless systems, free-flying insects   27, 28
– *Zophobas morio*   27, 35 36
Utah electrodes   97

v

Verichip® Corp   255
very large scale integrated circuits (VLSI)   395
VLSI. *See* very large scale integrated circuits (VLSI)

vocabulary module
- CAM 178
- hardware implementation 177
- wavelet engine 178

W

wavelet module
- biological signal processing 175
- DWT implementation 173
- parallel implementation 173, 174
- "processing element" 173, 174
- register minimization techniques 175
- register transfer logic 176
- thresholding schemes implementation 177

whole-cell biosensors 15–16
wideband telemetry link 228, 229
wireless communication
- device 146
- forward data link 51–54
- reduction, implants size 51
- reverse data link 54–55

wireless communication power delivery
- energy transfer 83
- far-field electromagnetic 82, 83
- implantable system 82
- near-field electromagnetic 82

wireless endoscopy
- data rate 232
- high-capacity wireless link 231
- intestine and real-time video images 230
- PillCam 231

wireless energy transfer 83
wireless power link
- approaches 227
- inductive link, class-E amplifier 226–227
- magnetic induction 226
- transistor device 227

World Diabetes Market Analysis 261

Z

Zophobas morio beetle, eye development process
- formation, adult organs 37, 38, 39
- hemolymph loss and damage, internal structures 36
- implantation, mid-pupation 37, 38
- interface and scanning electron micrograph 34, 35
- phototransduction mechanism 39
- survival rates and malformation, implants 37
- transparent cuticle 34, 36